AF836173

Fission and Properties of Neutron-Rich Nuclei

Fission and Properties of Neutron-Rich Nuclei

Proceedings of the Sixth International Conference on ICFN6

Sanibel Island, Florida, USA 6 – 12 November 2016

Editors

J. H. Hamilton
Vanderbilt University, USA

A. V. Ramayya
Vanderbilt University, USA

P. Talou
Los Alamos National Laboratory, USA

World Scientific

NEW JERSEY • LONDON • SINGAPORE • BEIJING • SHANGHAI • HONG KONG • TAIPEI • CHENNAI • TOKYO

Published by

World Scientific Publishing Co. Pte. Ltd.

5 Toh Tuck Link, Singapore 596224

USA office: 27 Warren Street, Suite 401-402, Hackensack, NJ 07601

UK office: 57 Shelton Street, Covent Garden, London WC2H 9HE

Library of Congress Cataloging-in-Publication Data

Names: International Conference on Fission and Properties of Neutron-Rich Nuclei (6th : 2016 : Sanibel Island, Fla.) | Hamilton, Joseph H., editor. | Ramayya, A.V., editor. | Talou, P., editor.

Title: Fission and properties of neutron-rich nuclei : proceedings of the sixth International Conference on ICFN6, Sanibel Island, Florida, USA, 6–12 November 2016 / editors, J.H. Hamilton (Vanderbilt University, USA), A.V. Ramayya (Vanderbilt University, USA), P. Talou (Los Alamos National Laboratory, USA).

Description: Singapore ; Hackensack, NJ : World Scientific, [2017] | Includes bibliographical references.

Identifiers: LCCN 2017026748| ISBN 9789813229419 (hardcover ; alk. paper) | ISBN 9813229411 (hardcover ; alk. paper)

Subjects: LCSH: Nuclear fission--Congresses. | Nuclear structure--Congresses. | Neutrons--Congresses.

Classification: LCC QC789.7 .I535 2016 | DDC 539.7/62--dc23

LC record available at https://lccn.loc.gov/2017026748

British Library Cataloguing-in-Publication Data

A catalogue record for this book is available from the British Library.

Copyright © 2018 by World Scientific Publishing Co. Pte. Ltd.

All rights reserved. This book, or parts thereof, may not be reproduced in any form or by any means, electronic or mechanical, including photocopying, recording or any information storage and retrieval system now known or to be invented, without written permission from the publisher.

For photocopying of material in this volume, please pay a copying fee through the Copyright Clearance Center, Inc., 222 Rosewood Drive, Danvers, MA 01923, USA. In this case permission to photocopy is not required from the publisher.

Printed in Singapore

Program Committee and Advisory Committee

International Advisory Committee

Joseph H. Hamilton (Chair), Vanderbilt University
A.V. Ramayya (Co-Chair), Vanderbilt University
Patrick Talou (Co-Chair), Los Alamos National Laboratory

Navin Alahari, GANIL
Andrei Andeyev, University of York
Juha Äystö, University of Helsinki
Maria Borge, CERN
Peter Butler, University of Liverpool
Jolie Cizewski, Rutgers University
Giacomo de Angelis, INFN Laboratori Nazionali di Legnaro
Christoph Duellmann, GSI
Amand Faessler, University of Tuebingen
Claes Fahlander, Lund University
Sydney Gales, IPN Orsay
Jacklyn Gates, Lawrence Berkeley National Laboratory
Paulo Gomes, Universidade Federal Fluminense
Robert Haight, Los Alamos National Laboratory
Koichi Hagino, Tohoku University
Franz-Josef Hambsch, EC-JRC-Dir G
David Hinde, Australian National University
Sigurd Hoffman, GSI
Jan Kliman, Slovak Academy of Sciences
Wojciech Krolas, Institute of Nuclear Physics
Reiner Kruecken, TRIUMF

Walter Loveland, Oregon State University
Yixiao Luo, Lawrence Berkeley National Laboratory
Witold Nazarewicz, FRIB/NSCL
Yuri Oganessian, Joint Institute for Nuclear Research
Dorin Poenariu, Horia Hulubei National Institute for Physics
Jorgen Randrup, Lawrence Berkeley National Laboratory
John O. Rasmussen, Lawrence Berkeley National Laboratory
Lee L. Riedinger, University of Tennessee
Mark Riley, Florida State University
Krzysztof Rykaczewski, Oak Ridge National Laboratory
Hiroyoshi Sakurai, RIKEN
Guy Savard, Argonne National Laboratory
Bradley M. Sherrill, NSCL, Michigan State University
Gavin Smith, University of Manchester
Michael S. Smith, Oak Ridge National Laboratory
Mark Stoyer, Lawrence Livermore National Laboratory
Julien Taieb, CEA DAM Bruyeres-le-Chatel
Gurgen Ter-Akopian, Joint Institute for Nuclear Research
Sait Umar, Vanderbilt University
Sherry Yennello, Texas A&M University
Michael Wiescher, University of Notre Dame
Furong Xu, Peking University
Shengjiang Zhu, Tsinghua University

Local Organizing Committee

David Dean, Oak Ridge National Laboratory
Alexandra Gade, Michigan State University
Robert Grzywacz, University of Tennessee
Joseph H. Hamilton, Vanderbilt University
A.V. Ramayya, Vanderbilt University
Patrick Talou, Los Alamos National Laboratory
Fredrik Tovesson, Los Alamos National Laboratory
Enhong Wang, Vanderbilt University
Edward Zganjar, Louisiana State University

Sponsors

Vanderbilt University
Los Alamos National Laboratory
NSCL/FRIB Michigan State University
Oak Ridge National Laboratory
Joint Institute for Nuclear Physics and its Applications

Preface

The Sixth International Conference on Fission and Properties of Neutron-Rich Nuclei was held on November 6–12, 2016 at Sanibel Island, Florida. The conference followed very closely the spirit set by the previous five conferences held in 1997 at Sanibel, 1999 at St. Andrews, Scotland, and 2002, 2007 and 2012 at Sanibel. With over 100 papers presented over five-and-a-half days, this sixth conference highlighted many exciting new insights and developments in studies of fission and neutron-rich nuclei.

At the opening of the Conference on Monday, we paused to honor the memory of our colleagues, Professor Walter Greiner, Professor Konstantin Gridnev and Professor Lakshman Chaturvedi who passed away since the last conference. They were regular speakers at our conferences and we missed their presence.

The conference assembled many of the world's experts in the fields of fission, properties of neutron-rich nuclei, and astrophysics, radioactive ion beams, and super heavy elements, both theoretical and experimental. Speakers presented overviews and the most recent data that provide the latest insights into these fields. The future of these fields is very bright as evidenced by the results presented including the plans and status of major new facilities being developed worldwide and by the presence of many young researchers and graduate students not only as attendees but also as speakers. Fission research is growing as seen in the large number of papers presented. A highlight was on November 8, 2016 during the conference when the International Union of Pure and Applied Chemistry announced the formal approval of the names of the four new elements 113, 115, 117 and 118: nihonium, moscovium, tennessine and oganesson.

The papers have been arranged essentially in the order of presentation in the program with the poster session papers at the end of the proceedings. Abstracts of conference talks are given in cases where full papers were not received.

We want to especially thank the International Advisory Committee and the Local Organizing Committee for all their diligent work. The scientific success was assured by many excellent suggestions for speakers and topics by the International Advisory Committee. As planning proceeded, several members of this committee took very active roles in helping to ensure that the most current and important results from all laboratories were presented and to help balance speakers and topics. In some cases, other co-authors on the paper presented their results. The conference setting, in addition to its natural beauty, provided convenient arrangements for informal discussions and for enjoyable outings. The Sundial Beach Resort's entire staff went beyond their job descriptions in providing excellent surroundings, food, and an excursion, and so are due special thanks in particular to David Irwin.

We want to thank each of our sponsors: Vanderbilt University, Los Alamos National Laboratory, NSCL/FRIB Michigan State University, Oak Ridge National Laboratory, and the Joint Institute for Nuclear Physics and its Applications for their financial support of the conference. Special thanks go to E. Wang, C. Zachary, J. Eldridge, A. Thibeault from Vanderbilt University and B. Musangu from Furman University for their many efforts in computer arrangements, website and the graphics for the conference.

The underpinning of any conference is the secretarial staff that sees to the endless details from long before to long after the conference including preparation of the proceedings. We are most appreciative for the dedicated work of Carol Soren, Vanderbilt University for her tireless efforts and very pleasant demeanor and calmness in the face of many demands in carrying out the work of the conference. We also want to thank Peggy Lucas-McGowan for designing the cover for the Abstract Book. We want to thank Y. X. Luo, D. Pickert and E. Wang for taking the conference photographs.

We are happy to announce that the FRIB Laboratory will take the lead role in planning the next Sanibel conference along with the support of Los Alamos National Laboratory and Vanderbilt University.

J. H. Hamilton
Vanderbilt University

A. V. Ramayya
Vanderbilt University

P. Talou
Los Alamos National Laboratory

Contents

Radioactive Ion Beams and New Facilities

Nuclear Structure II

Mass Measurements

Nuclear Structure III

Fission I

Nuclear Structure IV

Nuclear Astrophysics

Fission II

Posters

Opening Session

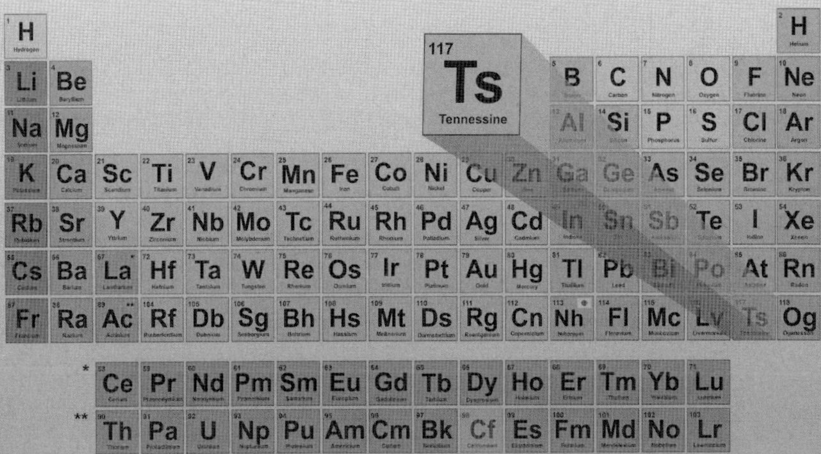

Prospects for Breakthroughs in Nuclear Theory

Witold Nazarewicz
Department of Physics and Astronomy and FRIB Laboratory
Michigan State University, East Lansing, Michigan 48824, USA

Understanding atomic nuclei is a quantum many-body problem of incredible richness and diversity and studies of nuclei address some of the great challenges that are common throughout modern science. In this talk, advances in nuclear theory will be reviewed in the context of the overarching scientific questions. Particular attention will be given to the science of rare isotopes and to theoretical studies of nuclei at the eve of extreme-scale computing.

Heaviest Elements

Yuri Oganessian
Flerov Laboratory of Nuclear Reactions,
Joint Institute for Nuclear Research

Experimental results on the synthesis of heaviest nuclei in the reactions induced by 48Ca are discussed in terms of further investigations of super-heavy elements.

From a comparison of the properties of super heavy and neutron-rich nuclei with theoretical predictions derive some consequences and consider the possibility of new experiments. It is shown that the further expansion of the scope of work associated with a significant increase in the sensitivity of experiments.

Heavy ion accelerator facilities and new generation of the experimental installation seems to be the key factors determining the effectiveness of future researches.

Nuclei in the Cosmos: Nuclear Astrophysics Experiments in the USA

A. Aprahamian

University of Notre Dame, Notre Dame, IN 46556

Nuclear Astrophysics is a broad field at the intersection of Nuclear Physics, Astrophysics, and Astronomy concerned with the energy production and evolution of stars and stellar explosions and the origin of the elements from the Big Bang to the present. Nuclei and nuclear processes experience and in turn define the exotic cosmic environments and yield what we see as the abundances of the elements. Stars evolve and die as a result of the nuclear reactions that take place in their interior using the available seed material. The associated burning processes during each evolutionary phase produce new elements that serve as fuel for subsequent burning processes at the specific environment, temperature, and density conditions. Explosive environments show similar evolution patterns only on much shorter time-scales, initiating a number of exotic burning scenarios that allow the synthesis of elements to the actinides and beyond. I will discuss the specific physical conditions of the astrophysical scenarios and discuss the experiments that we are building in the USA to access those parameters, FRIB at MSU and associated recoil mass separators, St. George at ND, CASPAR at Homestake Mine and others.

Recent Accomplishments in Nuclear Structure Research with Rare Isotopes and Future Challenges*

Alexandra Gade

National Superconducting Cyclotron Laboratory, Michigan State University, East Lansing, Michigan 48824, USA
Department of Physics and Astronomy, Michigan State University, East Lansing, Michigan 48824, USA

The often surprising properties of exotic nuclei have prompted extensive experimental and theoretical studies aimed at identifying the driving forces behind the dramatic changes encountered in the regime of extreme isospin. This talk will review some of the recent highlights in the field and identify upcoming challenges.

*This work was supported by the National Science Foundation under Grants No. PHY-0606007, No. PHY-1102511.

Overview of Nuclear Fission — Present Experiments*

Robert C. Haight

Los Alamos Neutron Science Center
Los Alamos National Laboratory
Los Alamos, NM 87545, USA
E-mail: haight@lanl.gov

After nearly 80 years of research on nuclear fission, we still do not fully understand the fission process. A combination of new requirements for accurate fission nuclear data, the ability to calculate sophisticated models of fission, and the development of new experimental techniques has re-energized the field. This paper concentrates on recent developments in the experimental realm and what the new results can contribute to the database for applications and to improving the underlying physical models. This overview is intended to be an introduction to fission presentations at this meeting.

Keywords: Fission; Experiments; Neutrons; Gamma Rays; Fission Fragments; Fission Products.

1. Introduction

Applications of nuclear fission and models of the underlying physical processes depend directly on measurements. Even after nearly 80 years of experimentation and a very large existing database of results, new measurements are now being made at many laboratories. Some of the activities focus on improving the accuracy of the data; others are exploring approaches to new observables. The laboratories where these measurements are being made, Fig. 1, indicate the extent of this renewed interest in fission.

* This work was performed under the auspices of the U.S. Department of Energy at Los Alamos National Laboratory by the Los Alamos National Security, LLC under Contract No. DE-AC52-06NA25396.

This paper will be limited to low-energy neutron-induced fission and spontaneous fission for two reasons, one being the author's interest in applications of fission nuclear data and the other being that these processes test many characteristics of the physics of fission including models of the fission barrier, the transition from saddle to scission, fission yields, and the energy distribution in the fission fragments that leads to prompt neutron and gamma-ray emission. This quantum mechanical process of transmission of correlated nuclei through or just above the fission barrier is unique in physics, but we note that it also has relevance to other processes in atomic and molecular physics and perhaps also in other realms. Properties of the fission products, such as beta-delayed neutron and gamma-ray emission will not be included here as they are part of the physics of neutron-rich nuclei, another focus of this meeting.

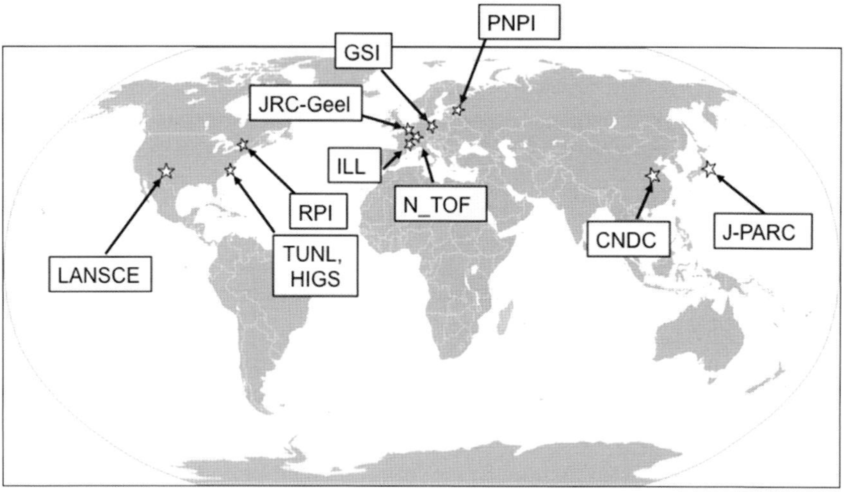

Fig. 1. Major laboratories where neutron-induced fission research is being carried out at present. Many institutions, not indicated in the figure, participate in many of the measurements. In addition, research is being carried out at universities and some other institutions not listed in the figure. The laboratories in the figure, from left to right, are the Los Alamos Neutron Science Center [LANSCE], Triangle Universities Nuclear Laboratory [TUNL] and the High-Intensity Gamma-ray Source [HIGS], Rensselaer Polytechnic Institute [RPI], Joint Research Center-Geel [JRC-Geel], Institut Laue Langevan [ILL], Neutron Time-of-Flight facility at CERN [N_TOF], Geschellschaft für Schwerion Forschung [GSI], Petersburg Nuclear Physics Institute [PNPI], Chinese Nuclear Data Center [CNDC] and the Japanese Proton Accelerator Research Complex [J-PARC].

Recent Measurements

Here we briefly describe some recent measurements, instrumentation and facility developments, and progress in improving the experimental nuclear database. The section highlights works that have come to the author's attention and is not meant to downplay other works that could be just as important. Most of works discussed have importance to applications (they pay the bills!), but I hope that this paper will also open doorways to better understanding the physics of fission.

The measurements outlined here are ordered somewhat in importance to applications but that also reflect, also in an approximate order, the path of the fission process from transition through or over the fission barrier, to scission, to prompt neutron and gamma-ray emission from the fully accelerated fission fragments. Recent innovative approaches are being made for the following observables:

- Fission cross sections
- Total kinetic energy released in fission
- Neutrons emitted promptly in fission, nu-bar, and neutron multiplicity distribution, P(nu)
- Prompt gamma-ray output
- Fission product yields

From the outset, we note that present-day experiments do not measure all of the possible observables simultaneously, event-by-event. What we would like is a spread sheet that, for each event, all of the above observables are recorded: the split into specific light and heavy fragments (Z_L, A_L, Z_H, A_H), the total kinetic energy released, the number of neutrons and their energies emitted by each fragment, the number of gamma rays and their energies emitted by each fragment, and the resulting fission products. With this information, average quantities and their variances could be calculated. Furthermore, correlations of quantities could be investigated. Correlated observables, such as total kinetic energy and mass yields, for example have been measured, but higher-order correlations that would include these observables correlated with prompt neutron and gamma-ray observables would provide even greater constraints on the models. These are measurements for the future.

1.1. *Fission cross sections*

Fission cross sections have obvious importance for applications as well as for understanding fission physics. Fission models must explain these data as the system passes through or over the fission barrier and also the competition between fission, gamma-ray decay (neutron capture), and inelastic scattering. Here the experimental database is huge, but there still there are many active efforts in this area. One can ask why. There are at least four reasons, in the opinion of this author:

- Improved accuracy is requested by the applications community; the resolution of discrepancies among previous works is essential.
- Uncertainties in the results need to be better understood so that covariance files, used in calculating design margins, will be realistic; many old measurements were not well documented.
- Known experimental limitations of previous works need to be confronted with new techniques: for example, the angular distribution of energetic fission fragments needs to be understood to quantify the effect of fissions that are not detected when the fission products are absorbed in the target foil. Further, non-uniformities of the beam and actinide sample need to be quantified in order to obtain accurate fission cross sections.
- Measuring the fission cross section of important minor actinides is complicated by the high radioactivity of the samples.

New experimental approaches that address the effect of the angular distribution of fission products on measurements of fission cross sections include:

- Time-projection chambers, a development from high-energy physics, have been scaled down to the size of a coffee can (LANSCE) [1,2]. This instrument also provides information on the sample and beam uniformities and can in principal investigate ternary fission.
- Micromegas detectors are used in a similar fashion (CERN) [3].
- A segmented-anode twin-Frisch-grid ionization chamber also gives position and angle information (JRC) [4].
- Multiwire proportional counters are also used to determine the angle of the energetic fission products (PNPI) [5].

- Simply tilting the fission foils is used to understand the effect of fission product angular distribution (CERN) [6].

It would be helpful to evaluators and fission experimenters if a comparison of approaches could be made with factors such as angular and energy resolution, fission-product mass resolution, counting rate, and scattering mass for future correlation experiments with prompt neutron and gamma output. A detailed comparison of results for the important $^{238}U/^{235}U$ fission ratio [2,6] is certainly in order.

For the challenges of highly radioactive minor actinides, approaches previously used with lead slowing-down spectrometers, e.g. [7] and single-pulse neutron sources from nuclear explosions, e.g. [8] are not much used now. The former approach is very useful but the energy resolution is at best 30% in $\Delta E/E$. And nuclear explosions are not being conducted by developed countries nowadays. Instead, new high-intensity neutron sources are being pursued at CERN [3], J-PARC [9], and LANSCE [10]. For special actinides that are encapsulated, an approach based on the detection of the fission neutrons or gamma rays rather than the energetic fission products is being developed at J-PARC [11].

For intense neutron sources, many comparisons have been made, usually by researchers at one of the facilities to show how their facility is superior. It would be helpful to evaluators and to the science community if the important factors of the facilities were written down and a comparison could be made in terms such as neutron flux at the flight paths that are actually used, repetition rate, pulse width at the source, neutron energy spectra for given times-of-flight (due to scattering in the production target), backgrounds from scattered neutrons, availability of beam time, and so forth.

1.2. *Total kinetic energy (TKE) released in fission*

Most of the energy released in fission is in the form of kinetic energy of the fission fragments (before prompt neutron emission). What is measured is the kinetic energy of the fission products (after prompt neutron emission) and, with fairly reliable corrections, the fission fragment kinetic energies can be inferred. Fission models depend on knowledge of the fragment kinetic energies, the variance of these energies, and their dependence on the split into (Z1, A1, Z2, A2) (hereafter called "mass splits"), as they reflect on the degree of deformation of the fissioning nucleus at scission. Several active experiments are in progress to improve the data base of this fundamental observable for neutron-induced fission of essential nuclei such as ^{235}U, ^{238}U and ^{239}Pu. These new experimental efforts are inspired by fission models that have different predictions, questions of the accuracy of literature results, the need for data at higher incident neutron energies where pre-equilibrium neutron emission can be important, and the challenge of obtaining more detailed data with better resolutions of the masses and charges of the fission products. Recent measurements include:

- Measurements with well calibrated surface-barrier detectors at LANSCE by Oregon State University [12].
- Double Frisch-Grid measurements (LANSCE, and JRC-Geel) [4,13].
- Time-projection chamber; data from this instrument can be analyzed for TKE, although the focus on this work is mainly on determining the fission cross section (LANSCE) [2].

A detailed comparison of the results with advantages and limitations of each seems appropriate at this time. The dependence of TKE results on incident neutron energy, fragment angles, and mass splits is desired.

1.3. *Neutrons emitted promptly in fission, nu-bar, neutron multiplicity distribution P(nu), and neutron energies*

Accurate values of nu-bar are essential in calculating criticality and great efforts were expended in the past to study this observable with scintillator tanks. Also the distribution of neutron multiplicities P(nu) was determined. Recently experiments have concentrated on the correlation in angle and energy of prompt fission neutrons, and its dependence on mass splits. Prompt fission neutron spectra are being addressed to try to resolve discrepancies in literature results as summarized in [14,15]. Past measurements, originally analyzed without neutron scattering corrections, are now being reanalyzed with advanced Monte Carlo codes [16]. Recent experiments include:

- The Chi-Nu project at LANSCE to obtain more accurate data on the energies and angles of the prompt fission neutrons [17].
- Neutron detector arrays around a fission chamber to determine the energy and angles of correlated neutrons emitted promptly in fission at Michigan and LANSCE [18].

- Measurements of the mass split of fission products from photonuclear fission where the mass resolution is sufficient to calculate nu = A(target) +1 –A1 – A2 at GSI [19]. This innovative approach uses actinides at relativistic energies excited by the Coulomb field of a heavy nucleus. Upcoming experiments will use the Neuland neutron detector to obtain information on the neutrons emitted from each fragment.

1.4. *Prompt gamma-ray output*

Gamma-rays that are emitted promptly after fission and prompt neutron emission give information on the distribution of excitation energies of the fission products. These data are also of importance in calculating the heating of components in a reactor. Recent experiments focus on:

- Highly segmented, high efficiency gamma-ray detector based on BaF_2 scintillators for fission induced by neutrons below 100 keV [20].
- Precise measurements with $LaBr_3$ gamma-ray detectors at JRC-Geel [21].

1.5. *Fission product yields*

The yields of fission products are of importance for considerations of the disposal of radioactive waste, for neutron dosimetry, for interpretation of the properties of neutrinos in basic science experiments such as the so-called "reactor neutrino problem" [22], and as a constraint on fission models, which must reproduce the mass splits. The status of needs and research efforts were summarized recently [23] and include the following experimental programs:

- Precision measurements of certain fission product yields, especially for dosimetry (TUNL) [24].
- 2E-2V spectrometers, of which there are at least four world-wide are summarized in [23] and go by the names of (LANSCE) SPIDER, (JRC-Geel) VERDI, and (ILL and n_TOF) STEFF, some of which are described at this meeting.
- Fission product yield measurements with excellent mass and charge resolution in photonuclear fission, the SOFIA experiments (GSI) [19].
- Use of Penning traps, which have excellent mass resolution so that, in favorable cases, the yield of an isomer can be compared with the yield of the ground state for fission products (Jyväskylä) [25].

1.6. *Correlated observables*

Correlated prompt observables can give further constraints on fission models, and data on correlations are being requested for applications, such as for identification of fissile material. Recent work includes:

- Gamma-ray and neutron-gamma-ray correlations with highly segmented detectors at DANCE (LANSCE) [26]
- Neutron-neutron and neutron-gamma-ray correlations with neutron and gamma-ray detector arrays (Michigan, LANSCE) [18].
- Correlations between the direction of the fission fragments and the directions and energies of the emitted neutrons (JRC-Geel, PNPI) [4,5].
- Correlations between the fission fragment direction, the charge and mass split and gamma-ray output, e.g. the STEFF detector [27].

Further instruments are under development to increase the number of correlated observables and to increase the energy and mass resolution. A closer coupling of these efforts to fission models would be helpful now, for it is difficult to imagine making a spread sheet, as described above, for all of the observables. There is also the experimental limitation in focusing on a small region of charge and mass splits, a small region of TKE, specific angle and energy ranges of emitted neutrons and gamma rays, and so forth. When the regions of each are small, that is, well focused, the number of events is a very small fraction of the total fissions and so the counting times can be unreasonably long.

Correlations of prompt fission neutron and gamma-ray emission (with multiplicities, energy and angular correlations) with mass splits and kinetic energies will provide further information and constraints on fission models. These measurements will be difficult and guidance on priorities could come from the models. For example, will it be necessary to look for rare events, such as mass splits in the tail of the distributions? Will excellent resolution be required, such as in Z and A, or can the observables be characterized by a few moments? How many and what correlations will be necessary to address the models and their basic assumptions?

Summary

In summary, experiments in neutron-induced fission are being actively pursued at many laboratories with researchers from a much larger list of institutions. Many of these programs are reported at this meeting. With new techniques and capabilities, better accuracy is being achieved, uncertainties are being better understood, and new measurements are now possible. The results from these efforts will have significant impact in applications and will provide new constraints on physical models. Finally, some suggestions are made for experimenters to analyze and compare the strengths and weaknesses of these new approaches and to interact with fission modelers to provide data that will best elucidate the fission process.

Acknowledgments

This work was performed under the auspices of the U.S. Department of Energy at Los Alamos National Laboratory by the Los Alamos National Security, LLC under contract No. DE-AC52-06NA25396.

References

1. M. Heffner *et al.*, A time projection chamber for high accuracy and precision fission cross-section measurements, *Nucl. Instr. Meth. in Phys. Res. A* **759**, 50 (2014).
2. B. Manning, High precision neutron-induced fission cross sections using a time projection chamber, this conference.
3. M. Diakaki *et al.*, Fission activities at the CERN n_TOF facility, this conference.
4. F.-J. Hambsch *et al.*, Prompt neutron emission correlations with fission fragment properties, this conference.
5. A. S. Vorobyev *et al.*, Measurements of angular and energy distributions of prompt neutrons from thermal neutron-induced fission, *Nuclear Instr. Meth. in Phys. Res.* **A598**, 795 (2009).
6. C. Paradela *et al.*, High-accuracy determination of the $^{238}U/^{235}U$ fission cross section ratio up to ≈ 1 GeVat n_TOF at CERN, *Phys. Rev.* **C91**, 024602 (2015).
7. D. Rochman *et al.*, Characteristics of a lead slowing-down spectrometer coupled to the LANSCE accelerator, *Nucl. Instr. Meth. Phys. Res.* **A 550**, 397–413 (2005).

8. M. S. Moore and G. A. Keyworth, *Phys. Rev.* **C3**, 1656 (1971).
9. J-PARC website: http://www.j-parc.jp/index-e.html
10. P. W. Lisowski and K. F. Schoenberg, *Nucl. Instrum. Meth. in Phys. Res.* **A 562**, 910 (2006).
11. K. Hirose *et al.*, Simultaneous measurement of neutron-induced fission and capture cross sections for ^{241}Am at neutron energies below fission threshold, *Nucl. Instr. Meth. Phys. Res.* A (accepted, in press).
12. W. Loveland and J. King, Total kinetic energy release in the fast neutron induced fission of ^{232}Th and ^{238}U, this conference.
13. D. L. Duke *et al.*, TKE measurements, this conference.
14. R. Capote *et al.*, Prompt Fission Neutron Spectra of Actinides, *Nucl. Data Sheets* **131**, 1 (2016).
15. D. Neudecker *et al.*, The Need for Precise and Well-documented Experimental Data on Prompt Fission Neutron Spectra from Neutron-induced Fission of ^{239}Pu, *Nuclear Data Sheets* **131**, 289 (2016).
16. T. N. Taddeucci, Multiple-scattering Corrections to Measurements of the Prompt Fission Neutron Spectrum, *Nuclear Data Sheets* **123**, 135 (2015).
17. J. A. Gomez *et al.*, Measuring PFNS a the Chi-Nu Experiment, this conference.
18. A. Enqvist, M. Flaska and S.A. Pozzi, Measurement and simulation of neutron/gamma-ray cross-correlation functions from spontaneous fission. *Nucl. Instr. & Meth. A* **595**, 426 (2008).
19. J. Taieb *et al.*, SOFIA studies on fission with Aladin, this conference.
20. J. L. Ullmann *et al.*, Prompt γ-ray production in neutron-induced fission of ^{239}Pu, *Phys. Rev.* **C 87**, 044607 (2013).
21. A. Oberstedt *et al.*, New prompt fission γ-ray spectral data and its implication on present evaluated nuclear data files, *Physics Procedia* **47**, 156 (2013).
22. A. A. Sonzogni, E. A. McCutchan, T. D. Johnson and P. Dimitriou, Effects of Fission Yield Data in the Calculation of Antineutrino Spectra for ^{235}U(n,fission) at Thermal and Fast Neutron Energies, *Phys. Rev. Lett.* **116**, 132502 (2016).
23. P. Dimitriou, F.-J. Hambsch and S. Pomp, Fission Product Yields Data, Current status and perspectives, INDC(NDS)-0713 (2016).
24. A. Tonchev, Fission product yield measurements using quasi-monoenergetic neutrons, this conference.
25. H. Penttila, Novel technique for fission yield measurements applying ion traps, this conference.

26. G. Rusev, B. Baramsai *et al.*, Measurements of correlated fission data with DANCE and NEUANCE, this conference.
27. J. Matarranza *et al.*, A multiparameter nuclear-fission experiment: Can all be obtained at once?, *Physics Procedia* **47,** 76 (2013).

Nuclear Structure I

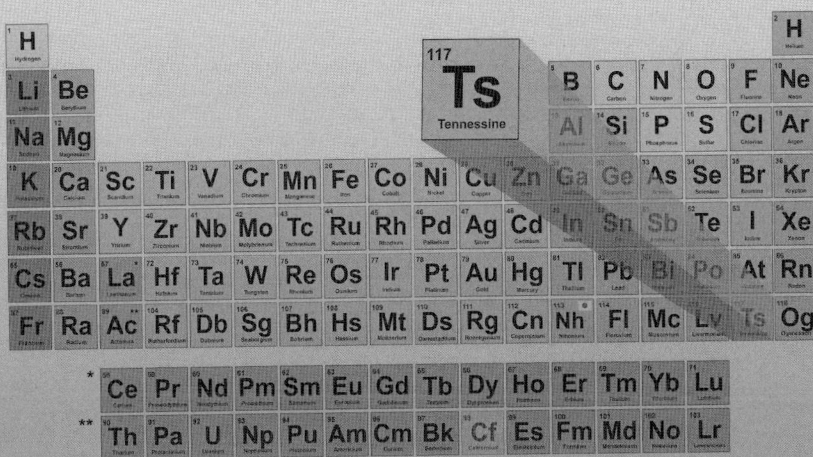

Fluctuations in Neutron-Induced Fission Cross Sections

G.F. Bertsch

Department of Physics and Institute of Nuclear Theory, Box 351560
University of Washington, Seattle, Washington 98915, USA

bertsch@uw.edu

David Brown

National Nuclear Data Center, Brookhaven National Laboratory, Upton NY, USA

We examine the published experimental data on the ^{235}U(n,f) reaction with a goal to quantify cross section fluctuations above the fission barrier. For neutron energies in the range 10 keV - 25 keV we find that there are significant fluctuations of the order of 5% measured at 0.25 keV energy resolution. Here we analyze the fluctuations by the autocorrelation function with the following conclusions: 1) The correlation width associated with peak at zero energy offset is smaller than the experimental energy resolution; 2) there are peaks in the autocorrelation function on an energy scale of several keV. Such a multi-peaked structure is incompatible with Ericson's statistical treatment of compound nucleus fluctuations but may arise in more detailed models of the fission dynamics.

1. Introduction

The dynamics of fission above the fission barrier is far from well understood. Present theory [1] is largely based on the Bohr-Wheeler framework [2] which introduced the concept of fission channels. The channels are characterized by their energy-dependent transmission coefficients $T(E)$ but there still no predictive theory to calculate them. Commonly T is parameterized by the Hill-Wheeler formula [3, p. 1140],

$$T(E) = (1 + \exp(2\pi(E - E_B)/\hbar\omega))^{-1} \tag{1}$$

where E_B is the barrier height and the ω is a frequency associated with the thickness of the barrier. However, there is no compelling experimental data to support the Hill-Wheeler parameterization. The transmission coefficients could equally well be described as resonant states at the barrier top, in analogy to quantum dots bridging conductors in the mesoscopic theory of electron transport [8]. The corresponding formula is

$$T(E) = \Gamma_R \Gamma_L / ((E - E_r)^2 + (\Gamma_R + \Gamma_L)^2/4). \tag{2}$$

This depends on three parameters, namely the resonance energy and the decay widths to the left and right. Measurement of the fluctuations in the cross section may provide a way to distinguish between these possibilities. In the Hill-Wheeler approach, the channels themselves introduce very little fluctuations–they open over an energy range of the order of $\hbar\omega \sim 1$ MeV, and once open they remain so for all higher energy. The only mechanism to introduce fluctuations on a finer scale is Ericson's theory [4] treating the statistical feeding of the channels from inside the defining barrier. In contrast, the picture of the channels as resonant states bridging the barrier permits arbitrary fluctuations associated with the widths Γ_R and Γ_L.

With this mind, we examine here the experimental evidence for fluctuations above the fission barrier. There is much data on neutron-induced reaction on ^{235}U including the energy dependence of fission, capture, and total cross sections. The fluctuations associated with different entrance channels are likely to be independent, and the best opportunity to observe them is when the number of entrance channels is small. For neutron energies below a few keV, cross sections are dominated by s-wave neutrons with two total angular momentum channels for an odd-A target such as ^{235}U. With this in mind, we focus on the data for the range of neutron energies from 10 to 25 keV.

2. The data

Three independent measurements of the reaction ^{235}U(n,f) in the energy range $E_n = 10 - 25$ keV are shown in Fig. 1. It is obvious that there is some fluctuation outside of the limits of experimental error. For example, note the prominent peak at 22 keV, present in all three measurements. The peak could arise from a doorway entrance channel or a transmission factor peak from in a resonant-state fission channel. The remaining possibility, a localized dip in the capture cross section is very unlikely due to the large number of capture channels.

We start with a very elementary test for fluctuations. This is the variance of the measured data with respect to a polynomial fit. Here we make

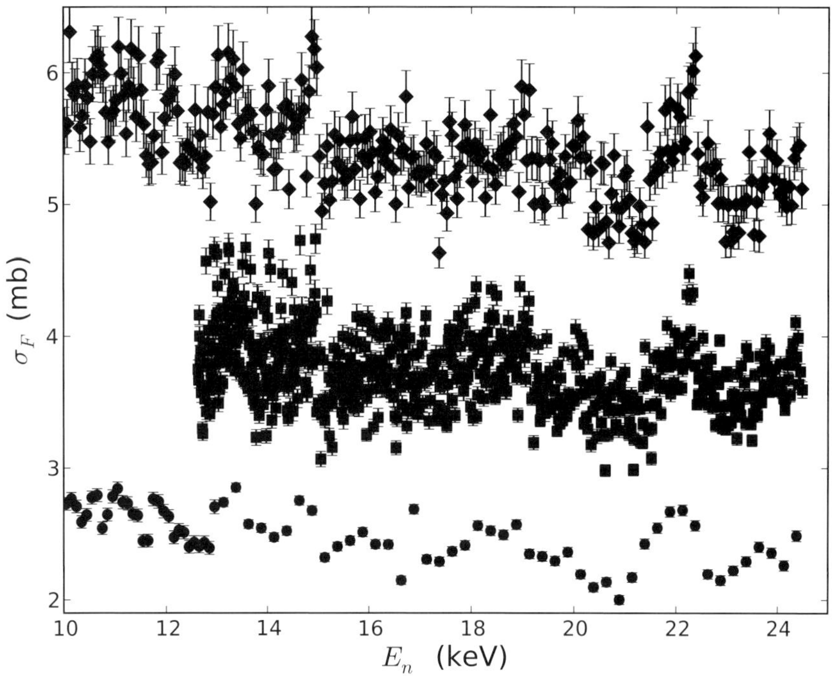

Fig. 1. Experimental fission cross sections in the range 10–25 keV. Circles: from Ref. [7]; squares: from [5]; diamonds: from [6]. The latter two data sets have been shifted upward for clarity in the plotting the figure.

a quadratic fit

$$\sigma_{fit}(E) = \bar{\sigma} + a(E - \bar{E}) + b(E - \bar{E})^2 \tag{3}$$

to each data set and compare the variance of the fit error

$$v = \left(\frac{1}{N_d} \sum_i^{D_d} (\sigma_{exp}(E_i) - \sigma_{fit}(E_i))^2 \right)^{1/2} \tag{4}$$

to the reported experimental error bar v_{exp}. The results are shown in Table 1. The variance for the data on the top line is nearly 4 times larger than the average quoted experimental error. The variance for the data on the other lines is even larger. We conclude that fluctuations in the cross sections are present at keV energy scales and can be measured. In conventional theory, the only way to explain them is by Ericson fluctuations associated with the compound nucleus states in either the first or second wells of the fission potential energy surface.

Table 1. Cross section variances for the data sets [5, 7] plotted in Fig. 1. See text for details.

data range keV	source	$\bar{\sigma}$ b	a b/keV	b b/keV^2	v b	v_{exp} b
10 - 24.5	[7]	2.4521	-0.0331	0.00250	0.15	0.04
12.6 - 24.5	[5]	2.2844	-0.0240	0.00348	0.25	0.06
10 -25	[6]	2.3415	-0.0457	0.00362	0.27	0.17

3. Autocorrelation analysis

More detailed information about the fluctuations can be obtained by an analysis of the autocorrelation function. The usual formulation assumes a constant average cross section, but this is not tenable over a large range of energies. We shall apply the analysis by examining the correlation in the reduced cross section $\tilde{\sigma}(E)$ obtained by normalizing to the fit function $\sigma_{fit}(E)$.

$$\tilde{\sigma}(E) = \frac{\sigma(E)}{\sigma_f(E)}. \tag{5}$$

The autocorrelation function is then defined as

$$C(\epsilon) = \langle \tilde{\sigma}(E)\tilde{\sigma}(E+\epsilon)\rangle \tag{6}$$

As shown by Ericson [4], the correlation function can be simply interpreted if the reaction is mediated by independent compound-nucleus resonances feeding the appropriate exit channels. Then it has the form

$$C(\epsilon) \approx 1 + \frac{1}{N}\frac{1}{1+(\epsilon/\Gamma)^2} \tag{7}$$

where Γ is the average decay width of the resonances and N is the effective number of exit channels.

We show in Fig. 2 the results for the data from [7] in the range 10–25 keV, using the quadratic fit for the average cross section. The experimental data was partially rebinned with a bin size $E_B = 0.25$ keV to calculate C in steps of E_B. The error bars on the graph were obtained by sampling data sets with random errors added to the experimental values. The added errors are Gaussian distributed with a variance given by the quoted experimental error. We see that the correlation function has a multi-peak structure. The first peak at $\epsilon = 0$ is very sharp, having an excess of 0.0043 and falling immediately to 1 in the next bin at $\epsilon = 0.25$ keV. It is clear that Γ in the

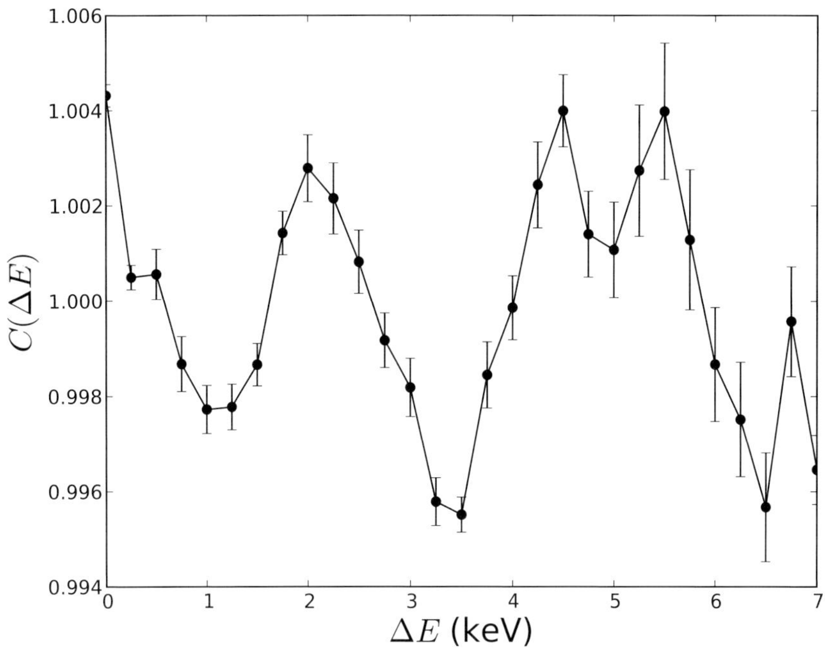

Fig. 2. Autocorrelation function for ^{235}U(n,f) in the range fission cross sections in the range 10–25 keV for the data in Ref. [7].

Ericson formula must be much less than the bin size. In that limit, the excess can be interpret by the formula [9]

$$C(0) - 1 = \frac{1}{N} \frac{1}{1 + (E_B/\pi\Gamma)}. \tag{8}$$

The average width Γ of the compound nucleus states in the first well can be reliably estimated as $\Gamma \approx 0.2$ eV. Eq. (8) then yields $N \approx 0.6$ which is much too small to be compatible with the experimental reaction cross section [10]. However, the existence of other peaks in C renders the entire Ericson fluctuation analysis problematic. The correlation function does not look anything like Eq. (7), deviating more from unity at finite ϵ than at $\epsilon = 0$. We have not shown the correlation functions extracted from the other data sets cited in Table I, but they also show a multi-peak structure similar to that exhibited in Fig. 2. Obviously, the dynamics are far more complicated than the compound nucleus picture permits.

4. Outlook

There are several possibilities to gain some understanding of the measured autocorrelation function. The first that comes to mind that doorway states in the entrance channel cause fluctuations in the reaction cross section. This can be dealt with by examining the ratio of fission to capture cross section, called α^{-1} in the literature. This analysis was attempted in Ref. [11] which found that large fluctuations still were present in the ratio. Another possibility is that the fluctuations are due to individual states in the second well acting like the resonant transition states in Eq. (2). Such resonance fluctuations are well known in fission under the barrier, but the common assumption is that the widths of these resonance become too large to play a role above the barrier. So far there is no theory to calculate these widths, but it is not out the question that theoretical tools can be developed to estimate these widths using phenomenological Hamiltonians based on nucleon-nucleon interactions.

Acknowledgments

This material is based upon work supported by the US Department of Energy, Office of Science, Office of Nuclear Physics, under Contract No. DE-SC0012704 (BNL).

References

[1] See for example: O. Bouland, J.E. Lynn and P. Talou, *Phys. Rev. C* **88** 054612 (2013).
[2] N. Bohr and J.A. Wheeler, *Phys. Rev.* **56** 426 (1939).
[3] D.A. Hill and J.A. Wheeler, *Phys. Rev.* **89** 1102 (1953).
[4] T. Ericson, *Ann. Phys.* (NY) **33** 390 (1963).
[5] C.D. Bowman, G.S. Sidhu, M.L. Stelts and J.C. Browne, 3rd Conf. Neutron Cross-Sections+Tech., Knoxville 1971, Vol. 2, p. 584.
[6] R.B. Perez *et al.*, *Nucl. Sci. Eng.* **55** 203 (1974).
[7] M.S. Moore *et al.*, *Phys. Rev. C* **18** 1328 (1978).
[8] Y. Alhassid, *Rev. Mod. Phys.* **72** 895 (2000).
[9] P. Fessenden, W.R. Gibbs and R.B. Leachman, *Phys. Rev. Lett.* **15** 796 (1965).
[10] M.S. Moore *et al.*, *Phys. Rev. C* **30** 214 (1984).
[11] G.F. Bertsch and T. Kawano, arXiv:1701.00276 (2017).

Binding Energies and Pairing Gaps in Semi-Magic Nuclei Obtained Using New Regularized Higher-Order EDF Generators

K. Bennaceur,[1,2,3] J. Dobaczewski,[2,3,4,5] and Y. Gao (高原)[6]

[1] *Univ Lyon, Université Lyon 1, CNRS/IN2P3, IPNL, F-69622 Villeurbanne, France*
[2] *Department of Physics, PO Box 35 (YFL), FI-40014 University of Jyväskylä, Finland*
[3] *Helsinki Institute of Physics, P.O. Box 64, FI-00014 University of Helsinki, Finland*
[4] *Department of Physics, University of York, Heslington, York YO10 5DD, United Kingdom*
[5] *Institute of Theoretical Physics, Faculty of Physics, University of Warsaw, ul. Pasteura 5, PL-02-093 Warsaw, Poland*
[6] *State Key Laboratory of Nuclear Physics and Technology, School of Physics, Peking University, Beijing 100871, China*

We present results of the Hartree-Fock-Bogolyubov calculations performed using nuclear energy density functionals based on regularized functional generators at next-to-leading and next-to-next-to-leading order. We discuss properties of binding energies and pairing gaps determined in semi-magic spherical nuclei. The results are compared with benchmark calculations performed for the functional generator SLyMR0 and functional UNEDF0.

Keywords: Nuclear Energy Density Functionals, Regularized Functional Generators, Nuclear Binding Energies, Pairing Gaps.

1. Introduction

A quest for energy density functionals (EDFs) that would precisely and accurately describe multitude of low-energy nuclear properties is at present one of the most important research avenues in nuclear physics. The problem has been addressed in quite a number of recent studies,[1–10] where various extensions of the standard EDFs, used for the last 60-odd years, were proposed.

In this conference communication, we present results of calculations performed for semi-magic nuclei across the mass chart, using the newly developed EDFs based on the regularized higher-order generators with pairing.[10] As discussed in Ref. 10, the proposed new parametrizations at next-to-leading order (NLO) REG2c.161026 and next-to-next-to-leading (N^2LO) REG4c.161026 correspond to a fairly low effective mass and overestimate pairing strength. Therefore, they are not good enough to embark for them on massive mass-table calculations. However, inexpensive calculations performed for spherical semi-magic nuclei, which are reported on in this paper, can constitute a useful illustration of the overall bulk properties corresponding to the newly developed EDFs.

2. Results

We performed the Hartree-Fock-Bogolyubov (HFB) calculations for all bound semi-magic nuclei across the mass chart, using the NLO REG2c.161026 and N^{2}LO REG4c.161026 EDFs,[10]) SLyMR0,[7]) and UNEDF0.[11]) For UNEDF0, the Lipkin-Nogami (LN) method was used to account for approximate particle-number restoration as in Ref. 12. For the finite-range NLO and N^{2}LO generators, we solved the non-local self-consistent equations using the newly developed code FINRES$_4$ (Finite-Range Self-consistent Spherical Space-coordinate Solver),[13] which is based on the method proposed by Hooverman.[14] For the zero-range generator SLyMR0, we obtained the solutions using the spherical solver LENTEUR,[15] and for the quasi-local functional UNEDF0, using the code HOSPHE,[16,17] similarly as in Ref. 18.

For NLO, N^{2}LO, and UNEDF0, where the fit covariance matrices are known, we determined statistical uncertainties of all calculated observables according to the methodology presented in Ref. 20. For SLyMR0, only the values of observables were determined.

In Figs. 1 and 2, for nuclei where the binding energies are known from experiment or systematics,[19]) we show the binding-energy residuals determined for the four studied EDFs. We note that for SLyMR0, the obtained differences between theory and experiment are significantly larger than those obtained for the other three EDFs, and therefore, in the figures they were divided by a factor of five. We also note that NLO, N^{2}LO, and SLyMR0 EDFs are built as exact averages of two-body or many-body generators, both in the particle-hole and pairing channels, and therefore, they are suitable without ambiguity for beyond-mean-field and symmetry-restoration calculations. On the other hand, EDF UNEDF0 is built as an average of a density-dependent two-body Skyrme-type generators that are different in particle-hole and pairing channels and with some omitted terms in the particle-hole channel.

Although the pattern of comparison with data is fairly different among all four studied EDFs, we clearly see that the new NLO and N^{2}LO EDFs

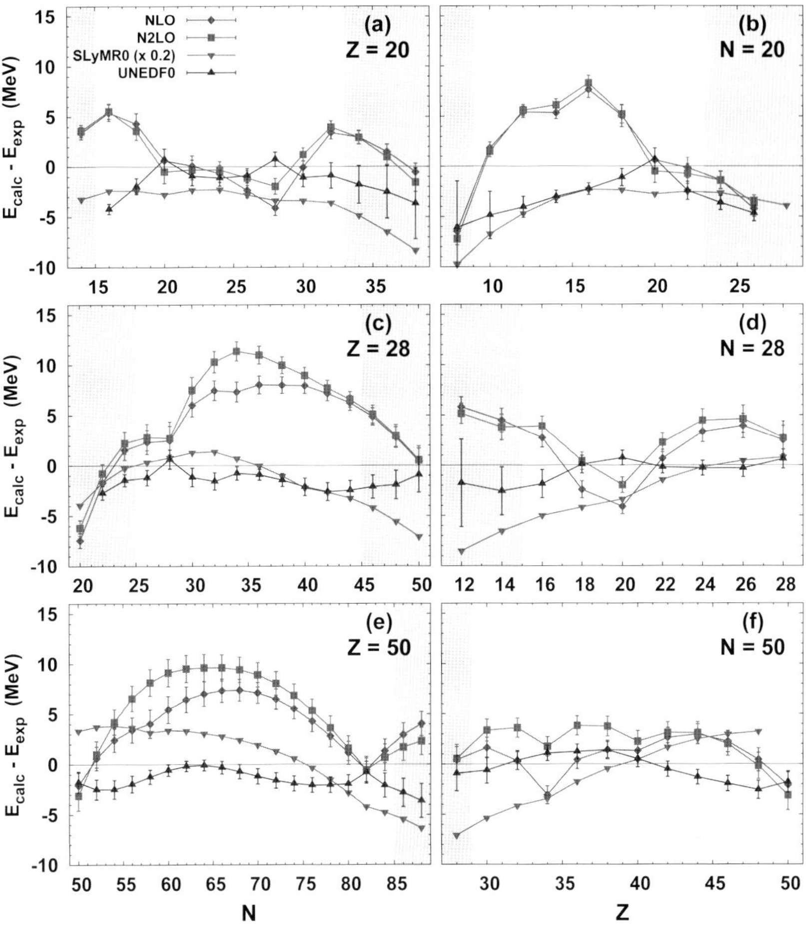

Fig. 1. Differences between calculated (HFB results) and experimental (AME12 [19]) ground-state energies of proton-magic (left panels) and neutron-magic nuclei with $Z, N = 20$, 28, or 50. Calculations were performed using EDFs NLO REG2c.161026 [10]) (diamonds), N^2LO REG4c.161026 [10]) (squares), SLyMR0 [7]) (down triangles), and UNEDF0 [11]) (up triangles). For SLyMR0, to fit in the figure, the differences were divided by a factor of five. Shaded zones correspond to the AME12 masses taken from systematics.

describe data better than SLyMR0 (recall the scaling factor of five used for SLyMR0). However, results obtained at NLO and N^2LO are still fairly worse than those obtained for the standard Skyrme-like EDF UNEDF0. In particular, for the NLO and N^2LO EDFs, we see conspicuous “arches” of

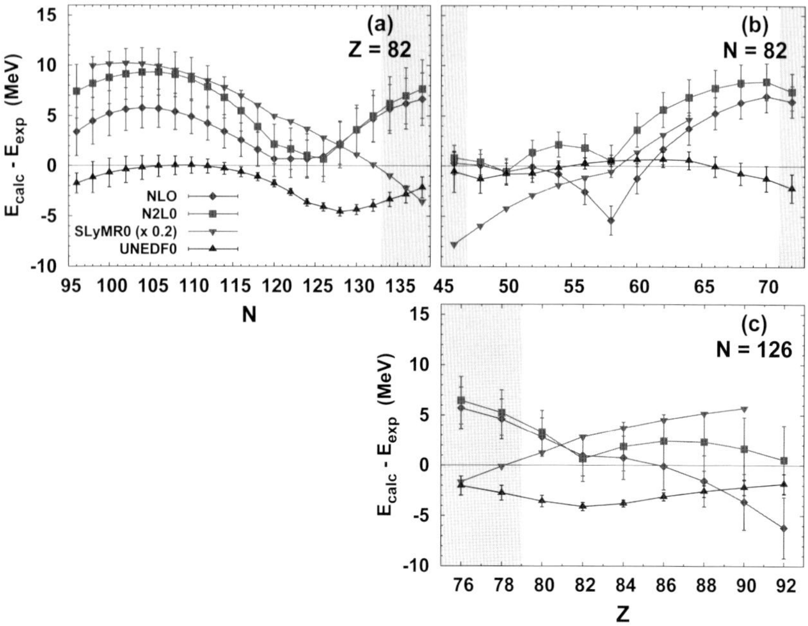

Fig. 2. Same as in Fig. 1 but for the semi-magic nuclei with $Z, N = 82$ and $N = 126$.

residuals between the doubly magic nuclei. Usually these feature of calculated ground-state energies was attributed to low effective mass, however, we note here that for SLyMR0, which has a low effective mass too, no such an effect is seen. We also note that the statistical uncertainties obtained for UNEDF0 significantly increase when going towards neutron-rich nuclei,[18] whereas those for the NLO and N^{2}LO EDFs depend on the neutron excess much less.

In Figs. 3 and 4, we show neutron and proton pairing gaps calculated as pairing fields averaged with density matrices,[12] and for the LN method (UNEDF0), corrected by adding the corresponding λ_2 LN parameters.[12] As discussed in Ref. 10, for the NLO and N^{2}LO EDFs, pairing correlations were adjusted to values largely overestimating experimental data. This feature is clearly visible in the figures, and should certainly be improved upon in future planned adjustment of parameters. We also see that the HFB results shown for the NLO, N^{2}LO, and SLyMR0 EDFs exhibit unphysical breaking of pairing at doubly magic gaps, which is a feature related to the lack of particle-number restoration, and thus is absent in the HFB+LN results shown for UNEDF0.

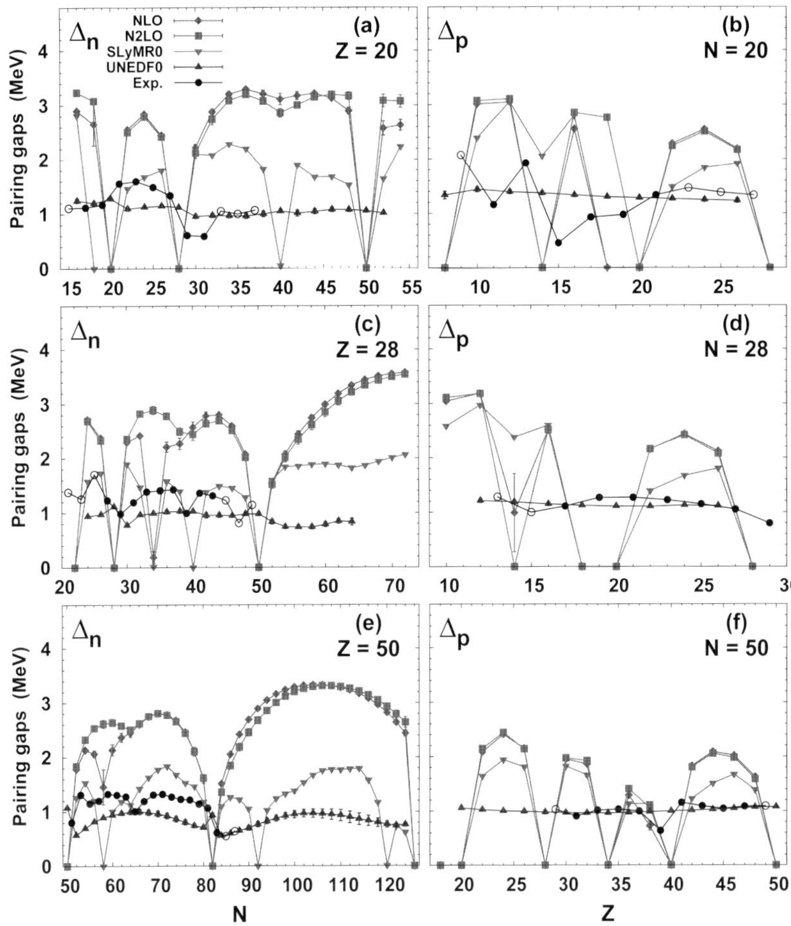

Fig. 3. Calculated (HFB results) and experimental (AME12[19]) neutron pairing gaps of proton-magic nuclei (left panels) and proton pairing gaps of neutron-magic nuclei with $Z, N = 20$, 28, or 50. Calculations were performed using EDFs NLO REG2c.161026[10]) (diamonds), N^{2}LO REG4c.161026[10]) (squares), SLyMR0[7]) (down triangles), and UNEDF0[11]) (up triangles). Experimental values (circles) correspond to the three-point mass staggering centered at odd particle numbers.[21]) Open circles indicate values, for which at least one of the three AME12 masses were taken from systematics.

3. Conclusions

In this article we presented the binding-energy residuals and average pairing gaps obtained for semi-magic nuclei using two recently adjusted finite-range

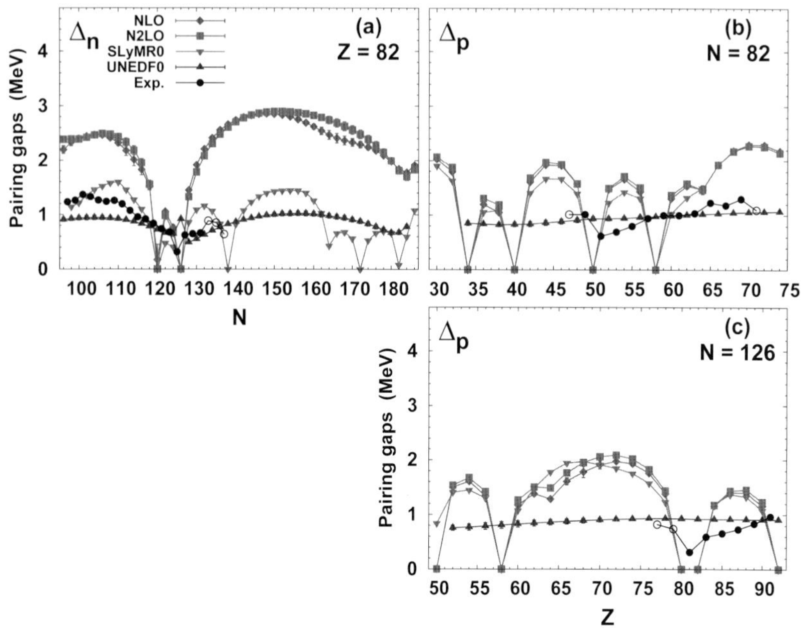

Fig. 4. Same as in Fig. 3 but for the semi-magic nuclei with $Z, N = 82$ and $N = 126$.

pseudopotentials at NLO and N^{2}LO, as well as the EDF UNEDF0 and zero-range pseudopotential SLyMR0. For all cases but SLyMR0, the propagated statistical errors of observables were calculated.

For the set of nuclei considered here, and for finite-range pseudopotentials, the average deviations between experimental and calculated binding energies are larger than those obtained for EDF UNEDF0, but they are significantly smaller than the ones obtained for the zero-range pseudopotential SLyMR0. For the NLO and N^{2}LO EDFs, the typical arches that appear in the binding-energy residuals between major shells might be due to their very low effective masses (close to 0.4). However, the fact that similar arches do not appear for SLyMR0, which has an effective mass of 0.47, questions this conjecture.

Based on the results obtained for binding-energy residuals and average pairing gaps, there was no clear improvement when going from the finite-range pseudopotential at NLO to the one at N^{2}LO. This does not necessarily mean that the additional degrees of freedom introduced at N^{2}LO are not relevant, but most likely reflects the fact that the penalty function based on spherical doubly magic nuclei did not allow for properly constraining them.

The next step in the development of this family of pseudopotentials will consist in increasing the effective mass. The obvious way to do this is to introduce three-body terms in the pseudopotential. The present computational ressources restrict this extension to the introduction of zero-range three-body terms in the spirit of the work of Onishi and Negele[22] and will require the use of a cut-off to prevent the divergence of the energy. The work along these lines is in progress.

Acknowledgments

This work was supported by the Academy of Finland, the University of Jyväskylä within the FIDIPRO program, by the CNRS/IN2P3 through PICS No. 6949.

References

1. B. G. Carlsson, J. Dobaczewski and M. Kortelainen, Local nuclear energy density functional at next-to-next-to-next-to-leading order, *Phys. Rev. C* **78**, p. 044326 (2008).
2. M. Zalewski, J. Dobaczewski, W. Satuła and T. R. Werner, Spin-orbit and tensor mean-field effects on spin-orbit splitting including self-consistent core polarizations, *Phys. Rev. C* **77**, p. 024316 (Feb 2008).
3. F. Raimondi, B. G. Carlsson and J. Dobaczewski, Effective pseudopotential for energy density functionals with higher-order derivatives, *Phys. Rev. C* **83**, p. 054311 (2011).
4. J. Dobaczewski, K. Bennaceur and F. Raimondi, Effective theory for low-energy nuclear energy density functionals, *J. Phys. G* **39**, p. 125103 (2012).
5. F. Raimondi, K. Bennaceur and J. Dobaczewski, Nonlocal energy density functionals for low-energy nuclear structure, *J. Phys. G: Nucl. Part. Phys.* **41**, p. 055112 (2014).
6. K. Bennaceur, J. Dobaczewski and F. Raimondi, Extended skyrme pseudopotential deduced from infinite nuclear matter properties, *EPJ Web of Conf.* **66**, p. 02031 (2014).
7. J. Sadoudi, M. Bender, K. Bennaceur, D. Davesne, R. Jodon and T. Duguet, Skyrme pseudo-potential-based EDF parametrization for spuriosity-free MR EDF calculations, *Phys. Scr.* **T154**, p. 014013 (2013).

8. J. Sadoudi, T. Duguet, J. Meyer and M. Bender, Skyrme functional from a three-body pseudopotential of second order in gradients: Formalism for central terms, *Phys. Rev. C* **88**, p. 064326 (2013).
9. D. Davesne, J. Navarro, P. Becker, R. Jodon, J. Meyer and A. Pastore, Extended skyrme pseudopotential deduced from infinite nuclear matter properties, *Phys. Rev. C* **91**, p. 064303 (Jun 2015).
10. K. Bennaceur, A. Idini, J. Dobaczewski, P. Dobaczewski, M. Kortelainen and F. Raimondi, arXiv:1611.09311.
11. M. Kortelainen, T. Lesinski, J. Moré, W. Nazarewicz, J. Sarich, N. Schunck, M. V. Stoitsov and S. Wild, Nuclear energy density optimization, *Phys. Rev. C* **82**, p. 024313 (Aug 2010).
12. M. V. Stoitsov, J. Dobaczewski, W. Nazarewicz, S. Pittel and D. J. Dean, Systematic study of deformed nuclei at the drip lines and beyond, *Phys. Rev. C* **68**, p. 054312 (2003).
13. Bennaceur K. *et al.* 2017, to be submitted to Computer Physics Communications.
14. R. H. Hooverman, A technique for numerical solution of the Schroedinger equation with non-local potentials, *Nuclear Physics A* **189**, 155 (1972).
15. K. Bennaceur, LENTEUR HFB code unpublished.
16. B. Carlsson, J. Dobaczewski, J. Toivanen and P. Veselý, Solution of self-consistent equations for the N3LO nuclear energy density functional in spherical symmetry. The program HOSPHE (v1.02), *Comput. Phys. Comm.* **181**, p. 1641 (2010).
17. B. G. Carlsson, J. Toivanen, P. Veselý and Y. Gao, to be published.
18. Y. Gao, J. Dobaczewski, M. Kortelainen, J. Toivanen and D. Tarpanov, Propagation of uncertainties in the skyrme energy-density-functional model, *Phys. Rev. C* **87**, p. 034324 (2013).
19. M. Wang, G. Audi, A. H. Wapstra, F. G. Kondev, M. MacCormick, X. Xu and B. Pfeiffer, The AME2012 atomic mass evaluation (ii). tables, graphs and references, *Chin. Phys. C* **36**, 1603 (2012).
20. J. Dobaczewski, W. Nazarewicz and P.-G. Reinhard, Error Estimates of Theoretical Models: a Guide, *J. Phys. G: Nucl. Part. Phys.* **41**, p. 074001 (2014).
21. W. Satuła, J. Dobaczewski and W. Nazarewicz, Odd-even staggering of nuclear masses: Pairing or shape effect?, *Phys. Rev. Lett.* **81**, 3599 (1998).
22. N. Onishi and J. Negele, Two-body and three-body effective interactions in nuclei, *Nuclear Physics A* **301**, 336 (1978).

Covariant Density Functional Theory and Beyond for Deformed Nuclei*

Peter Ring

Physics Department, Technical University Munich, D85738 Garching, Germany

Density functional theory (DFT) provides a very successful description of nuclear structure in medium heavy and heavy nuclei. Covariant versions (CDFT) take into account full Lorentz invariance, a basic symmetry of QCD. It puts stringent restrictions on the number of parameters without loosing the good agreement with experimental data. The concept of DFT is a mapping of the complicated quantum mechanical many-body problem to a one-body problem, which can be solved relatively easily.

Large-scale axially symmetric relativistic Hartree-Bogoliubov calculations are presented for all $Z \leq 104$ even-even nuclei between the two-proton and two-neutron drip lines for several covariant density functionals. The spreads of predictions within these functionals are discussed in known regions of the nuclear chart and their propagation towards the neutron drip line.

On the mean field level the one-body densities for the ground states correspond to a fixed deformation, i.e. to a single point in deformation space. Therefore the success of static DFT is limited to the ground states of nuclei with well pronounced minima.

*Supported by the DFG cluster of excellence "Origin and Structure of the Universe" (www.universe-cluster.de).

In order to describe shape coexistence and to take into account fluctuations in deformation space the generator coordinate method (GCM) is used. Here the wave functions are linear combinations of Slater determinants with various deformations and the symmetries broken on the mean field level are restored by projection techniques. We discuss recent covariant applications of these methods for the description of shape co-existence, quantum-phase transitions and neutrino-less double beta decay.

These methods are powerful but they require considerable numerical efforts. For heavy nuclei, the Gaussian overlap approximation allows a completely microscopic derivation of a collective hamiltonian, which is relatively easy to solve. The validity of this approximation is investigated in a benchmark calculation for the transitional nucleus 76Kr.

The advantage of all these methods is the fact that there are no new parameters necessary. All the results are derived from well known and successful universal covariant density functionals.

Low Energy Magnetic Radiation (LEMAR) of Warm Nuclei

S. Frauendorf
University Notre Dame,
Notre Dame, IN 46556, USA
E-mail: sfrauend@nd.edu

B. A. Brown
National Superconducting Cyclotron Laboratory and Department of Physics and Astronomy, Michigan State University,
East Lansing, Michigan 48824, USA

R. Schwengner
Institut für Strahlenphysik, Helmholtz Zentrum Dresden Rossendorf,
01328 Dresden, Germany

The enhancement observed below 2 MeV in the radiative strength function of nuclei near closed shells is explained by shell model calculations as M1 transitions between excited states. In the open-shell a change to a bimodal structure composed of the zero energy spike and a scissors resonance is found. The features are caused by realignment of high-j orbitals.

Keywords: Radiative Strength, Shell Model, High-j Orbitals.

1. Introduction

Photonuclear reactions and the inverse radiative-capture reactions between nuclear states in the region of high excitation energy and large level density are of considerable interest in many applications. Radiative neutron capture, for example, plays a central role in the synthesis of the elements in various stellar environments, for next-generation nuclear technologies, and as the transmutation of long-lived nuclear waste.

A critical input to calculations of the reaction rates is the average strength of the cascade of γ-transitions de-exciting the nucleus, which is described by the radiative strength function. An increase of the dipole strength function below 3 MeV toward low γ-ray energy has recently been

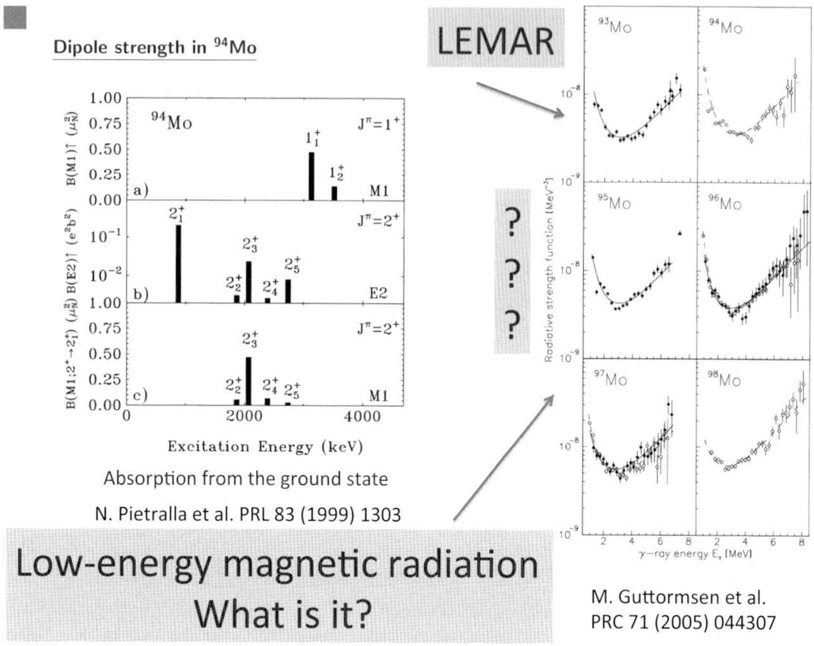

Fig. 1. Left: Reduced absorption probabilities from the ground state by (γ, γ') experiments[1]. Right: Strength function for γ de-excitation after the (^{3}He,^{3}He') reaction[2].

observed in nuclides in the mass range from $A \approx 40$ to 100. As an example, Fig. 1 shows the γ strength function obtained from (^{3}He,^{3}He') reactions on various Mo isotopes. The enhancement is not observed in the inverse process of absorbing γ-quanta by nuclei in the ground state. Only few discrete lines are found within the interval of the first 4 MeV. The question arises which is the origin of the radiation enhancement and how does it depend on the N and Z. Refs.[3,4] showed that the low-energy enhancement may increase the neutron capture rate in the stellar r-process environment up to a factor 100-1000. The dominant dipole character of the low-energy strength was demonstrated in Ref.[5] and an indication for an M1 character was discussed for the case of ^{60}Ni[6].

2. The LEMAR spike

Our Shell Model studies of the Mo isotopes explained the enhancement as Low Energy MAgnetic Radiation (LEMAR) generated by transitions

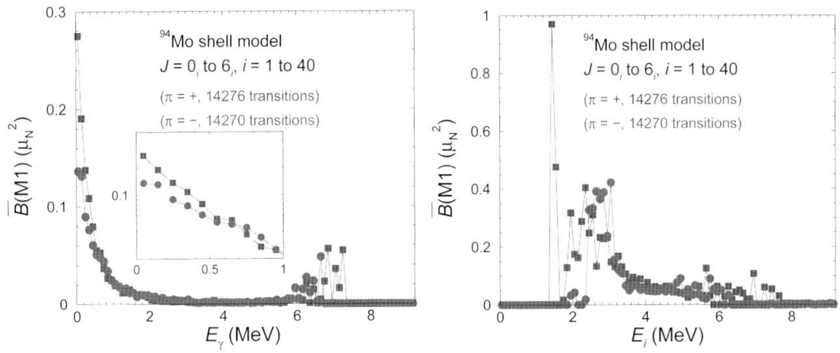

Fig. 2. Left: Average $B(M1)$ values in 100 keV bins of transition energy calculated for positive-parity (squares) and negative-parity (circles) states in ^{94}Mo. The inset shows the low-energy part in logarithmic scale. Right: Average $B(M1)$ values in 100 keV bins of excitation energy calculated for positive-parity (squares) and negative-parity (circles) states in ^{94}Mo. From [7].

between excited states few MeV above yrast [7]. The Shell Model calculations included the active proton orbits $\pi(0f_{5/2}, 1p_{3/2}, 1p_{1/2}, 0g_{9/2})$ and the neutron orbits $\nu(0g_{9/2}, 1d_{5/2}, 0g_{7/2})$ relative to a ^{68}Ni core. A set of empirical matrix elements for the effective interaction and effective g-factors of $g_s^{\rm eff} = 0.7g_s^{\rm free}$ have been used. The calculations included states with spins from $J = 0$ to 6 for ^{90}Zr and ^{94}Mo and from $J = 1/2$ to $13/2$ for ^{95}Mo. For each spin the lowest 40 states were calculated. The reduced transition probabilities $B(M1)$ were calculated for all transitions from initial to final states with energies $E_f < E_i$ and spins $J_f = J_i, J_i \pm 1$. This resulted in more than 14000 $M1$ transitions for each parity $\pi = +$ and $\pi = -$, which were sorted into 100 keV bins according to their transition energy $E_\gamma = E_i - E_f$. The average $\overline{B}(M1, E_\gamma)$ value for one energy bin was obtained as the sum of all $B(M1)$ values divided by the number of transitions within this bin. The results for ^{94}Mo are shown in Fig. 2. The left panel demonstrates appearance of the LEMAR spike, which exponentially decreases with the transition energy E_γ. The exponential dependence on the transition energy is retained by the M1 strength functions, which are defined by the relation

$$f_{M1}(E_\gamma) = 16\pi/9(\hbar c)^{-3}\overline{B}(M1, E_\gamma)\rho(E_i), \tag{1}$$

where the level density at the initial state $\rho(E_i)$ is obtained from the Shell Model calculations.

As seen in the right panel of Fig. 2, LEMAR is generated by transition originating from states between 2 and 6 MeV. Inspecting the composition

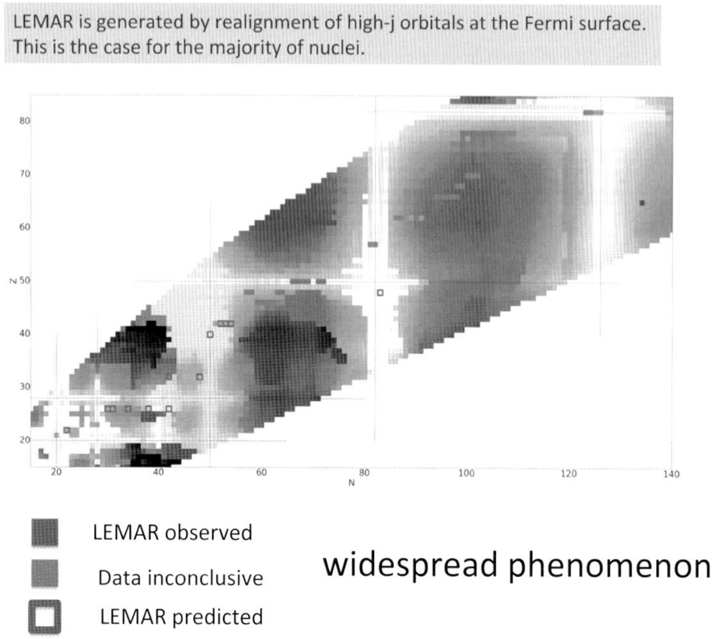

Fig. 3. Experimental presence of the LEMAR spike and predictions by Shell Model calculations.

of initial and final states, we found large $B(M1)$ values for transitions between states that contain a large component (up to about 50%) of the same configuration with broken pairs of both protons and neutrons in high-j orbits. The largest $M1$ matrix elements connect configurations with the spins of $g_{9/2}$ protons re-coupled with respect to $g_{9/2}$ neutrons to the total spin $J_f = J_i, J_i \pm 1$. These orbitals have large magnetic moments. Large M1 matrix elements connect configurations which differ by only the angular momentum projection of these high-j orbitals. Our Shell Model studies of ^{56,57}Fe[8] and ^{60,64,68}Fe[9] found also the LEMAR spike, which is generated by reorientation of $f_{7/2}$ proton holes and $g_{9/2}$ neutrons. High-j particle or hole orbitals with $j = l + 1/2$ have the largest magnetic moments. As they appear near the Fermi level all over the nuclear chart one expects that LEMAR spike is omnipresent as well.

The left panel of Fig. 4 shows the logarithm of the level density. At low energy the level density follows the constant temperature expression $\rho(E) = \rho_0 \exp(E/T)$. Above 4 MeV the level density is underestimated because of

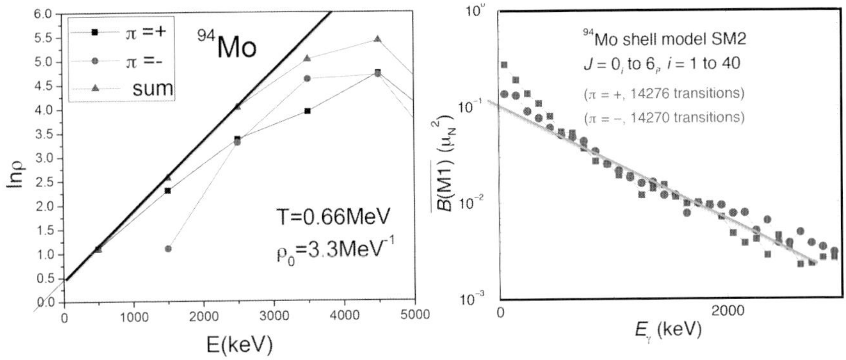

Fig. 4. Left: Shell Model calculation[7] of the level density (left) and average $B(M1)$ values of ^{94}Mo. The straight lines have the slope $\pm 1/T$, $T = 0.66$ MeV.

the truncation of the configuration space. The right panel demonstrates that the exponential decrease of the LEMAR spike is consistent with T derived from the level density. For the Mo isotopes there is consistency between the value of $T \approx 0.8$ MeV derived from the experimental level densities[10] and from the LEMAR spike of the γ strength functions[2]. As illustrated by Fig. 7 below, the consistency also holds for the experimental level densities and γ strength functions of ^{151,153}Sm.

3. Consequences of deformation and properties of the scissors resonance

In our Shell Model studies of ^{60,64,68}Fe[9] we investigated the influence of quadrupole correlation on LEMAR. The calculations were carried out in the CA48PN model space with the CA48MH1 Hamiltonian using the code NuShellX@MSU. The model space included the $\pi(0f_{7/2}^{(6-t)}, 0f_{5/2}^{t}, 1p_{3/2}^{t}, 1p_{1/2}^{t})$ proton orbits with $t = 0, 1, 2$, and the $\nu(0f_{5/2}^{n5}, 1p_{3/2}^{p3}, 1p_{1/2}^{p1}, 0g_{9/2}^{g9})$ neutron orbits. The calculations of M1 strengths were carried out in the same way as for the Mo isotopes. They included the lowest 40 states each with spins from $J_i, J_f = 0$ to 10, which resulted in more than 22000 M1 transitions for each parity. Effective g factors of $g_s^{\mathrm{eff}} = 0.9 g_s^{\mathrm{free}}$ were applied.

The comparison with the experimental $B(E2, 2_1^+ \rightarrow 0_1^+)$ values in the Table included in Fig. 5 shows that the calculations reproduce the quadrupole collectivity in the considered isotopic chain, in particular the

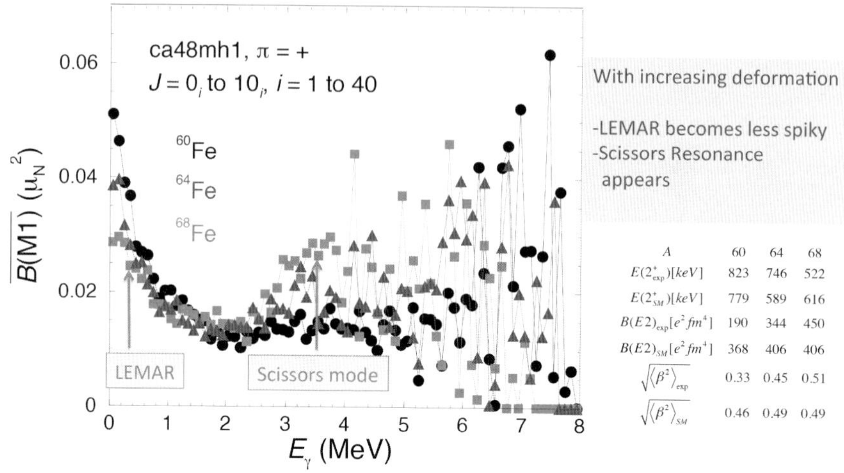

A	60	64	68
$E(2^+_{\text{exp}})[keV]$	823	746	522
$E(2^+_{SM})[keV]$	779	589	616
$B(E2)_{\text{exp}}[e^2 fm^4]$	190	344	450
$B(E2)_{SM}[e^2 fm^4]$	368	406	406
$\sqrt{\langle\beta^2\rangle_{\text{exp}}}$	0.33	0.45	0.51
$\sqrt{\langle\beta^2\rangle_{SM}}$	0.46	0.49	0.49

Fig. 5. Calculated average B(M1) values for the positive parity states in ^{60,64,68}Fe. The table compares the calculated properties of the first 2^+ state with experiment.

enhancement observed around $N = 40$. The respective experimental ratios $E(4^+)/E(2^+)$=2.57, 2.36, 2.66, which are well accounted for by the calculations $E(4^+)/E(2^+)$=2.56, 2.68, 2.43, indicate the transitional character of the ground state bands of all considered isotopes.

The comparison of the $\overline{B}(M1)(E_\gamma)$ values of the various isotopes shows that the shape of the distributions changes when going from $N = 34$, four neutrons above the closed shell, to $N = 42$, the middle of the fpg shell. One observes a weakening of the LEMAR spike and the development of a bump in the range from about 2 to 5 MeV, which is most pronounced in ^{68}Fe. We interpret this bump as the Scissors Resonance (SR) built on excited states.

Fig. 6 shows the M1 strength functions for ^{60,68}Fe, which were calculated using level densities from the Shell Model calculations. Like the average $\overline{B}(M1)$ values they develop the bimodal LEMAR-SR structure when going to the middle of the open shell. The integrated strength deduced from the strength functions below 5 MeV varies by only 8% at most from an average of 9.80 μ_N^2. The slight increase with N is attributed to the progressive occupation of the $g_{9/2}$ shell for $N = 34$, 38, 42. That is, the bimodal structure develops with increasing N by shifting strength from the LEMAR spike to the SR while the sum stays nearly constant.

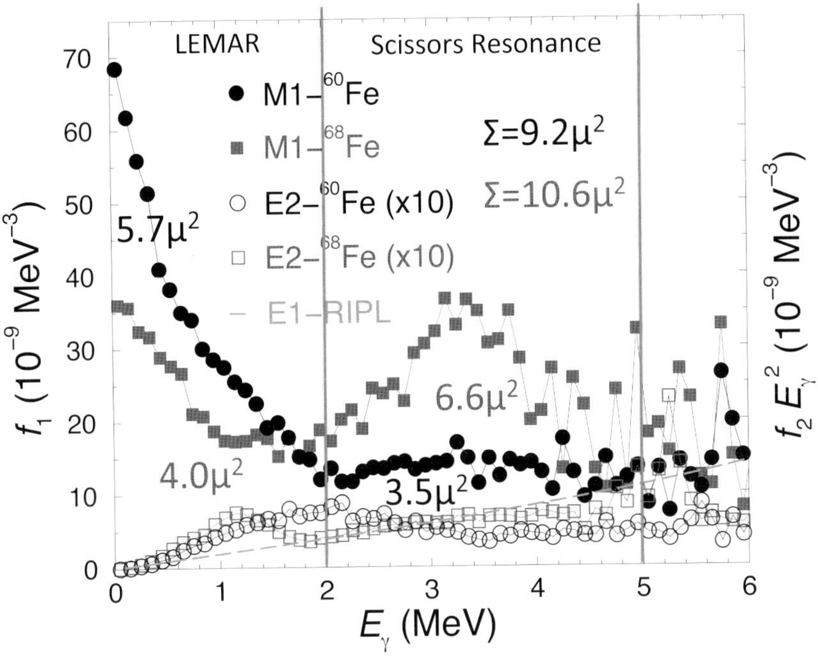

Fig. 6. Calculated M1 strength functions (f_1) and E2 strength functions $\left(f_2 E_\gamma^2\right)$ in ^{60,68}Fe. Note the factor of 10 for E2. The numbers quote the integrated strength for the LEMAR spike $\left(\int_0^2 f_1(E)dE\right)$ and the Scissors Resonance $\left(\int_2^5 f_1(E)dE\right)$. The vertical lines indicate the respective integration regions. For ^{68}Fe the sum of the M1 strength from the ground state to all 1^+ states is $\sum B(M1, 0_1^+ \to 1^+) = 1.7\mu^2$.

For comparison, the standard E1 strength function with parameters taken from the RIPL data base and E2 strength functions f_2 obtained from the present calculations are also shown in Fig. 6. The latter were multiplied with E_γ^2 to be directly comparable with the dipole strength functions f_1. The E2 strength $f_2 E_\gamma^2$ is more than one order of magnitude smaller and the E1 strength exceeds the M1 strength only at energies greater than 6 MeV.

The strength functions deduced by means of the Oslo method from high-resolution experiments by A. Simon *et al.*[4] showed for the first time the LEMAR spike down to about $E_\gamma = 1$ MeV as well as a SR around 3 MeV in one nuclide. Fig. 7 displays the results together with a table of the properties of the 2^+ states of the neighbors, which demonstrate the well deformed nature of the nuclides. Thus we conclude that the bimodal LEMAR-SR structure is a general feature expected for all deformed nuclei.

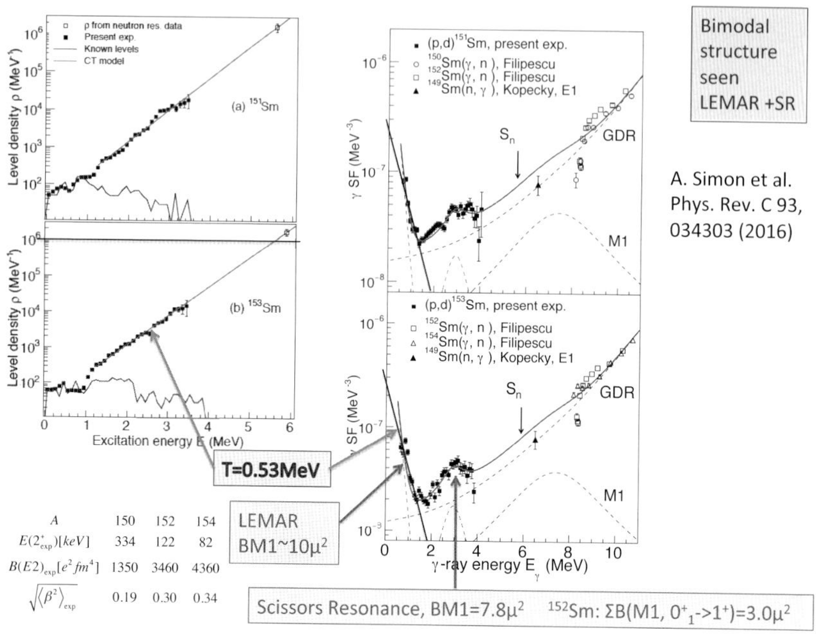

Fig. 7. Experimental level densities (left) and γ strength functions (right) of ^{151,153}Sm measured and derived by means of the Oslo method[4]. The dashed lines show a fit of three Lorentzians (GDR, M1GR, SR) to the γ strength functions. The temperature T derived from the slope of the level density (left) is used to extrapolate the LEMAR spike to $E_\gamma = 0$. The sum of the M1 strength from the ground state to all 1^+ states is quoted at the bottom.

Fig. 8 illustrates how the development of the bimodal LEMAR-SR structure can be attributed to the onset of stable quadrupole deformation above the yrast line when entering the open shell. The mechanism causing large $B(M1)$ values is reorientation of the high-j single-particle angular momenta. Without deformation this occurs between the various configurations of the nucleons in an incompletely filled j-shell, $h_{11/2}$ in the example, which generates the LEMAR spike as suggested in Ref.[7] and discussed above. The magnetic (m) substates of the high-j multiplet split with the onset of deformation. Reorientation occurs in two different ways. First, there are the real transitions between the m-substates. Particle-hole excitations of this type are known to generate the SR[11]. As seen in Fig. 8, the splitting between the m-states corresponds to the position of the SR for $\varepsilon = 0.3$. Second, there are the virtual excitations from the occupied to the empty

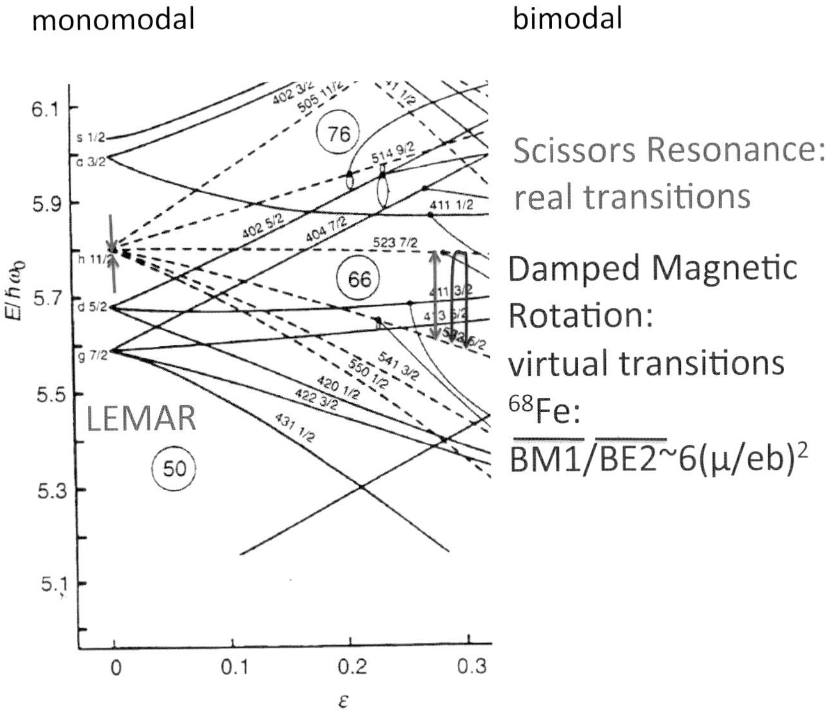

Fig. 8. Transitions that generate the LEMAR spike and the Scissors Resonance. The energy scale is $\hbar\omega_0$=7.7 MeV for A =153.

m-states. These are known generating the rotational motion of deformed nuclei. Along a rotational band the angular momentum increases by a coherent gradual alignment of the spins of the nucleons in the deformed orbitals, which is described by the virtual particle-hole excitations between the m-states. As discussed in our paper[9], the rotational coherence is only partially realized for the transition between the states several MeV above yrast. With increasing deformation, the LEMAR spike changes from incoherent thermal-like radiation to partially coherent rotational radiation. The onset of damped rotation can be seen in the E2 strength function for ^{68}Fe shown in Fig. 6. The bump at 1.2 MeV is generated by stretched E2 transitions with $E_\gamma = 2I/\mathcal{J}$, where $\mathcal{J} \approx 12$ MeV^{-1} is close to the rigid body moment of inertia. The LEMAR spike of ^{68}Fe in Figs. 5 and 6 deviates from the exponential form at the lowest energies, which can be traced back to the presence of an rotational component with $E_\gamma = I/\mathcal{J}$.

In (γ, γ') experiments 1^+ states are excited from the ground state that group to the SR in deformed nuclei. In the case of ^{68}Fe the summed strength of these transitions is calculated to be $\sum B(M1, 0_1^+ \rightarrow 1^+) = 1.7\ \mu^2$. The integral of the strength function between 2 and 5 MeV amounts to 6.6 μ^2, which is three and a half times larger (see Fig. 6). From their experiment on ^{151,153}Sm, Ref.[4] assigned an integrated strength of 7.8 μ^2 to the SR, which is two and a half times larger that the summed strength of the excitation of the SR from the ground state of ^{152}Sm (see Fig. 7). A similar enhancement of the SR strength in the γ decay as compared with the excitation of the SR from the ground state by (γ, γ') experiments has been reported before (see Ref.[4] for references).

The origin of the enhancement of the transition strength is an increase of transitions from excited states as compared to $0_1^+ \rightarrow 1^+$. For example in ^{68}Fe, $\sum B(M1, 0_2^+ \rightarrow 1^+) = 3.4\mu^2$. We attribute the larger summed strengths between excited states to a suppression of the pair correlations with increasing excitation energy, i.e. the thermal quenching of pairing. Pair correlations tend to couple the high-j orbits to zero spin, which obstructs the reorientation that generates the M1 radiation.

S. F. acknowledges support by the DOE Grant DE-FG02-95ER4093, B. A. B. support by the NSF Grant PHY-1404442, and all authors acknowledge the importance of the ECT* workshop on "Statistical properties of nuclei", July 11 - 15, 2016, for advancing this research.

References

1. N. Pietralla *et al.*, *Phys. Rev. Lett.* **83**, 1303 (1999).
2. M. Guttormsen *et al.*, *Phys. Rev. C* **71**, 044307 (2005).
3. A. C. Larsen and S. Goriely, *Phys. Rev. C* **82**, 014318 (2010).
4. A. Simon *et al.*, *Phys. Rev. C* **93**, 034303 (2016).
5. A. C. Larsen *et al.*, *Phys. Rev. Lett.* **111**, 242504 (2013).
6. A. Voinov *et al.*, *Phys. Rev. C* **81**, 024319 (2010).
7. R. Schwengner, S. Frauendorf and A. C. Larsen, *Phys. Rev. Lett.* **111**, 232504 (2013).
8. B. Alex Brown and A. C. Larsen, *Phys. Rev. Lett.* **113**, 252502 (2014).
9. R. Schwengner, S. Frauendorf and A. Brown, *Phys. Rev. Lett.* in print.
10. R. Chankova *et al.*, *Phys. Rev. C* **73**, 034311 (2006).
11. I. Hamamoto and S. Åberg, *Phys. Lett. B* **145**, 164 (1984).

Studies of Neutron-Rich Nuclei at RIBF — From Tetraneutron to Fission

Tomohiro Uesaka

RIKEN Nishina Center for Accelerator-Based Science

The RI Beam Factory at RIKEN is producing beams of highly neutron-rich nuclei. New experimental data obtained with the beams will definitely provide a sound basis for better understanding of the nuclear structure of exotic nuclei.

Highlights of the experimental studies of neutron-rich nuclei are presented in the conference.

In the last part of the talk, an idea of future fission barrier height measurements for neutron-rich heavy nuclei will be given.

Nuclei, Quantum Entanglement, and Qubits

David J. Dean

ORNL

All closed quantum systems are governed by interactions among the particles of that system. The correlations produced require one to describe the overall system as a single many-body wave function. It is interesting to understand how different parts of the Hilbert space contribute to the overall wave function and how these spaces are interconnected. The interconnections can be calculated through an application of Ranyi entropy methods. I will describe some initial Auxiliary Field Monte Carlo calculations of Ranyi entropies in a simple pairing model to demonstrate entanglement of particle-hole states. I will also discuss the formulation of time-dependent coupled-cluster theory to investigate the entanglement.

Understanding the Structure of Medium-Mass Nuclei from First Principles

H. Hergert

NSCL/FRIB Laboratory and Department of Physics & Astronomy,
Michigan State University,
East Lansing, MI 48824-1321, USA
E-mail: hergert@nscl.msu.edu

We present applications of the In-Medium Similarity Renormalization Group (IMSRG) framework in *ab initio* nuclear structure theory. We present two complementary strategies for the computation of nuclear ground- and excited-state properties, namely direct evaluations and the derivation of valence-space interactions for the nuclear configuration interaction (Shell model) approach. We highlight recent results for ground- and excited state observables of *sd*- and *pf*-shell nuclei, and look ahead at the next steps towards a predictive *ab initio* theory of medium-mass nuclei.

Keywords: Nuclear Structure, *Ab Initio* Methods, Renormalization Group.

1. Introduction

Efforts to describe nuclear structure and dynamics *ab initio*, i.e., from first principles, have made tremendous progress in recent years. Computationally efficient many-body methods like Coupled Cluster[1,2], Self-Consistent Greens Function theory[3,4], and the In-Medium Similarity Renormalization Group (IMSRG)[5,6] are nowadays routinely applied for nuclei in the pf-shell and even beyond[7,8]. This allows us to confront modern nuclear interactions and transition operators from Chiral Effective Field Theory (EFT) with a wealth of experimental data.

Here, we present results from recent applications of IMSRG methods, namely direct calculations of the ground-state properties of open-shell nuclei in the Multi-Reference IMSRG (MR-IMSRG), as well as nuclear configuration interaction (Shell model) calculations of excitation spectra based on interactions derived in the targeted valence-space IMSRG (TVS-IMSRG) — these approaches are discussed in detail in Refs. 5,6,15. We consider the implications of our findings for the present status of chiral two- (NN) and three-nucleon ($3N$) interactions serve as inputs for our calculations, and look ahead at efforts towards their improvement.

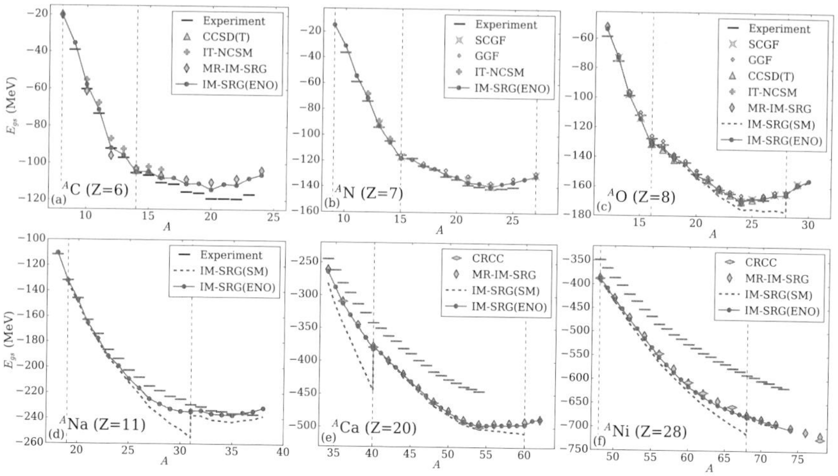

Fig. 1. Ground-state energies for multiple isotopic chains from MR-IMSRG, TVS-IMSRG (referred to as IMSRG(ENO) here), and a variety of other methods, compared to experimental data[13,14] (see Ref. 15 for details). All calculations are based on the same chiral the $NN+3N(400)$ interaction at $\lambda = 1.88\,\mathrm{fm}^{-1}$. Figure reproduced with permission from Ref. 15, Copyright American Physical Society, 2017.

2. Ground-State Properties of Medium-Mass Nuclei

In Fig. 1, we show the ground-state energies for isotopic chains in the p, sd, and pf shells, obtained with the MR-IMSRG and TVS-IMSRG approaches[5,6,15]. We employ a Hamiltonian that we refer to as $NN + 3N(400)$, consisting of the N^{3}LO interaction by Entem and Machleidt[16,17], with cutoff $\Lambda_{NN} = 500$ MeV, and a local NNLO interaction with a reduced cutoff $\Lambda_{NN} = 400$ MeV[18,19]. This Hamiltonian is subsequently evolved to a lower resolution scale λ with the help of a free-space SRG transformation (see Refs. 5,9).

Ground-state energies obtained with the $NN + 3N(400)$ Hamiltonian at are in excellent agreement with experimental data for the nitrogen and oxygen isotopes, and results are not sensitive to variations of the resolution scale λ in a window around 2.0 fm^{-1} (see Refs. 6,10,11 for more details). This is especially impressive since the parameters of the Hamiltonian are entirely fixed by data in $A = 2, 3, 4$ nuclei. Moreover, if we use the same Hamiltonian as input in other many-body methods, we obtain consistent results: Discrepancies are explained by slightly different truncations in the many-body expansion.

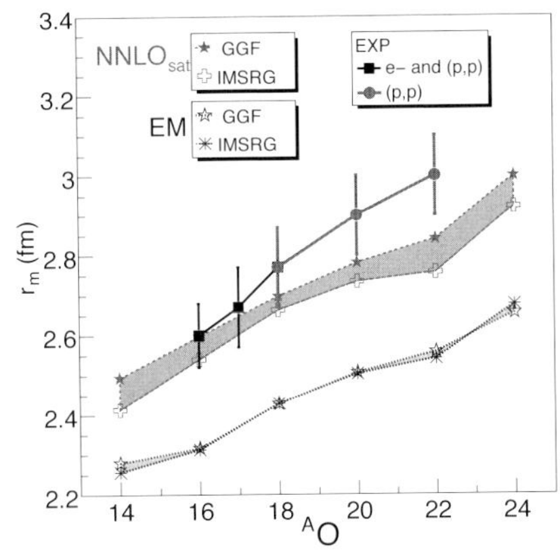

Fig. 2. Oxygen matter radii from MR-IMSRG(2) and the self-consistent Gor'kov Greens Function approach (GGF) for NN+$3N$(400) with $\lambda = 1.88, \ldots, 2.24\,\mathrm{fm}^{-1}$ (indicated by shading, curves labeled "EM"), and NNLO$_{\mathrm{sat}}$, compared to an analysis of experimental data. See Ref.[20] for more details.

In the neutron-rich carbon isotopes, we obtain notable underbinding which must be due to a deficiency in the $NN + 3N(400)$ Hamiltonian, because we replicate essentially exact results from the importance-truncated No-Core Shell Model[21,22]. Moving into the $sd-$shell and eventually to the calcium and nickel chains, the Hamiltonian causes an increasing amount of overbinding (also see Ref. 7), indicating issues with its saturation behavior. This problem is reflected in results for other ground-state observables. Despite the good reproduction of the ground-state energies in the oxygen isotopes, the $NN+3N(400)$ Hamiltonian yields matter radii that are about 10% too small, as shown in Fig. 2.

In response to this issue, Ekström *et al.* have developed an optimization protocol for the parameters of chiral interactions that explicitly includes certain many-body data as constraints. By including selected carbon and oxygen radii, the saturation behavior of chiral interactions can be improved. The first $NN + 3N$ interaction optimized with this protocol is NNLO$_{\mathrm{sat}}$[23], which is able to accurately describe the ground-state energies and charge radii of ^{40,48}Ca at the same time[24]. Figure 2 shows that NNLO$_{\mathrm{sat}}$ improves the agreement between theoretical and experimental oxygen radii as well, although the deviation grows with neutron number. Similar discrepancies were found in Coupled Cluster calculations for the charge radii of ^{50,52}Ca radii[25].

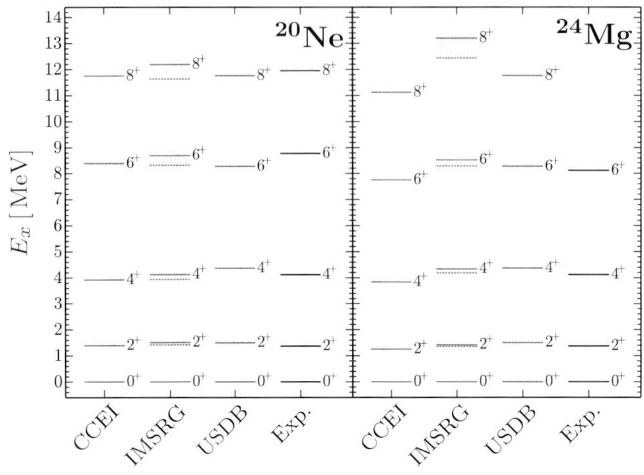

Fig. 5. Ground-state rotational bands of ^{20}Ne and ^{24}Mg. Here, we compare effective interactions derived by TVS-IMSRG decoupling ($e_{\max} = 14, E_{3\max} = 14, \hbar\omega = 20$ MeV (dashed lines) and 24 MeV (solid lines)), the A-independent CCEI interaction of Ref. 34 ($e_{\max} = 12, E_{3\max} = 14, \hbar\omega = 20$ MeV), and the phenomenological USDB interaction[31] to experimental levels[33].

For the shown nuclei, the agreement between theory and experiment is quite satisfactory given that the chiral input Hamiltonian for TVS-IMSRG and CCEI is entirely fixed by $A \leq 4$ data. In ^{24}F, states below 3 MeV are described well, and aside from the 0^+ and 1^+ states, results are well converged. The TVS-IMSRG interaction exhibits a gap in the spectrum between the 3_2^+ and 2_2^+ states, which USDB fills with a group of states that has not been observed at the corresponding energy in the experiment.

Expanding our calculations to the entire *sd* shell, we find that the TVS-IMSRG interactions perform quite well[6,15,35], provided chiral $3N$ interactions are included in the input Hamiltonian — these forces are of crucial importance for the correct reproduction of level orderings and spacings.

Through Shell model calculations with TVS-IMSRG interactions we can also immediately access nuclei with intrinsic deformation. The ground-state rotational bands of ^{20}Ne and ^{24}Mg are shown in Fig. 5. The levels obtained with the TVS-IMSRG and the A-independent CCEI interactions[34], both based on the $NN + 3N(400)$ Hamiltonian ($\lambda = 1.88$, fm^{-1}), are in good agreement with each other as well as the USDB interaction. While rotational bands emerge naturally in these nuclei even without an initial chiral $3N$ force[36], its inclusion markedly improves the agreement of the theoretical excitation energies and level spacings with experimental data, as noted before[15,35].

4. Conclusions and Outlook

In this contribution, we have attempted to illustrate the current reach and capabilities of modern *ab initio* nuclear structure theory. The discussion is by no means exhaustive, given the rapid rate at which new advances are being introduced. The presented results show that the chiral $NN+3N$ itneractions that serve as input to our many-body calculations are currently the dominant source of uncertainty. Several efforts to better understand the deficiencies are underway, and will hopefully lead to a new generation of improved chiral interactions. As the *ab initio* structure community works towards that goal, it is crucially important that theoretical results for many-body observables are confronted with current and forthcoming experimental data. Methods like the MR-IMSRG and TVS-IMSRG that can be used to systematically investigate the evolution of observables along entire isotopic and isotonic chains are ideally suited for that purpose.

A multitude of efforts are underway to further extend the capabilities of the MR-IMSRG and TVS-IMSRG approaches[12,15,37,38], especially towards electroweak transitions. In Chiral EFT, the low-energy constants that parameterize short-range appear in interactions as well as electroweak operators and currents in a consistent fashion. Therefore, measurements of lifetimes and transition rates can provide important additional constraints in the construction of chiral forces (see, e.g., Ref. 19).

In summary, we hope that this contribution has convinced our readers that the developments in *ab initio* nuclear structure theory in recent years have borne fruit, and that there are many exciting opportunities for collaboration in the domain of medium-mass and even heavy nuclei.

Acknowledgments

First and foremost, I am indebted to all my collaborators on the IMSRG framework underlying the present work: S. K. Bogner, K. Tsukiyama, A. Schwenk, T. D. Morris, N. M. Parzuchowski, J. D. Holt, and S. R. Stroberg. Furthermore, I would like to thank C. Barbieri, S. Binder, A. Calci, T. Duguet, F. Evangelista, R. J. Furnstahl, E. Gebrerufael, G. Hagen, K. Hebeler, M. Hjorth-Jensen, G. R. Jansen, R. Roth, J. Simonis, V. Somà, and K. A. Wendt for many useful discussions, as well as providing inputs and/or theoretical results for comparison.

I am grateful to the National Superconducting Cyclotron Laboratory (NSCL)/Facility for Rare Isotope Beams (FRIB) and Michigan State University for startup support. Computing resources were provided by the Ohio Supercomputing Center (OSC), the Michigan State University High Performance Computing Center (HPCC)/Institute for Cyber-Enabled Research (iCER), and the National Energy Research Scientific Computing Center (NERSC), a DOE Office of Science User Facility supported by the Office of Science of the U.S. Department of Energy under Contract No. DE-AC02-05CH11231.

References

1. G. Hagen, T. Papenbrock, M. Hjorth-Jensen and D. J. Dean, *Rept. Prog. Phys.* **77**, 096302 (2014).
2. G. Hagen, M. Hjorth-Jensen, G. R. Jansen and T. Papenbrock, *Phys. Scripta* **91**, 063006 (2016).
3. W. Dickhoff and C. Barbieri, *Prog. Part. Nucl. Phys.* **52**, 377 (2004).
4. V. Somà, A. Cipollone, C. Barbieri, P. Navrátil and T. Duguet, *Phys. Rev. C* **89**, 061301 (2014).
5. H. Hergert, S. K. Bogner, T. D. Morris, A. Schwenk and K. Tsukiyama, *Phys. Rept.* **621**, 165 (2016).
6. H. Hergert, *Phys. Scripta* **92**, 023002 (2017).
7. S. Binder, J. Langhammer, A. Calci and R. Roth, *Phys. Lett. B* **736**, 119 (2014).
8. G. Hagen, G. R. Jansen and T. Papenbrock, *Phys. Rev. Lett.* **117**, 172501 (2016).
9. S. K. Bogner, R. J. Furnstahl and A. Schwenk, *Prog. Part. Nucl. Phys.* **65**, 94 (2010).
10. H. Hergert, S. Binder, A. Calci, J. Langhammer and R. Roth, *Phys. Rev. Lett.* **110**, 242501 (2013).
11. H. Hergert, S. K. Bogner, T. D. Morris, S. Binder, A. Calci, J. Langhammer and R. Roth, *Phys. Rev. C* **90**, 041302 (2014).
12. N. M. Parzuchowski, T. D. Morris and S. K. Bogner, Ab Initio Excited States from the In-Medium Similarity Renormalization Group, *Phys. Rev. C*, in press (2017), `arXiv:1611.00661`.
13. M. Wang, G. Audi, A. Wapstra, F. Kondev, M. MacCormick, X. Xu and B. Pfeiffer, *Chin. Phys. C* **36**, 1603 (2012).
14. F. Wienholtz *et al.*, *Nature* **498**, 346 (2013).

15. S. R. Stroberg, A. Calci, H. Hergert, J. D. Holt, S. K. Bogner, R. Roth and A. Schwenk, *Phys. Rev. Lett.* **118**, 032502 (2017).
16. D. R. Entem and R. Machleidt, *Phys. Rev. C* **68**, 041001 (2003).
17. R. Machleidt and D. Entem, *Phys. Rept.* **503**, 1 (2011).
18. R. Roth, J. Langhammer, A. Calci, S. Binder and P. Navrátil, *Phys. Rev. Lett.* **107**, 072501 (2011).
19. D. Gazit, S. Quaglioni and P. Navrátil, *Phys. Rev. Lett.* **103**, 102502 (2009).
20. V. Lapoux, V. Somà, C. Barbieri, H. Hergert, J. D. Holt and S. R. Stroberg, *Phys. Rev. Lett.* **117**, 052501 (2016).
21. R. Roth, *Phys. Rev. C* **79**, 064324 (2009).
22. C. Forssén, R. Roth and P. Navrátil, *J. Phys. G* **40**, 055105 (2013).
23. A. Ekström, G. R. Jansen, K. A. Wendt, G. Hagen, T. Papenbrock, B. D. Carlsson, C. Forssén, M. Hjorth-Jensen, P. Navrátil and W. Nazarewicz, *Phys. Rev. C* **91**, 051301 (2015).
24. G. Hagen *et al.*, *Nat. Phys.* **12**, 186 (2015).
25. R. F. Garcia Ruiz *et al.*, *Nat. Phys.* **12**, 594 (2016).
26. J. D. Holt, T. Otsuka, A. Schwenk and T. Suzuki, *J. Phys. G* **39**, 085111 (2012).
27. J. D. Holt, J. Menéndez, J. Simonis and A. Schwenk, *Phys. Rev. C* **90**, 024312 (2014).
28. A. T. Gallant *et al.*, *Phys. Rev. Lett.* **109**, 032506 (2012).
29. D. Steppenbeck *et al.*, *Nature* **502**, 207 (2013).
30. G. R. Jansen, J. Engel, G. Hagen, P. Navratil and A. Signoracci, *Phys. Rev. Lett.* **113**, 142502 (2014).
31. B. A. Brown and W. A. Richter, *Phys. Rev. C* **74**, 034315 (2006).
32. L. Cáceres *et al.*, *Phys. Rev. C* **92**, 014327 (2015).
33. National Nuclear Data Center, information extracted from the NuDat 2 database, http://www.nndc.bnl.gov/nudat2/.
34. G. R. Jansen, M. D. Schuster, A. Signoracci, G. Hagen and P. Navrátil, *Phys. Rev. C* **94**, 011301 (2016).
35. S. R. Stroberg, H. Hergert, J. D. Holt, S. K. Bogner and A. Schwenk, *Phys. Rev. C* **93**, 051301 (2016).
36. P. Maris, M. A. Caprio and J. P. Vary, *Phys. Rev. C* **91**, 014310 (2015).
37. T. D. Morris, N. M. Parzuchowski and S. K. Bogner, *Phys. Rev. C* **92**, 034331 (2015).
38. E. Gebrerufael, K. Vobig, H. Hergert and R. Roth, *Phys. Rev. Lett.*, in press (2017), `arXiv:1610.05254`

Laser Spectroscopic Studies of Fission Products and Nobelium

B. Cheal

Department of Physics, University of Liverpool,
Liverpool, L69 7ZE, UK
Bradley.Cheal@liverpool.ac.uk
www.liverpool.ac.uk/physics

Laser spectroscopy is a prominent tool in the study of nuclear ground and isomeric state properties. Being able to measure nuclear spins, magnetic dipole and electric quadrupole moments, and changes in mean-square charge radii, make it an effective probe of both single particle and collective phenomena. Topics of recent investigation have included: evolution of single particle levels and changing magicity in the vicinity of shell closures, emerging collectivity, establishing new isomeric states, and the first optical measurements of a transfermium element.

Keywords: Laser Spectroscopy; Nuclear Spins; Moments; Charge Radii.

1. Techniques of Laser Spectroscopy

Most high resolution optical studies of radioactive nuclei use the technique of *collinear laser spectroscopy*. The technique is usually to be found at ISOL facilities, where the ~30keV radioactive beam is overlapped in parallel with a narrow-linewidth laser beam. This gives a long interaction length and a suppression of the velocity spread in the forward direction due to the fast beam and that the energy spread is a constant under acceleration. Doppler broadening is therefore minimized or even reduced to below the natural linewidth. Applying a tuning potential to the region of interaction Doppler shifts the effective laser frequency over the resonances. In most cases, resonant excitation leads to photon emission, detected using photomultiplier tubes, and an optical spectrum is formed. Optical transitions depend most of all on the atomic number, hence why laser spectroscopy studies isotopic chains. An atomic (ionic) transition, usually found in exhaustive tables, is chosen for study, and then high resolution scans reveal the isotope shift and hyperfine structure. These allow measurement of the nuclear spins, moments and charge radii. More than one overview of the field has recently been published [1,2,3].

Collinear laser spectroscopy experiments take place at: COLLAPS and CRIS at ISOLDE, CERN; at JYFL, Jyväskylä, Finland; TRIUMF, Canada; and more recently at the new BECOLA line at MSU, Michigan, USA. Experimental developments in recent years have included the widespread adoption of gas-filled RFQs for cooling and bunching, first at JYFL [4], and now at all four of these facilities. For collinear fluorescence spectroscopy, where photons are counted, the dominant source of background is the scattering of laser light into the photomultiplier tubes at a constant rate. By bunching the beam into ~5μs every ~50ms and applying a gate to the photon detection, the resonant signal is preserved while the background is suppressed by four orders of magnitude. This is the most common variant of the technique, and at ISOLDE is the predominant form used at the COLLAPS beamline [5]. For the Collinear Resonance Ionization Spectroscopy (CRIS) setup, multi-step excitations are used to resonantly re-ionize neutralized beams, and methods of particle detection rather than photon detection are employed. Here, the use of bunched beams reduces duty cycle losses originating from the use of pulsed lasers [6]. At ISOLDE, since the primary beam is pulsed, proton-triggering techniques also can be used to include or exclude isobars which have a fast release from the target or short half-life [7]. In many cases, the use of a Laser Ion Source also provides a high degree of Z-selectivity. Multi-step excitation schemes are used to resonantly ionize a particular element, which is then mass filtered in the usual way [8]. With some modification, this gives rise to an alternative form of spectroscopy, which is generally more sensitive but lower resolution. As with CRIS, the spectrum is then formed either using particle count rates or decay spectroscopy.

2. Calcium to Nickel Regions

Studies in the nickel region originally began with long-chain measurements of copper and gallium at COLLAPS. The main interest had been the evolution of the proton single-particle energy levels with neutron number, due to the tensor interaction. This revealed itself not only in the inversion in the ground state spin for both elements, but the level migrations could also be tracked by comparing the nuclear moments (which are a particularly sensitive probe of the nuclear wave function) with shell model calculations [9].

Isotopes of copper up to ^{75}Cu and of gallium up to ^{82}Ga were measured. Since those studies, the copper measurements have been extended by the CRIS collaboration, while at COLLAPS, the photon detection region has been significantly upgraded and measurements have been performed on zinc and nickel isotopes. For zinc, a highlight was to firmly establish the existence of an isomeric state in ^{79}Zn, recently inferred from decay spectroscopy measurements, and to determine its character [10]. With the hyperfine structure showing a characteristic I = ½ structure, the magnetic moment revealed this to be a $½^+$ intruder state, requiring the $\nu s_{1/2}$ orbital to be included in the model space. Furthermore, a significant isomer shift was observed between the ground and isomeric states, indicating a large change in the mean-square charge radius. This is consistent with shape coexistence between the ground and isomeric states. The moments of the other isotopes were studied with a variety of model spaces and shell model interactions but did not give any grounds to doubt the anticipated magicity of ^{78}Ni [11].

Production of nickel isotopes past $\sim^{72}Ni$ is notoriously difficult at conventional thick-target ISOL facilities, although there is a possibility that these could become feasible at the JYFL IGISOL. However, closer to stability, another area of physics interest is the character of ^{68}Ni. Many properties, such as the high 2^+ energy, are normally signatures of shell closure. In the mean-square charge radii, a slight upward kink at $N = 40$, also seen at shell closures, is observed in copper and gallium [12]. It will be interesting to see if this effect is maximized in the nickel chain.

While the microscopic behavior of the radii in copper and gallium will remain clear, the global trends are subject to a reliable calibration. Mean-square charge radii are determined from isotope shifts using a simple formula requiring two atomic factors to be determined, which depend on the optical transition only. Theoretical (MCDF) calculations of these proves to be a challenge [13]. However, where three or more stable isotopes exist, either (i) non-optical measurements of the mean-square charge radii exist and can be used as a calibration, or (ii) the isotope shifts can be studied on a variety of optical transitions without the need for radioactive beam time, including transitions which are perhaps spectroscopically weak, but computationally more convenient. Unlike copper and gallium, this *is* the case for zinc and nickel.

In the proton mid-shell region below $Z = 28$, far from observing a subshell effect at $N = 40$, a study of manganese isotopes at COLLAPS show an increase in deformation [14]. A comparison of moments using the GXPF1A and LNPS interactions showed a widening deviation as $N = 40$ was approached, with the LNPS calculations reproducing the experimentally observed trend. This indicated the increasing role of neutron excitations past $N = 40$, and protons excitations across $Z = 28$. Quadrupole moments are a more direct probe of emerging collectivity, and the LNPS calculations of these proved sensitive specifically to the inclusion of the neutron $d_{5/2}$ orbital. Unfortunately, the ground state atomic transition studied provided insufficient sensitivity to the quadrupole moment. In order to remedy this issue, a strong transition in the manganese ion was selected, which shows much greater sensitivity, but is an excitation from a metastable state. Following a program of upgrades and realignment of the device [15], optical pumping in the cooler/buncher ISCOOL was used for the first time at ISOLDE [16], following pioneering work at JYFL [17,18]. This enabled the population of the chosen metastable state to be enhanced (and more selectively than the population of individual atomic states that followed charge exchange) to enable efficient collinear spectroscopic study of ionic manganese. The increasing quadrupole deformation proved the specific need for the $d_{5/2}$ orbital in addition to the $g_{9/2}$. Extracted from both transitions as a cross check, the charge radii also showed a sudden upward deviation from $N = 35$ [19].

Another physics motivation was the possible existence of a subshell effect at $N = 32$. Although no sign of this was observed for the manganese isotopes, it is expected to be most prominent in the calcium chain. An investigation at COLLAPS revealed not only that there seemed to be no indication of a subshell effect in the charge radii, but on the contrary, the radius was significantly *enlarged* [20].

3. IGISOL 4 at JYFL

At the JYFL Accelerator Laboratory, University of Jyväskylä, Finland, the new IGISOL 4 laboratory has now been commissioned since the 2012 conference. Rather than use thick targets of conventional ISOL facilities, like at ISOLDE and at TRIUMF, thin foil targets of a few mg/cm^2 are used, allowing reaction products (both fission and fusion-evaporation) to recoil out directly. These are then stopped in ~200 mbar of helium gas, and extracted with the gas flow through a nozzle in a supersonic jet. While the gas is then pumped out of the outer chamber using Roots pumps, the (predominantly) singly charge ions are captured by a sextupole ion guide and extracted. This allows sub-millisecond extraction times irrespective of the physical or chemical properties. The IGISOL therefore gains an advantage for refractory elements, for example.

A brand new experimental hall now houses the IGISOL, together with a dedicated MCC30 cyclotron and experimental facilities, as shown in Figure 1. The MCC30 is able to provide protons at up to 30 MeV (ideal for fission) and 100μA (too much for thin foil targets but opening up opportunities for neutron

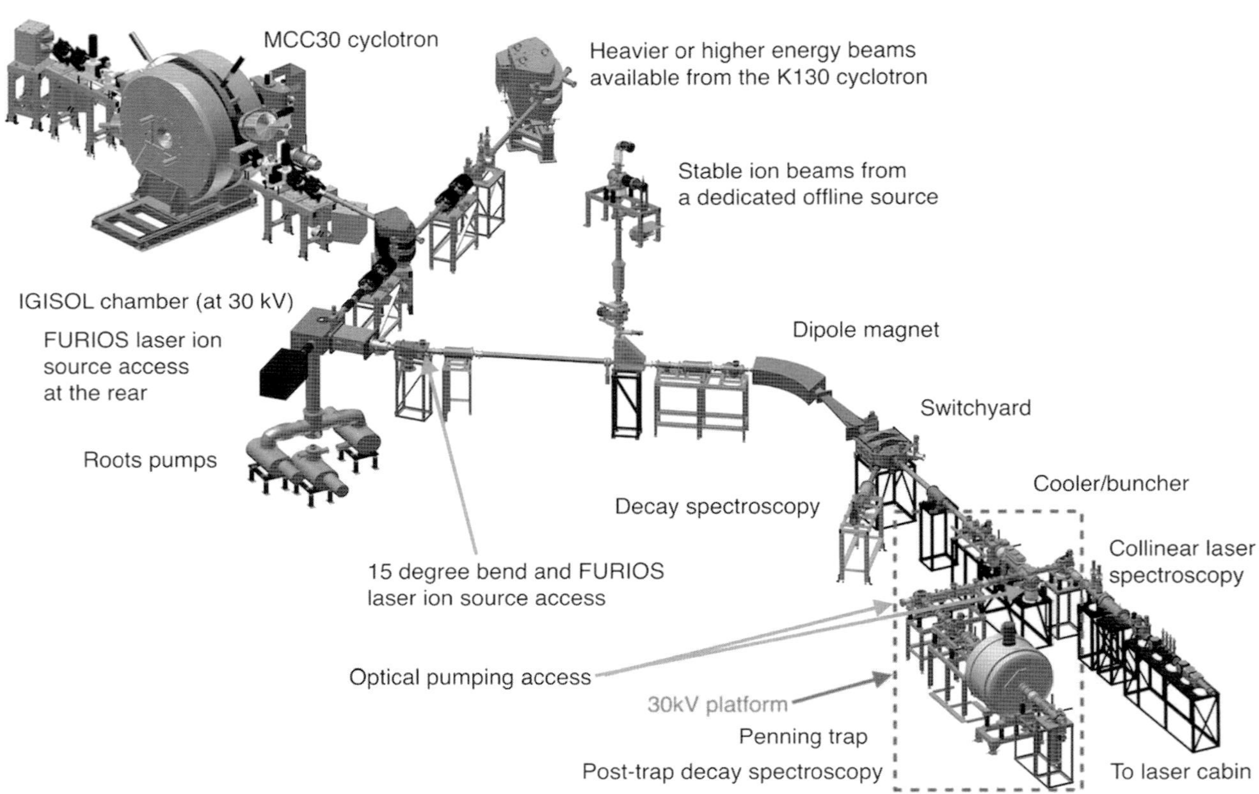

Fig. 1. The new IGISOL 4 laboratory, at the JYFL Accelerator Laboratory, Jyväskylä, Finland. Figure courtesy of Ilkka Pohjalainen.

conversion) or deuterons at up to 15 MeV. Heavier ions for fusion evaporation reactions, and higher primary beam energies, are still available from the existing K130 cyclotron. However, the MCC30 will provide an increase in beam time available for IGISOL [21].

At present, when stable isotope beams are required for testing spectroscopic schemes, a sample of the element is placed in the IGISOL on the cathode of an electrical discharge source. With long experimental campaigns envisaged, high radiation levels and significant cooling times have prompted the installation of a separate offline "mini-IGISOL" placed upstairs. Stable beams, also required to provide a laser frequency calibration during online running, can then be switched in with electrostatic deflectors.

One of the new technical developments which took place at IGISOL 3 was the use of optical pumping in the cooler (later used at ISOLDE for the manganese work, as mentioned earlier). Sending pulsed lasers (which while generally unsuited to high resolution spectroscopy, can also more easily access a greater range of wavelengths) along the trapping axis of the device, optically pumps electronic populations from the ground state to a selected metastable state. This population enhancement persists as the ions are released and arrive for high resolution collinear spectroscopy. This enables spectroscopy to be performed from metastable states rather than being limited to the (otherwise well populated) ground state, opening up the choice of transitions for study. At IGISOL 4, the layout has been optimized to allow easy optical access to the cooler. In addition, a longer low-energy (800 eV) transfer section provides opportunities for optical pumping, but in vacuum. This may help preserve excited level populations in some problem cases.

4. Laser Spectroscopy of Nobelium

In all of the collinear laser spectroscopy experiments above, a reasonably precise value for the frequency of the chosen transition can be found in tables. That is to say, almost as "precise" as can be, before high resolution scans concentrated over a limited range reveal the isotopic dependence of the optical structure (the isotope shift and hyperfine splitting). Finding these resonances for the first time, especially with a high resolution technique, would not be possible without knowing where to start digging.

For spectroscopy of nobelium, several problems present themselves for the field of laser spectroscopy. Since no naturally occurring isotopes exist, and even the longest-lived isotope cannot be extracted from reactors and made into samples, even the "gross" *atomic* physics which we rely upon is entirely absent.

We must first find the optical transitions ourselves (and in doing so determine the energies of the electronic energy levels precisely) before turning our attention to the nuclear physics effects. Additionally, production of nobelium requires an in-flight technique, whereas laser spectroscopy has traditionally been performed at ISOL facilities, where the beam quality and energy is more suitable. Even then, the yields are too low for the collinear technique, even if we knew precisely where to start looking frequency-wise.

To overcome these difficulties, a new experimental technique was developed [22]. Isotopes of nobelium were produced from enriched lead targets using a ^{48}Ca beam from the GSI UNILAC. These were filtered using the velocity filter, SHIP, degraded, and collected onto a filament using electric fields. During each measurement cycle, the beam was then switched off and the filament was heated, producing a cloud of nobelium atoms within a helium buffer gas. A laser was then used to search for an optical transition, which, if resonantly excited, would be energetically sufficient to cause ionization by a second, powerful, non-resonant laser. Ionization is detected by using fields to transport any ions to an alpha detector, and nobelium ions identified by their characteristic decay energy. In the first phase of this experiment, the resolution of the lasers was deliberately kept low during the initial search for a resonance. Even then, the frequency range would be prohibitive within the beam time available. Therefore, five lasers simultaneously scanned complementary regions until a resonance was found, and the laser responsible was determined [23].

This achievement, published in *Nature*, then paves the way to proceed in two ways. Firstly, replacing the lasers for the first (resonant) step, with a single narrow-linewidth laser beam, enables the transition to be scanned with a higher resolution. With different nobelium isotopes available from the use of different isotopically enriched targets, the resolution is sufficient to extract nuclear properties (at these high atomic numbers). Secondly, the non-resonant broadband laser for the second step can be replaced with a tunable laser, able to measure a series of transitions, from the electronic state(s) now identified, to the Rydberg levels. This converging sequence of frequencies enables the ionization potential to be precisely determined.

Further analysis and publications are in progress. Preparations for similar studies, but of lawrencium, are also underway.

Acknowledgments

The author would like to thank: the COLLAPS collaboration of ISOLDE, CERN; the laser spectroscopy collaborations of JYFL, Finland; the RADRIS collaboration at GSI, Germany; the collinear laser spectroscopy collaboration at TRIUMF, Canada. The author would also like to thank Dr David Joss (University of Liverpool) for kindly covering his PHYS107 problems class, enabling him to attend this fine conference.

References

1. B. Cheal and K. T. Flanagan, *J. Phys. G* **37**, 113101 (2010).
2. K. Blaum, J. Dilling and W. Nörtershäuser, *Phys. Scr.* **T152**, 014017 (2013).
3. P. Campbell, I. D. Moore and M. R. Pearson, *Prog. Part. Nucl. Phys.* **86**, 127 (2016).
4. A. Nieminen *et al.*, *Phys. Rev. Lett.* 88, 094801 (2002).
5. K. Kreim *et al.*, *Phys. Lett. B* **731**, 97 (2014).
6. T. J. Procter *et al.*, *J. Phys.: Conf. Ser.* **381**, 012070 (2012).
7. H. Heylen, C. Babcock *et al.*, *Phys. Rev.* C **92**, 044311 (2015).
8. V. N. Fedosseev, Y. Kudryavtsev and V. I. Mishin, *Phys. Scr.* **85** 058104 (2012).
9. B. Cheal *et al.*, *Phys. Rev. Lett.* **104**, 252502 (2010).
10. X. F. Yang, C. Wraith *et al.*, *Phys. Rev. Lett.* **116**, 182502 (2016).
11. C. Wraith, X. F. Yang *et al.*, Submitted to *Phys. Lett.* B (2017).
12. M. L. Bissell *et al.*, *Phys. Rev.* C **93**, 064318 (2016).
13. B. Cheal, T. E. Cocolios and S. Fritzche, *Phys. Rev.* A **86**, 024501 (2012).
14. C. Babcock, H. Heylen *et al.*, *Phys. Lett.* B **750**, 176 (2015).
15. C. Babcock, PhD Thesis, University of Liverpool, UK (2015).
16. C. Babcock, H. Heylen *et al.*, *Phys. Lett.* B **760**, 387 (2016).
17. B. Cheal *et al.*, *Phys. Rev. Lett.* **102**, 222501 (2009).
18. F. Charlwood *et al.*, *Phys. Lett.* B **690**, 346 (2010).
19. H. Heylen, C. Babcock *et al.*, *Phys. Rev.* C **94**, 054321 (2016).
20. R. F. Garcia Ruiz *et al.*, *Nat. Phys.* **12**, 594 (2016).
21. B. Cheal and D. H. Forest, Hyperfine Interact. **223**, 223 (2014).
22. F. Lautenschläger *et al.*, *Nucl. Instrum. Meth.* B **383**, 115 (2016).
23. M. Laatiaoui *et al.*, *Nature* **538**, 495 (2016).

Overview on What We Have Learned from GRETINA*

Shaofei Zhu

Argonne National Laboratory

During the first campaign of GRETINA at ATLAS, a total of roughly 130 days of beam time was devoted to measurements with GRETINA for 18 PAC-approved experiments over a little more than a year. A large fraction of the research program was dedicated to Coulomb excitation measurements where the device was coupled with the CHICO-2 array of PPAC counters for particle detection. In addition, a number of measurements focusing on studies of the intrinsic capabilities of GRETINA itself were carried out.

*This work was supported by the DOE Office of Nuclear Physics under Contracts Nos. DE-AC02-06CH11357 and DE-AC02-06CH11231.

Nuclear Structure at High Isospin and Spin: The Next Step

A. Navin

Grand Accélérateur National d'Ions Lourds, 14000 Caen, France

The fission process is one of the ways to create short lived radioactive nuclei not existing on earth. These are necessary both to search and understand the regular and simple patterns that emerge in the internal structure of complex microscopic nuclei and for understanding macroscopic astrophysical objects. The characteristic γ-decay of the excited states of the long chains of isotopes of these fission fragments provides fingerprints of the evolution of the underlying nuclear configurations of individual and collective motion of the nucleons as a function neutron-proton asymmetry and angular momentum. To access the most neutron rich nuclei produced in fission reactions, typically with a fractional fission yields of $1/10^5$, would require an improvement in the available sensitivity to carry out spectroscopy of these exotic nuclei produced with low cross sections. Recently the characterization of the prompt γ-decay of the excited states of these nuclei, especially those far from stability received a boost due to increased selectivity and sensitivity, primarily arising from the isotopic identification of the fission fragments produced in fusion-fission and transfer-induced fission, at energies around the Coulomb barrier @GANIL using a large acceptance spectrometer coupled with a large gamma-ray detector array [1, 2].

The talk will give an overview of recent results obtained using the ^{9}Be(^{238}U,ff) @6.2 MeV/u using the VAMOS++ spectrometer and the EXOGAM γ-array at GANIL, focusing on a possible new frontier for nuclear physics, namely, BOTH high spin ***and*** large neutron proton asymmetry [1-7]. Further improvements in both selectivity and sensitivity, with a further upgraded VAMOS++ spectrometer and its combination with the next generation AGATA array, for prompt and delayed γ-rays will also be presented.

References

1. A. Navin *et al.*, *Phys. Lett. B* **728**, 136 (2014).
2. A. Navin and M. Rejmund, McGraw-Hill Yearbook of Encyclopedia of Science & Technology, Access Science (2014) p. 137.
3. E.H. Wang *et al.*, *Phys. Rev. C* **92**, 034317 (2015).
4. M. Rejmund *et al.*, *Phys. Rev. C* **93**, 024312 (2016).
5. M. Rejmund *et al.*, *Phys. Lett. B* **753**, 86 (2016).
6. S. Biswas *et al.*, *Phys. Rev. C* **93**, 024312 (2016).
7. A. Navin *et al.*, submitted to *Phys. Lett. B* (2016).

Experiments and Theory of Superheavy Elements

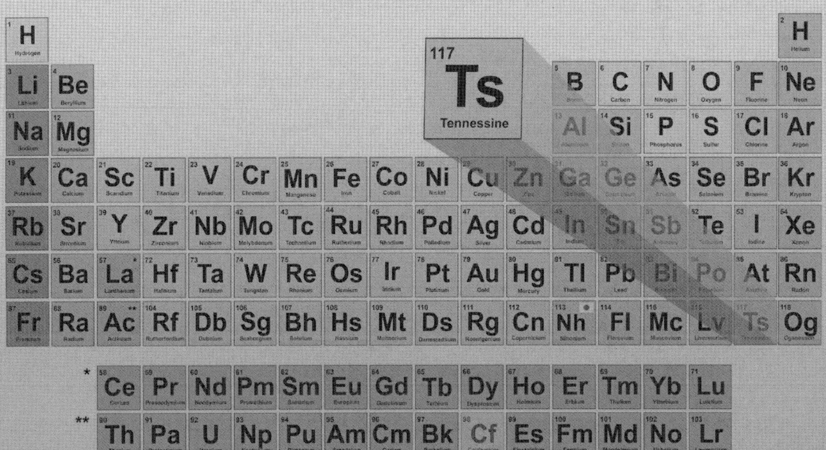

Discovery of Element 113 and Perspectives

K. Morita[1], K. Morimoto, D. Kaji and H. Haba

Nishina Center, RIKEN,
2-1 Hirosawa, Wako, Saitama 351-0198, Japan
[1]Department of Physics, Kyushu University,
Higashi-ku, Fukuoka 812-8581, Japan

H. Kudo

Department of Chemistry, Niigata University,
8050 Ikarashi, 2-no-cho, Nishi-ku, Niigata 950-2181, Japan
E-mail: hkudo@sc.niigata-u.ac.jp

Experiments for producing heaviest elements by a cold-fusion type reaction were performed using a gas-filled recoil ion separator, GARIS, at RIKEN. After confirmation experiments of the reported nuclides, ^{271}Ds, ^{272}Rg, and ^{277}Cn, a new superheavy nuclide ^{278}113 has been searched by the reaction of ^{70}Zn on ^{209}Bi. Totally 3 decay chains due to ^{278}113 were observed during the net irradiation time of 576 days. The 1st and 2nd decay chains consist of four alpha decays from ^{278}113 to ^{266}Bh, and terminate by spontaneous fission of ^{262}Db. The 3rd chain consists of 6 consecutive alpha decays down to ^{254}Md. Observed decay properties from 3 decay chains were consistent with each other. We also examined the decay properties of ^{266}Bh produced by the reaction of ^{23}Na on ^{248}Cm to establish the cross-reaction of ^{278}113 decay chain. It was found that ^{266}Bh can be regarded as an anchor nuclide of ^{278}113 decay chain, and thus the production of ^{278}113 was confirmed.

Keywords: Cold-Fusion Type Reaction; ^{278}113; Gas-Filled Recoil Ion Separator.

1. Introduction

In 2002 we started to investigate production of very heavy elements by a gas-filled recoil ion separator, GARIS, at the up-graded RIKEN Linear Accelerator Facility, RILAC. So-called cold-fusion type reactions were employed for the production of heavy elements, because produced new nuclides can be connected to known nuclides by α-decay chains. Hofmann et al. reported the production of

132 MeV which correspond to 121, 124, and 126 MeV, respectively, at the middle of the target.

The ^{23}Na beam was pulsed in two ways, micro and macro pulse structures. A purpose of the micro pulse structure was to prevent the irradiation of the target frame. A macro pulse structure was 3 s beam ON and 3 s beam OFF independently of the micro pulse structure. A logical AND signal of the micro and macro pulse structures was sent to the beam pulsing system of the accelerator. All the measurement was done only in the beam OFF period. A counting duty in the macro beam OFF period was 100%, while that in the macro beam ON period was 45%. The beam duty factor was 27.5%. A typical beam intensity was 4.4 particle-μA in the micro beam ON period.

GARIS was used to separate evaporation residues (ERs) from the beam particles and other unwanted charged particles. GARIS was filled with helium gas at a pressure of 33 Pa. The value of magnetic rigidity of GARIS was set at 2.07 Tm for 132 and 130 MeV beam energies, and 2.19 Tm for 126 MeV and a part of 132 MeV to reduce a counting rate of the focal plane detector. The detection system is about the same with the case of 113 experiment except for the timing detectors. Because the kinetic energy of ER is very low in the reaction of ^{23}Na on ^{248}Cm, the timing detectors were removed. Energy calibration of the focal plane detectors was made using long-lived transfer products, ^{248}Cf, ^{252}Fm, and ^{254}Fm which were implanted in the PSD. A total counting rate of the focal plane detector was 5–10 kHz in the beam ON period, while that was 5–10 Hz in the beam OFF period. Because of the high implantation rate to the PSD in the beam ON period, the PSD was gradually damaged in the course of the measurement. Therefore, the PSD was replaced three times during the experimental period of one and a half months.

We have done off-line analyses to search for mother-daughter correlation, i.e. α-α and α-SF event based on position in PSD, energy, and time difference. Because we took data only at beam OFF period, a signal of implantation of ER was not recorded. Therefore, time difference between ER implantation and its decay event was not obtained. An average counting rate of particles having energies of greater than 50 MeV for each strip of the PSD was 2.7×10^{-4} /s. A counting rate in the positional window of 2 mm is calculated to be 9.3×10^{-6} /s. Considering a time window of 300 s, a probability of an accidental correlation is calculated to be 2.8×10^{-3} per event. This low value of the accidental probability was realized by using a combination of GARIS and PSD, which enables a low background α-SF correlation measurement compared with a gas-jet transport system.

3. Experimental Results

The first event of the candidate of $^{278}113$ was observed on July 23, 2004. The net irradiation time was 79 days and the total dose of ^{70}Zn was 1.7×10^{19}. The event consisted of an implantation signal in the PSD followed by four consecutive α-decays and a spontaneous fission as a terminal signal. The measured positions of all six sequential signals were within the spatial resolution of the PSD. The observed energies and the time differences between decay signals are indicated in Fig. 1 as a label of "2004". The energy was calibrated for α-particles, so that the energy for the spontaneous fission may be not accurate because of possible pulse height defect. The α-particles of the 1st and 4th events were detected only by the PSD whereas those of the 2nd and 3rd were partly measured by the PSD and partly by the SSD. The energy resolution measured by only the PSD was 39 keV in full width at half maximum (FWHM) and that measured by both the detectors was 66 keV in FWHM. For all the five decay events, no TOF signal registered from the timing detectors was associated. Considering the counting rate of decay events and the time difference, the probabilities of accidental coincidence between ER and individual decays were evaluated to be 2.9×10^{-9}, 8.2×10^{-8}, 1.4×10^{-7}, 2.1×10^{-5}, and 9.5×10^{-6}, for $\alpha 1$, $\alpha 2$, $\alpha 3$, $\alpha 4$, and SF, respectively. The mass number of ER was roughly estimated to be 280+/-15 from the data of TOF between the timing counters of 44.6 ns and the energy measured by the PSD of 36.75 MeV using a method similar to that described in Ref. 5.

The second candidate was observed on April 2, 2005. The net irradiation time was 161 days and the total dose of ^{70}Zn was 4.45×10^{19}. The second decay sequence is about the same with that observed in the previous work, that is, four consecutive α-decays followed by a spontaneous fission after the implantation of ER. The decay energies and the time differences between decay signals are indicated in Fig. 1 as a label of "2005". The decay energy of the 3rd α-decay was measured only in part owing to the limited solid angle coverage (85%) of the detector box. The mass number are almost the same with that of the previous one, because the values of TOF data and the energies at the PSD are in good agreement with each other.

The third decay chain was observed on August 12, 2012. The observed decay chain consisted of six consecutive α-decays after the implantation of an ER in the PSD. The associated α-energies and the time intervals are summarized as "2012" in Fig. 1. The α-decay mode was first observed for the fifth decay in the third chain, while the modes observed in the first and second decay chains were the SF decay. The observed α-energy of E_α = 8.63+/−0.06 MeV is in good agreement with the adopted values for ^{262}Db[13]. In the present decay chain, the sixth α-decay (α_6) of E_α=8.66+/−0.06 MeV was also observed with a decay time of

3.78 s, which corresponds to $T_{1/2}=2.6^{+12}_{-1.1}$ s. These decay properties are in good agreement with the adopted properties for ^{258}Lr which is the α-decay daughter of ^{262}Db. The mass number of ER was roughly estimated to be 291+/−15.

In the off-line analysis of the ^{266}Bh experiment, the maximum correlation time was set to 300 s and the position window was set to +/−2mm for the correlation events measured only in the PSD. We have assigned total of 32 correlations to true ones. Four of the events are α_1-α_2-α_3 correlations, one is an α_1-α_2-SF, nine are α_1-α_2 and eighteen are α_1-SF correlations. Because ER cross sections of asymmetric reactions in the heaviest region exhibit a broad peak around the optimum energies, ^{267}Bh, which is a product of the 4n evaporation channel of the reaction, could be produced simultaneously with ^{266}Bh. We focused on these two isotopes in the analyses. The direct productions of ^{262}Db and ^{263}Db, which are the α-decay daughters of ^{266}Bh and ^{267}Bh by α5n and α4n evaporation channels, respectively, were not considered because of the reaction in the sub-barrier energy region. As a result of correlation analysis, 11 chains which contain α-decay of ^{262}Db and 3 chains which include SF of ^{262}Db were conclusively assigned to ^{266}Bh decay. The obtained decay properties are in agreement with the reported ones[14,15]. As for the half-life of ^{266}Bh, we could only state that it is longer than 1 s because of the small counting statistics.

4. Discussion

The experiment was designed to produce the isotope, ^{278}113, by the one-neutron evaporation channel in the ^{209}Bi + ^{70}Zn reaction. On the basis of a systematic study of the most probable reaction energies for the one-neutron evaporation channel in the ^{208}Pb(^{64}Ni, n)^{271}Ds (*Z*=110)[4], ^{209}Bi(^{64}Ni, n)^{272}Rg (*Z*=111)[5], and ^{208}Pb(^{70}Zn, n)^{277}Cn (*Z*=112)[6] reactions, a reaction energy of 349.0 MeV at the half-depth of the targets was adopted to produce ^{278}113. The corresponding energy of the center of mass frame is 261.4 +/− 2.0 MeV, where +/-2.0 MeV indicates the range of reaction energy due to the energy loss of the beam in the bismuth targets. The excitation energy of the compound nucleus, ^{279}113, is calculated to be 14.1+/−2.0 MeV using the predicted mass for a compound nucleus [16] and the experimental masses of the beam and target. This excitation energy is considered to be too high for a radiative capture process (zero neutron emission channel) and too low for a two-neutron emission channel. The possibility of a one-proton evaporation channel leading to the product of ^{278}112 is excluded by comparing the observed decay time of SF with that of ^{262}Rf which decays by 100% spontaneous fission with a half-life of 47+/−5 ms [17]. As a result, the members of the decay chains

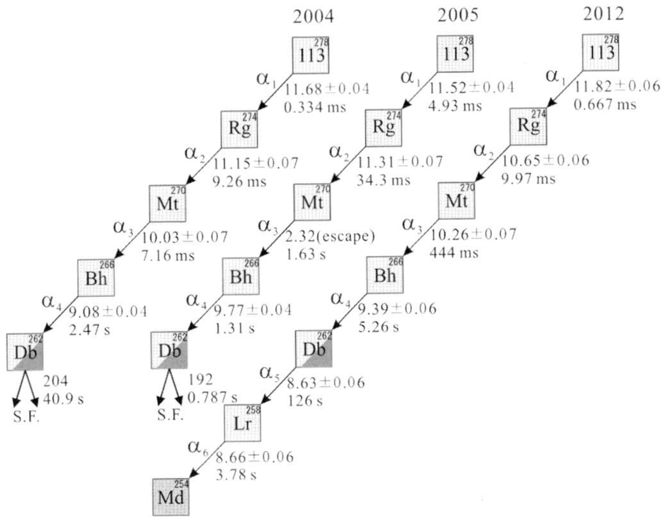

Fig. 2. Observed Z=113 decay chains. Values are E_α(MeV) and lifetimes.

were assigned as ^{278}113, ^{274}Rg, ^{270}Mt, ^{266}Bh, ^{262}Db, and ^{258}Lr. The observed decay chains are displayed in Fig. 1.

The decay time distributions of individual generations in the three decay chains are shown in Fig. 3, where the logarithm of the decay time is taken as the abscissa. It was found that the decay time of each generation closely follows a single exponential decay curve. This indicates that the observed decays originated from nuclides with the same mean life. The α-decay energies of ^{278}113, ^{274}Rg, ^{270}Mt, and ^{266}Bh show some discrepancies among the three chains. However, the observation in the rather widely distributed decay energies of the decay chains of odd-odd nuclei is a natural feature due to the decays to many excited states in their daughters, and due to the possible summing effect of α-energy with a conversion electron or γ-ray energy, which is emitted simultaneously with α-decay inside the detector.

The total beam dose was 1.4x10^{20}. Combining all three events, the production cross section of ^{278}113 was determined to be 22^{+20}_{-13} fb with 1σ error. The error includes only a statistical one. To deduce the cross section, the values of the transmission of the GARIS was assumed to be 0.8.

In conclusion, the isotope of the 113th element, i.e., ^{278}113, was produced in the ^{209}Bi(^{70}Zn, n)^{278}113 reaction and was unambiguously identified by the firm connection to the well-known daughter nuclides ^{266}Bh, ^{262}Db, and ^{258}Lr.

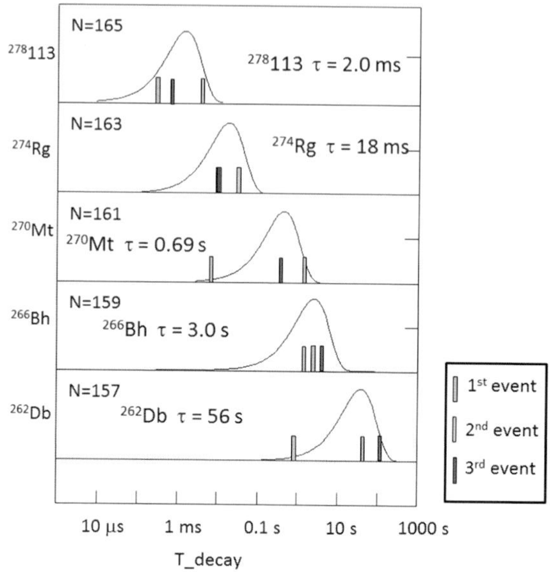

Fig. 3. Decay time distributions of individual generations in the three decay chains.

References

1. S. Hofmann *et al.*, *Z. Phys.* A350 (1995) 281.
2. S. Hofmann *et al.*, *Z. Phys.* A354 (1996) 229.
3. S. Hofmann *et al.*, *Eur. Phys. J.* A14 (2002) 147.
4. K. Morita *et al.*, *Eur. Phys. J.* A21 (2004) 257.
5. K. Morita *et al.*, *J. Phys. Soc. Jpn.* 73 (2004) 1738.
6. K. Morita *et al.*, *J. Phys. Soc. Jpn.* 76 (2007) 043201.
7. K. Morita *et al.*, *J. Phys. Soc. Jpn.* 73 (2004) 2593.
8. K. Morita *et al.*, *J. Phys. Soc. Jpn.* 76 (2007) 045001.
9. K. Morita *et al.*, *J. Phys. Soc. Jpn.* 81 (2012) 103201.
10. K. Morita *et al.*, *J. Phys. Soc. Jpn.* 78 (2009) 064201.
11. Y. Ts. Oganessian *et al.*, *Phys. Rev.* C69 (2004) 021601(R).
12. L. C. Northcliffe and R. F. Schilling, *Nucl. Data Tables* A7 (1970) 233.
13. R. Dressler *et al.*, *Phys. Rev.* C59 (1999) 3433.
14. P. A. Wilk *et al.*, *Phys. Rev. Lett.* 85 (2000) 2697.
15. Z. Qin *et al.*, *Nucl. Phys. Rev.* 23 (2006) 400 (Chinese journal in English).
16. W. D. Myers and W. J. Swiatecki, *Nucl. Phys.* A601 (1996) 141.
17. L. P. Somerville *et al.*, *Phys. Rev.* C31 (1985) 1801.

Discovery of Elements 113–118

V. K. Utyonkov†,1, Yu. Ts. Oganessian[1], S. N. Dmitriev[1], M. G. Itkis[1], K. J. Moody[2], M. A. Stoyer[2], D. A. Shaughnessy[2], J. B. Roberto[3], K. P. Rykaczewski[4], J. H. Hamilton[5] for the Collaboration[1,2,3,4,5,6,7]

1Flerov Laboratory of Nuclear Reactions, Joint Institute for Nuclear Research, RU-141980 Dubna, Russian Federation
2Nuclear and Chemical Sciences Division, Lawrence Livermore National Laboratory, Livermore, California 94551, USA
3Science and Technology Partnerships Directorate, Oak Ridge National Laboratory, Oak Ridge, Tennessee 37831, USA
4Physics Division, Oak Ridge National Laboratory, Oak Ridge, Tennessee 37831, USA
5Department of Physics and Astronomy, Vanderbilt University, Nashville, Tennessee 37235, USA
6Department of Physics and Astronomy, University of Tennessee, Knoxville, Tennessee 37996, USA
7Research Institute of Atomic Reactors, RU-433510 Dimitrovgrad, Russian Federation
†E-mail: utyonkov@jinr.ru

Review of discovery of isotopes of elements 113–118 produced in the ^{48}Ca-induced reactions with actinide target nuclei is presented. These superheavy nuclei were synthesized at DGFRS; the decay properties of the majority of these nuclei were subsequently reproduced in experiments carried out at SHIP, BGS, TASCA, and GARIS. The increased role of shell effects in the stability of superheavy nuclei with rise of their neutron number is demonstrated by comparison of the experimental data and results of theoretical calculations.

Keywords: Superheavy Elements; α Decay; Spontaneous Fission.

1. Introduction

Since 1998, six new superheavy elements with atomic numbers 113 through 118 were synthesized for the first time [1,2]. Radioactive properties of more than fifty new heaviest nuclides were explored in the ^{238}U–^{249}Cf+^{48}Ca reactions in experiments with the DGFRS and later at the chemical setups and at the separators SHIP, BGS, TASCA, and GARIS. These superheavy nuclei (SHN) demonstrate a consistent pattern of nuclear properties in the region of the heaviest nuclides, reveal a substantial increase of nuclear stability with increase of neutron number and closer approach to the predicted shell N=184, and provide experimental proof of the existence of the "Island of stability" of SHN predicted by various theoretical models for half a century.

2. Results

The first SHN ^{289}Fl was discovered on June 25, 1999, in the ^{244}Pu+^{48}Ca reaction studied at DGFRS (here and after we refer to reviews [1,2] containing references to most of earlier experimental data). Two identical decay chains, both consisting of signals from evaporation residue (ER), then two consecutive α decays of ^{289}Fl and ^{285}Cn, and spontaneous fission (SF) of ^{281}Ds, were observed in 1999, in experiment carried out at the lowest excitation energy E^* of the compound nucleus (CN). With gradual increase of E^*, the lighter neighboring isotopes ^{288}Fl and ^{287}Fl were produced. On one hand, the number of nuclei and their decay properties in the decay chains of $^{287\text{-}289}$Fl are different which proves observation of different isotopes. On the other hand, the difference in E_α and $T_{1/2}$ values of parent nuclei $^{287\text{-}289}$Fl is rather small (by about 0.1 MeV and by factor of about 2, respectively); this observation excludes the assignment of the observed nuclides to *differing types of reactions* (e.g., considering that one isotope is produced in the *xn* channel and others – in the *pxn* or αxn channels).

The same isotope ^{287}Fl and two lighter ones, ^{285,286}Fl, were produced in the reactions with the lighter Pu isotopes, ^{242}Pu+^{48}Ca and ^{240}Pu+^{48}Ca studied also at different excitation energies. The excitation functions together with decay properties of the observed nuclides demonstrate *the production of the neighboring isotopes of the same element* in each reaction. In these experiments, two Fl isotopes were observed in cross bombardments ^{244}Pu(^{48}Ca,5n), ^{242}Pu(^{48}Ca,3n) $\rightarrow$ ^{287}Fl and ^{242}Pu(^{48}Ca,5n), ^{240}Pu(^{48}Ca,3n) $\rightarrow$ ^{285}Fl. Moreover, daughter nuclei ^{282}Cn and ^{283}Cn were produced in another cross-bombardment reaction ^{238}U+^{48}Ca. The identity of the decay properties observed in cross bombardments [3] definitely proves that the same nuclei were produced in different reactions. This proves that all of these nuclei were produced in *the same type of reaction*.

The isotopes of Lv were synthesized at DGFRS in the ^{248}Cm+^{48}Ca and ^{245}Cm+^{48}Ca reactions at different excitation energies of CNs [1,2]. Again, the decay properties of descendants of the isotopes ^{290}Lv-^{293}Lv were in full agreement with those found in the ^{242,244}Pu, ^{238}U+^{48}Ca cross-bombardment reactions. The heaviest known element Og was synthesized in the ^{249}Cf(^{48}Ca,3n)^{294}Og reaction studied at DGFRS [1,2,4]. Note, the descendant SF isotope ^{282}Cn was produced in four cross reactions ^{238}U(^{48}Ca,4n)^{282}Cn, ^{242}Pu(^{48}Ca,4n)^{286}Fl, ^{245}Cm(^{48}Ca,3n)^{290}Lv, and ^{249}Cf(^{48}Ca,3n)^{294}Og. Furthermore, in accordance with [3], the decay properties of several parent nuclides and their descendants were reproduced in experiments at separators SHIP, TASCA, BGS, GARIS, and chemistry setups [1,2,4,5].

Assignment of all even-Z nuclides to the products of the xn channel is justified mainly by excitation-function measurements, production of most of the nuclides in cross bombardments, as well as by decay properties of nuclei, especially their measured α-particle energies [3] (see Figure 1a). The $Q_\alpha(N)$ systematics for isotopes of neighboring elements never intersect, the same behavior is known for various lighter nuclei [9]). The Q_α values for isotopes of Sg and Hs perfectly follow the trend of variation of $Q_\alpha(N)$ at the crossing of the magic neutron number $N = 162$ (identical to what is observed for the isotopes of Ds and Rg as well as for numerous nuclides with $N \approx 152$ and 126). Therefore, the systematics of α-decay energies of even-Z nuclei provides proof of identification of the Z and N of these elements. Moreover, the results of the chemical experiments [13] "establish element 112 as a typical element of group 12". Thus, the atomic numbers of all of the isotopes in the decay ^{291}Lv-^{287}Fl-^{283}Cn-^{279}Ds-^{275}Hs-^{271}Sg- ^{267}Rf were determined by the genetic relation [3].

The nuclei with $Z = 113$, 115, and 117 were first synthesized at DGFRS in the ^{243}Am(^{48}Ca,2-4n)^{287,288,289}Mc and ^{249}Bk(^{48}Ca,3-4n)^{293,294}Ts reactions [1,2]. Similarly to the reactions with even-Z target nuclei, two or three different parent nuclides were produced in the 2n- to 4n-evaporation channels. The excitation functions have been measured for the both these reactions. Only at high excitation

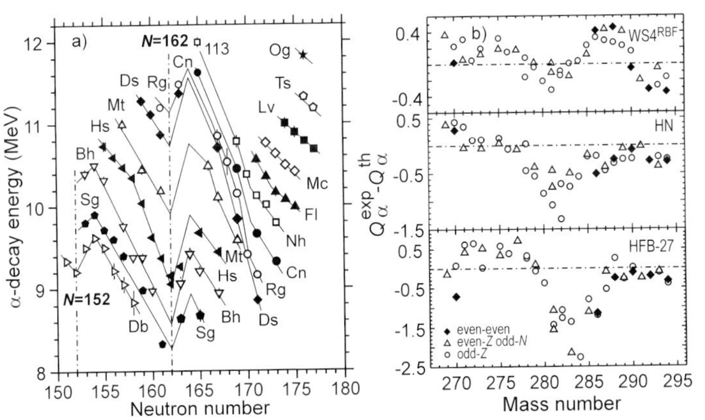

Fig. 1. (a) Measured α-decay energy vs. neutron number for the isotopes of elements 105-118 (filled and open symbols refer to even-Z and odd-Z nuclei, respectively [1,2,4-9]. The lines are drawn to guide the eye. (b) Deviation between experimental and calculated α-decay energies of nuclei Q_α. Theoretical Q_α values were taken from [10,11] for macroscopic-microscopic models WS4^{RBF} and HN, respectively, and Skyrme-Hartree-Fock-Bogoliubov model HFB-27 [12].

energies of the $^{249}Bk+^{48}Ca$ reaction we observed short decay chains of the even-N isotope $^{293}117$; each consisted of three consecutive α decays terminated by SF of ^{281}Rg or 5-ms ^{277}Mt instead of α decay observed for their odd-N neighbors $^{280,282}Rg$ and $^{276,278}Mt$. Similar α-α-SF decay chains were observed at the lowest ^{48}Ca energies in the $^{243}Am(^{48}Ca,2n)^{289}Mc$ reaction. Therefore, assignment of these chains to the product of the $2n$-reaction channel, ^{289}Mc, seems to be the most reasonable.

The same reactions were studied at TASCA [6,8] and BGS [7]. In [8], the decay times of nuclei in the short decay chains were re-analyzed with a newly suggested method which led to the interpretation that some of these chains might start from the isotope ^{288}Mc. However, the analysis of data by methods of conventional statistics demonstrate reasonable agreement of α-particle energies and decay times of nuclei assigned to ^{289}Mc, ^{285}Nh, and ^{281}Rg [14].

Therefore, similar to even-Z elements, the assignment of all of the odd-Z nuclides to the products of the xn-reaction channel is based on excitation-function measurements for the reactions with ^{243}Am and ^{249}Bk, production of ^{289}Mc in cross bombardments, as well as decay properties of nuclei. The decay properties of all of the odd-Z nuclei clearly differ from those observed for $^{290\text{-}293}Lv$, $^{285\text{-}289}Fl$, etc. (potential products of the pxn-reaction channels or electron capture (EC) of parent nuclei). The Q_α values for these odd-Z nuclei, estimated from the highest measured α-particle energies [1,2] and measured for ^{288}Mc-^{272}Bh from the α-γ coincidences [6,7], are in good agreement with systematics for other nuclei. The α-decay energies of all of the odd-Z nuclei are located in between and are in good agreement with the interpolations/extrapolations from neighboring even-Z nuclides with (Z±1) and the same neutron number as shown in Figure 1a. The Q_α values for Bh isotopes are located between $Q_\alpha(N)$ systematics for Sg and Hs and perfectly correspond to the pattern of variation of $Q_\alpha(N)$ for Bh isotopes at the crossing of $N = 162$. Moreover, the terminal SF nuclide ^{268}Db in the decay chain of ^{288}Mc produced in the $^{243}Am+^{48}Ca$ reaction was chemically characterized as a transactinide element [1,2]. Thus, all of the precursors ^{288}Mc-^{272}Bh discovered in [1,2] were identified by establishing the genetic link between the ancestor and the descendant. Therefore, conformity with criteria developed by IUPAC/IUPAP in 1991 [3] for the discovery of a new chemical element demonstrates, beyond a reasonable doubt, that the observed SHN originate from the complete-fusion reactions followed by the evaporation of the neutrons.

3. Discovery of Island of Stability

Moving from the magic number $N = 162$ and approaching the hypothetical magic neutron number $N = 184$, the isotopes of the elements produced in the U-Cf+^{48}Ca reactions exhibit a decrease in α-decay energy (Figure 1a). This results in a considerable increase of their half-lives (Figure 2).

Since the mid-1960s, different versions of the macroscopic-microscopic (MM) theory predict magic numbers $Z = 114$ and $N = 184$. Subsequent self-consistent microscopic calculations suggest that the magic shells should appear at $Z = 120$-126 and $N = 184$ in the most models (see, e.g., [10]). In spite of this disagreement, the resulting calculated nuclear masses do not strongly deviate from experimental masses. In Figure 1b, the experimental Q_α data are compared with the results of calculations within two MM models [10] and [11], which show minimum root mean square discrepancies with the measured mass values and with the purely microscopic Hartree-Fock-Bogoliubov approach [12]. The deviation between theory and experiment is given as the difference in decay energies $\Delta Q_\alpha = Q_\alpha^{exp} - Q_\alpha^{th}$ for nuclei produced in the ^{48}Ca-induced reactions. The ΔQ_α values for even-even nuclei are within approximately ±0.5 MeV; however, the HFB-27 data systematically exceed the experimental Q_α values by approximately 0.5 MeV. For all of the nuclei, the discrepancies ΔQ_α are within +0.5 to −0.4 MeV (WS4^{RBF}), +0.5 to −1.4 MeV (HN), and +0.8 to −2.3 MeV (HFB-27). Note, results of Q_α calculations within other MM models and purely microscopic approaches also agree with experimental data ([10] and references therein). Therefore, the hypothesis of existence of superheavy nuclei has received experimental confirmation.

Moreover, the effect of nuclear shells in the domain of superheavy elements results not only in substantially higher stability to various decay modes, but also

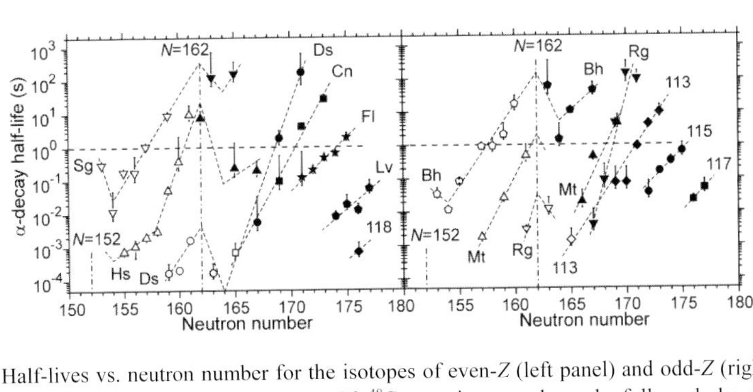

Fig. 2. Half-lives vs. neutron number for the isotopes of even-Z (left panel) and odd-Z (right panel) elements with $Z = 106$–118 (results from Ra-Cf+^{48}Ca reactions are shown by full symbols; other data are taken from [9]. Lines are drawn to guide the eye.

in an increase of the production cross section for complete-fusion reactions with ^{48}Ca projectiles. From numerous experiments it is known that in the synthesis of heavy nuclei with $Z \geq 102$, both in hot and cold fusion reactions, the cross section σ_{ER} decreases rapidly with increasing Z_{CN}. Extrapolating the dependence $\sigma_{ER}(Z_{CN})$ to $Z > 110$, we would arrive at extremely low cross sections for the production of isotopes of elements Fl and Mc. However, experimental values of $\sigma_{ER}(Z_{CN} = 114, 115)$ appeared to be about three orders of magnitude higher (Figure 3a). A strong change in $\sigma_{ER}(Z_{CN})$ with increasing Z could be caused by the survival probability of excited nuclei, which is directly related to the change in the fission barrier height B_f of the nuclei. The calculated values of difference of B_f and neutron binding energies B_n are shown in Figure 3b for the compound nuclei having the

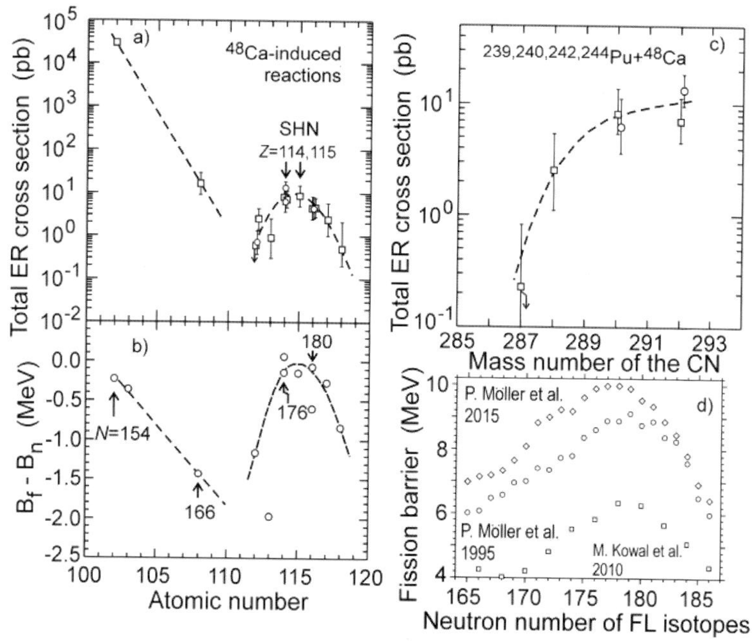

Fig. 3. (a) Maximum cross sections of the production of the nuclei in ^{48}Ca-induced reactions (E*=35-40MeV). Data measured at DGFRS are shown by squares, results obtained at SHIP, BGS, and TASCA are shown by circles. (b) Difference of fission barrier heights and neutron binding energies of the CNs in these reactions calculated in the MM model [10] are shown. Arrows show number of neutrons in the CN with the given atomic number. Lines are drawn to guide the eye. (c) Maximum of the production cross sections of Fl isotopes in the ^{48}Ca-induced reactions with targets shown in figure vs mass number of the CN. Data from DGFRS are shown by squares as well as from BGS (^{290}Fl) and TASCA (^{292}Fl) by circles. The dashed line is drawn to guide the eye. (d) Fission barrier heights vs neutron number of Fl isotopes calculated in the MM models [15-17].

production cross sections given in Figure 3a. Comparing these results, one can see that the relatively high cross sections for the production of evaporation residues in the fusion reactions with ^{48}Ca are caused by the large survivability of the compound nuclei, which is directly related to the high fission barriers in nuclei that appear due to nuclear shell effects.

On the other hand, a strong decrease in the production cross sections of the Pu+^{48}Ca reactions was observed with decrease of mass number of the compound nucleus (Figure 3c). With decreasing number of neutrons, the fission barriers are predicted to be lower (see, e.g., Refs. [15-17]) while neutron-binding-energy values steadily (if neutron magic number is not crossed) increase undergoing 1 MeV fluctuations due to the odd-even effect for nuclei in this mass region [9,10]. Therefore, a large decrease of the production cross sections of neutron-deficient Fl isotopes would be caused by a strong decrease in fission barriers (Figure 3d).

In accordance with reducing cross sections, the SF half-lives rapidly decrease for neutron-deficient even-even isotopes of Cn and Fl. This property of the nucleus is also governed by the fission barrier. The partial SF half-lives of even-even nuclei with $N \geq 162$, produced in the ^{48}Ca-induced reactions with ^{238}U-^{249}Cf, together with the half-lives of SF nuclides with $N < 162$, are shown in Figure 4. The influence of the $N = 152$ and 162 shells on the stability of nuclei against SF is clearly seen for isotopes of No and Rf-Hs, respectively. A similar stabilizing

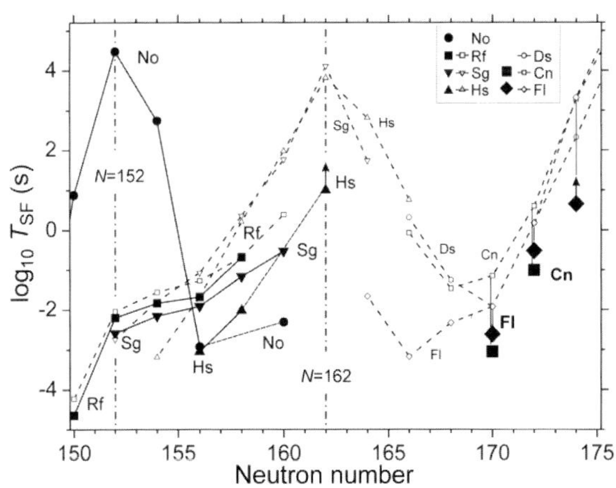

Fig. 4. Common logarithm of experimental partial spontaneous fission half-life vs neutron number for even-even isotopes of elements with Z = 102 to 114 (data are from Refs. [1,2,9], larger symbols). Short-dashed lines connecting smaller symbols show the theoretical T_{SF} values (see [10] and references therein) for even-even Z=104 to 114 isotopes. Long-dashed lines connect T_{SF} values for isotopes $^{258,262}No$ and $^{266,270}Hs$.

influence of the $N = 184$ shell seems apparent for Cn and Fl. These variations of T_{SF} are in agreement with theoretical calculations within MM models [10], as well as with a recent self-consistent Skyrme-HFB approach [18]. Thus, the considerable decrease of the evaporation cross sections of the Pu+^{48}Ca reactions with decreasing mass number of the compound nucleus as well as the decline of the half-life of spontaneously fissioning isotopes of Cn and Fl may indicate one is approaching the neutron-deficient border of this region of enhanced stability of SHN.

Acknowledgments

We acknowledge the support by the JINR Directorate grant and the RFBR Grant No. 16-52-55002. Research at ORNL was supported by the U.S. DOE Office of Nuclear Physics under DOE Contract No. DE-AC05-00OR22725 with UT-Battelle, LLC. Research at LLNL was supported by LDRD Program Project No. 08-ERD-030, under DOE Contract No. DE-AC52-07NA27344 with Lawrence Livermore National Security, LLC. This work was also supported by the U.S. Department of Energy through Grant No. DE-FG-05-88ER40407 (Vanderbilt University). Most of the actinide target materials used in the DGFRS experiments were made at ORNL and provided through the DOE Isotope Program and at RIAR.

References

1. Yu.Ts. Oganessian, *J. Phys. G: Nucl. Part. Phys.* **34**, R165 (2007).
2. Yu.Ts. Oganessian and V.K. Utyonkov, *Rep. Prog. Phys.* **78**, 036301 (2015); *Nucl. Phys. A* **944**, 62 (2015).
3. A.H. Wapstra, *Pure Appl. Chem.* **63**, 879 (1991).
4. K.P. Rykaczewski *et al.*, *EPJ Web of Conferences* **131**, 05005 (2016).
5. K. Morita *et al.*, *RIKEN Accel. Prog. Rep.* **47** (2014); D. Raji *et al.*, *RIKEN Accel. Prog. Rep.* **48**, 69 (2015).
6. D. Rudolph *et al.*, *Phys. Rev. Lett.* **111**, 112502 (2013).
7. J.M. Gates *et al.*, *Phys. Rev. C* **92**, 021301(R) (2015).
8. U. Forsberg *et al.*, *Nucl. Phys. A* **953**, 117 (2016).
9. M. Wang *et al.*, *Chin. Phys.* **36**, 1603 (2012); G. Audi *et al.*, ibid p. 1287; G. Audi *et al.*, ibid p. 1157.
10. A. Sobiczewski and K. Pomorski, *Prog. Part. Nucl. Phys.* **58**, 292 (2007).
11. N. Wang *et al.*, *Phys. Lett. B* **734**, 215 (2014).
12. S. Goriely, N. Chamel and J.M. Pearson, *Phys. Rev. C* **88**, 061302(R) (2013).

13. R. Eichler *et al.*, *Nature* **447**, 72 (2007).
14. V.B. Zlokazov and V.K. Utyonkov, to be published.
15. P. Möller *et al.*, *At. Data Nucl/ Data Tables* **59**, 185 (1995).
16. M. Kowal, P. Jachimowicz and A. Sobiczewski, *Phys. Rev. C* **82**, 014303 (2010).
17. P. Möller *et al.*, *Phys. Rev. C* **91**, 024310 (2015).
18. A. Staszczak, A. Baran and W. Nazarewicz, *Phys. Rev. C* **87**, 024320 (2013).

Fission Barriers of Super-Heavy Nuclei and Search for Element 120

S. Hofmann[1,2,*], S. Heinz[1], R. Mann[1], J. Maurer[1], G. Münzenberg[1,3], S. Antalic[4], W. Barth[1], H.G. Burkhard[1], L. Dahl[1], K. Eberhardt[5], R. Grzywacz[6,7], J.H. Hamilton[8], R.A. Henderson[9], J.M. Kenneally[9], B. Kindler[1], I. Kojouharov[1], R. Lang[1], B. Lommel[1], K. Miernik[6,10], D. Miller[7], K.J. Moody[9], K. Morita[11], K. Nishio[12], A.G. Popeko[13], J.B. Roberto[6], J. Runke[1], K.P. Rykaczewski[6], S. Saro[4], C. Scheidenberger[1], H.J. Schött[1], D.A. Shaughnessy[9], M.A. Stoyer[9], P. Thörle-Pospiech[5], K. Tinschert[1], N. Trautmann[5], J. Uusitalo[14] and A.V. Yeremin[13]

[1]GSI, 64291 Darmstadt, Germany
[2]Goethe-Universität Frankfurt, 60438 Frankfurt, Germany
[3]Manipal University, Manipal 576104, Karnataka, India
[4]Comenius University in Bratislava, 84248 Bratislava, Slovakia
[5]Johannes Gutenberg-Universität Mainz, 55128 Mainz, Germany
[6]Oak Ridge National Laboratory, Oak Ridge, Tennessee 37831, USA
[7]University of Tennessee, Knoxville, TN 37996, USA
[8]Vanderbuilt University, Nashville, Tennessee 37235, USA
[9]Lawrence Livermore National Laboratory, Livermore, California 94551, USA
[10]University of Warsaw, Warsaw, Poland
[11]RIKEN, 2-1 Hirosawa, Wako, Saitama 351-0198, Japan
[12]Japan Atomic Energy Agency, Tokai, Ibaraki, 319-1195, Japan
[13]Joint Institute for Nuclear Research, 141980 Dubna, Russian Federation
[14]Department of Physics, University of Jyväskylä, 40351 Jyväskylä, Finland
[*]Email: S.Hofmann@gsi.de

After a successful production of isotopes of element 116 in the reaction ^{48}Ca + ^{248}Cm, the reaction ^{54}Cr + ^{248}Cm → ^{302}120* was investigated at the velocity filter SHIP at GSI, Darmstadt, aiming to search for isotopes of element 120. One chain of events was observed, which is not created by chance with high probability. Parts of the chain agree with model predictions and measured data from a decay chain which could start at the isotope ^{299}120. In a complementary study, model dependent shell-correction energies and related heights of fission barriers were deduced from measured Q_α values. The results are compared with predictions of macroscopic-microscopic models. The consequences for calculations of cross-sections are discussed.

Keywords: Super-Heavy Nuclei; Fission Barriers; Cross-Sections.

1. Introduction

The nuclei presently known in the region at the upper end of the chart of nuclei are plotted in Figure 1. These heavy and super-heavy nuclei (SHN) were produced in so called cold fusion reactions using targets of lead and bismuth and in hot fusion reactions using actinide targets. Excitation energies of the produced compound nuclei (CN), shown as circles in Figure 1, are 10–20 MeV in the case of cold fusion and 35–55 MeV for hot fusion. Therefore, the highest cross-sections are measured for evaporation of 1 and 2–5 neutrons in the case of cold and hot fusion, respectively. The experimental results and descriptions of the experimental set-ups are presented in detail in review articles [1–4].

Definitely, the attempts to search for and produce SHN were triggered by theoretical predictions. These started with the introduction of the nuclear shell model, they culminated with the invention of the macroscopic-microscopic (MM) model in the 1960s [8,9] and were continuously improved up to now [10,11]. On the other hand, erroneous theoretical developments were corrected on the basis of

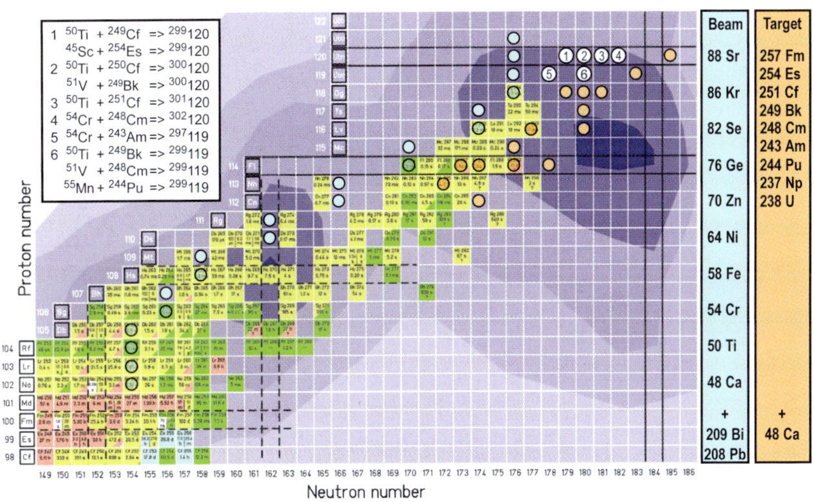

Fig. 1. Upper end of the chart of nuclei showing the presently known isotopes. Circles mark the CN produced in cold (blue) and hot (orange) fusion reactions. Beams are given on left of the two columns for cold fusion and targets for production of the most neutron-rich isotopes using hot fusion on the right. CN of possible reactions for production of isotopes of elements 119 and 120 are marked with numbers 1–6. For completeness, Cn produced in reactions with a ^{48}Ca beam and targets of ^{254}Es and ^{257}Fm are also shown. For a tentative assignment of an electron capture (EC) branch of ^{287}Fl and ^{290}Fl and tentative re-assignment of the decay chains measured in [5–7], see discussion in [4]. The background structure shows the calculated shell-correction energies according to the MM model [10] from –7 MeV (dark blue) in steps of +1 MeV.

new experimental findings. Examples are the predictions of a so called extra-push energy for fusion and the underestimation of quasi-fission in early cross-section calculations. A review on theoretical work has been presented in [10].

The reason for the stability of heavy and super-heavy nuclei are shell effects. In MM models [10,11], these are added as shell correction energies (SCE) to the macroscopic binding energy determined from the liquid drop model. The background color in Figure 1 shows those SCE for a wide range of heavy nuclei. Clearly, two regions of minimal SCE are recognized, one for deformed nuclei located at proton number Z = 108 and neutron number N = 162, the other for spherical nuclei near the double shell closure at Z = 114 and N = 184. In both cases, the experimental data related to the stability of these nuclei, like half-lives and decay modes, are well in agreement with the theoretical predictions.

However, uncertainty exists between model predictions for the stability of nuclei at Z = 120 and beyond. In addition or instead of the closed shells predicted by the MM models at Z = 114 and N = 184, relativistic mean field models obtain increased stability also at Z = 120 and 126 and N = 172 and 178. In terms of the shell model, subshells are closed at these proton or neutron numbers. Therefore, measured data for isotopes of element 120 are urgently required. As a by-product, the production of element 120 and already that of 119 will open a new row in the periodic table of elements where the last row was filled by the identification of element 118, a noble gas, now named 'oganesson' with the abbreviation 'Og'.

2. Search for Isotopes of Element 120 at SHIP

For the production of isotopes of element 120, the reaction ^{54}Cr + ^{248}Cm => ^{302}120* was proposed in November 2010 [13]. This reaction is more asymmetric than the reactions ^{58}Fe + ^{244}Pu [14] and ^{64}Ni + ^{238}U [15] studied in earlier experiments, and thus less Coulomb repulsion exists in the entrance channel. Although the reaction ^{50}Ti + ^{249}Cf studied in [16] is even more asymmetric, in our choice we expected to profit from being three neutrons closer to the N = 184 shell closure, see Figure 1. So far, higher cross-sections were measured when more neutron-rich projectiles were used.

Technical arguments were an additional reason for using a ^{248}Cm target. This isotope is an α emitter with a long half-life of 3.4×10^5 years and a low spontaneous fission (SF) branching of 8 %. Due to the resulting low specific activity, target wheels can be produced and handled without using special manipulators and thick radiation shielding.

Safe operation under experimental conditions was tested in a preparatory experiment in July 2010 [17]. In the reaction $^{48}Ca + ^{248}Cm \rightarrow ^{296}Lv^*$ previously measured data [3] on the decay of ^{293}Lv and ^{292}Lv were confirmed.

A total of 37 scientists from laboratories of various countries gathered for contributing to the experiment with experience, equipment, and material. Most substantial was the contribution of enriched ^{248}Cm target material by Lawrence Livermore National Laboratory (LLNL) and the digital pulse processing by Oak Ridge National Laboratory (ORNL). The accelerator UNILAC and the separator SHIP and its position-sensitive detector system were suggested for delivering the ^{54}Cr beam and for separation and identification of the fusion reaction products, respectively.

A beam time of 140 days was proposed. We estimated that within this time we could reach a cross-section limit of 100 fb. In the first part of the element-120 experiment lasting for 38 days from April 24th to June 1st in 2011, we reached a beam dose of 7.0×10^{18}, which would correspond to a one event cross-section limit of 0.58 pb.

One chain of correlated signals was measured for which a probability of 4×10^{-8} was calculated that this chain occurred by chance. A possible assignment of the signals to a decay chain starting at $^{299}120$ is shown on the left side in Figure 2. Not shown in the figure are the signals from the implantations of the produced nuclei into the detector, which serve as the start for the measurement of the life-times of the parent nuclei of the chains.

Arguments for the given assignment are, firstly the agreement of the third signal with measured values for both energy and half-life of ^{291}Lv [3]. Secondly, the good agreement of the energies of the first two signals with predictions of the MM model [11,12]. Thirdly, the agreement of the fourth and fifth signal with data measured in [5–7]. In those experiments, the reactions $^{48}Ca + ^{238}U \rightarrow ^{286}Cn^*$ and $^{48}Ca + ^{242}Pu \rightarrow ^{290}Lv^*$ were studied, and the measured data were assigned to the isotopes ^{283}Cn and ^{287}Fl, respectively. However, the measured relatively long half-lives of 5.1 min and 5.5 s, respectively, are not in agreement with established data for these nuclei measured in other experiments, see review articles [1,3,4]. Therefore, tentatively, a new interpretation was given in [4], which considers a possible EC branch of ^{287}Fl. Within this scenario, a reasonable agreement with the decay chain measured at SHIP is obtained. However, instead of EC, population and decay of isomeric states cannot be excluded as an alternative possibility.

As open questions to be discussed remain the relatively long half-lives of 3.7 s and 181 ms determined for the decays of $^{299}120$ and $^{295}118$, respectively. At α energies of 13.14 and 11.81 MeV, a WKB calculation (see [4]) results in 1.1 μs and 0.16 ms, respectively, for unhindered transitions. The measured half-lives

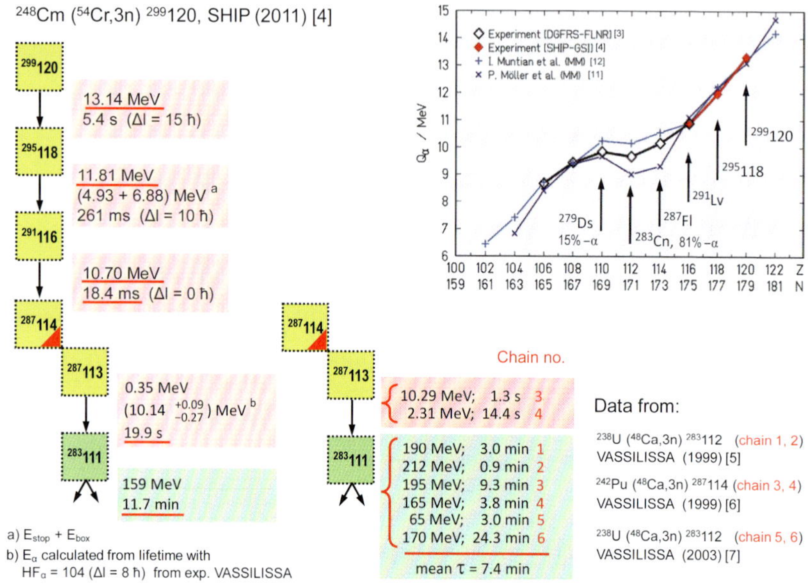

Fig. 2. Properties of the event chain measured at SHIP on May 18, 2011, and its tentative assignment (left). Properties of events measured in [5–7] and tentative re-assignment (bottom right). Comparison of measured [3] and calculated [11,12] Q_α values of the decay chain passing through ^{291}Lv (top right). See [4] for details of the experiment at SHIP and arguments for the suggested re-assignment.

could be explained assuming transitions hindered by angular momenta of $\Delta l = 15$ ħ and 10 ħ, respectively. Interestingly, in discussions on production probabilities of SHN, an often used argument is that the increase of the fission barrier by an additional centrifugal barrier could result in an increase of the cross-section for production of such high spin isomers.

Due to the existing uncertainties of measured data, we cannot definitely assign the new event chain to an isotope of element 120. Under these circumstances, a continuation of the experiment would be the best and only solution. However, beam time was not available for this experiment at the UNILAC. Therefore, we undertook a review of known experimental data of even element SHN. In particular we tried to derive information on the trend of the binding energies of the nuclei within the decay chains from measured Q_α values. This information was used to extract ‘experimental’ but model dependent SCE and related heights of fission barriers which are used in models for calculations of fusion-evaporation cross-sections.

3. Review of Known Data and Comparison with Predictions

3.1. *Masses from measured Q_α values*

Investigated decay chains of nuclei of even elements are plotted in Figure 3 in Columns 1, 2, 4, 6, and 7. For details of the priority of discovery and of the assignment of the measured data, see review articles [1,3,4]. Not considered in the evaluation are the chains in Columns 3, 5, and 8. In these cases, additional assumptions as EC or isomeric states are needed in order to place them into the systematics of known data.

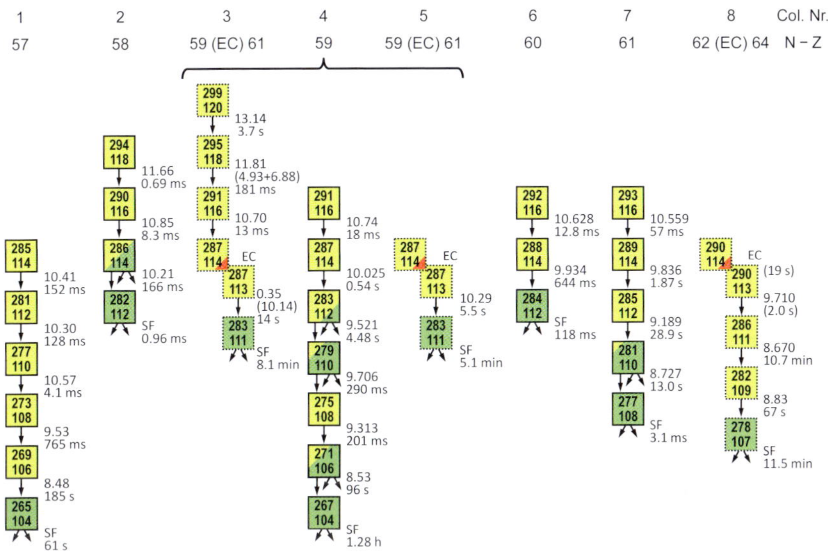

Fig. 3. Known and established α-decay chains of even element SHN are shown in Columns 1, 2, 4, 6, and 7. The chains are ordered according to their difference N – Z between neutron and proton number. The data are taken from review articles [1,3,4]. Energies are given in MeV. For a tentative assignment of the chain shown in Column 3 and a tentative re-assignment of the decay chains shown in Columns 5 and 8 considering a possible EC branch, see discussion in [4].

In a first step, the measured Q_α values were used to determine relative mass values. For none of the nuclei shown in Figure 3 an experimental mass value is known. Therefore, we used the estimates of the AME 2012 mass evaluation [18] for normalizing the masses of the nuclei at the end of the chains, filled symbols in Figure 4.

In Figure 4, we plotted the difference of the so determined relative masses and the predictions of the MM model of Sobiczweski *et al.* (MM-S, upper part)

[10] and Möller *et al.* (MM-M, lower part) [11]. Although both models are based on the same principle, the MM model, significant differences are observed in the region of SHN extending up to values of 4.5 MeV.

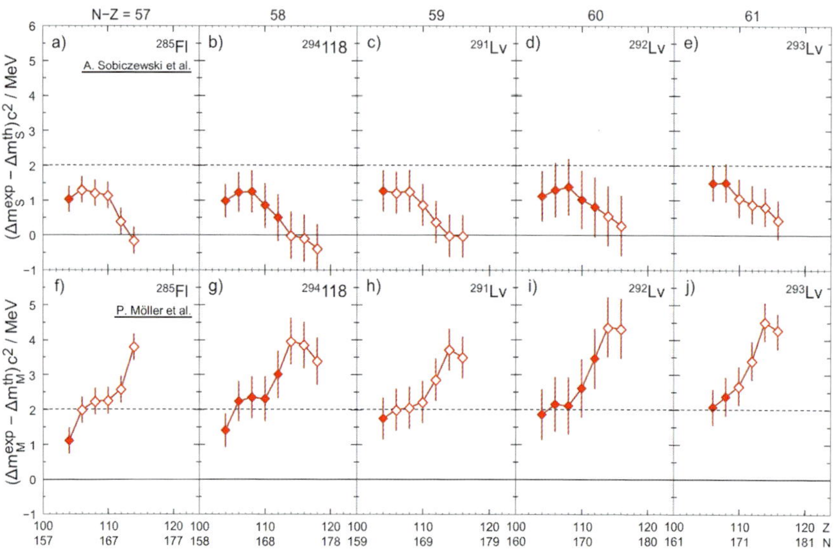

Fig. 4. Differences of experimental and theoretical masses of five neighboring decay chains of even-element SHN. The data are compared with theoretical masses of the MM-S model in (a)–(e) [10] and with the MM-M model in (f)–(j) [11]. Filled symbols mark differences obtained with extrapolated masses given in the AME2012 evaluation [18]. These data were used for normalizing the masses of the nuclei at the end of the α-decay chains. Open symbols mark those where the experimental masses were determined from the measured Q_α values [1,3,4]. The horizontal dashed lines at 2 MeV are drawn to guide the eye. The figure was taken from [19].

3.2. *Shell-correction energies and fission barriers*

The relative masses were used determining trends of SCE and fission barriers. This is achieved subtracting the liquid-drop mass used in the specific model from the experimental mass. The so determined model dependent 'experimental' SCE are plotted in Figure 5.

The upper row shows the data using the MM-S model [10], the lower row those using the MM-M model [11]. In addition to SCE based on the AME mass predictions, we also normalized the masses of the nuclei at the end of the chains to the theoretical masses (data without error bars). In this case, differences in the trends between theoretical and experimental SCE become more visible.

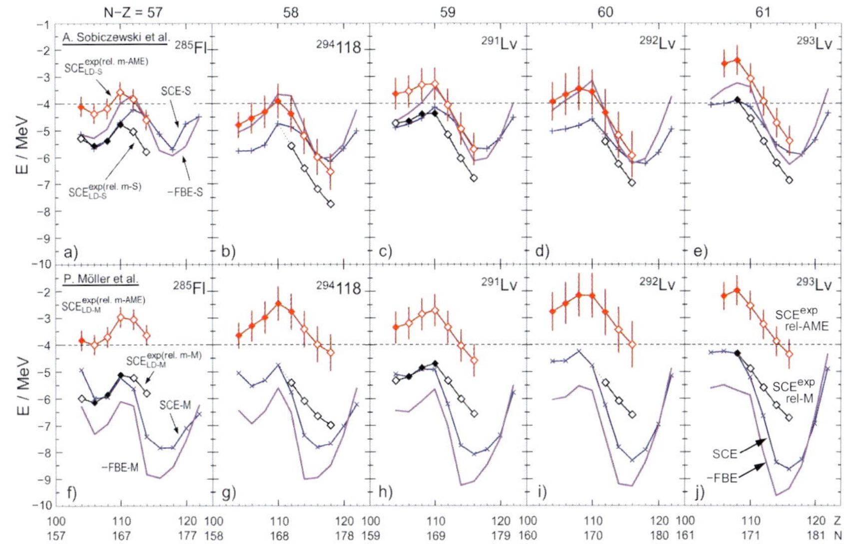

Fig. 5. Experimental shell-correction energies of nuclei of five neighboring decay chains are compared with theoretical values of the MM-S model [10] in (a)–(e) and of the MM-M model [11] in (f)–(j). Experimental data with error bars are based on the AME mass estimates, see also Figure 4, from which the theoretical liquid drop masses used in MM-S and MM-M, respectively, were subtracted. Experimental data without error bars were normalized to the theoretical masses of the nuclei at the end of the decay chains. Curves without symbols show the negative values of the heights of the fission barriers, −FBE-S [10] and −FBE-M [20], obtained in models MM-S and MM-M. The horizontal dashed lines at −4 MeV are drawn as reference lines to guide the eye. The figure was taken from [19].

Common in both models is the minimum of SCE at Z = 114–116. However, the minimum is more pronounced in MM-M. There, it is in the range from –8.0 to –8.5 MeV, whereas in MM-S it is at about –6.0 MeV.

The trend of the experimental SCE values normalized to the theoretical masses indicates a slightly stronger decrease from isotopes of element 110 into direction of 114–116 in MM-S, whereas significantly less shell strength is obtained in the comparison with MM-M. For the more neutron rich chains of N – Z = 60 and 61, the difference amounts to about 2.0 MeV.

Fission barriers are dominantly determined by SCE in the region of SHN. Therefore, it is reasonable to compare the experimental SCE also with the calculated heights of the fission barriers. In fact, –SCE values were often used in calculations of cross-sections as approximation to the heights of fission barriers, as long as these were not available. In Figure 5, the negative values of the heights of fission barriers determined in MM-S and MM-M are also plotted. Only small differences are observed in the case of MM-S, whereas the fission barriers are about 1 MeV higher than the corresponding SCE values in the case of MM-M. Obviously, a pronounced positive saddle-point SCE is responsible for the increase of the heights of the fission barriers.

3.3. *Cross-sections*

The height of the fission barrier is primarily responsible for the survival of the CN and is, thus, a significant factor contributing to the cross-section of the fusion-evaporation reaction. In the case of hot fusion reactions for production of SHN, a rough dependence of the cross-section from the height of the fission barrier is given by the rule of thumb that a 1 MeV increase of the fission barrier (in the ground-state) increases the cross-section by a factor of ten and vice versa [21]. Considering the uncertainties of fission barriers suggested by the data shown in Figure 5, we have to conclude that calculated cross-sections will differ by many orders of magnitude when different predictions of fission barriers are used.

In many of those calculations the values –SCE-M were used as approximation to the height of the fission barrier. This has two reasons. Firstly, the MM-M model is a global model which reproduces well the masses of known nuclei and predicted well the masses of nuclei which were discovered after its publication. Secondly, the binding energies together with their macroscopic and microscopic contributions are well tabulated for a wide region of nuclei, in particular also for SHN [11]. On the other hand, the parameters of the MM-S model [10] are adjusted to measured data of nuclei located in the limited region beyond lead. Consequently, it is expected that the MM-S model will reproduce well the experimental binding energies of SHN, whereas the global MM-M model reveals good results also for lighter nuclei. However, this model exhibits larger differences for nuclei in the vicinity of closed shells [24].

In cross-section calculations for production of isotopes of elements 114 and 116, which used the high fission barriers of MM-M, the experimental data were well reproduced due to using a correspondingly low fusion probability. If, in reality, as indicated by SCE deduced from experimental Q_α values, the heights of the fission barriers are lower, similar as predicted by MM-S, then the lower survival probability of the CN has to be compensated by an increase of the fusion probability. A rough quantitative estimate for the change of the calculated cross-sections for production of isotopes of elements 114 and 120 is shown in Figure 6. The rule of thumb was used as given before. The result is obtained that the cross-sections for production of $^{299}120$ and $^{298}120$ in the reaction $^{54}Cr + ^{248}Cm$ =>

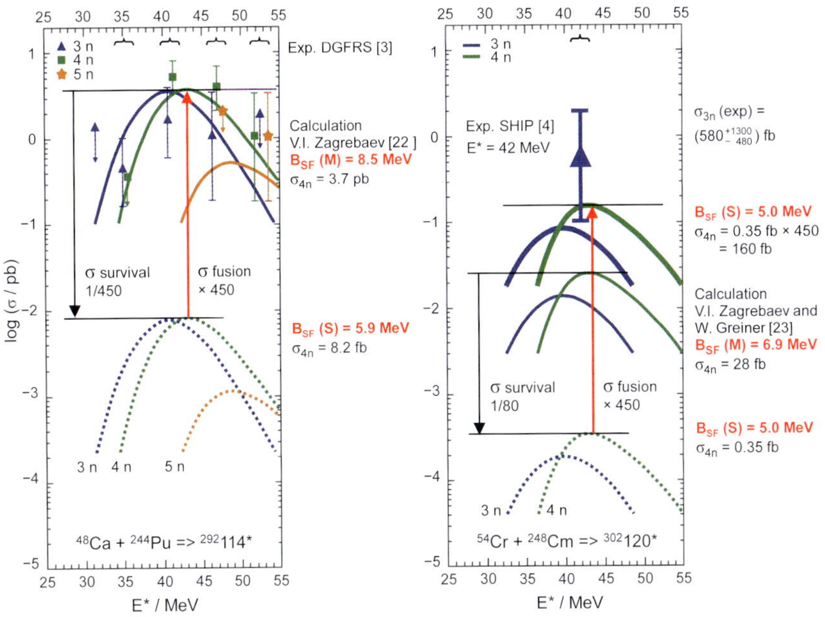

Fig. 6. Measured [3] and calculated [22] (lines) evaporation-residue cross-sections (3n–5n) for the reaction ^{48}Ca + ^{244}Pu => ^{292}Fl* (left). A reduction of the fission barrier from 8.5 to 5.9 MeV results in a reduction of the cross-section by a factor of 450 (dotted lines) assuming a factor of 10 reduction by a 1 MeV lower height of the fission barrier [21]. An increase of the fusion probability by the same factor is needed for reproducing the experimental values. Similarly, the calculated cross-sections for the reaction ^{54}Cr + ^{248}Cm => ^{302}120* [23] (right, thin lines) are reduced by a factor of 80 when the fission barrier is reduced from 6.9 to 5.0 MeV. Applying the same factor of 450 for the increase of the fusion probability, a maximum cross-section of 160 fb is obtained for the 4n channel (thick lines). A lower fusion probability due to the higher proton number of the ^{54}Cr beam compared to the ^{48}Ca beam on the left is already included in the original calculations [23]. The symbol with error bars marks the cross-section of the event chain measured at SHIP.

^{302}120* could be higher than the predicted values of 14 and 28 fb [23] for the 3n and 4n channels, respectively. This surprising and at a first glance contradictory result that lower fission barriers reveal higher cross-sections, is due to the locally pronounced shell effects for isotopes of elements 114 and 116 of the MM-M model used in calculations [22,23], which are, as elaborated, less pronounced in reality demanding for a higher fusion probability.

4. Conclusion

We investigated the reaction ^{54}Cr+^{248}Cm in an attempt to produce isotopes of the new element 120. Correlated α-decay–like signals and an also correlated SF event were observed. The probability is low that this chain of signals was produced by chance. Part of the data are in agreement with previously measured data assigned to decay chains starting at ^{291}Lv and ^{287}Fl, respectively, and with predictions for the α energies of ^{295}Og and $^{299}120$. However, uncertainties still exist, which hamper a definite assignment.

Experimental data on isotopes of even element SHN measured since 1998 were reviewed, in order to prepare a comprehensive basis for comparison of experimental data among each other and with results of model calculations. Good agreement was observed between experimental data and the results of the MM models of Sobiczewski et al. [10] and Möller et al. [11] concerning properties of α decay and SF. In particular, the much higher probability for α decay compared to SF of nuclei in the island of SHN is verified.

However, from a detailed analysis of measured Q_α values aiming to extract shell-correction energies and related heights of fission barriers, we deduced lower heights for Fl and Lv isotopes than predicted in [11,20] where particularly high fission barriers were calculated for isotopes of these elements due to strong shell effects at $Z = 114$ and $N = 184$. Consequences for cross-section calculations were discussed, see also [4,19].

The results of our search for isotopes of element 120 including those of the review may be of interest for search experiments for that element being in preparation at other laboratories. At the Flerov Laboratory in Dubna it is planned to irradiate a target of ^{249}Cf, ^{250}Cf, and ^{251}Cf, first with ^{48}Ca ions for production of more isotopes of Og. In a second step, it is planned to switch the beam from ^{48}Ca to ^{50}Ti for synthesis of element 120 [25].

At GARIS at RIKEN, it is planned to produce isotopes of Og using the reaction $^{50}Ti + ^{248}Cm \rightarrow ^{298}Og^*$ and, in a second step, to switch the beam to ^{54}Cr for synthesis of element 120 [26] as in our experiment at SHIP.

The perspectives for a successful production of the new elements 120 and 119 which could be produced also with a ^{248}Cm target and a beam of ^{51}V, are quite promising. The discovery of these two elements, which would open a new row in the periodic table of the elements, will be the most remarkable event in future research of SHN. However, for a better understanding of the properties and the extension of the island of SHN, experimental results on lighter elements are still waiting to be performed. Among those are the measurements of small branching ratios for both α decay and SF of nuclei for which only SF or α decay, respectively, is known so far. Similarly, EC or pxn channels, both occurring also with small probability, open possibilities producing nuclei of odd elements with neutron numbers slightly higher than known so far. Two candidates for EC, ^{287}Fl and ^{290}Fl were discussed in this work and in [4]. A tentative re-assignment of previously measured decay chains of nuclei with exceptional long lifetimes led to this conclusion. Confirmation of this assumption as well as the reproduction of the result of our search experiment for isotopes of element 120 belongs to the challenging but promising tasks of future experiments.

Acknowledgments

We acknowledge support of the U.S. Department of Energy through Contracts No. DE-AC05-00OR2272 (ORNL) and No. DE-AC52-07NA27344 (LLNL), and Grants No. DE-FG-05- 88ER40407 (Vanderbilt University). SA acknowledges support from the Slovak grant agency VEGA (contract No. 1/0532/17) and the Slovak Research and Development Agency (contract No. APVV-14-0524).

References

1. J.H. Hamilton *et al.*, *Annu. Rev. Nucl. Part. Sci.* **63**, 383 (2013).
2. S. Hofmann and G. Münzenberg, *Rev. Mod. Phys.* **72**, 733 (2000).
3. Yu.Ts. Oganessian and V.K. Utyonkov, *Nucl. Phys. A* **944**, 62 (2015).
4. S. Hofmann *et al.*, *Eur. Phys. J. A* **52**, 180 (2016).
5. Yu.Ts. Oganessian *et al.*, *Eur. Phys. J. A* **5**, 63 (1999).
6. Yu.Ts. Oganessian *et al.*, *Nature* **400,** 242 (1999).
7. Yu.Ts. Oganessian *et al.*, *Eur. Phys. J. A* **19**, 3 (2004).
8. V.M. Strutinsky, *Nucl. Phys. A* **95**, 420 (1967).
9. S.G. Nilsson *et al.*, *Nucl. Phys., A* **131**, 1 (1969).
10. A. Sobiczewski and K. Pomorski, *Prog. Part. Nucl. Phys.* **58**, 292 (2007).
11. P. Möller *et al.*, *At. Data Nucl. Data Tables* **59**, 185 (1995).
12. I. Muntian *et al.*, *Phys. At. Nucl.* **66**, 1015 (2003).
13. S. Hofmann *et al.*, *Study of superheavy elements using ^{248}Cm targets, Part II: the reaction ^{54}Cr + ^{248}Cm → ^{302}120*, GSI Proposal* **U-248**, (2010).
14. Yu.Ts. Oganessian *et al.*, *Phys. Rev. C* **79**, 024603 (2009).
15. S. Hofmann *et al.*, *GSI Scient. Rep. 2008, GSI Report* **2009-1**, 131 (2009).
16. J. Khuyagbaatar *et al.*, *GSI Scient. Rep., GSI Report* **2013-1**, 131 (2013).
17. S. Hofmann *et al.*, *Eur. Phys. J. A* **48**, 62 (2012).
18. M. Wang *et al.*, *Chinese Phys.* **36**, 1603 (2012).
19. S. Hofmann *et al.*, *Eur. Phys. J. A* **52**, 116 (2016).
20. P. Möller *et al.*, *Phys. Rev. C* **79**, 064304 (2009).

21. M.G. Itkis *et al.*, *Phys. Rev. C* **65**, 044602 (2002).
22. V.I. Zagrebaev, *Nucl. Phys. A* **734**, 164 (2004).
23. V.I. Zagrebaev and W. Greiner, *Phys. Rev. C* **78**, 034610 (2008).
24. P. Möller *et al.*, *At. Data Nucl. Data Tables* **109-110**, 1 (2016).
25. V.K. Utyonkov, contribution to this conference (2016).
26. H. Kudo, contribution to this conference (2016).

On the Search for Elements Beyond Z = 118 — An Outlook Based on Lessons from the Heaviest Known Elements

Christoph E. Düllmann

PRISMA Cluster of Excellence & Institute of Nuclear Chemistry, Johannes Gutenberg University, 55099 Mainz, Germany
GSI Helmholtzzentrum für Schwerionenforschung, Planckstr. 2, 64291 Darmstadt, Germany
Helmholtz Institute Mainz, 55099 Mainz, Germany
E-mail: duellmann@uni-mainz.de
www.superheavies.de

All elements up to Z = 118 are discovered, and several search experiments for elements beyond have been carried out. None of them, however, have reported the discovery of a heavier element. Relevant aspects for a successful search for elements with $Z \geq 119$ are discussed.

Keywords: Superheavy Elements; Elements with $Z \geq 119$.

1. Introduction: Discovery of the elements up to Og (Z = 118)

The search for new, artificial elements has been a driving force in the nuclear sciences since the 1930s. After the development of the nuclear shell model, the motivation to find and explore a region of enhanced nuclear stability, predicted to be centered around pronounced shell effects at atomic number Z = 114 and neutron number N = 184 [1, 2] fueled such experiments even further. Modern theoretical approaches suggest spherical shell closures likely to be located at Z = 114, 120, or 126 and N = 172, 184, [3-5], forming an extended region of enhanced stability [6]. The first transuranium elements up to fermium (Z = 100) were accessible in neutron capture reactions. Heavier elements are accessible in fusion reactions of heavy-ion beams with suitable target nuclei. Reactions induced by projectiles with mass number A ~ 6-30 with neutron-rich transuranium isotopes lead to the formation of compound nuclei (CN) with rather high excitation energies, E*, of at least ~ 40-50 MeV, and are thus called hot fusion reactions. Elements up to seaborgium (Z = 106) were first produced this way, as evaporation residues (ER) after the fusion of the corresponding nuclei followed by the evaporation of typically 3-5 neutrons, with each evaporation step being in competition with fission. Experimental ER cross sections, σ_{ER}, were found to decrease exponentially with increasing Z [7]. After the realization that fusion reactions involving shell-stabilized nuclei feature lower minimum accessible E* of ~10-20 MeV [8], thus reducing the number of competitive fission vs. neutron evaporation steps in the exit channel, the next elements bohrium (Z = 107) to nihonium (Z = 113) were discovered using such cold fusion reactions, based on targets of (doubly-)magic ^{208}Pb and ^{209}Bi [7, 9, 10]. Steeply exponentially decreasing σ_{ER} with increasing Z were also found for cold fusion reactions, reaching the level of only a few ten fb at Z = 113 [11].

In contrast to hot fusion reactions, where the σ_{ER} are small predominantly due to losses in the exit channel, cold fusion reactions with their massive partners inherently suffer from finite probabilities, P_{CN}, of the projectile and target to form a CN in the first place, with the dominant flux going to quasi-fission [12] (QF), where $P_{QF} = 1 - P_{CN}$. Based on measured data, the general impossibility to produce elements beyond Z $\approx$ 112 was concluded [13]. Nevertheless, experimental efforts did not cease. Beyond Z = 113, a new twist in heavy-ion fusion reactions proved successful: fusion reactions involving a less massive projectile, ^{48}Ca, which is doubly-magic and very neutron-rich, with the heaviest target nuclei available in sufficient (~mg) quantities. Reactions of that type have been studied for a long time [14], but only in the past ~15 years has the experimental sensitivity been high enough to allow the discovery of new elements up to oganesson (Z = 118) [10, 15, 16]. As Cf is the highest-Z element available in mg-amounts [17], as are needed to make targets for such experiments, the successful ^{48}Ca route to new superheavy elements is exhausted at Z = 118. Hence, yet again, the search for new, heavier elements requires the identification of the most promising route.

2. Synthesizing elements with Z ≥ 118

2.1. *Nuclear reaction aspects*

Fusion reactions induced by high-intensity stable beams still appear to be most promising approach to synthesize elements beyond Z = 118 [18]. Basic features of reactions with actinide targets leading to the yet unknown element with Z = 120 are listed in [19] and are discussed in more detail in [20]. Alternatively, employing reactions between partners that are more equal in mass (symmetric reactions) has been suggested [21]. Commonly, though, the following four reactions are considered most promising for reaching Z = 120: ^{64}Ni + ^{238}U, ^{58}Fe + ^{244}Pu, ^{54}Cr + ^{248}Cm, and ^{50}Ti+^{249}Cf. The former three all lead to $^{302}120^{*}$ as

CN, whereas the most asymmetric reaction leads to $^{299}120^*$. Fig. 1 shows the location of the landing points and decay chains on the current chart of nuclei for these four reactions. For reaching Z = 119, the ^{50}Ti+^{249}Bk reaction leading to $^{299}119^*$ appears promising.

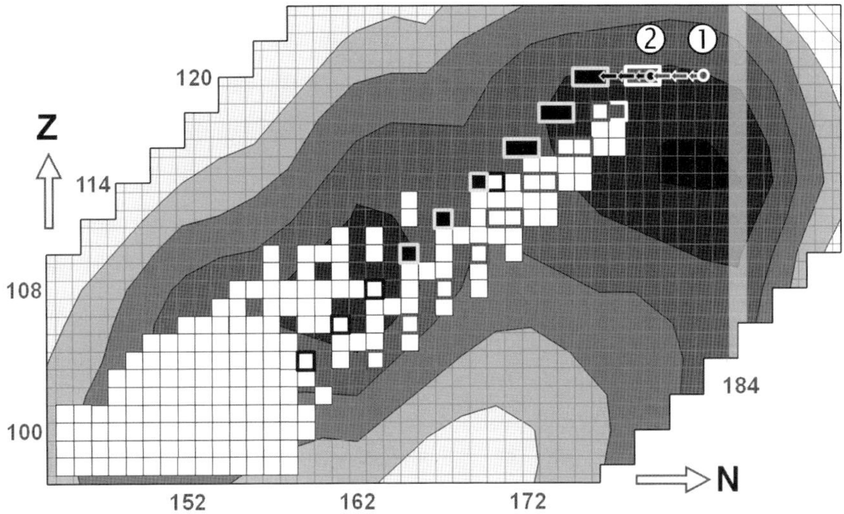

Fig. 1. Cut-out of the chart of nuclei showing the region of the heaviest elements. The underlying contour plot shows calculated shell stabilization energies according to a macroscopic-microscopic model [22]. The predicted neutron-shell closure at N = 184 is indicated. White boxes denote known nuclei. The compound nuclei (CN), the 3n and 4n evaporation residues and their α-decay chains are highlighted, where the reactions ^{64}Ni + ^{238}U, ^{58}Fe + ^{244}Pu, and ^{54}Cr + ^{248}Cm start with the CN indicated by ① and the reaction ^{50}Ti+^{249}Cf with the CN indicated by ②.

When selecting the most promising reaction, i.e., the one with the maximum σ_{ER}, a survey of theoretical predictions reveals an inhomogeneous picture. Predicted values for σ_{ER} for each reaction span orders of magnitude, with lower values approaching single fb. As discussed, e.g., in [19, 20], the most asymmetric reaction ^{50}Ti+^{249}Cf is generally favored, but estimates of the magnitude of σ_{ER} appear uncertain. The often-stated expectation that the most asymmetric reaction might be the optimum one reiterates the importance of the entrance channel, where the CN formation probability p_{CN} generally scales with reaction asymmetry. On the other hand, this most asymmetric reaction differs from all others in the neutron number of the CN. It features three fewer neutrons and thus populates a CN that is approaching the generally expected next spherical shell closure less closely. This might be expected to be dis-advantageous in the exit channel, as the fission barrier might be expected to be highest at the shell closure. Also here, though, the situation is not quite as clear; take as an example Fig. 6 in [23], where calculated fission barrier heights in the relevant region of the chart of nuclei are shown. In the approach of [23], highest fission barriers in the Z = 120 isotopic sequence occur at neutron numbers substantially smaller than N = 184. Guidance that theoretical predictions give to experimental searches of new elements appears thus somewhat unsatisfactory. A comprehensive analysis of current predictions is summarized in [20]. Table 1 below shows some relevant predictions for the $^{50}Ti + ^{249}Cf$ reaction.

Table 1. Overview of predicted maximum cross section for producing Z = 120 ER in the 3n and 4n evaporation channels.

σ_{ER} (3n) / fb	σ_{ER} (4n) / fb	Ref
750	300	[24]
130	50	[25]
100	65	[26]
100	4	[27]
45	4	[28]
40	45	[29]
34	15	[30]
20	5	[31]
6	6	[32]

To assess the credibility of individual theoretical models, a comparison to the $^{48}Ca+^{244}Pu$ reaction, leading to $^{288,289}Fl$ in the 3n and 4n evaporation channels, respectively, where experimental data are available [33], is contained in [20], along with a comparison of the predicted optimum bombarding energies. The latter vary substantially among the different predictions, with the differences being much larger than the range of energies that can be accommodated in a single experiment.

2.2. *Nuclear structure aspects of Z = 120*

Proton number 120 is of special interest as it is sometimes predicted to correspond to a spherical shell closure, most pronouncedly in microscopic relativistic-mean-field (MicRMF) based calculations. Properties of Z = 120 nuclei have thus received special attention from nuclear theory, e.g., [34]. Occurrence of a nuclear shell closure at this Z would manifest, e.g., in reduced α-decay energies, Q_α, when compared to results from models, in which 120 is no magic proton number, with direct impact on the magnitude of partial α-decay half-lives. A nuclear shell closure at Z = 120 is also predicted to dramatically – up to several orders of magnitude – affect σ_{ER} of fusion reactions leading to

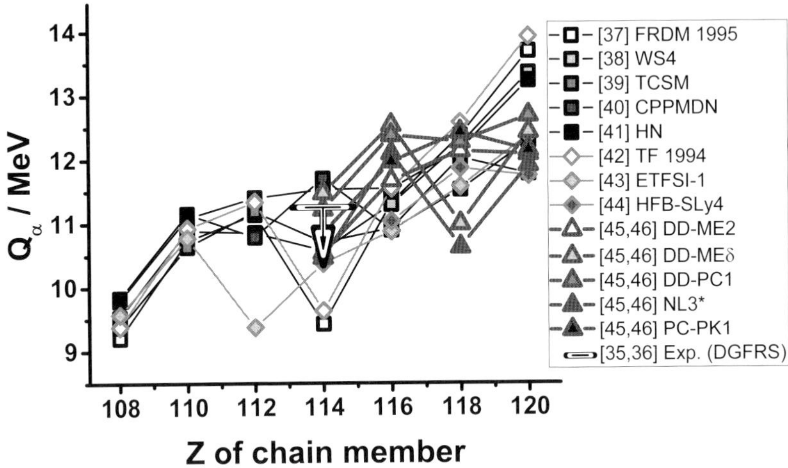

Fig. 2. Predicted Q_α values for the α-decay chain originating from $^{296}120$, the 4n evaporation channel of the ^{50}Ti + ^{249}Cf reaction. Data are grouped according to theoretical framework. MM: squares; MicNR: diamonds; MicRMF: triangles. An upper limit derived from the experimental non-observation of α decay is indicated for ^{284}Fl.

Z = 120 (e.g., [24]). A comprehensive summary of Q_α values predicted for $^{295,296}120$ (the 3n and 4n evaporation channels of the ^{50}Ti + ^{249}Cf reaction) is given in [20]. Generally, MicRMF-based approaches, suggesting Z = 120 to be a shell closure, yield lower Q_α values and correspondingly longer partial half-lives for α decay than other approaches like macroscopic-microscopic (MM) or microscopic non-relativistic (MicNR) ones preferring magic numbers at Z = 114 or Z = 126.

Experiments searching for Z = 120 rely on identifying the decay-chain populated by the ER decay. Therefore, not only decay properties of the ER are of interest, but also those of their daughters. In Fig. 2, Q_α values of $^{296}120$, ^{292}Og (Z = 118), ^{288}Lv (Z = 116), ^{284}Fl (Z = 114), ^{280}Ds (Z = 110), and ^{276}Hs (Z = 108) calculated in different theoretical frameworks are given; among these, the ^{284}Fl is known [35] to decay by SF with $T_{1/2}$ = 2.5 ms. Based on the observation of several SF decays, but no α decay, a conservative "experimental" Q_α value of about ≥ 11.25 MeV (which would correspond to a partial α-decay half-life of about 2.5 ms in the formalism of [36]) can be inferred.

Half-lives obtained from the data given in Fig. 2 via [36], i.e., deriving from predictions for Q_α values from various mass tables, span several orders of magnitude; a comparison of the performance of the different frameworks in neighboring isotopes, where experimental data are available, will allow to narrow this range.

Several models suggest the Q_α value of ^{284}Fl to be substantially lower than the inferred upper limit; others, though, locate it in its vicinity, including Ref. 41 (11.53 MeV, corresponding to 0.5 ms for $T_{1/2}(\alpha)$) and Ref. 47 (11.61 MeV; corresponding to 0.3 ms for $T_{1/2}(\alpha)$ according to [36] or 8.8 ms according to [47]). This may hint at an α branch large enough for its detection in a more sensitive experiment, as discussed in [35, 47].

2.3. *Status of efforts to discover elements beyond Z = 118*

Several experiments have been performed with the goal to discover elements with Z > 118. Table 2 [48] summarizes these efforts.

Table 2. Summary of search experiments for elements with Z > 118 performed in the past 10 years.

Reaction	Compound nucleus	Beamtime invested	Cross section limit[1]	Lab.	Ref.
^{50}Ti + ^{249}Bk	$^{299}119^*$	4.5 months	[2]	GSI	[20]
^{64}Ni + ^{238}U	$^{302}120^*$	4 months	≥ 90 fb	GSI	[49]
^{58}Fe + ^{244}Pu	$^{302}120^*$	2 months	≥ 400 fb	FLNR	[50]
^{54}Cr + ^{248}Cm	$^{302}120^*$	> 1 month	≥ 580 fb	GSI	[47]
^{50}Ti + ^{249}Cf	$^{299}120^*$	> 1 month	[2]	GSI	[20]

[1] 63.2% confidence level ("one-event limit").

[2] Under final analysis.

More than one year of accelerator beamtime has been devoted to search for elements beyond Z = 118, but the reached experimental sensitivity was insufficient for a positive identification. Before a new effort on the search for new elements is undertaken, it appears necessary to gain an improved understanding of the above listed nuclear reactions, including the determination of the height of the Coulomb barrier, which is one of the necessary ingredients for more reliable predictions, e.g., of the optimum bombarding energies. For this, further developments on the experimental as well as on the theoretical side will complement each other.

Acknowledgment

I like to thank A. Afanasjev for providing data that entered Figure 2.

References

1. A. Sobiczewski *et al.*, *Phys. Lett.* **22**, 500 (1966).
2. W. D. Myers and W. J. Swiatecki, *Nucl. Phys.* **81**, 1 (1966).
3. A. Sobiczewski and K. Pomorski, *Prog. Part. Nucl. Phys.* **58**, 292 (2007).
4. W. Zhang *et al.*, *Nucl. Phys. A* **753**, 106 (2005).
5. S. Ćwiok, P.-H. Heenen and W. Nazarewicz, *Nature* **433**, 705 (2005).
6. M. Bender *et al.*, *Phys. Lett. B* **515**, 42 (2001).
7. S. Hofmann, *Rev. Mod. Phys.* **72**, 733 (2000).
8. Y. T. Oganessian *et al.*, *Nucl. Phys. A* **A239**, 353 (1975).
9. K. Morita *et al.*, *J. Phys. Soc.* Japan **73**, 2593 (2004).
10. P. J. Karol *et al.*, *Pure Appl. Chem.* **88**, 139 (2016).
11. K. Morita *et al.*, *J. Phys. Soc.* Japan **81**, 103201 (2012).
12. J. Tōke *et al.*, *Nucl. Phys. A* **440**, 327 (1985).
13. P. Armbruster, *Acta Phys. Pol. B* **34**, 1825 (2003).
14. P. Armbruster *et al.*, *Phys. Rev. Lett.* **54**, 406 (1985).
15. R. C. Barber *et al.*, *Pure Appl. Chem.* **83**, 1485 (2011).
16. P. J. Karol *et al.*, *Pure Appl. Chem.* **88**, 155 (2016).
17. J. B. Roberto *et al.*, *Nucl. Phys. A* **944**, 99 (2015).
18. W. Loveland, *Phys. Rev. C* **76**, 014612 (2007).
19. Ch. E. Düllmann, in 5th International Conf. on Fission and Neutron-Rich Nuclei, edited by J. Hamilton, Sanibel Island, FL, USA, 2013.
20. Ch.E. Düllmann *et al.*, in preparation.
21. K. Choudhury and Y. K. Gupta, *Phys. Lett. B* **731**, 168 (2014).
22. A. Sobiczewski *et al.*, *Phys. Rev. C* **63**, 034306 (2001).
23. M. Kowal *et al.*, *Phys. Rev. C* **82**, 014303 (2010).
24. G. G. Adamian *et al.*, *Eur. Phys. J. A* **41**, 235 (2009).
25 J. Zhang, C. Wang and Z. Ren, *Nucl. Phys. A* **909**, 36 (2013).
26 Z.-H. Liu and J.-D. Bao, *Phys. Rev. C* **83**, 044613 (2011).
27 A. K. Nasirov *et al.*, *Phys. Rev. C* **84**, 044612 (2011).
28 N. Wang *et al.*, *Phys. Rev. C* **85**, 041601 (2012).
29. V. Zagrebaev and W. Greiner, *Phys. Rev. C* **78**, 034610 (2008).
30. Z.-H. Liu and J.-D. Bao, *Phys. Rev. C* **87**, 034616 (2013).
31. N. Wang, J. Tian and N. Scheid, *Phys. Rev. C* **84**, 061601 (2011).
32. K. Siwek-Wilczynska *et al.*, *Phys. Rev. C* **86**, 014611 (2012).
33. Ch.E. Düllmann *et al.*, *Phys. Rev. Lett.* **104**, 252701 (2010).
34. E. Litvinova, *Phys. Rev. C* **85**, 021303(R) (2012).
35. V. K. Utyonkov *et al.*, *Phys. Rev. C* **92**, 034609 (2015).
36. A. Parkhomenko and A. Sobiczewski, *Acta Phys. Pol. B* **36**, 3095 (2005).
37. P. Möller *et al.*, *At. Data Nucl. Data Tab.* **59**, 185 (1995).

38. N. Wang *et al.*, *Phys. Lett. B* **734**, 215 (2014).
39. A. N. Kuzmina *et al.*, *Phys. Rev. C* **85**, 014319 (2012).
40. K. P. Santhosh *et al.*, *Nucl. Phys. A* **955**, 156 (2016).
41. I. Muntian *et al.*, *Acta Phys. Pol. B* **34**, 2073 (2003); *Phys. At. Nucl.* **66**, 1015 (2003).
42. W. D. Myers and W. Swiatecki, *Nucl. Phys. A* **601**, 141 (1996).
43. Y. Aboussir *et al.*, *At. Data Nucl. Data Tab.* **61**, 127 (1995).
44. D. E. Ward *et al.*, *Phys. Rev. C* **92**, 014314 (2015).
45. S. E. Agbemava *et al.*, *Phys. Rev. C* **92**, 054310 (2015).
46. A. Afanasjev, priv. comm.
47. S. Hofmann *et al.*, *Eur. Phys. J. A* **52**, 180 (2016).
48. Ch. E. Düllmann, *EPJ Web Conf.* **131**, 08004 (2016).
49. S. Hofmann, *J. Phys. G* **42**, 114001 (2015).
50. Yu. Ts. Oganessian *et al.*, *Phys. Rev. C* **79**, 024603 (2009).

Superheavy Element Factory

Dmitriev S.N.
Flerov Laboratory of Nuclear Reactions, JINR, 141980, Dubna, Moscow Region, Jolio Curie 6,
dmitriev@jinr.ru

The synthesis of superheavy elements (SHE) with atomic numbers 113-118 has been achieved and consequently the seventh period of the D.I. Mendeleev's Periodic Table has been completed. Unique data on the nuclear-physical properties of isotopes of the new elements have been obtained [1]. Pioneering experiments on the chemical properties of elements 112 (Cn), 113 and 114(Fl) have been carried out [2].

The pursuit of the next level of these investigations will only be possible through the creation of a new experimental framework capable of delivering significant increases in the efficiency of the experiments. As a response to this complex challenge, a new project has been launched at FLNR/JINR — the creation of the first in the world superheavy elements "factory" [3].

The project includes the construction of a new accelerator of stable and long-living isotopes in the mass range $A = 10 - 100$, with intensities up to 10 pμA and energies up to 8 MeV/nucleon; the construction of a new building for housing the accelerator and the experimental infrastructure, including five channels for the transportation of beams to a 1200 sqm experimental hall equipped with systems of shielding and control for operations with radioactive beams and materials; development and construction of new separators of reaction products along with upgrades of existing separators, as well as development and construction of new detection modules for the study of the nuclear, atomic and chemical properties of the new elements.

Commissioning of the facility and the first experiments are planned for 2018.

References

1. Oganessian Yu., Utyonkov V., *Rep. Prog. Phys.*, **78**, 036301 (2015).
2. Oganessian Yu.Ts., Dmitriev S.N., *Russ. Chem. Rev.*, 2016, **85** (9), 901-916, DOI: 10.1070/RCR4607.
3. Oganessian Yu.Ts., Dmitriev S.N. and Itkis M.G., Proceedings of International Symposium on Exotic Nuclei, World Scientific, Singapore, 2015, pp. 483–490.

Recent Results and Future Prospects at LBNL

J.M. Gates

Lawerence Berkeley National Laboratory, Berkeley, CA, USA, 94720

Despite knowing about SHE for more than a decade, several important questions about these nuclei remain unanswered, including what are the atomic numbers, Z, and masses of these nuclei. Recently, our group has been working towards answering this intriguing question through one of three methods: (i) linking the SHE to known nuclides (ii) Z identification through observation of characteristic x-rays (iii) mass determination through mass analysis. This talk will review the progress that has been made for each of the three methods and discuss possible avenues for the future.

*This work was supported by the U.S. Department of Energy, Office of Science, Nuclear Physics, Low Energy Physics Program, through the Lawrence Berkeley National Laboratory under Contract No. DE-AC02-05CH11231

Impact of Quasifission on SHE Production

M. G. Itkis, I. M. Itkis, G. N. Knyazheva and E. M. Kozulin

Flerov Laboratory of Nuclear Reactions, Joint Institute for Nuclear Research, Dubna,141980, Russia
E-mail: itkis@jinr.ru

The properties of the mass and energy distributions of fissionlike fragments formed in the reactions ^{48}Ca,^{58}Fe + ^{208}Pb, ^{36}S,^{48}Ca,^{48}Ti,^{64}Ni + ^{238}U, ^{48}Ca + ^{232}Th,^{244}Pu,^{248}Cm at energies around the Coulomb barrier have been analyzed to define the systematic trend of compound nucleus fission and quasifission in cold and hot fusion reactions. The measurements have been carried out at the U400 cyclotron of the FLNR, JINR using the double-arm time-of-flight spectrometer CORSET. The fusion probabilities have been deduced from the analysis of mass and energy distributions. It was found that for the studied reactions fusion probability depends exponentially on mean fissility parameter of the system. For the reactions with actinide nuclei leading to the formation of superheavy elements the fusion probabilities are of several orders of magnitude higher than in the case of cold fusion reactions.

Keywords: Fusion Probability; Fusion-Fission; Quasifission.

1. Introduction

The existence of an island of stability in the region of nuclei with $Z = 114$ and $N = 184$ predicted theoretically fifty years ago [1] caused the experimental investigation in the field of the superheavy element (SHE) synthesis. Nowadays more than 30 isotopes of SHE with $Z = 108$–118 have been synthesized both in cold and in hot fusion reactions. In the Flerov Laboratory of Nuclear Reactions great success was achieved in the synthesis of superheavy elements in the hot fusion reactions of ^{48}Ca ions with actinide target nuclei. Unfortunately, nuclei with $Z > 118$ cannot be synthesized in ^{48}Ca-induced reactions since ^{249}Cf is the heaviest target material available for these purposes. One of the possible pathways to synthesize new isotopes of SHE with $Z = 118$–126 is the reactions of actinide targets with Ti, Fe, and Ni ions.

At energies around the Coulomb barrier the complete fusion of two massive nuclei is suppressed by competing binary process, the quasifission (QF), in which the composite system separates in two main fragments without forming a compound nucleus (CN). QF happens to be the most important mechanism that prevents the formation of SHE in the fusion of heavy nuclei.

The cross section for forming a heavy evaporation residue (ER) in fusion reactions depends on the capture cross section σ_{ER}, the fusion probability P_{CN}, and the survival of the completely fused system against fission $W_{survival}$. P_{CN} is the least known of these quantities. The experimental measurements of fusion probability show the general dependence of P_{CN} upon fissility, nevertheless, they strongly vary from each other as well as the theoretical predictions.

Experimentally, the capture cross section is defined as the sum of the QF, CN-fission, and ER cross sections. In the case of reactions leading to the formation of SHEs the ER cross section is considerably lower compared with CN fission and QF, and can be neglected. So, to determine P_{CN} it is necessary to know the relative contributions of QF and CN-fission components into the capture cross section. Since both processes are characterized by large nucleon exchange and energy dissipation, the experimental separation between QF and CN fission appeared to be not a trivial procedure. The identification of the reaction mechanism (fusion or QF) is usually based upon the following criteria,

- The reaction Coulomb factor Z_1Z_2 (charge product of reaction partners). This parameter relates to the Coulomb energy in the entrance channel. According to the calculations in the frame of macroscopic-microscopic model of Swiatecki [2] the Z_1Z_2 threshold value for the QF appearance is about 1600.
- Entrance channel mass asymmetry $\alpha_0 = (A_{projectile} - A_{target})/(A_{projectile} + A_{target})$. Unexpected fusion suppression observed in quite asymmetric combinations of colliding nuclei can be qualitatively explained by an effect of the conditional barrier arising along the mass-asymmetry coordinate on a path to the formation of a spherical CN. According to this criterion QF appears for systems with α_0 lower than the Businaro-Gallone mass asymmetry [3].
- Effective fissility parameter x_{eff} connected with repulsive and attractive forces in the reaction entrance channel (or mean fissility parameter x_m, defined as a linear combination between x_{eff} and the true fissility parameter x_{CN}, reflecting the stability of CN with respect to fission: $x_m = 0.75\, x_{eff} + 0.25\, x_{CN}$). From the analysis of a large set of experimental data it was found that the QF takes place for reactions with $x_m > 0.68$ and becomes dominant at $x_m > 0.765$ [4].

This paper presents the results of the study of mass-energy distributions of binary fragments for a wide range of nuclei with Z = 102-122 produced in the reactions with heavy ions at energies close and below the Coulomb barrier.

Reaction products were registered using a double-arm time-of-flight spectrometer CORSET [5] with a high time and position resolution. The wide set of the measured reactions made it possible to investigate the influence of the different properties of the entrance channel (mass-asymmetry, interaction energy, angular momentum, deformation of nuclei) on the competition between the quasifission and fusion-fission processes.

2. Results and discussion

In Figure 1 the mass distributions of fissionlike fragments for the ^{36}S, ^{48}Ca, ^{48}Ti, ^{64}Ni + ^{238}U reactions at energies near and above the Coulomb barrier are shown. At first sight the distributions for the reaction ^{48}Ti + ^{238}U are similar to those for the reactions with ^{48}Ca ions: the wide two-humped shape with the large QF component peaked around the mass 208–210 u. In the case of ^{36}S + ^{238}U reaction the asymmetric QF (QF_{asym}) peaks around masses of 200 u, and for ^{64}Ni + ^{238}U shifts to the heavier mass 215 u. The position of the QF_{asym} seems to be defined by the proton and neutron closed shells at Z = 28, 82 and N = 50, 126, and the deviation of the QF fragment masses from 208 u (double magic lead is expected to affect the formation of QF fragments strongly) is due to the influence of the shells Z = 28 and N = 50 in the light fragment [6, 7, 8]. The driving potentials as a function of mass, shown in Fig. 1, have been calculated at contact configuration in the diabatic approximation within the proximity model with the aid of the Nuclear Reaction Vision Project (NRV) [9]. For the ^{36}S-ions-induced reaction the positions of the minima of the driving potential agree with the positions of the peaks in the experimental asymmetric QF mass distribution, whereas for the reactions with ^{48}Ti and ^{64}Ni the QF_{asym} peak is shifted to the more asymmetric masses compared to the minimum of the driving potential.

As a first step to evaluate the CN-fission cross section the contribution of symmetric fragments with masses $A_{CN}/2 \pm 20$ u can be considered. TKE distributions of these fragments for the reactions ^{36}S, ^{48}Ca, ^{48}Ti, ^{64}Ni + ^{238}U are shown in the right panel of Fig. 1. It is clearly seen that the TKE distributions change markedly showing nearly Gaussian shape in the case of ^{36}S and ^{48}Ca ions and two-humped shape for the reactions with ^{48}Ti and ^{64}Ni.

To evaluate the contribution of the CN-fission process in the symmetric mass region the TKE distributions are decomposed as a sum of three Gaussians. One of them is associated with the CN-fission process (filled region). The low-energy component is attributed to QF_{asym} whereas the high-energy one is connected with symmetric QF (QF_{sym}). Since for these reactions the ER cross section is a

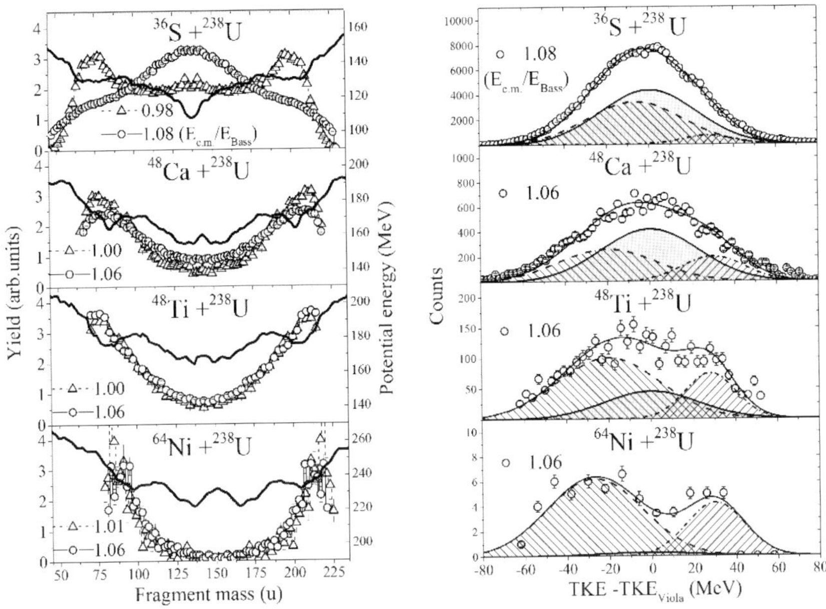

Fig. 1. Left panel: Mass distributions of fissionlike fragments obtained in the reactions ^{36}S,^{48}Ca,^{48}Ti,^{64}Ni+^{238}U at energies near (the triangles) and above (the circles) the Coulomb barrier together with potential energies of the systems (the thick lines). Right panel: TKE distributions of fragments with masses $A_{CN}/2\pm20$ u at energies above the Coulomb barrier. The filled region corresponds to the TKE distributions for CN fission. The dashed and dashed-dotted curves are associated with asymmetric and symmetric QF, respectively.

negligible part of the capture cross section, we can deduce the fusion probability P_{CN} as the ratio between the number of events attributed to CN fission in the framework of the present analysis of the measured mass-energy distributions and all fissionlike fragments.

The absolute differential cross sections for all fissionlike events observed in the studied reactions were measured at an angle $\theta_{c.m.}\approx90^{o}$ at energies near the Coulomb barrier. Capture cross sections $\sigma_{capture}$ for all fissionlike events were estimated assuming that the angular distribution is proportional to $\sin^{-1}\theta_{c.m.}$ since the detailed experimental data as well as any model to describe the QF fragments angular distribution are not available.

In Fig. 2 the measured capture cross sections $\sigma_{capture}$ for the reactions with uranium target together with the cross sections for the formation of symmetric fragments with masses $A_{CN}/2\pm20$ u and the CN-fission cross sections $\sigma_{CN\text{-}fission}$, deduced from the obtained fusion probabilities and capture cross sections together

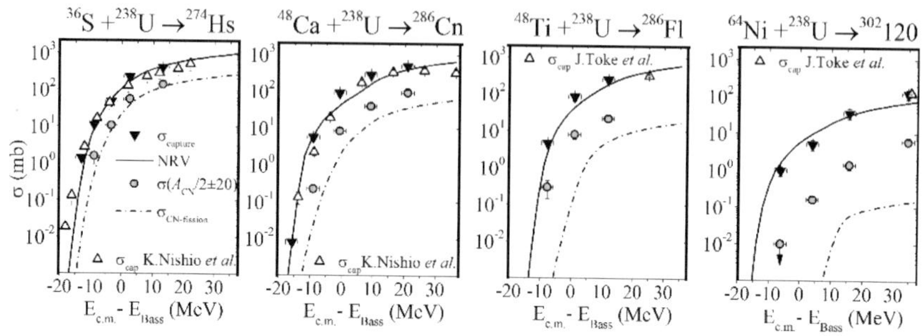

Fig. 2. The cross sections for fissionlike fragments obtained in the reactions ^{36}S,^{48}Ca,^{48}Ti,^{64}Ni+^{238}U (the solid triangles) compared with the data from Refs. [10, 11, 12] (the open triangles). The solid lines are the empirical model calculations [9]. The circles are the cross sections for fragments with masses $A_{CN}/2\pm20$ u. The dashed-dotted lines are the estimated CN-fission cross sections.

with the data from Refs. [10, 11, 12] are presented. One can clearly see that the capture cross sections decrease at the transition from ^{36}S to ^{64}Ni ions. At energies above the barrier for the ^{36}S + ^{238}U and ^{48}Ca + ^{238}U reactions $\sigma_{capture}$ are close to the geometrical cross sections, while for the ^{48}Ti + ^{238}U reaction it is ~60% of the geometrical cross section and ~15% in the case of the ^{64}Ni-ions-induced reaction. The experimental capture and the upper limits for CN-fission cross sections for the reactions with ^{48}Ca ions are presented on Fig. 3. The lower limits of the fission barriers of the for No, Cn, Fl, and Lv compound nuclei were deduced in the frame of Fermi gas model using the obtained CN-fission and ER cross sections. The obtained values of fission barriers are 5.9 MeV for No, 5.2 MeV for Cn, 6.7 MeV for Fl, and 6.9 MeV for Lv.

The dependence of fusion probability on the mean fissility parameter for the hot and cold fusion reactions at above-barrier energy, deduced from the analysis of mass-energy distributions, is shown in Fig. 4. The P_{CN} has been obtained as the ratio between the yield of CN-fission fragments and the experimental values of capture cross sections. However, the capture cross section is ambiguously defined both experimentally and theoretically. From an experimental point of view it is not so easy to separate elastic and quasielastic processes from capture events. Moreover, since the angular distribution for QF is often not known, it is assumed to be proportional to $\sin^{-1}\theta_{c.m.}$ for all fissionlike events. To avoid the uncertainties connected with the definition of the capture cross section we renormalized the estimated values of P_{CN} to the contact cross section proposed by Zagrebaev [17].

The deduced fusion probabilities P^0_{CN} for hot and cold fusion reactions were described according to an equation proposed by Zagrebaev [17],

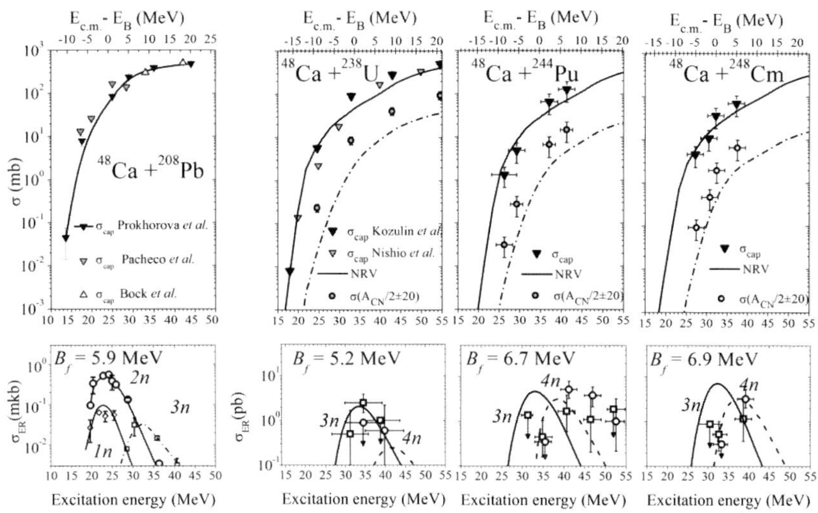

Fig. 3. Top: Triangles are the cross sections for fission-like fragments of the reactions $^{48}Ca+^{208}Pb$ [13, 14], ^{238}U [11, 15],^{244}Pu, and ^{248}Cm. Solid lines are the coupled channel calculations. Circles are the cross sections for fragments with masses $A_{CN}/2 \pm 20$ u, the dashed-dotted lines are estimated CN-fission cross sections $\sigma_{CNfission}$. Bottom: excitation functions for $2n$ (circles), $3n$ (squares) and $4n$ (circles) evaporation channels taken from Ref. [16]. Solid and dotted curves are the ER cross sections for the $3n$ and $4n$ channels, respectively, evaluated using the values of $\sigma_{CNfission}$ obtained in the present work.

$$P_{CN}^{0} = \frac{1}{1+\exp\left(\frac{x_m - \xi}{\tau}\right)}, \tag{1}$$

where τ and ξ are just the fitting parameters.

The fusion probabilities decrease exponentially with increasing the mean fissility parameter both for cold and for hot fusion reactions but with different rates. Such different behaviors of the P_{CN} dependence on the mean fissility parameter for the reactions with spherical and strongly deformed target nuclei may be explained by the orientation effect which plays an important role in the evolution of the dinuclear system. The relative orientation of deformed nuclei changes the Coulomb barrier that affects the balance between repulsive and attractive forces. It was shown experimentally that at near-side collisions a high CN formation probability is expected, whereas the elongated configurations (near-tip collisions) favor the QF process. In the case of spherical target nuclei the fusion probability is higher up to the values of mean fissility parameter $x_m \approx 0.8$, and for the larger values of x_m the fusion probability sharply decreases as

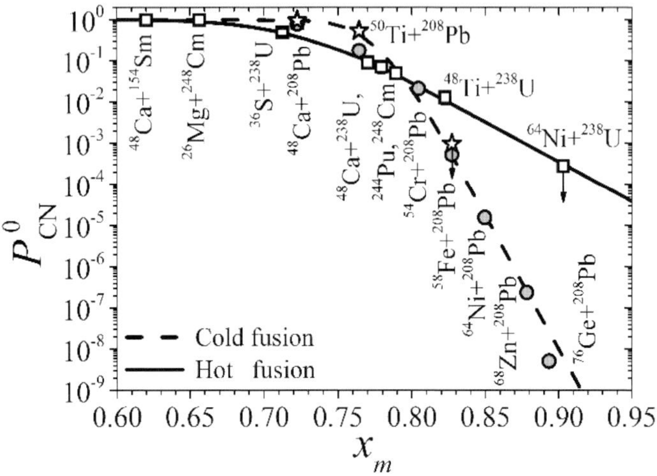

Fig. 4. Fusion probability in cold (spherical target nuclei) and hot fusion (strongly deformed target nuclei) reactions at energies above the Coulomb barrier in dependence on the mean fissility parameter. The deduced fusion probabilities for the reactions with well-deformed target nuclei (the squares) and with lead (the stars). The circles are the fusion probabilities calculated for cold fusion reactions [17].

compared to the reactions with strongly deformed target nuclei. Therefore, in the reactions with strongly deformed nuclei, QF starts to suppress CN formation at the lower values of x_m as compared to the reactions with spherical targets due to the contribution of near-tip collisions, whereas for $x_m > 0.8$ the near-side collisions give a gain in fusion probabilities even at large values of Z_1Z_2.

Notice that the fusion probabilities obtained from the analysis of mass-energy distributions of fissionlike fragments refer only to the upper limit. Nevertheless, with the help of Eq. (1) we can estimate the fusion probability decrease for production of the $Z = 120$ isotopes in the reactions with ^{50}Ti, ^{54}Cr, ^{58}Fe, and ^{64}Ni at above-barrier energies in comparison with the ^{48}Ca + ^{249}Cf reaction (the last reaction suitable for SHE synthesis using ^{48}Ca ions).

3. Summary

Fusion probabilities P_{CN} have been estimated for the reactions with actinide nuclei. At sub-barrier energies P_{CN} is strongly suppressed for these reactions. This suppression is the larger the heavier a system. At above-barrier energies P_{CN} decreases exponentially when the mean fissility parameter x_m increases. It was found that the fusion probability in the reactions with actinide target nuclei (strongly deformed) is several orders of magnitude higher than in the case of cold fusion reactions at the same value of mean fissility parameter for the systems with $x_{\mathrm{m}} > 0.8$.

The capture cross section decreases with increasing effective fissility parameter for the systems with $x_{\mathrm{eff}} > 0.75$ due to the large Coulomb repulsion in the entrance channel. The transfer process becomes dominant for the systems with $x_{\mathrm{eff}} > 0.8$. Besides, the capture cross sections decrease more slowly for the reactions with U target nucleus compared with Pb nucleus.

From the analysis of mass and energy distributions of fragments formed in the reactions of ^{48}Ca with actinide targets fusion probabilities have been estimated and the lower limits for survival probabilities of $\sim 10^{-10}$ for Cn, Fl and Lv compound nuclei have been found at the excitation energies of CN of about 35-40 MeV.

The fusion probability strongly decreases with increasing Coulomb factor of the reaction. At the transition from ^{48}Ca to heavier ions the fusion probability drops down by orders of magnitude. A possible alternative pathway of the production of long-lived neutron-rich SHE is represented by using QF or deep-inelastic reactions in the collisions of transuranium ions with actinide targets. The strong influence of $Z = 82$ and $N = 126$ nuclear shells in these processes can result in large mass and charge rearrangement and may noticeably increase the yield of surviving neutron-rich superheavy nuclei complementary to the projectile-like fragments around lead.

Acknowledgments

We thank the staff of the U400 cyclotron for their careful work. This work was supported by the Russian Foundation for Basic Research (Grants No. 13-02-01282a, 15-52-45098).

References

1. A. Sobiczewski, F. A. Gareev and B. N. Kalinkin, *Phys. Lett.* **22**, 500 (1966).
2. W. J. Swiatecki, *Phys. Scr.* **24**, 113 (1981).
3. H. Abe, KEK Preprint 8-26, KEK TH-128 (1986).
4. R. du Rietz *et al.*, *Phys. Rev. C* **88**, 054618 (2013).
5. E. M. Kozulin *et al.*, *Instrum. Exp. Tech.* **51**, 44 (2008).
6. E. M. Kozulin *et al.*, *Phys. Rev. C* **90**, 054608 (2014).
7. I. M. Itkis *et al.*, *Phys. Rev. C* **83**, 064613 (2011).
8. E. M. Kozulin *et al.*, *Phys. Rev. C* **94**, 054613 (2016).

9. V. I. Zagrebaev *et al.*, http://nrv.jinr.ru/nrv.
10. J. Toke *et al.*, *Nucl. Phys. A* **440**, 327 (1985).
11. K. Nishio *et al.*, *Phys. Rev. C* **86**, 034608 (2012).
12. K. Nishio *et al.*, *Phys. Rev. C* **77**, 064607 (2008).
13. E. V. Prokhorova *et al.*, *Nucl. Phys. A* **802**, 45 (2008).
14. A. J. Pacheco *et al.*, *Phys. Rev. C* **45**, 2861 (1992).
15. E. M. Kozulin *et al.*, *Phys. Lett. B* **686**, 227 (2010).
16. Yu. Ts. Oganessian *et al.*, *Phys. Rev. C* **70,** 064609 (2004); **74,** 044602 (2006).
17. V. I. Zagrebaev and W. Greiner, *Nucl. Phys. A* **944**, 257 (2015).

Using TDHF to Study Quasifission Dynamics

A. S. Umar
Physics and Astronomy, Vanderbilt University, Nashville, TN 37235, USA
E-mail: umar@compsci.cas.vanderbilt.edu

C. Simenel
Department of Nuclear Physics, The Australian National University, Canberra ACT 2601, Australia

We show that the microscopic TDHF approach provides an important tool to shed some light on the nuclear dynamics leading to the formation of superheavy elements. In particular, we discuss studying quasifission dynamics and calculating ingredients for compound nucleus formation probability calculations.

Keywords: TDHF; Superheavy Nuclei; Quasifission.

1. Introduction

The time-dependent Hartree-Fock (TDHF) theory has provided a possible means to study the diverse phenomena observed in low energy nuclear physics[1]. As a result of theoretical approximations (single Slater determinant), TDHF describes reaction channels in a common mean-field, usually corresponding to the dominant reaction channel. For instance, it describes above-barrier fusion and average transfer dynamics[2]. To obtain multiple reaction channels or widths of observables one must go beyond TDHF[3–5]. In connection with superheavy element formation, the theory predicts best the cross-section for a particular process which dominates the reaction mechanism. This is certainly the case for studying capture cross-sections and quasifission.

In recent years has it become numerically feasible to perform TDHF calculations on a 3D Cartesian grid without any symmetry restrictions and with much more accurate numerical methods[6,7]. During the past several years, a novel approach based on TDHF called the density constrained time-dependent Hartree-Fock (DC-TDHF) method was developed to compute heavy-ion potentials[8,9] and excitation energies[10] directly from TDHF time-evolution. This method was applied to calculate capture cross sections for fusion reactions leading to superheavy element $Z = 112$[11]. Furtermore, within the last few years the TDHF approach has been utilized for studying the dynamics of quasifission[12–18]. The study of quasifission is showing a great promise to provide insight based on very favorable comparisons with experimental data.

2. Recent quasifission studies

One of the major questions that is asked by the experimental superheavy element community is why a ^{48}Ca beam is so crucial in forming such systems and whether one could produce new superheavy nuclei using projectiles different than ^{48}Ca and actinide targets. Our first work in this area focused on the quasifission studies for the ^{40,48}Ca+^{238}U system [12,13], showing that for neutron-rich ^{48}Ca beams quasifission is substantially reduced. While, the above work is also in the time span of this proposal, below we discuss the more recent studies of the ^{48}Ca+^{249}Bk and ^{50}Ti+^{249}Bk systems [14]. The reaction ^{48}Ca +^{249}Bk creates superheavy isotopes of element 117 with cross-sections of 2–3 picobarns. However, the ^{50}Ti +^{249}Bk reaction so far has not produced any superheavy isotopes of element 119, with an upper cross-section limit of 50 fb [19].

We have calculated the microscopic DC-TDHF nucleus-nucleus potential barriers for the ^{48}Ca+^{249}Bk and ^{50}Ti+^{249}Bk systems for two extreme orientations of the ^{249}Bk nucleus (tip and side). For the ^{48}Ca+^{249}Bk system, the tip orientation of ^{249}Bk results in a significantly lower barrier, E_B(tip)= 191.22 MeV, as compared to the side orientation, E_B(side)= 204.36 MeV. This reaction has been studied at $E_{\text{c.m.}} = 204$–218 MeV in Dubna [20] and $E_{\text{c.m.}} = 211$–218 MeV with GSI-TASCA [21]. Thus, the highest experimental energy $E_{\text{c.m.}} = 218$ MeV is above both barriers but the lowest experimental energy $E_{\text{c.m.}} = 204$ MeV is slightly below the barrier for the side orientation of ^{249}Bk. Similarly, the corresponding potential barriers for the ^{50}Ti +^{249} Bk system are E_B(tip)= 211.2 MeV and E_B(side)= 224.6 MeV. Experimentally, $E_{\text{c.m.}} = 233.2$ MeV was used in the GSI-TASCA experiment [21], well above both barriers.

Figure 1(a) (left) shows the contact time as a function of center-of-mass energy for central collisions of ^{48}Ca with ^{249}Bk. For the tip orientation of the ^{249}Bk nucleus (dashed line) we observe contact times of order 10–12 zs which are essentially constant over a wide range of energies, $E_{\text{c.m.}} = 191$–230 MeV. Only at energies below the potential barrier, E_B(tip)= 192.2 MeV, do the contact times drop off very rapidly

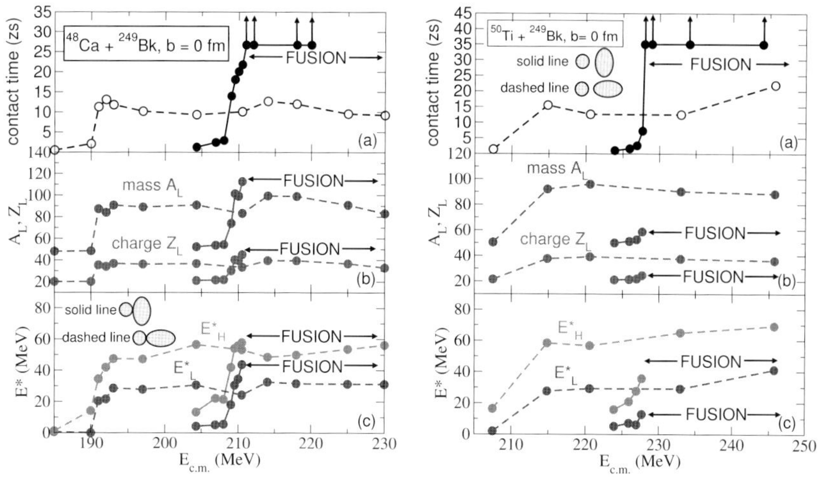

Fig. 1. (a) Contact time, (b) mass and charge of the light fragment, and (c) excitation energy E^* of the heavy and light fragments as a function of $E_{\mathrm{c.m.}}$ for central collisions of ^{48}Ca with ^{249}Bk (left) and ^{50}Ti with ^{249}Bk (right). Solid lines are for the side orientation of the deformed ^{249}Bk nucleus, and dashed lines are for the tip orientations.

because these events correspond to inelastic scattering and few-nucleon transfer reactions. A dramatically different picture emerges for the side orientation of the ^{249}Bk nucleus (solid line): At energies above the barrier E_B(side)$=$ 205.4 MeV, the contact times rise very steeply with energy and reach values up to 22 zs at $E_{\mathrm{c.m.}} = 210$ MeV. For energies above this value, fusion is observed which we define by a large contact time exceeding 35 zs and a mononuclear shape without a neck.

Figure 1(b) (left) shows the corresponding mass and charge of the light fragment. We observe that the mass and charge transfer to the light fragment are roughly proportional to the nuclear contact time. In particular, for the side orientation of ^{249}Bk, we find quasielastic collisions at energies below $E_{\mathrm{c.m.}} = 204$ MeV. Quasifission is limited to a small range of energies $E_{\mathrm{c.m.}} = 209$–211 MeV, whereas for energies above 211 MeV we find fusion. Naturally, non-central impact parameters can show quasifission in the range where we see fusion. The quasifission results are very different for the tip orientation of ^{249}Bk, ranging over a much wider energy domain $E_{\mathrm{c.m.}} = 191$–230 MeV with a lower maximum mass and charge transfer compared to the side orientation of ^{249}Bk.

We have also used DCTDHF to calculate the excitation energy of *each fragment* directly from the TDHF density evolution. This gives us new information on the repartition of the excitation energy between the heavy and light fragments which is not available in standard TDHF calculations. In Fig. 1 (c) we show the excitation energies of the heavy and light fragments which contain approximately 55–60 MeV and 30–45 MeV of excitation energy (side orientation) and 50 MeV and 30 MeV (tip orientation), respectively, for c.m. energies corresponding to quasifission.

Right panel of Fig. 1 shows the corresponding results for central collisions of ^{50}Ti with ^{249}Bk. The contact times and the masses and charges of the light fragment show a similar behavior as a function of energy as compared to the ^{48}Ca $+^{249}$Bk reaction. For the tip orientation, we find quasifission for $E_{\text{c.m.}} \geq 214$ MeV, with excitation energies of $E_H^* = 57$–69 MeV for the heavy fragment and $E_L^* = 27$–41 MeV for the light fragment, respectively. The mass and charge of the fragments indicate a strong influence of the shell effects in the ^{208}Pb region, as in reactions with ^{48}Ca. However, $N = 50$ does not seem to play a role here. For the side orientation, we find inelastic and multi-nucleon transfer reactions at energies $E_{\text{c.m.}} = 223$–227 MeV. Quasifission is confined to an extremely narrow energy window around $E_{\text{c.m.}} = 227.4$–227.7 MeV, with excitation energies of $E_H^* \simeq 36$ MeV and $E_L^* \simeq 13$ MeV. At energies $E_{\text{c.m.}} > 228$ MeV, fusion sets in.

3. Shape evolution and collective dynamics of quasifission

The proper characterization of fusion-fission and quasifission is one of the most important tasks in analyzing reactions leading to superheavy elements. Experimental analysis of fusion-fission and quasifission fragment angular distributions $W(\theta)$ is commonly expressed in terms of a two-component expression [22],

$$W(\theta) = \sum_{J=0}^{J_{\text{CN}}} \mathcal{F}_J^{(FF)}(\theta, K_0(FF)) + \sum_{J=J_{\text{CN}}}^{J_{\max}} \mathcal{F}_J^{(QF)}(\theta, K_0(QF)) \,. \tag{1}$$

Here, J_{CN} defines the boundary between fusion-fission and quasifission, assuming a sharp cutoff between the angular momentum distributions of each mechanism. The quantum number K is known to play an important role in fission. It is the projection of the total angular momentum onto the deformation axis. In the Transition State Model (TSM), the characteristics of the fission fragments are determined by the K distribution at scission. The argument K_0 entering Eq. (1) is the width of this distribution which is assumed to be Gaussian. It obeys $K_0^2 = T\Im_{eff}/\hbar^2$, where the effective moment of inertia, $\Im_{eff}$, is computed from the moments of inertia for rotations around the axis parallel and perpendicular to the principal deformation axis $\frac{1}{\Im_{eff}} = \frac{1}{\Im_{\parallel}} - \frac{1}{\Im_{\perp}}$, and T is the nuclear temperature at the saddle point. The physical parameters of the fusion-fission part are relatively well known from the liquid-drop model [23]. In contrast, the quasifission process never reaches statistical equilibrium. In principle, it has to be treated dynamically, while Eq. (1) is based on a statistical approximation. In addition, the usual choice for the nuclear moment of inertia for the quasifission component, $\Im_0/\Im_{eff} = 1.5$ [24], is somewhat arbitrary. Here, $\Im_0$ is the moment of inertia of an equivalent spherical nucleus.

We have developed methods to extract the moment of inertia of the system (the main collective observable of interest for fission and quasifission) directly from TDHF time-evolution of collisions resulting in quasifission. The proper way to calculate the moment-of-inertia for such time-dependent densities (particularly for non-zero impact parameters) is to directly diagonalize the moment-of-inertia tensor represented by a 3×3 matrix with elements

$$\Im_{ij}(t)/m = \int d^3r\ \rho(\mathbf{r},t)(r^2\delta_{ij} - x_i x_j)\ , \tag{2}$$

where ρ is the local number-density calculated from TDHF evolution, m is the nucleon mass, and $x_{i=1,2,3}$ denote the Cartesian coordinates. Numerical diagonalization the matrix $\Im$ gives three eigenvalues. One eigenvalue corresponds to the moment-of-inertia $\Im_{\parallel}$ for the nuclear system rotating about the principal axis. The other two eigenvalues define the moments of inertia for rotations about axes perpendicular to the principal axis. Using the time-dependent moment-of-inertia obtained from the TDHF collision one can calculate the so-called effective moment-of-inertia defined above. We have calculated the moment-of-inertia ratio for the ^{48}Ca + ^{249}Bk non-central collisions at $E_{\mathrm{c.m.}} = 218$ MeV. At the point of final touching configuration the moment-of-inertia ratios are in the range 1.4-1.8, suggesting a relatively strong impact parameter dependence which should be accounted for in future extensions to the TSM.

Acknowledgments

This work has been supported by the U.S. Department of Energy under grant No. DE-SC0013847 with Vanderbilt University and by the Australian Research Councils Future Fellowship (project number FT120100760) and Discovery Projects (project number DP160101254) funding schemes.

References

1. C. Simenel, *Eur. Phys. J. A* **48**, p. 152 (2012).
2. C. Simenel and B. Avez, *Int. J. Mod. Phys. E* **17**, 31 (2008).
3. M. Tohyama and A. S. Umar, *Phys. Lett. B* **516**, 415 (2001).
4. D. Lacroix and S. Ayik, *Eur. Phys. J. A* **50**, p. 95 (2014).
5. C. Simenel, *Phys. Rev. Lett.* **106**, p. 112502 (2011).
6. A. S. Umar, M. R. Strayer, J. S. Wu, D. J. Dean and M. C. Güçlü, *Phys. Rev. C* **44**, 2512 (1991).
7. J. A. Maruhn, P.-G. Reinhard, P. D. Stevenson and A. S. Umar, *Comp. Phys. Comm.* **185**, 2195 (2014).
8. A. S. Umar and V. E. Oberacker, *Phys. Rev. C* **74**, p. 061601 (2006).
9. A. S. Umar, C. Simenel and V. E. Oberacker, *Phys. Rev. C* **89**, p. 034611 (2014).
10. A. S. Umar, V. E. Oberacker, J. A. Maruhn and P.-G. Reinhard, *Phys. Rev. C* **80**, p. 041601 (2009).
11. A. S. Umar, V. E. Oberacker, J. A. Maruhn and P.-G. Reinhard, *Phys. Rev. C* **81**, p. 064607 (2010).
12. A. Wakhle *et al.*, *Phys. Rev. Lett.* **113**, p. 182502 (2014).
13. V. E. Oberacker, A. S. Umar and C. Simenel, *Phys. Rev. C* **90**, p. 054605 (2014).
14. A. S. Umar, V. E. Oberacker and C. Simenel, *Phys. Rev. C* **94**, p. 024605 (2016).
15. A. S. Umar, V. E. Oberacker and C. Simenel, *Phys. Rev. C* **92**, p. 024621 (2015).
16. A. S. Umar and V. E. Oberacker, *Nucl. Phys. A* **944**, 238 (2015).
17. K. Hammerton *et al.*, *Phys. Rev. C* **91**, p. 041602(R) (2015).
18. K. Sekizawa and K. Yabana, *Phys. Rev. C* **93**, p. 054616 (2016).
19. C. E. Düllmann and the TASCA collaboration, in *Fission and Properties of Neutron-Rich Nuclei*, (World Scientific, 2013).
20. Y. T. Oganessian *et al.*, *Phys. Rev. C* **87**, p. 054621 (2013).
21. J. Khuyagbaatar *et al.*, *Phys. Rev. Lett.* **112**, p. 172501 (2014).
22. B. B. Back, *Phys. Rev. C* **31**, 2104 (1985).
23. A. J. Sierk, *Phys. Rev. C* **33**, 2039 (1986).
24. R. Yanez *et al.*, *Phys. Rev. C* **88**, p. 014606 (2013).

Superheavy Nuclei (SHN) — Stability Beyond the Liquid Drop

D. Ackermann
Grand Accélérateur National d'Ions Lourds - GANIL, 14076 Caen, France
GSI Helmholtzzentrum für Schwerionenforschung GmbH, 64846 Darmstadt, Germany

The synthesis of new elements has reached up to Z = 118 with the IUPAC/IUPAP assigning recently the naming rights for the elements 114,115,116,117 and 118 to groups at the FLNR and at RIKEN [1,2,3,4]. The A and Z assignment, however is in some cases still based on indirect evidence. A direct experimental prove of Z could be the measurement of characteristic X-rays where attempts have been made recently with limited success [5,6].

The nuclei beyond fermium are solely stabilized by quantum mechanics. A charged liquid drop would fall apart due to Coulomb repulsion. Therefore those species provide an ideal laboratory for nuclear structure studies.

Decay spectroscopy after separation of deformed nuclei around Z = 108 and N = 152-162, provides direct links to the next heavier spherical closed shell nuclei, by investigating single particle levels which are, according to model calculations, relevant in both regions [7]. Another particularly interesting feature, accessible for the this experimental approach, are meta-stable states created by nuclear deformation, so called K-isomers. Following the trend of vanishing deformation they can be used to trace the spherical SHN and to locate the island of stability [8].

Improved experimental means like new high intensity ion accelerators, efficient ionoptical separators and spectrometers, and highly efficient detector systems with fast electronics are the essential ingredients for the success of the field. The new SPIRAL2 facility presently under construction at the accelerator laboratory GANIL in Caen, France, will offer great perspectives for the field [8].

An overview of the recent developments in SHN research, including a detailed discussion of the nuclear structure studies will be given.

References

[1] Yu.Ts. Oganessian *et al.*, *Nucl. Phys. A* **944**, 62, (2015).
[2] K. Morita *et al.*, *J. Phys. Soc. Jpn.* **73**, 2593 (2004).
[3] P.J. Carol *et al.*, Pure and Applied Chemistry 88, 139-153, ISSN (Online) 1365-3075, ISSN (Print) 0033-4545, DOI: 10.1515/pac-2015-0502, January 2016.
[4] P.J. Carol *et al.*, Pure and Applied Chemistry 88, 155-160, ISSN (Online) 1365-3075, ISSN (Print) 0033-4545, DOI: 10.1515/pac-2015-0501, January 2016.
[5] D. Rudolph *et al.*, *Phys. Rev. Lett.* 111, 112502 (2013).
[6] J.M. Gates *et al.*, *Phys. Rev. C* **92**, 021301 (2015).
[7] M. Asai *et al.*, *Nucl. Phys. A* **944**, 308, (2015).
[8] D. Ackermann, *Nucl. Phys. A* **944**, 376, (2015).

Actinide Targets for the Synthesis of Superheavy Nuclei: Current Priorities and Future Opportunities*

J. B. Roberto† and K. P. Rykaczewski

Oak Ridge National Laboratory,
Oak Ridge, Tennessee 37831, USA
†E-mail: robertojb@ornl.gov
www.ornl.gov

Recent experiments using actinide + ^{48}Ca reactions have led to the discovery of five new elements and more than 50 new isotopes on the "hot fusion" island. These new nuclei, all having a large neutron excess over protons, provide strong evidence for the existence of the proposed island of stability for superheavy nuclei. As atomic number Z increases beyond Z = 115, the reaction cross sections are decreasing, requiring new approaches for continued progress. Ion beams heavier than ^{48}Ca will be required to reach elements above Z = 118, and new actinide target technologies and separation capabilities will be needed to accommodate higher beam intensities and optimize the use of heavier ^{251}Ca targets in experiments. In this paper, we review the availability of actinide materials, the use of new target/projectile combinations, and related experimental opportunities, including prospects for new element discoveries and for approaching the presumed closed shell at N = 184. Research opportunities include ongoing experiments to produce the heaviest nuclei to date using mixed Cf targets, and measurements of excitation functions for ^{48}Ca and ^{50}Ti on Pu and Cm targets to estimate cross sections for planned experiments using ^{50}Ti beams on ^{249}Bk and ^{251}Cf for synthesis of elements 119 and 120. New developments for the production of ^{249}Bk and for increasing the availability of ^{251}Cf and possibly ^{254}Es are described.

Keywords: Actinide; Superheavy Element; Berkelium; Californium; Einsteinium.

* This work supported by the U. S. DOE Office of Science, Office of Nuclear Physics, through the Isotope Development and Production for Research and Applications Program.

1. Introduction

Over the past decade, five new superheavy elements with Z = 114-118 have been discovered in actinide + ^{48}Ca hot fusion reactions at the Joint Institute for Nuclear Physics in Dubna, Russia [1,2]. In total, more than 50 new heaviest nuclei have been produced in related decay chains. These nuclei exhibit enhanced lifetimes as the closed nuclear shell at N = 184 is approached, providing evidence for the existence of the "island of stability" for superheavy elements. Fig. 1 summarizes the nuclear chart for elements 114-118.

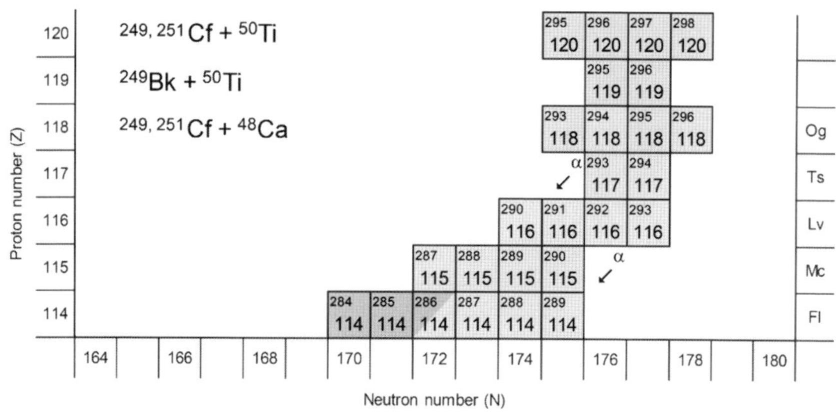

Fig. 1. Section of the nuclear chart for new elements Z = 114-118. New nuclei ^{293,295,296}Og and Z = 119 and 120 can in principle be reached via the 3n and 4n channels in the hot fusion reactions indicated at left. An ^{254}Es target (if available) could potentially yield ^{299,298}119 and ^{301,300}121 in similar reactions using ^{48}Ca and ^{50}Ti beams, respectively.

Neutron rich actinide targets, including Pu, Am, Cm, Bk, and Cf, have been essential to these discoveries. These heavy actinides are produced in nuclear reactors and require separation and purification in heavily-shielded hot cells [3]. Progress in superheavy element discovery and continued exploration of the island of stability will require new beam/target combinations to reach higher Z and N. This paper reviews the availability of actinide targets, the role of actinides in superheavy element research, and related opportunities for new element discoveries and for approaching the next magic number N = 184 at the island of stability.

2. Actinide targets

Actinide targets have been used for the discovery of all superheavy elements with $Z \geq 114$. Production and processing of actinides heavier than uranium requires high-flux reactors and heavily-shielded chemical processing facilities available in only a few locations worldwide [3]. Actinide target materials have been provided by Oak Ridge National Laboratory (ORNL), through the DOE Isotope Development and Production for Research and Applications Program, and by the Research Institute of Atomic Reactors at Dimitrovgrad, Russia. Table 1 shows the actinide targets that have been used to synthesize elements 114 through 118, and the total number of these nuclei produced to date.

Table 1. New elements produced since 2000 using actinide + ^{48}Ca reactions.

Element (Z)	Year	Target	Nuclei produced (directly/total)	Reference
Flerovium (114)	2000	$^{239,240}Pu$ $^{242,244}Pu$	82/124	[1,4-7]
Moscovium (115)	2004	^{243}Am	113/155	[1,8-10]
Livermorium (116)	2004	$^{245,248}Cm$	37/42	[1,4,11-12]
Tennessine (117)	2010	^{249}Bk	22	[1,13-15]
Oganesson (118)	2006	^{249}Cf	5	[1,16,17]

The High Flux Isotope Reactor (HFIR) and Radiochemical Engineering Development Center (REDC) at ORNL provide a unique resource [3] for the production and processing of actinide materials. Current inventories include ~gram quantities for many long-lived heavy actinides, including $^{240,242}Pu$, $^{241,243}Am$, $^{244,248}Cm$, and hundred mg quantities of ^{249}Cf. Facilities and processes are available for production of shorter-lived actinides, including up to 200 mg of ^{252}Cf, 20 mg of ^{249}Bk, and 1-2 μg of ^{254}Es.

A typical Cf production campaign begins with intense thermal neutron irradiation of ~40 g of Cm/Am in HFIR for several 24-day reactor cycles, followed by three months of cooling to reduce short-lived activation products. The irradiated material is then transferred to REDC for several months of chemical processing and purification performed remotely in specially-designed hot cells. Bk is typically produced as a byproduct of Cf production, and Es and ^{257}Fm (picograms) can also be produced in these campaigns.

For superheavy element (SHE) synthesis, the actinide material is typically electrodeposited on thin titanium foils at a thickness of ~350 μg/cm^2. These foils are mounted on rotating target wheels for production of superheavy nuclei, typically by irradiation with an intense ^{48}Ca beam from a heavy ion cyclotron.

3. Superheavy element research using actinide targets

The actinide + ^{48}Ca hot fusion process was first successfully implemented at JINR in the late 1990s [2], and resulted in a rapid succession of superheavy element discoveries, elements 114 (flerovium, Fl) and 116 (livermorium, Lv) in 2004 [4], element 115 (moscovium, Mc) in 2004 [8], element 118 (oganesson, Og) in 2006 [16], and element 117 (tennessine, Ts) in 2010 [13]. Decreasing production cross sections with increasing Z for the then-used cold fusion process had reduced production rates to one atom/yr for $Z = 113$. Hot fusion reversed this trend, increasing production cross sections for superheavy nuclei by one of more orders of magnitude, enabling continued progress in new element discovery.

3.1. *Plutonium, americium, and curium*

These actinides have been used with ^{48}Ca beams at JINR to produce Fl, Mc, and Lv [4,8], at GSI to produce Fl [6], Mc [9] and Lv [11], and at Berkeley to produce Fl [5] and Mc [10], and at RIKEN to produce Lv [12]. Current experiments [7,17] with these target materials are focused on extending the produced nuclei to lower neutron numbers, defining the western boundary of the island of stability and connecting it to the nuclear mainland.

3.2. *Berkelium*

Element 117 (Ts) was discovered in 2010 [13] using ^{249}Bk target material from ORNL irradiated with a ^{48}Ca beam at JINR. ^{249}Bk has a 327-day half-life and must be produced fresh for each experiment. Approximately 22 mg of ^{249}Bk was produced in HFIR and separated and purified at REDC to less than a nanogram of ^{252}Cf and other contaminants. The experiment was performed at the excitation energies of ^{297}Ts* compound nucleus of 39 and 35 MeV corresponding to the expected peaks in the production cross sections for the $A = 293$ and 294 isotopes of Ts. During 140 days of ^{48}Ca irradiation at JINR, the experiment produced two isotopes of element 117 and eleven new heaviest isotopes in related decay chains, all closer to the presumed center of the island of stability. Half-lives increase substantially with increasing neutron number for all observed isotopes of a given element consistent with the expected influence of a closer approach to the closed shell at $N = 184$. To date 22 atoms of Ts have been produced [13-15], 16 of ^{293}117 and 6 of ^{294}117.

Bk production in HFIR reaches 90% of its maximum after two reactor cycles (~2 months). Total yields can in principle be increased [18] by a factor of two or more by using neutron filters to decrease the thermal to epithermal neutron flux ratio.

3.3. *Californium*

Element 118 (Og) was first produced in 2006 [16] at JINR using ^{249}Cf target material from ORNL. A total of five ^{294}Og nuclei have been produced to date, three in 2006, one in 2012 [14], and one in 2015 [17]. The 2012 observation was during a ^{249}Bk + ^{48}Ca experiment at JINR to confirm earlier Ts results. During the experiment, the target acquired a significant ^{249}Cf concentration from the radioactive decay of short-lived ^{249}Bk, enabling production of both Og and Ts.

^{251}Cf is a long-lived isotope of Cf, two neutrons heavier than ^{249}Cf. This is an ideal target for heavier isotopes of oganesson, as well as element 120 using a ^{50}Ti beam. ^{251} Cf can be recovered from old ^{252}Cf targets that have decayed over several decades. The resulting material is a mixed Cf target containing approximately 35% ^{251}Cf, 15% ^{250}Cf, and 50% ^{249}Cf. Approximately 15 mg of mixed Cf was produced at ORNL in 2015 by chemical purification of substantially decayed ^{252}Cf sources [19]. This material was fabricated into targets at ORNL [19] for use at JINR in a collaborative search for the A = 295 and 296 isotopes of Og. These would be the heaviest nuclei produced to date. Early results include the fifth atom of ^{294}Og, most likely produced from the ^{249}Cf fraction of the mixed Cf target [17]. An actinide separator could produce separated ^{251}Cf, increasing the production rate for heavier Og isotopes by a factor of three.

3.4. *Einsteinium*

The extension of ^{48}Ca + actinide hot fusion experiments beyond Og will require an einsteinium target. This is an attractive path to element 119 if sufficient quantities of ^{254}Es (half-life 276 days) can be produced. The result for the 3n reaction channel would be ^{299}119. At N = 180, this nucleus would be three neutrons closer to N = 184 than the currently achieved N = 177 for ^{294}Ts and ^{293}Lv.

Current production methods for ^{254}Es as a byproduct during Cf campaigns yield ~1-2 μg of ^{254}Es, more than two orders of magnitude lower than typically required for superheavy element (SHE) experiments.

In the 1980s, a dedicated approach, the Large Einsteinium Activation Project (LEAP), was proposed [20] to produce up to 40μg of ^{254}Es using a sequence of tailored irradiations in HFIR. The resulting quantity would be sufficient to cover a 0.1 cm^2 area at the thickness required for hot fusion SHE experiments. Such a target would need to be designed to survive continuous irradiation at high intensity with the full accelerator beam.

4. Future opportunities using actinide targets

Continued progress toward higher Z and N will require new facilities with higher beam currents, as well as heavier ion beams, such as ^{50}Ti and ^{51}V. The Superheavy Element Factory, nearing completion at JINR, offers the potential to increase the discovery potential by factors of 10-100 assuming availability of sufficient quantities of the required actinides. RIKEN and GANIL are also undergoing major upgrades with significantly enhanced production capability.

Elements 119 and 120 are currently accessible using ^{50}Ti beams on Bk and Cf targets. A key factor in these experiments will be a better understanding of the excitation function for ^{50}Ti + actinide reactions. Experiments such as $^{48}Ca + ^{244}Cm$ and $^{50}Ti + ^{242}Pu$, both of which yield ^{292}Lv, as well as $^{50}Ti + ^{244}Pu$, can help us estimate cross sections and optimum beam energies for planned element 119/120 experiments using ^{50}Ti beams. Continued progress toward higher Z and N will require new beams such as ^{54}Cr and $^{58,60}Fe$ with the heaviest actinide targets, ^{251}Cf and (possibly) ^{254}Es.

Acknowledgments

This research was supported by the U.S. DOE Office of Nuclear Physics, Isotope Development and Production for Research and Applications Program. We are grateful to ORNL's Radiochemical Engineering Development Center and High Flux Isotope Reactor (HFIR) for production and chemical separation of Bk and Cf, and to the DOE Office of Basic Energy Sciences for support of HFIR. We are indebted to our many collaborators at JINR, GSI, University of Mainz, Lawrence Livermore National Laboratory, Vanderbilt University, and the University of Tennessee, Knoxville.

References

1. Yu. Ts. Oganessian and V. K. Utyonkov, *Rep. Prog. Phys.* **78**, 036301 (2015).
2. Yu. Ts. Oganessian, *J. Phys. G* **34**, R165 (2007).
3. J. B. Roberto, C. W. Alexander, R. A. Boll, J. D. Burns, J. G. Ezold, L. K. Felker, S. L. Hogle and K. P. Rykaczewski, *Nucl. Phys. A* **944**, 99 (2015).
4. Yu. Ts. Oganessian *et al.*, *Phys. Rev. C* **69**, 054607 (2004).

5. L. Stavsetra *et al.*, *Phys. Rev. Lett.* **103**, 132502 (2009).
6. Ch. E. Duellmann *et al.*, *Phys. Rev. Lett.* **104**, 252701 (2010).
7. V. K. Utyonkov *et al.*, *Phys. Rev. C* **92**, 034609 (2015).
8. Yu. Ts. Oganessian *et al.*, *Phys. Rev. C* **69**, 021601 (2004).
9. D. Rudolph *et al.*, *Phys. Rev. Lett.* **111**, 112502 (2013).
10. J. M. Gates *et al.*, *Phys. Rev. C* **92**, 021301R (2015).
11. S. Hofmann *et al.*, *Eur. Phys. J. A* **48**, 62 (2012).
12. K. Morita *et al.*, RIKEN *Accel. Prog. Rep.* 47, xi (2014).
13. Yu. Ts. Oganessian *et al.*, *Phys. Rev. Lett.* **104**, 142502 (2010).
14. Yu. Ts. Oganessian *et al.*, *Phys. Rev. Lett.* **109**, 162501 (2012).
15. J. Khuyagbaatar *et al.*, *Phys. Rev. Lett.* **112**, 172501 (2014).
16. Yu. Ts. Oganessian *et al.*, *Phys. Rev. C* **74**, 044602 (2006).
17. K. P. Rykaczewski, J. B. Roberto, N. T. Brewer and V. K. Utyonkov, ORNL actinide materials and a new detection system for superheavy nuclei, in Proceedings of Nobel Symposium 160, Chemistry and Physics of Heavy and Superheavy Elements, Bäckaskog Castle, Sweden, May 29–June 3, 2016, *Eur. Phys. J.* (in press).
18. S. Hogle, C. W. Alexander, J. D. Burns and J. G. Ezold, Experimental validation for optimization of transcurium isotope production, in 8th International Conference on Isotopes, Chicago, IL (2014).
19. R. A. Boll, J. B. Roberto, N. J. Sims, K. P. Rykaczewski, S. M. Van Cleve and V. K. Utyonkov, Electrodeposition of decay-enriched ^{251}Cf, in 16th International Workshop on Targetry and Target Chemistry, Santa Fe, New Mexico (August 29–September 1, 2016).
20. J. E. Bigelow, C. W. Alexander, and L. J. King, Proposed production of a large (40 μg) sample of ^{254}Es, in Symposium on Heavy Element Research, AIChE Summer National Meeting, Seattle, WA (1985).

performed at the Dubna Gas-Filled Recoil Separator (DGFRS) using the ^{48}Ca beam accelerated at the U400 cyclotron of the Flerov Laboratory of Nuclear Reactions, JINR. The ^{48}Ca ion beam was delivered to the target with a maximum intensity of 1.3 particle μA and the beam energy was determined with a systematic uncertainty of 1 MeV by a time-of-flight system placed in front of the DGFRS.

The target materials were provided by Oak Ridge National Laboratory (ORNL) salvaged from old ^{252}Cf sources and contained multiple Cf isotopes with the following weight percentages at the time of target manufacture: 50.7% ^{249}Cf, 12.9% ^{250}Cf, 36.4% ^{251}Cf, and 0.2% ^{252}Cf. Target sectors were manufactured at ORNL and shipped to Dubna to assemble a standard target wheel.

The irradiation with ^{48}Ca was performed at a beam energy of 252 MeV and occurred between Oct. 2015 and April 2016. A total beam dose of approximately 1×10^{19} particles was obtained. The experiment was discontinued because beam-induced target contamination altered the thickness of the target thus reducing recoils of products into the DGFRS.

Evaporation residues (ERs) recoiling from the target were separated in flight from ^{48}Ca beam ions, scattered particles, and transfer-reaction products by the DGFRS. The transmission efficiency of the separator for $Z = 118$ nuclei was estimated to be about $35\% \pm 5\%$. Recoils passed through a time-of-flight (TOF) system and were implanted in the focal plane detectors.

The array of silicon detectors at the DGFRS final focus was a 0.3-mm-thick double-sided silicon strip detector (DSSD). This large DSSD has 1-mm-wide strips, 48 on the front side and 128 on the back side. The recoil implantation counter was surrounded by six single Si detectors, each 500 microns thick and having an active area of 65 mm by 120 mm, forming a detector system akin to a shoebox with the lid removed.

One correlated decay chain consisting of an EVR, two alpha-decays, terminated by fission was observed[7]. This decay chain is consistent with the four other decay chains observed for ^{294}Og from past ^{48}Ca irradiations of ^{249}Cf[8] and ^{249}Bk[9] targets (see Figure 1 for decay chain properties) and most likely occurred from the ^{249}Cf portion of the target as the $3n-$reaction channel from ^{249}Cf would be expected to have a larger cross-section than the $5n-$reaction channel from ^{251}Cf at this beam energy.

3. Fission Properties

In the region of enhanced stability, as one moves farther away from the $N = 184$ closed neutron shell, a rapid decrease in the Fl production cross-section is shown in Figure 2. This indicates a western border of the region where spontaneous fission most likely dominates, thus decreasing the production cross-section. Exploration and production of lighter isotopes might be severely limited, however, characterization of fission properties of nuclei in this region will establish the boundaries of enhanced stability and may give key reaction dynamics understanding. Currently, experiments are being run at DGFRS to try to produce lighter isotopes of Fl, and to observe any alpha-decay branch of ^{284}Fl.

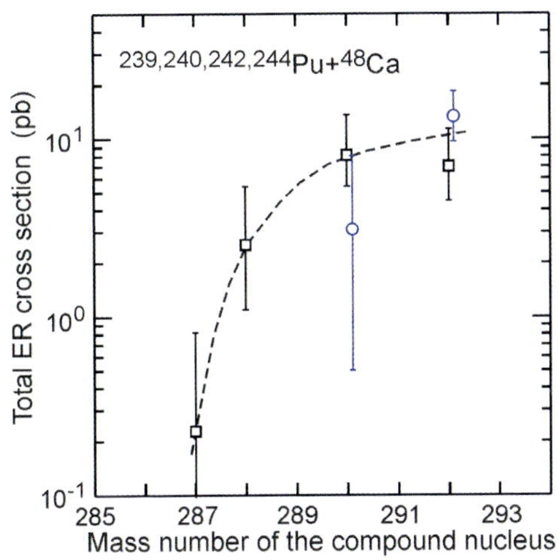

Fig. 2. Maximum of the production cross-sections of Fl isotopes in ^{48}Ca-induced reactions shown in figure as a function of mass number of the compound nucleus. Squares are data from DGFRS and circles are data from BGS and TASCA. Figure is reproduced from Utyonkov, *et al.*[10]. The dashed line is drawn to guide the eye and shows a rapid decrease in cross-section with decreasing N – farther away from the $N = 184$ neutron shell.

4. Monte Carlo Random Probability (MCRP) Analysis

The Monte Carlo Random Probability (MCRP) method is described in more detail in references[5,6] and recent results are discussed in more detail in Stoyer, *et al.*[7]. This method uses the actual data collected during an experiment and the same search technique used to identify correlated events making an interesting decay chain or sequence. Random events are inserted into the data, and the search algorithm is run again to find any correlation that might be in the data, but now correlated with a random event. The events must be within the same front and back strips (vertical and horizontal position) and within a given time window in order to be correlated. Because the actual data is used, this method automatically includes fluctuations in background, counting rates, detector performance, beam rates, cosmic ray rates, high voltage drifts, gain shifts, etc. that may occur during an experiment running 24 hours a day for months at a time. The MCRP method was tested with a variety of random number generators, distributions of random events in the detector system, and a variety of other factors which are detailed in references[5,6] and found to be relatively insensitive to these factors. For this study, a flat distribution of random events was inserted into the detector with no attempt made to mimic the distribution of EVRs implanted into the detector. Typically 1×10^9 or more random events were inserted into the data. Random probabilities below 1×10^{-9} were obtained by extrapolation or multiplication of various pairs of correlated random probabilities.

The MCRP results for about half of the data for the ^{48}Ca + ^{251}Cf experiment are shown in Table 1 as a function of correlation time. Note the increase in probability as the correlation time increases from 0.1 s to 5 s. The MCRP probability that the observed R-A-A-F decay chain in the ^{48}Ca + ^{251}Cf experiment, assigned to the previously observed ^{294}Og, is 1.8×10^{-14} and indicates that this decay chain is highly unlikely to be due to random events.

Table 1. MCRP method calculated for varying correlation times for a ^{294}Og-like decay chain consistent with the one shown in Figure 1 using about half of the ^{48}Ca + ^{251}Cf experimental data. Evaporation residues (EVR), alpha-particles (A), fission fragments (F) and random (R) events are correlated in various combinations.

Correlation	0.1 s	1 s	5 s
R-F	6.3×10^{-9}	8.3×10^{-8}	3.8×10^{-7}
R-A	2.5×10^{-6}	2.4×10^{-5}	1.0×10^{-4}
A-R	3.2×10^{-6}	2.7×10^{-5}	1.4×10^{-4}
R-A-A-F*	1.8×10^{-14}	2.2×10^{-12}	4.6×10^{-11}

*These values were calculated from random probabilities determined for decay chain fragments and multiplication of the appropriate probabilities together for this longer decay chain.

5. Conclusions and Future Work

A total of five atoms of ^{294}Og have been observed in several experiments (three using a ^{249}Cf target, one using a ^{249}Bk target that had significant in-growth of ^{249}Cf, and one using the ^{251}Cf target, most likely from the ^{249}Cf impurity in the target). Monte Carlo Random Probability analysis indicates that these observed ^{294}Og decay chains are highly unlikely to be due to random events. Future experiments aimed at the synthesis of lighter Fl isotopes are underway at DGFRS.

Acknowledgments

These studies were supported by the Russian Foundation for Basic Research, including recent Grants No. 13-02-12052 and No. 13-03-12205, and by the Moscow Region Government through Grant No. 736/36-16.09.2014. Research at ORNL was supported by the U.S. DOE Office of Nuclear Physics under DOE Contract No. DE-AC05-00OR22725 with UTBattelle, LLC. Research at LLNL was supported by LDRD Program Project No. 08-ERD-030, under DOE Contract No. DE-AC52-07NA27344 with Lawrence Livermore National Security, LLC. This work was also supported by the U.S. DOE through Grant No. DE-FG-05-88ER40407 (Vanderbilt University).

References

1. Yu. Ts. Oganessian *et al.*, *J. Phys. G* **34**, R165 (2007), and references therein.
2. Yu. Ts. Oganessian *et al.*, *Phys. Rev. Lett.* **109**, 162501 (2012).
3. Yu. Ts. Oganessian *et al.*, *Prog. Theor. Phys. Supp.* **154**, 406 (2004).
4. Yu. Ts. Oganessian *et al.*, *Phys. Rev. C* **62**, 041604 (2000).
5. N.J. Stoyer *et al.*, *Nucl. Instrum. Methods Phys. Res. A* **455**, 433 (2000).
6. M.A. Stoyer *et al.*, *Eur. J. Phys. A* **25**, 595 (2005).
7. M. A. Stoyer *et al.*, to be published in the Proceedings of the International Nuclear Physics Conference held in Adelaide, Australia Sep. 11-16, 2016, (2017).
8. Yu. Ts. Oganessian *et al.*, *Phys. Rev. C* **74**, 044602 (2006).
9. Yu. Ts. Oganessian *et al.*, *Phys. Rev. C* **83**, 054315 (2011).
10. V.K. Utyonkov *et al.*, *Phys. Rev. C* **92**, 034609 (2015).

Spontaneous Fission of Superheavy Nuclei

D. N. Poenaru* and R. A. Gherghescu

Horia Hulubei National Institute of Physics and Nuclear Engineering (IFIN-HH), P.O. Box MG-6, RO-077125 Bucharest-Magurele, Romania and Frankfurt Institute for Advanced Studies, Johann Wolfgang Goethe University, Ruth-Moufang-Str. 1, D-60438 Frankfurt am Main, Germany

**E-mail: poenaru@fias.uni-frankfurt.de*

http://www.theory.nipne.ro/~poenaru/

Microscopic theories of alpha decay and cluster radioactivity explain these decay modes as a quantum tunneling of a preformed cluster at the nuclear surface. In the present work we show for the first time that in a spontaneous cold fission process the shell plus pairing corrections calculated with Strutinsky's procedure may give a strong argument for preformation of a light fission fragment near the nuclear surface. It is obtained when the radius of the light fragment, R_2, is increased linearly with the separation distance, R, of the two fragments, while for R_2 = constant one gets the well known two hump potential barrier. Nuclear physics community also contributed to nanocluster physics by applying the macroscopic-microscopic method to explain the shell effects experimentally observed since 1984.

Keywords: Spontaneous Fission, Lifetimes, Properties of Superheavy Nuclei.

In 1928 G. Gamow[1] as well as Condon and Gurney[2] gave the first explanation of alpha decay as a quantum tunneling of a preformed particle at the nuclear surface. Soon after experimental discovery in 1984 by Rose and Jones[3] of cluster radioactivity, confirming earlier (1980) predictions by Sandulescu, Poenaru and W. Greiner[4], a microscopic theory[5] also explained the phenomenon in a similar way.

Among the almost 2450 nuclides (nuclear species) known up to now only 288 are stable to occur primordially: their half-lives are comparable to, or longer than the Earth's age (4.5 billion years), hence a significant amount survived since the formation of the Solar System. The metastable nuclides are decaying toward the stable ones.

The first informations about nuclei have been obtained in 1896, when Antoine Henri Becquerel discovered a "mysterious" radiation of a uranium salt (potassium uranyl sulfate) which was bent by a magnetic field. The term radioactivity was coined by Marie Curie. Together with her husband, Pierre Curie, they discovered radium (symbol Ra) and polonium (Po), which possess a million times much stronger radioactivity.

Ernest Rutherford (ER) gave the names α (^{4}He nuclei), β (electrons) and γ (electromagnetic radiation with frequencies of 3×10^{19} Hz or higher and wavelengths of 10^{-11} m (10 pm) or lower). Three of the four fundamental forces (strong, weak, and electro-magnetic) are responsible for them. They are produced by the decay of excited nuclei of radioactive elements. Gamma rays can penetrate through several centimeters of lead and large doses of them are harmful. From scattering experiments (1911) ER deduced that atomic particles consisted primarily of empty space surrounding a central core called nucleus. He transmuted one element into another, elucidated the concepts of the half-life and decay constant. By bombarding nitrogen with α-particles produced oxygen. The atomic nucleus was discovered around 1911.

After 1928 the microscopic theories of α decay have been developed, see e.g.[6]. Simple relationships are also very useful[7,8] to estimate the half-lives.

The liquid drop model (LDM) was introduced by Sir John William Strutt, Lord Rayleigh. His book *Theory of Sound*, was published in 1878. Niels Bohr applied the LDM to Nuclear Physics[9]. It was used by Lise Meitner and her nephew O.R. Frisch[10] to explain the *induced fission* discovered by Otto Hahn and Fritz Strassmann, who identified by chemical means among the fission fragments a barium isotope[11]. N. Bohr and J.A. Wheeler[12] published a theoretical paper, based on his LDM; they showed that fission was more likely to occur with ^{235}U than ^{238}U. An interesting historical account of the discovery of induced fission was written in 1984 by the famous Edoardo Amaldi, former member of the Enrico Fermi's (EF) team who did the first experiment two years before Otto Hahn, but was wrong in interpreting the data. Even a genius like EF was a human being, hence could be sometimes wrong. The energetic (nuclear power plants) and military applications of the induced fission changed completely our world.

Spontaneous fission was discovered in 1940 by G.N. Flerov and K.A. Petrzhak[13]. Usually the fission fragments are deformed and excited; they decay by neutron emission and/or γ rays, so that the total kinetic energy (TKE) of the fragments is smaller by about 25-35 MeV than the released energy, or Q-value. The asymmetric mass distributions of the fission fragments and the spontaneously fissioning shape isomers[14] could not be explained until 1967, when V.M. Strutinsky reported[15] his macroscopic-microscopic method. His calculations gave for the first time a two hump potential barrier. Shape isomers occupied the second minimum.

Besides α, β, γ decay and fission there are other types of nuclear disintegrations sometimes referred to "exotic decay modes" as beta-delayed particle emissions, particle-accompanied fission (or ternary fission), fissioning shape-isomers, proton radioactivity, heavy particle radioactivities (HPR)[16,17], etc. A brief presentation, at a level of non-specialist, of the large diversity of nuclear decay modes may be found in the Ref.[18]. We re-

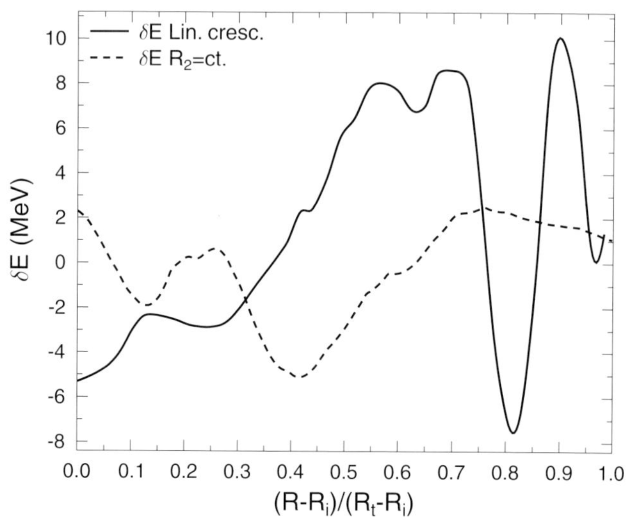

Fig. 1. Comparison of absolute values of shell and pairing correction energies for symmetrical fission of ^{282}Cn with R_2 constant (dashed line) and linearly increasing R_2 (solid line).

ported[19,20] results obtained within macroscopic-microscopic method using cranking inertia[21] and the best two-center shell model[22] in the plane of two independent variables (R, η), where R is the separation distance of the fragments and $\eta = (A_1 - A_2)/A$ is the mass asymmetry with A, A_1, A_2 the mass numbers of the parent and nuclear fragments. Phenomenological deformation energy, E_{Y+E}, was given by Yukawa-plus-exponential model[23], and the shell plus pairing corrections, $\delta E = \delta U + \delta P$ are based on the asymmetric two center shell model (ATCSM). This time we give more detailed arguments for the neighboring nucleus ^{282}Cn. The deep minimum of total deformation energy near the surface is shown for the first time as a strong argument for light fragment preformation in a cold fission[24]. It is obtained when the radius of the light fragment, R_2, is increased linearly with the separation distance, R, of the two fragments, while for R_2 = constant one gets the well known two hump potential barrier.

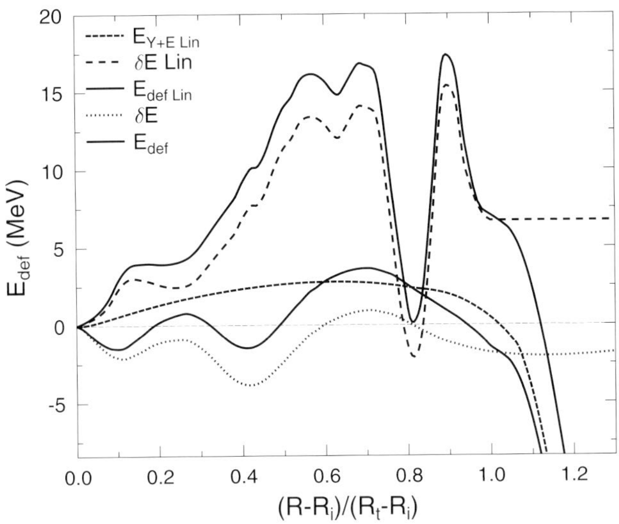

Fig. 2. Comparison of deformation energies with respect to spherical shapes for symmetrical fission of ^{282}Cn with R_2 constant and linearly increasing R_2 (Lin). The contribution of the macroscopic deformation energy is included.

In our paper[19] we reported results obtained within macroscopic-microscopic method[15] using cranking inertia[21,25] and the best two-center shell model[22] in the plane of two independent variables (R, η), or (R, R_2) where R_2 is the radius of the light fragment (emitted particle), R is the separation distance of the fragments and $\eta = (A_1 - A_2)/A$ is the mass asymmetry with A, A_1, A_2 the mass numbers of the parent and nuclear fragments. Phenomenological deformation energy, E_{Y+E}, was given by Yukawa-plus-exponential model[23,26], and the shell plus pairing corrections, $\delta E = \delta U + \delta P$ are based on the asymmetric two center shell model (ATCSM)[22]. The variation of R_2 with R during the fission process is very important, allowing to find the least action trajectory. Previously we have tried four different trajectories: (1) $R_2 = R_{2f} = r_0 A_2^{1/3}$ constant, with $r_0 = 1.16$ fm; (2) exponentially decreasing law: $R_2 = R_{2f} + (R_{20} - R_{2f})e^{-k_2(R-R_i)/(R_t-R_i)}$, where $R_{20} = R_0 = r_0 A^{1/3}$, $R_i = 0$ is the initial separation distance, $R_t = R_{1f} + R_{2f}$ is the same value at the touching point, $k_2 = 4$, and $R_{1f} = r_0 A_1^{1/3}$; (3) linearly decreasing law from $R_{20} = R_0$ to R_{2f}:

$R_2 = R_{2f} + (R_{20} - R_{2f})(R_t - R)/(R_t - R_i)$, and (4) linearly increasing law from 0 to $R_{2f} = R_e$ $R_2 = R_{2f}(R - R_i)/(R_t - R_i)$.

The best result (least action trajectory) was obtained with the linearly increasing law. As one can see in figure 1, this kind of variation is essentially different from the three other laws, which are more or less similar. Our results are confirming the expectations. The half-life against the most important decay modes of any parent nucleus is one of the most important measured characteristics. Consequently there is an increased interest to know details about the mechanism implied and to make a progress toward a unified treatment of the three decay modes.

This time we give more detailed arguments. The deep minimum of total deformation energy near the surface is shown for the first time as a strong argument for cluster preformation.

References

1. G. Gamow, *Zeitschrift für Physik* **51**, 204 (1928).
2. R. W. Gurney and E. U. Condon, *Nature* **122**, p. 439 (1928).
3. H. J. Rose and G. A. Jones, *Nature* **307**, 245 (1984).
4. Encyclopaedia Britannica Online, 2011. Web http://www.britannica.com/EBchecked/topic/465998/.
5. R. G. Lovas, R. J. Liotta, A. Insolia, K. Varga and D. S. Delion, *Physics Reports* **294**, 265 (1998).
6. A. M. Lane, *Reviews of Modern Physics* **32**, 519 (1960).
7. D. N. Poenaru, M. Ivaşcu and D. Mazilu, *Journal de Physique Lettres* **41**, L589 (1980).
8. Y. Wang, S. Wang, Z. Hou and J. Gu, *Phys. Rev. C* **92**, p. 064301 (June 2015).
9. N. Bohr, *Nature* **137**, 344 (1936).
10. L. Meitner and O. R. Frisch, *Nature* **143**, 239 (1939).
11. O. Hahn and F. Strassmann, *Naturwissenschaften* **27**, 11 (1939).
12. N. Bohr and J. Wheeler, *Physical Review* **56**, 426 (1939).
13. G. N. Flerov and K. A. Petrjak, *Physical Review* **58**, 89 (1940).
14. S. M. Polikanov, V. Druin, V. Karnaukhov, V. Mikheev, A. Pleve, N. Skobelev, V. Subotin, G. Ter Akopian and V. Fomichev, *Soviet Physics JETP* **15**, 1016 (1962).
15. V. M. Strutinsky, *Nuclear Physics, A* **95**, 420 (1967).
16. D. N. Poenaru and M. Ivaşcu (eds.), *Particle Emission from Nuclei*, 3 volumes (CRC Press, Boca Raton, Florida, 1989).

17. D. N. Poenaru (ed.), *Nuclear Decay Modes* (Institute of Physics Publishing, Bristol, UK, 1996).
18. W. Greiner and D. N. Poenaru, Radioactivity, in *Encyclopedia of Condensed Matter Physics, Vol. 5*, eds. F. Bassani, G. L. Liedl and P. Wyder (Elsevier, Oxford, 2005) pp. 106–116, 0123694019.
19. D. N. Poenaru and R. A. Gherghescu, *Phys. Rev. C* **94**, 014309 (2016).
20. D. N. Poenaru and R. A. Gherghescu, *European Physical Journal A* **52**, 349.
21. M. Brack, J. Damgaard, A. Jensen, H. C. Pauli, V. M. Strutinsky and G. Y. Wong, *Review of Modern Physics* **44**, 320 (1972).
22. R. A. Gherghescu, *Physical Review* **C 67**, 014309 (2003).
23. H. J. Krappe, J. R. Nix and A. J. Sierk, *Physical Review, C* **20**, 992 (1979).
24. C. Signarbieux, M. Montoya, M. Ribrag, C. Mazur, C. Guet, P. Perrin and M. Maurel, *Journal de Physique (Paris) Lettres* **42**, L437 (1981).
25. D. R. Inglis, *Physical Review* **96**, 1059 (1954).
26. D. N. Poenaru, M. Ivaşcu and D. Mazilu, *Computer Physics Communications* **19**, 205 (1980).

Radioactive Ion Beams and New Facilities

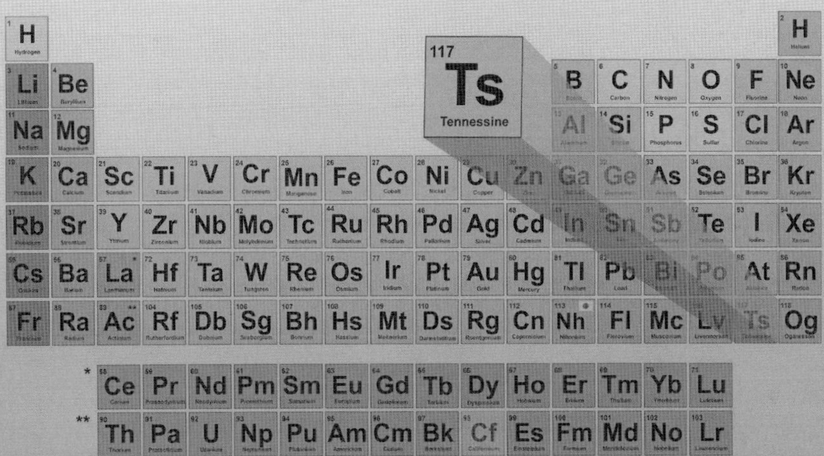

Overview of the Gamma-Ray Energy Tracking: GRETA

P. Fallon

Nuclear Science Division, Lawrence Berkeley National Laboratory
Berkeley, CA 94720, USA
E-mail: pfallon@lbl.gov

Gamma-ray spectroscopy continues to be an important tool for the study of nuclei. Excitation energies can be measured directly and in model-independent ways, and thus are among the key observables that can guide our understanding of atomic nuclei. The Gamma-Ray Energy Tracking Array (GRETA) is a full 4π γ-ray tracking detector, capable of reconstructing the energy and three-dimensional position of γ-ray interactions within a compact sphere of high-purity germanium crystals. GRETA, with its combination of high efficiency, good background rejection, and excellent energy and position resolution, will be a key instrument at FRIB, and will add new capabilities to existing facilities.

Keywords: Nuclear Structure, Gamma-Ray Spectroscopy, Rare Isotopes, Germanium Detectors.

1. Introduction

The study of atomic nuclei – their structure and reactions – directly relates to fundamental questions in physics, such as understanding the origin of the elements and how complex many-body quantum systems organize. Much of what we know about nuclei comes from the measurement and characterization of excited states, and many experimental tools and techniques have been developed to aid these studies. Among them, high-resolution γ-ray spectroscopy has proven to be the most powerful, and γ-ray spectrometers play a central and vital role in nuclear science.

Planned as a key instrument for the Facility for Rare Isotope Beams (FRIB), which is being constructed at Michigan State University, the Gamma-Ray Energy Tracking Array (GRETA) is the realization of a full 4π γ-ray tracking detector, capable of reconstructing the energy and three-dimensional position of γ-ray interactions within a compact sphere of high-purity germanium crystals (Figure 1). It consists of a total of 120 highly segmented large-volume, coaxial germanium crystals, with four crystals combined to form a total of 30 Quad Detector Modules, designed to cover

2.1. ^{60}Ca: Establishing the location of the neutron dripline

Predicting the location of the neutron dripline is a fundamental benchmark of our understanding of nuclei far from stability. Currently there is very limited data that can constrain and guide theory and predictions can widely vary. For example, calculations with realistic two- and three-body forces[8,9] predict the neutron dripline for $Z = 20$ to be located around ^{60}Ca, while many mean-field and density-functional calculations have the (even A) Ca isotopes bound out to $A = 68 - 76$[10], where often a fine interplay between single-particle and many-body degrees of freedom decides the fate of which calcium isotopes that can exist[11]. Information on the structure of neutron-rich nuclei around ^{60}Ca is therefore an important test of calculations, which can differ in their prediction of the location of the Ca dripline by more than 10 nucleons. First results on ^{60}Ti ($N = 38$) have been obtained[12] with GRETINA at the NSCL, however, spectroscopy of the key nuclei ^{60,61}Ca remains out of reach for current instrumemts and facilities.

With GRETA at FRIB spectroscopy of ^{60}Ca will be possible and in Figure 2 we show a simulation of the γ-ray spectrum that one could obtain for ^{60}Ca produced in a one-proton knockout ("fast-beam") reaction induced by a ^{61}Sc projectile beam. The detailed study of neutron-rich calcium out to ^{60}Ca will help establish the significance and role of three nucleon forces and the continuum in describing these nuclei.

2.2. Octople collectivity

Deformation and the associated degrees of freedom (e.g. rotations, vibrations) plays an important role in our understanding of nuclear structure[16]. Using reaccelerated beams at FRIB together with GRETA, it will be possible to carry out high-statistics, multi-step Coulomb excitation experiments to provide a precise determination of nuclear static moments and excited state transition strengths used to study a broad range of collective phenomena and deformation. The 4π angular coverage and tracking ability of GRETA will also provide linear polarization and angular distribution information to enable spin and parity assignments of excited nuclear states.

One interesting case for future studies involves octupole deformation and searches for a permanent electric dipole moment (EDM). The observation of a permanent EDM would indicate time-reversal (T) or charge-parity (CP) violation, most likely due to physics beyond the Standard Model[17,18]. Nuclear EDMs are proportional to a quantity called the Schiff moment, the residual response proportional to the square of the nuclear radius, after

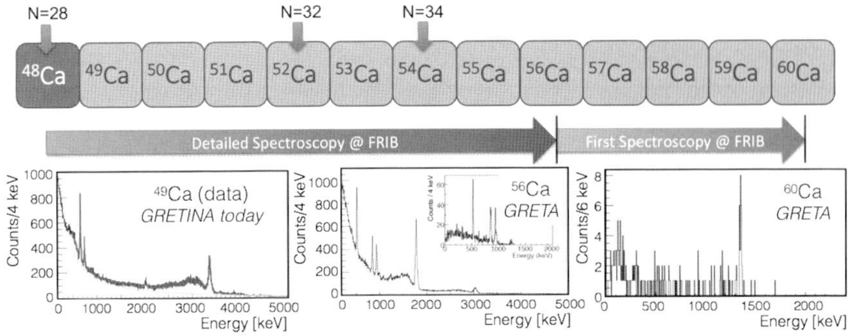

Fig. 2. Overview of the spectroscopic reach in the Ca isotopes today, and in the future with GRETA at FRIB. (Left) Experimental GRETINA spectrum for neutron knockout from ^{50}Ca to ^{49}Ca (whole range), fit with a GEANT4 simulation (starting from 500 keV) and a smooth background (bottom line). (Center) Simulated GRETA spectrum for neutron knockout from ^{57}Ca into ^{56}Ca at FRIB, based on a theoretical level scheme[13]. Inset shows simulated $\gamma\gamma$ coincidence spectrum for the 2^+ to 0^+ transition at 1.8 MeV. (Right) Simulation for ^{9}Be(^{61}Sc, ^{60}Ca + γ) with GRETA at the proposed high-rigidity spectrometer (HRS)[14] at FRIB; this combined setup will enable first spectroscopy in ^{60}Ca.

cancellations due to the screening of the nucleus by its electron cloud are taken into account. The occurance of a nuclear octupole deformation, which not only breaks rotational symmetry but also reflection symmetry in the intrinsic frame, can lead to considerably larger Schiff moments; and because a CP-violating Schiff moment induces a contribution to the atomic EDM, a large enhancement due to nuclear octupole effects translates into an improved sensitivity to an atomic EDM when compared to systems without this deformation. Enhancement factors of $10^2 - 10^3$ have been calculated[17,19,20]. As an example, detailed measurements can be performed to characterize the octupole collectivity and properties of ^{225}Ra and ^{223}Rn, where efforts are underway to use these nuclei in experiments to observe an EDM. In addition, new candidate nuclei for EDM searches can also be studied for the presence or absence of octupole deformation as inferred from properties of the excited levels, including spin-parity assignments and electromagnetic transition rates. One such possible candidate is ^{229}Pa, for which the required level of detailed study is currently not possible. The most sensitive EDM search to date come from measurements on ^{199}Hg[21]. In ^{229}Pa, the EDM contribution induced by the Schiff moment is predicted to be 3 x 10^4 times larger than the one for (spherical) ^{199}Hg and 40 times larger than the one for ^{225}Ra[22].

3. γ-Ray Tracking: Underlying Principles

The idea to construct a γ-ray detector array from a "shell of germanium" stemmed from the goal to improve on the performance of the, then current, large scale detector arrays based on individual Compton (escape) suppressed detectors[23], which were constructed during the 1980's and 1990's and culminated in the Gammasphere[24] and Euroball[25] 4π arrays. It was realized that such a new detector would require a different approach and the concept of a γ-ray tracking array soon emerged as the way forward. In order to realize a γ-ray tracking array, a number of technologies were developed, including electrically segmentated Ge detectors, digital signal processing, and signal pulse shape analysis algorithms. Two such tracking arrays have since been constructed, GRETINA in the United States and the AGATA[26] demonstartor in Europe. Further technical details on GRETINA are given in[27–29].

In a Ge detector, γ rays in the energy range between of 0.2 to 5 MeV will generally undergo several Compton scattering processes before they are absorbed by a photo electric interaction. If the position and the energy deposited of each of the γ-ray interactions can be measured, the sequence of the interactions can be determined or "tracked" by requiring that each interaction satisfies the relation between the scattering angle and the energy deposition of the Compton scattering process given by the the Klein-Nishina formula.

Each HPGe detector in GRETA is cut from a single crystal into a tapered irregular hexagon shape (9 cm long and approximately 300 cm^3 in volume after being cut), so that 120 of them could be packed into a spherical shell. The detector is segmented electrically into 36 segments with six longitudinal segmentation lines along the flat tapered surface and five transverse segmentation lines. The signals that are produced when a γ-ray ray interacts and deposits energy in a crystal are then digitized locally in dedicated ADC/digitizer boards and transmitted to signal filter boards, which calculate the energy deposited by the γ ray and extract a time-windowed portion of the waveform. System wide trigger and timing modules determine when a desired event has occurred and synchronize the timing information for all the electronics modules. A time-stamped event (comprising the sampled waveform, energy, time, as well as trigger information) is then passed to the computing system. The primary task of the computing system is to decompose the time-correlated waveforms into a linear combination of signals from individual γ-ray interactions; this process provides the number of interactions and their position and energy. Next, the positions and energies from all detectors are time correlated and input into a tracking algorithm. The algorithm identifies the interactions that belong to individual γ rays, selects γ rays with full energy deposition in the array and rejects those with partial energy deposition (Compton rejection), and determines the scattering sequence of the full-energy γ-ray events. The results are further analyzed to obtain the spectrum and extract physics results.

The array geometry for GRETA is the same as GRETINA and is based on a spherical geodesic design, comprised of 120 hexagons and 12 pentagons that tile the complete sphere. The design ensures a close-packed geometry to maximize efficiency and allow seamless integration with the existing GRETINA detector modules. The 30 quad modules will be held in place precisely and accurately around a target chamber by two hemispherical aluminum frames. Both frames independently translate and rotate to allow access to the target chamber and to allow detectors to be installed and removed in any module location.

Acknowledgements

GRETINA was funded by the US Department of Energy (DOE), Office of Science. The writing of this article was supported under DOE grant DE-AC02-05CH11231 (LBNL), and this work is the result of the contributions and effort of the many people involved in the GRETINA project and science campaigns.

References

1. IY. Lee, *Nucl. Instrum. Methods Phys. Res., Sect. A* 422:195 (1999).
2. D. Bazin *et al.*, *Nucl. Instrum. Methods Phys. Res., Sect. B* 204:629 (2003).
3. P. Fallon, A. Gade and IY. Lee, *Annu. Rev. Nucl. Part. Sci.* 66:321 (2016).
4. G. Savard *et al.*, *JPS Conf. Proc* 6:010008 (2015).
5. G. Savard *et al.*, *Nucl. Instrum. Methods Phys. Res., Sect. B* 266:4086 (2015).
6. *The Gamma-ray Energy Tracking Array: GRETA*, LBNL Report 1004136 (2015) `http://greta.lbl.gov/documents/GRETA-WhitePaper.pdf`

7. *The Nuclear Science Advisory Committee 2015 Long Range Plan for Nuclear Science*
`http://science.energy.gov/~/media/np/nsac/pdf/2015LRP/2015_LRPNS_091815.pdf`
8. L.D. Holt, T. Otsuka, A. Schwenk and T. Suzuki, *J. Phys. G* 39:085111 (2012).
9. G. Hagen *et al.*, *Phys. Rev. Lett.* 109:032502 (2012).
10. J. Erler *et al.*, *Nature* 486:509 (2012).
11. C. Forssen *et al.*, *Phys. Scr.* T152:014022 (2013).
12. A. Gade *et al.*, *Phys. Rev. Lett.* 112:112503 (2014).
13. J.D. Holt *et al.*, *Phys. Rev. C* 90:024312 (2014).
14. *HRS: A High Rigidity Spectrometer for FRIB* (2015), `https://people.nscl.msu.edu/~zegers/HRS_draft.pdf`
15. S.M. Lenzi, F. Nowacki, A. Poves and K. Sieja, *Phys. Rev. C* 82:054301 (2010).
16. A. Bohr and B. Mottelson, New York: W. A. Benjamin (1975).
17. J. Engel, M. Ramsey-Musolf and U. van Kolck, *Prog. Part. Nucl. Phys.* 71:21 (2013).
18. M. Pospelov and A. Ritz, *Ann. Phys.* 318:119 (2005).
19. J. Dobaczewski and J. Engel, *Phys. Rev. Lett.* :232502 (2005).
20. V. Spevak, N. Auerbach and V.V. Flambaum, *Phys. Rev. C* 56:1357 (1997).
21. W.C. Griffith *et al.*, *Phys. Rev. Lett.* 102:101601 (2009).
22. V.V. Flambaum, *Phys. Rev. A* 77:024501 (2008).
23. J. Eberth and J. Simpson, *Prog. Part. Nucl. Phys.* 60:283 (2008).
24. IY. Lee, *Nuclear Physics A* 520:641 (1990).
25. Simpson J., *Zeitschrift fur Physik A Hadrons and Nuclei* 358:138 (1997).
26. Akkoyun S. *et al.*, *Nucl. Instrum. Methods Phys. Res., Sect. A* 668:26 (2012).
27. S. Paschalis *et al.*, *Nucl. Instrum. Methods Phys. Res., Sect. A* 709:44 (2013).
28. T. Lauritsen *et al.*, *Nucl. Instrum. Methods Phys. Res., Sect. A* 836:46 (2016).
29. D. Weisshaar *et al.*, *Nucl. Instrum. Methods Phys. Res., Sect. A* 847:187 (2017).

Commissioning and Spectroscopic Studies of Fission Products with IGISOL-4*

Ari Jokinen

Department of Physics, PB 35, FI-40014 University of Jyväskylä

An ion guide isotope separator on-line (IGISOL)-facility in the Department of Physics, University of Jyväskylä (JYFL) has been moved to the new extension of the Accelerator Laboratory. The present IGISOL installation is already fourth generation of IGISOL apparatus with various technical upgrades and new installations, which will be introduced in this contribution.

A combination of IGISOL technique and ion trap techniques has provided extended set of precise atomic mass data on neutron-rich nuclei [1]. A mass measurement program at IGISOL-4 has just started and the most recent results will be presented.

Variety of spectroscopic studies have been performed at IGISOL-4 by using different detector setups both in the decay station and after the Penning trap. Those include total absorption spectrometer DTAS, neutron counter setup BELEN and a new decay station with Ge-array. Highlights of the recent studies will be reviewed and further installations at IGISOL-4 will be illustrated.

Reference

1. A. Kankainen, J. Äystöja and A. Jokinen, *J. Phys. G.* **39** (2012) 093101.

*This work has been supported by the Academy of Finland under the Finnish Centre of Excellence Programme 2012-2017 (Finnish Centre of Excellence in Nuclear and Accelerator-Based Physics Research).

FRIB/NSCL Laboratory Update

C. Konrad Gelbke
FRIB/NSCL Laboratory, Michigan State University,
East Lansing, Michigan 48824-1321, USA

Michigan State University is designing and establishing the Facility for Rare Isotope Beams (FRIB) as a DOE Office of Science National User Facility in support of the mission of the Office of Nuclear Physics. FRIB will be a cutting-edge research facility to advance understanding of rare nuclear isotopes and the evolution of the cosmos. In this talk I will provide a high-level update of the DOE funded FRIB project, of the NSF-funded NSCL user program and the envisaged transition from NSCL to FRIB user operations, and additional instrumentation required to fully exploit FRIB capabilities for the user-driven nuclear science research.

Recent Results with Low-Energy and Reaccelerated Neutron-Rich Radioactive Beams at CARIBU*

Guy Savard

Argonne National Laboratory, Argonne, Illinois 60439, USA
University of Chicago, Chicago, Illinois 60637, USA

CARIBU provides neutron-rich low-energy and reaccelerated beams to the ATLAS user's community. It is unique in that it relies on spontaneous fission from a ^{252}Cf source to produce the neutron-rich isotopes that are cooled and extracted as a low-energy beam using a large gas catcher and guiding radio-frequency structures. These beams are accelerated to 50 kV, purified by a compact high-resolution separator, and subsequently delivered to either a low-energy experimental area or reaccelerated to above the Coulomb barrier using the ATLAS superconducting linac. Furthermore, the beam extraction and selection techniques developed for CARIBU are universal, essentially independent of the chemical properties of the isotopes, so that beams of short-lived refractory and reactive isotopes not typically extracted from ISOL-type systems are also available.

Recent highlights from the CARIBU programs, including the Coulomb excitation reaccelerated beam campaign using neutron-rich beams from CARIBU with the combination of GRETINA and Chico-II, and the new mass measurement results at low-energy using an improved Penning trap mass measurement technique, will be presented.

Recent CARIBU upgrades to reduce beam contamination of both low-energy and reaccelerated beams using a new MRTOF spectrometer and an EBIS charge-breeder will be briefly described, together with the installation of a new low-energy experimental area with much lower radioactivity background for decay spectroscopy.

*This work was supported by the US DOE, Office of Nuclear Physics, under contract DE-AC02-06CH11357.

Results of the ISOLDE Facility and the HIE-ISOLDE Project

M.J.G. Borge

EP-ISOLDE, CERN,
Geneva, Switzerland
Instituto de Estructura de la Materia, CSIC,
Serrano 113bis, E-28002-Madrid, Spain
E-mail: mgb@cern.ch
http://isolde.web.cern.ch

ISOLDE is the CERN facility dedicated to the production of radioactive ion beams for many different experiments in the fields of nuclear and atomic physics, materials science and life sciences. By a clever combination of target and ion source units pure beams of over 1000 different nuclei of 74 elements have been produced and delivered to experiments. Since more than ten years ISOLDE offers the largest variety of post-accelerated radioactive beams in the world today. The HIE-ISOLDE intensity and energy upgrade has finished its phase 1 dedicated to upgrade the energy up to 5.5 MeV/u, producing the first radioactive beams at this energy in September 9th 2016. The results from this campaign of post-accelerated beams will be described together with very recent ISOLDE highlights of neutron-rich nuclei.

Keywords: Radioactive Beams; Post-Accelerated Beams; Neutron-Rich Nuclei.

1. The ISOLDE Facility

ISOLDE's production of radioactive nuclei began in 1967, and half a century later, ISOLDE is the oldest experiment still in operation at CERN. Over the years, it has developed into a Facility dedicated to fundamental science and its applications. ISOLDE's success can be traced to two key elements. One is the continuous development of new radioactive ion beams and the steady improvement of experimental conditions and devices. ISOLDE produces radioactive ion beams in spallation, fragmentation and fission reactions between stable nuclei from thick targets and high energy and high intensity proton beam accelerated in the first ring of the CERN accelerator complex, the Proton Synchrotron Booster (PSB). The proton beams hits the hot target typically with an energy of 1.4 GeV and an intensity of up to 2 μA.

The heart of ISOLDE is its target and ion source system, which provide the selective production of the desired isotopes. The production process begins when the target, heated between 1000°C and 2000°C depending on the material, is bombarded by the proton beam and in the collision, the already mentioned nuclear reactions occur either by boiling off a few protons or neutrons or fragmenting them in lighter species. The high temperature allows the fast release of radioactive nuclei before their decay. Release times vary from a few dozens of seconds for targets with temperatures in between 700°C and 1400°C to a few tenths of a second for the fastest targets that reach more than 2000°C. Target materials partly determine the radioactive nuclei production rate and their release time, thus a careful selection is of paramount importance. ISOLDE offers more than 25 target materials, with uranium carbide being the most requested because of its versatility.

ISOLDE uses three types of ion sources: surface, plasma and laser. The simplest method — and highly efficient in the case of chemical elements with low ionization potential — is to have the nuclei interact with a hot surface of a metal tube, also called *line*. The line's material should have a higher work function than the atom that is to be ionized and can reach temperatures of up to 2400°C. The second type of ion sources, plasma sources, can be used in combination with most target materials and are especially suited for elements that cannot be surface-ionized. In this method, a mixture of gases — argon and xenon are most often used — is ionized by accelerated electrons with a voltage of approximately 130 V to produce the plasma, while an extra magnetic field confines the plasma and optimizes the ionization process.

Although both these methods are efficient, a higher degree of chemical selectivity can be achieved with the Resonance Ionisation Laser Ion Source (RILIS), a method pioneered by ISOLDE. RILIS takes advantage of resonant excitation of atomic transitions using tunable laser radiation and allows a high degree of isobar-free selectivity. Presently the used of RILIS has expanded up to 85 % of the beam time. In total ISOLDE is able to produce up to 1300 nuclei of 74 elements [1]. The ISOLDE beams is then delivered to the different experimental stations or post-accelerated.

For the production of neutron-rich fission fragments a solid tungsten proton-to-neutron converter has been used for many years. This reduces the neutron deficient isobaric contaminants. An improved converter design has been obtained based on simulations using energy-dependent cross-sections for proton and neutron induced fission. Recently this new type of converter was successfully tested experimentally [2].

For many experiments the beam quality is of major importance. To achieve reduced emittance and energy spread an ion cooler and buncher, ISCOOL [3], was designed and installed after the high resolution separator (HRS) of ISOLDE. This resulted in strongly reduced emittance and energy spreads for all experiments downstream the beam line. Using its bunching capabilities, ISCOOL permits an increase in the sensitivity for experiments such as those devoted to collinear laser spectroscopy.

The importance and interest on post-acceleration of the large palette of ISOL beams can be trace back to 1966. The final project, called REX and approved in 1994, was based in the previous expertise existing in the Facility, and it was designed in the following way: the radioactive ions produced in the ISOLDE target-ion-source are mass separated and collected in an ion-trap, after cooling and bunching in the trap the ions are brought into a charge-state breeder (REXEBIS) using the technique of electron-beam ionization, the charge-bred ions are then brought into a room temperature linear accelerator. In the linear accelerator the ions are first accelerated with a radio-frequency quadrupole (RFQ), followed by an Interdigital H-type (IH) structure and finally three 7-gap resonators increase the energy of the radioactive beams to 2.2 MeV/u. From the very start in 2001 an upgrade was conceived by adding a 9-gap acceleration stage behind the 7-gap section. This allowed already in 2004 for a maximal energy of 3 MeV/u for ions with $A/Q = 3.5$. The availability of post-accelerated beams opened up new fields of research within the ISOLDE program. REX-ISOLDE, originally planned for accelerating light isotopes up to mass around A~50, has gone far beyond this goal and has provided post-accelerated beams covering the entire mass range from ^{6}He up to ^{224}Ra for reaction studies. The physics program have mainly profited from the opportunities offered by Coulomb excitation studies in inverse kinematics and transfer and inelastic reaction studies of light nuclei. A detailed account for technical details of REX-ISOLDE and its physics results is given in Ref. [4].

2. Physics at ISOLDE with n-rich nuclei

In the following I present some recent highlights with focus on n-rich nuclei.

2.1. *Exotic decay modes: The case of ^{11}Be*

A characteristic feature of nuclei approaching the drip-line regions is the increase of Q-values available for beta decay combined with decreasing separation energies for one or several particles. This allows the study of beta-delayed particle-emission processes for the most exotic nuclei on both sides of the beta-stability line. Several of the more rare beta-delayed decay modes were first observed at ISOLDE like β2n, β3n, βt and βα [5]. In addition beta-delayed deuteron emission was search for as consequence of the halo structure in some near drip line 2n-halo nuclei and it was first observed ISOLDE [6] in the decay of ^{6}He ($T_{1/2}$ = 807 ms).

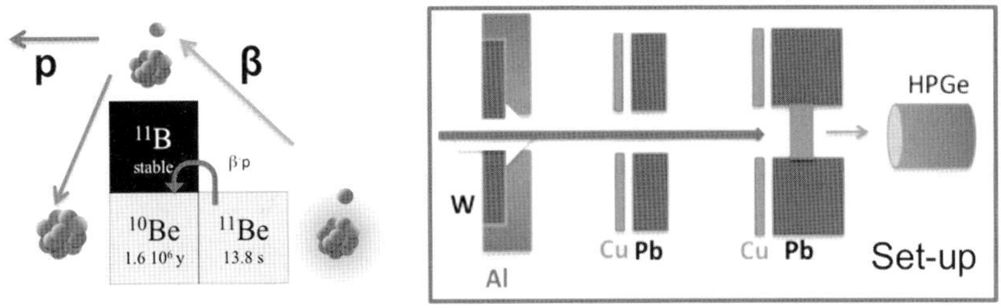

Fig. 1. On the left hand side the βp decay mode observed for first time in the β-decay of the 1n-halo ^{11}Be. On the right hand side the setup used at ISOLDE to count the ^{11}Be ions using a HPGe detector to determine the number of decays from the 2125 keV γ-ray transition.

The one-neutron halo nucleus ^{11}Be has a neutron-separation energy as low as S_n = 504 keV, which means that β-delayed proton emission is energetically allowed in this n-rich nucleus, as illustrated in Fig. 1. The energy required for this process to occur is Q_p(MeV) = 0.782-S_n (MeV). The proton spectrum will expand only up to 280 keV, making difficult the use of conventional counting techniques to observe this decay. Further, theoretical estimates pointed to a branching ratio of the order of 10^{-8}. To determine the branching ratio of this "new" decay mode atomic mass spectroscopy was applied profiting from the large difference in half-life between the parent ($T_{1/2}$ = 13.8 s) and the β-delayed proton daughter ^{10}Be ($T_{1/2}$ = 1.39 × 10^6 y) and the high purity of the ISOLDE beams. The ^{11}Be was collected at ISOLDE with the setup shown on the right hand side of Fig. 1. The amount of ^{10}Be produced by βp in the collected sample, measured at the VERA facility in Vienna, was found to correspond to a branching ratio of 8.3(9) × 10^{-6} [7]. To understand the branching ratio, which is a factor of 100 higher than predicted, and that points to the presence of an unknown resonance in ^{11}B, further work is needed. To pin-down whether the decay is through a resonance or directly to a continuum a new experiment has been approved to determine the proton spectrum in the 110-287 keV range by the used of an optical time projection chamber.

2.2. *The crossing between spherical and deformed configurations in $N = 20$ nuclei finally identified in ^{34}Al*

The evolution of the N = 20 magic number has been widely studied experimentally from the Ca down to the Na nuclei using different techniques. It has been suggested that the "magicity" persists along the N = 20 isotonic chain, from ^{40}Ca (Z = 20) down to ^{34}Si (Z = 14), and suddenly disappears only two protons below, in the deformed nuclide ^{32}Mg [8]. Recent experimental studies of ^{32}Mg and ^{34}Si have provided support to the hypothesis that a crossing between normal and intruder regime occurs between these two nuclei [9]. However the ground state of ^{33}Al (N = 20) is outside the island of inversion [10, 11]. Therefore, ^{34}Al (N = 21) should offer the possibility to explore this crossing. First, two β-decaying states with spin values 4^- and 1^+ have been identified, corresponding to normal and intruder configurations, respectively. Second, an abnormal crossover on the S_{2n} surface has been found between the Al and Mg chains at N = 21 [12], suggesting a structural change there.

Therefore the ^{34}Al excited states were studied by the β-decay of ^{34}Mg produced at ISOLDE through fragmentation reactions between a 1.4 GeV pulsed proton beam delivered by the PS-Booster and a UC_x target. The ^{34}Mg production rate was estimated from βγ measurements to be of 700 ions/s. The β-decay half-life of ^{34}Mg, $T_{1/2}$ = 44.9(4) ms, was obtained from the weighted average values extracted by gating on the most intense transitions in ^{34}Al at 364.5, 441.2, 975.0 and 1051.7 keV. This new value is twice as long as the previously measured half-life of $T_{1/2}$ = 20(10) ms [13] obtained from β-delayed neutron emission. Study of the γγ-coincidences and the decay pattern allowed to construct a detailed level scheme of ^{34}Al. The vast majority of γ-ray cascades lead to the first 1^+ state and their summed energies consistently amount to 1416(1) keV. However, three of them, have a summed energy 46.6 keV below. Assuming that the 4^- state lies 46.6 keV below the 1^+state, we are able to build the level scheme consistently. Large scale shell-model calculations with the recently developed sdpf-u-mix interaction are compared with the new data and used to interpret the mechanisms at play at the very border of the N = 20 Island of Inversion, for more details see Ref. [14].

2.3. *Exploring the nature of the $N = 32$ magic number*

The calcium isotopic chain (Z = 20) is a unique nuclear system to study how protons and neutrons interact inside the atomic nucleusas two of its stable isotopes are double magic (^{40}Ca and ^{48}Ca). In addition experimental evidence of doubly-magic features in the exotic ^{52}Ca are found based on nuclear spectroscopy experiments as well as in mass measurements [15].

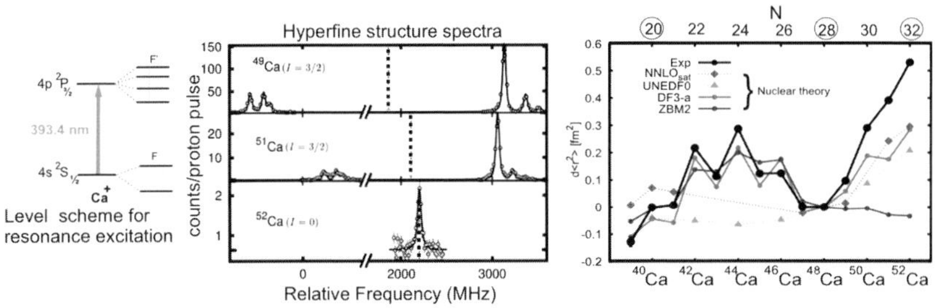

Fig. 2. On the left hand side the hyperfine structure spectrum of ^{48,51,52}Ca and on the right hand side the meansquarecharged radius from $^{39\text{-}52}$Ca. Illustrated the unexpected increase of charged radius of ^{52}Ca and the comparison with different models, for more details see [16].

Despite an excess of eight neutrons, ^{48}Ca exhibits the striking feature that it has essentially the same charge radius as ^{40}Ca. A fundamental explanation of these anomalous features has been a longstanding problem for nuclear theory. To determine the radius beyond ^{48}Ca is therefore crucial both from an experimental and theoretical point of view mainly considering the recently determined magic character of ^{52}Ca mentioned above. The measurements were done by using high-resolution bunched-beam collinear laser spectroscopy. Changes in the charge radii for $^{40\text{-}52}$Ca isotopes were obtained from the measured optical isotope shifts determined from hyperfine structure as shown in Fig. 2 for ^{48,51,52}Ca. With a production yield of only a few hundred ions per second, the measurement of ^{52}Ca represents one of the highest sensitivities ever reached by using fluorescence detection techniques. The isotope shifts were extracted from the fit of the hyperfine experimental spectra. The resulting charge radius of ^{52}Ca is found to be much larger than expected for a doubly magic nucleus, see Fig. 2. The large and unexpected increase of the size of the neutron-rich calcium isotopes beyond N = 28 challenges the doubly-magic nature of ^{52}Ca and opens new intriguing questions on the evolution of nuclear sizes away from stability, which are of importance for our understanding of neutron-rich atomic nuclei [16].

2.4. *Cadmium isotopes: Too simple or too complex?*

The region around the doubly-magic pair, Z = 50 and N = 82, attracts considerable interest. In this circumstances ISOLDE is able to attack the problem with all our existing tools: measurements of ground states properties: masses, spins, electromagnetic moments, beta decay studies and reactions.

The observation of an anomalous behavior of the first 2^+ state with a decrease of its energy passing from ^{126}Cd to ^{128}Cd prompted a "safe" Coulomb excitation experiment at REX-ISOLDE with 2.85 MeV/u beams of ^{122,124,126}Cd. The reduced transition probabilities, B(E2), between ground states and first 2+ states and limits for the quadrupole moments for the three isotopes were determined [57]. The results were discussed in terms of onset of collectivity in this region.

Measurements of the masses for the three isotopes $^{129\text{-}131}$Cd were performed at the ISOLTRAP spectrometer [17] to determine the r-process path in this region. The beta decay strength distribution of waiting point nuclei is extremely valuable to anchor nuclear models necessary to calculate half-lives farther away from stability. The beta decay of the r-process waiting point nucleus ^{132}Cd was recently measured at the ISOLDE Decay Station (IDS) [18].

The spin and electromagnetic moments of $^{107\text{-}129}$Cd isotopes reaching N = 82 shell closure have been measured also [19]. Deep ultraviolet excitation and radioactive beam bunching provided the required sensitivity, even allowing for the discovery of long lived isomers in ^{127,129}Cd. The quadrupole moments on isotopes with spin 11/2 related to the $h_{11/2}$ orbital show a lineal increase with the neutron occupation as expected in an extreme shell model description. The linear trend extends beyond the filling of this orbit indicating a certain degeneracy with the neighboring orbitals. It is amazing to find such a simplicity in these complex systems. In addition, the difference in radii between the ground state and the isomeric one shows an unusual simple parabolic behavior when moving from a slightly oblate to a prolate shape [20].

3. The HIE-ISOLDE Project and first beams

The HIE-ISOLDE Project will be a major staged upgrade of the existing ISOLDE facility [21]. It includes an energy increase of the REX post-accelerator up to 5.5 MeV/u, which was already operative in 2016, followed by a further increase to 7.5 MeV/u in 2017 and 10 MeV/u planned for 2018. An upgrade of the REX low energy stage is also expected. The beam quality will be improved e.g. with the installation of a RFQ cooler. The beam purification upgrade foresees to include a new high resolution separator and a Multi-Reflection Time-of-Flight, MR-ToF, for mass separation. The beam intensity will be increased by two independent parts of the injectors. First the production will linearly increase with the enhanced proton intensity delivered by the new ion source and pre-injector called Linac-4. Second the increase in energy of the PS Booster to 2

GeV will enhance the cross section for fragmentation and spallation in factor of two to 10 and with them the production [22].

The energy increase became possible with the construction of a superconducting linear accelerator, a high-energy beam transfer line and three beam lines to the experimental areas. For the energy upgrade up to 10 MeV/u four cryo-modules will be used, each containing five high-beta ($\beta = v/c = 10.3\%$) superconducting cavities and solenoid. The latter is also integrated in the common vacuum system and provides the transverse beam focusing to the accelerator. Three identically designed beam lines connect the linac with the experimental devices. In each of them, the beam is bent by 90 degrees by two dipole magnets and at least a doublet-quadrupole to focus the beam into the experimental station. The HEBT is designed to reach an absolute transport efficiency close to 100%.

In the first beam line, MINIBALL, a 8 cluster HPGe array, is located with three alternative setups: the CD type Double-sided Silicon Detector for detecting ejectiles, the SPEDE spectrometer for electron conversion detection both for Coulomb excitation studies and T-REX for transfer reaction studies. The second beam line will have the ISOL Solenoidal Spectrometer (ISS) that will be operative at the end of 2017 for high efficiency transfer reaction studies. The third beamline is connected to a ScattEring chamber, SEC, widely used as it can host different charged particle detector arrays for elastic, inelastic and transfer studies, and the corset setup for quasi-fission measurements, i.e. for reaction studies where gamma-ray information is not necessary. Other movable setups such as the active target ACTAR, the optical time projection chamber and the tilted foil β-NMR setup will be also installed at the end of this third beamline.

3.1. *The physics opportunities and the first beams*

The HIE-ISOLDE Project was formally approved September 1st, 2009 and the first radioactive beams with one cryomodule were accelerated the 22nd of October 2015 and with two cryomodules accomplishing the first stage of HIE-ISOLDE the 9th of September 2016.

The higher energy allows for the exploration of interesting regions in the nuclear chart, increases the cross section in most cases and improves accessibility to detailed nuclear structure information. Many HIE-ISOLDE experiments plan to use Coulomb excitation to address the nuclear collective degree of freedom, exotic shapes and shape coexistence. A number of transfer reaction experiments have also been proposed, as well as a few studies on elastic scattering reactions as well as studies of astrophysical interest.

During the 2016 campaign 685h of post-accelerated beams were provided allowing for six experiments to be realized, Table 1 summarizes the radioactive beams used, the A/q, the transmission and the energy. The ^{9}Li beam at 6.8 MeV/u was used for transfer reaction studies to explore the unbound nucleus ^{10}Li using the SEC chamber. The ^{66}Ni beam at 4.5 MeV/u was used to study whether the unexpected increase in the gamma strength function at low energy observed in several stable nuclei using the Oslo method was confirmed for exotic beams and using a very different setup composed of MINIBALL together with three large $LaBr_3$ detectors. The ^{78}Zn beam at 4.3 MeV/u was used to study via multi-step Coulomb excitation the evolution of the nuclear structure near ^{78}Ni. The ^{100}Sn at 4.5 MeV/u used multi-step Coulomb excitation to determine the B(E2) between the $4^+ \rightarrow 2^+$ and $2^+ \rightarrow 0^+$ as the results obtained for light tin isotopes in experiments done at various energies differ greatly. Moving from the light tin to the heavy ones, the ^{132}Sn beam accelerated at 5.5 MeV/u was used to study by multi-step Coulomb the reduced transition probabilities $4^+ \rightarrow 2^+$ and $2^+ \rightarrow 0^+$ in this double magic nucleus. The multi-step Coulomb study of the ^{142}Xe beam at 4.5 MeV/u aimed to study the octupole correlations in this nucleus. Population up to the 8+ excited state was identified while running the experiment.

Table 1. Exotic beams (RIB) post-accelerated at ISOLDE in 2016. Six experiments were done properties of the beams used and their transmission are stated.

RIB	A/q	Beam Energy (MeV/u)[a]	Beam Transmission REX/HIE Linac
$^{9}Li^{3+}$	3.0	6.8	75 %
$^{66}Ni^{16+}$	4.125	4.5	68 %
$^{78}Zn^{20+}$	4.3	4.3	68 %
$^{100}Sn^{26+}$	4.23	4.5	68 %
$^{132}Sn^{31+}$	4.26	5.5	68 %
$^{142}Xe^{33+}$	4.5	4.5	60 %

[a]The accelerating energy of 5.5 MeV/u was calculated for A/q = 4.5, higher energies can be achieved for lower A/q.

4. Summary and Outlook

Almost fifty years since the beginning of its operation, in October 1967, ISOLDE remains a leading facility both in terms of technical infrastructure and production of frontier physics.

In this contribution I have presented very recent results obtained at ISOLDE centered in n-rich nuclei: the discovery of a new decay mode, βp in n-rich nuclei due the 1-halo structure of ^{11}Be. The crossing between spherical and deformed structure in the region of N = 20 has been identified in ^{34}Al. The different manifestations of the new magic number N = 32, and the special character that the Cd-isotopes show as well as the relevance of these data for the r-process.

The success of the different post-accelerated beam experiments has demonstrated the capacities of HIE-ISOLDE and the physics opportunities that lie ahead thanks to the upgrade. At the start of 2018, HIE-ISOLDE is expected to complete the energy upgrade, reaching 10 MeV/u and becoming an even more attractive facility for a wide variety of experiments.

References

1. M. J. G. Borge, *Nucl. Instrum. Methods Phys. Res., Sect. B* **376**, 408 (2016).
2. A. Gottberg *et al.*, *Nucl. Instr. and Methods Phys. Res., Sect. B* **336**, 143 (2015).
3. E. Mane *et al.*, *Eur. Phys. J A* **42**, 503 (2009).
4. P. Van Duppen and K. Riisager, *J. Phys. G.* **38**, 024005 (2011).
5. M. J. G. Borge, *Phys. Scr.* **T152**, 014013 (2013); B. Blank and M. J .G. Borge, *Prog in Part and Nucl. Phys.* **60**, 403 (2008).
6. K. Riisager *et al.*, *Phys. Lett. B* **235**, 30 (1990).
7. K. Riisager *et al.*, *Phys. Lett. B* **732**, 305 (2014).
8. H. L. Crawford *et al.*, *Phys. Rev. C* **93**, 031303(R) (2016).
9. A. O. Macchiavelli *et al.*, *Phys. Rev. C* **94**, 051303(R) (2016).
10. E. Caurier, F. Nowacki and A. Poves, *Phys. Rev. C* **90**, 014302 (2014).
11. H. Heylen *et al.*, *Phys. Rev. C* **94**, 034312 (2016).
12. A. A. Kwiatkowski *et al.*, *Phys. Rev. C* **92**, 061301(R) (2015).
13. M. Langevin *et al.*, *Nucl. Phys. A* **414**, 151 (1984).
14. R. Lica *et al.*, *Phys Rev C* **95**, 021301(R) (2017).
15. F. Wienholtz *et al.*, *Nature* **498**, 346 (2013).
16. R. F. Garcia Ruiz *et al.*, *Nature Physics* **12**, 594 (2016).
17. D. Atanasov *et al.*, *Phys. Rev. Lett.* **115**, 232501 (2015).
18. See contribution of S. Paulauskas to this issue.
19. D. T. Yordanov *et al.*, *Phys. Rev. Lett.* **110**, 192501 (2013).
20. D. T. Yordanov *et al.*, *Phys. Rev. Lett.* **116**, 032501 (2016).

21. M. Lindroos and T. Nilsson (ed.), *"HIE-ISOLDE: the technical options"*, CERN-2006-013; K. Riisager, P. Butler, M. Huyse and R. Kruecken (ed.), "HIE-ISOLDE: the Scientific Opportunities", CERN-2007-008.
22. M. J. G. Borge and K. Riisager, *Eur. Phys. J. A* **52**, 334 (2016).

The SPES Radioactive Ion Beam Project: Status and Perspectives

Giacomo de Angelis

Laboratori Nazionali di Legnaro, Istituto Nazionale di Fisica Nucleare
Viale dell'Universit 2, I-35020, Legnaro, Italy
E-mail: giacomo.deangelis@lnl.infn.it

SPES is a new infrastructure of the Legnaro National Laboratories (Italy) dedicated to nuclear physics and nuclear physics based applications. The facility will provide high intensity and high-quality beams of unstable neutron-rich nuclei with the aim to perform forefront research in nuclear structure, reaction dynamics and interdisciplinary fields like medical, biological and material sciences. SPES represents a second generation ISOL radioactive ion beam facility with technical solutions which define it as an intermediate step toward the future European ISOL facility EURISOL [1]. Due to its specific technical solutions, it is part of the network of ISOL-facilities presently under construction in Europe constituting the so-called EURISOL distributed facility (EURISOL-DF). To produce radioactive beams it uses the ISOL method with a uranium carbide (UCx) Direct Target able to sustain a maximum power of 8 kW. The primary proton beam impinging on the target is delivered by a Cyclotron accelerator with an energy up to 70 MeV and a beam current up to 750 μA. Neutron-rich radioactive ions are produced by fission at an expected maximum rate in the target of the order of 10^{13} fissions per second. The exotic isotopes, after extraction and separation, are re-accelerated by the ALPI superconducting linac of the Legnaro National Laboratories at energies of 10 MeV/A and higher, for masses up to A=200 amu, with an expected rate on the secondary target of 10^7-10^9 pps for the most prolific species.

Keywords: Radiactive Beams, Nuclear Structure, Nuclear Astrophysics.

1. Challenges of the SPES project

To achieve a full understanding of the properties of a nucleus it is essential to determine the interaction between its components as well as to establish the arrangement of the nucleons, i.e. the structure. Only recently unexplored regions of the nuclear chart, especially on the neutron-rich side, have become accessible due the availability of beams of unstable ions. Starting from a nucleus located on the stability line and adding successively neutrons, the binding energy of the last neutron decreases steadily until it vanishes and the system decays by neutron emission. The position in the nuclear chart where the neutron emission starts defines the neutron drip line. Due to the absence of electrical repulsion between neutrons, the neutron drip line lies much farther away from the valley of stability than the corresponding drip line associated with protons. Nuclei with large neutron excess are of interest for studying the location of the neutron drip line but also for the investigation of the density dependence of the effective interaction between the nucleons. Variations of the nuclear density and size in nuclei with increasing N/Z ratios are expected to lead to different nuclear symmetries and new excitation modes. While a halo structure has been identified in the case of some very light nuclei, for heavier nuclei one expects the formation of a skin of neutrons.

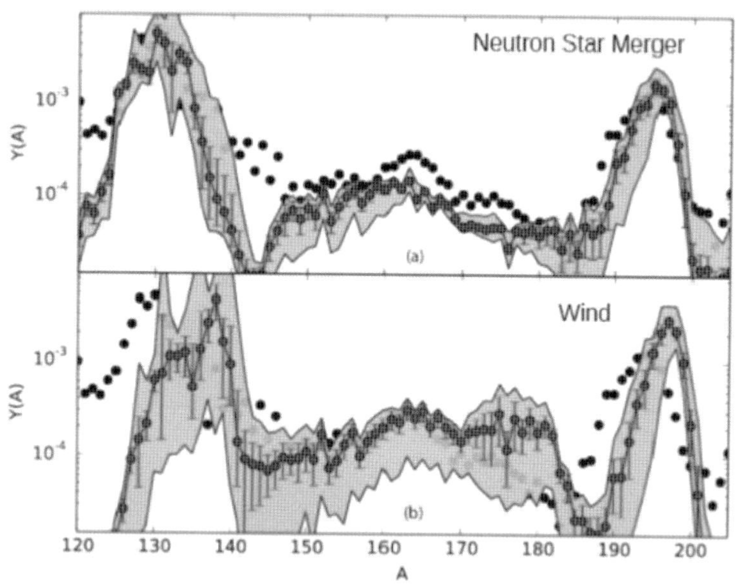

Fig. 1. Calculated elemental abundancies for explosive nucleo-synthesis based on different astrophysical scenarios (Neutron star merger or neutrino driven wind).

The modification of nuclear properties when moving towards the neutron drip line depends on the shell structure changes for increasing neutron excess. This evolution has consequences on the ground state properties of the nuclei and on the nature of single-particle and collective excitations. As an example, studies of neutron-rich nuclei beyond the doubly magic ^{132}Sn are of key importance to investigate the single-particle structure above the N=82 shell closure, the stability of the shell closure and to find out how the effective interaction between valence nucleons behaves far from stability. In connection with the formation of a neutron skin new modes of collective motion are also expected, namely oscillation of the skin against the core, similar to the soft dipole mode already identified in the case of very light halo nuclei. Neither the thickness nor the detailed properties of the neutron skin of exotic nuclei are presently known. This information is essential to allow a quantitative description of compact systems like neutron stars, where exotic nuclei forming a Coulomb lattice are immersed in a sea of free neutrons, a system displaying the properties of both finite and infinite (nuclear matter) objects. At the energies of the secondary beams provided by the SPES facility, it will be possible to address important questions related to the study of neutron-rich matter such as nuclear forces, level density, viscosity, barrier, neutron pairing and collective modes. Challenging and broader range studies in nuclear spectroscopy and reaction mechanism will be performed using the high intensity secondary beams delivered by SPES. Interesting areas where new data will be collected are those in the very neutron rich regions, where shell evolution is a major issue. Investigations of the modification of the pairing interaction in the nuclear medium will receive significant inputs by measurements of multi-nucleon transfer reactions to specific nuclear states. Beams of neutron rich nuclei will be used to investigate the tunneling process in sub-barrier fusion reactions in presence of very positive Q-values, an issue interesting also in the field of nuclear astrophysics. Another important issue is the study of the basic nuclear properties, like lifetimes, masses and neutron capture probabilities, for systems involved in the element formation in the universe. Those quantities can allow to probe the predictive power of the nuclear models used to predict the element abundancies in different astrophysical scenarios, Fig. 1 (for a detailed discussion see[2]).

2. The SPES radioactive ion beam facility

SPES is the new radioactive ion beam infrastructure of INFN. Its realization strongly involves two national laboratories, the Legnaro National Laboratories (LNL) and the Laboratori Nazionali del Sud (LNS), and other INFN sites located all over Italy. It is presently under construction at LNL and makes use of the Isol (Isotope separation on line) technique for the production of radioactive beams. A primary proton beam provided by a driver accelerator, a cyclotron, induces nuclear reactions inside a thick, mainly uranium carbide, target. The reaction products are extracted from the target by thermal process due to the high temperature of the target-ion source system (exceeding 2000^0 C). Once reached the 1^+ source, the products are ionized and extracted. After an isotopic selection obtained using an high resolution spectrometer and a further ionization to an higher n^+ charge state, the exotic beam is injected into a re-accelerator. The SPES facility is based on a proton beam to induce fission on a UCx target (Direct Target). The driver is a 70 MeV proton cyclotron with two exit ports providing a total current of 750 μA. Two production Isol targets are planned to be installed and operated alternatively even if only one is presently under construction for budgetary reason. As the cyclotron can supply two beams at the same time, a second independent facility can be operated in parallel. In particular the high intensity proton beam will be used to produce radioisotopes for nuclear medicine as well as a neutron source. The proton beam of the SPES cyclotron is suited for the production of new radioisotopes for medical applications. Taking advantage of the high energy and current, it opens up the possibility to produce ^{82}Sr, ^{67}Cu, and ^{47}Sc isotopes to be used in therapy and diagnostics.

The Isol front-end: it is the system coupling the proton beam with the UCx target, the ion source and the first part of the transport line of the exotic beam. The whole system has been completed and is now in operation with stable beams at LNL for off-line tests. The most critical element of the SPES Isol facility is the Direct Target. It has been constructed following the basic design of the CERN ISOLDE target but represents an innovation in term of capability to sustain an higher primary beam power. The design is carefully oriented to optimize the radiative cooling, taking advantage of the target system's high operating temperature, which is of the order of 2000^0C. The SPES target is designed in order to maximize the release efficiency and the heat dissipation. It consists of multiple thin disks housed in a cylindrical graphite box. The presently installed one is composed of 7 UCx disks (diameter and thickness of 40 and 1.3 mm, respectively), appropriately spaced in the axial direction in order to dissipate, by thermal radiation, the average power of 8 kW originated by the energy release of the proton beam which, passing through them, induces nuclear reactions. The graphite box is housed inside a tubular hollow tungsten ohmic heater. In fact, due to the intense heat exchange by radiation, the energy released by the proton beam is not sufficient to make the system reaching the operating temperature. This allows for a better thermal control of the target operation. At the basis of the target-ion-source design there is an extensive simulation study of the target behavior for thermal and release properties. Experimental work to benchmark the simulations was carried out in collaboration with HRIBF[3], the Oak Ridge National Laboratory Isol facility (USA) and with the i-Themba Nuclear Physics Laboratory in Stellenbosch (South - Africa). Partial results on the measured production rates (and on the expected rates of the reaccelerated SPES beams) are shown in Fig. 2.

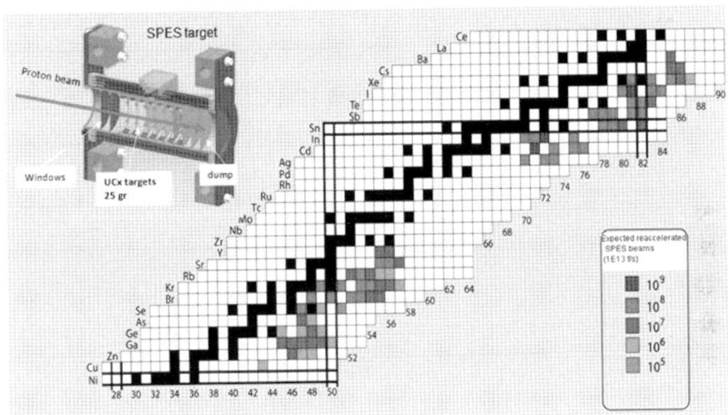

Fig. 2. Expected beam on target intensities. Experimental data scaled to: 0.2 mA proton current, 2-5 % transport efficiency. ISOL target operated at 2000^0 C, ionization and extraction with 1^+ plasma source.

The ion-source system: neutron-rich fission products are produced through the interaction of the proton beam with the UCx target. Such neutron rich nuclei are subsequently extracted by thermal motion and ionized at 1^+ charge state by the Isol source, directly connected to the production target. The hot-cavity ion source chosen for the SPES facility was designed at CERN (ISOLDE)[4] and adapted to the LNL environment. The source has the basic structure of the standard high temperature radioactive ion sources employed for on-line operation. The ionizer cavity is a tungsten tube (34 mm length, 3 mm inner diameter and 1 mm wall thickness) resistively heated to near 2000 C. Due to the high temperature the isotopes produced in the target diffuse in the target material and then effuse through the transfer tube into the ionizer cavity where they undergo surface or laser ionization. The surface ionization process occurs when an atom comes into contact with a hot metal surface. In the positive surface ionization, the transfer of a valence electron from the atom to the metal surface is energetically favorable for elements with an ionization potential lower than the work function of the metal. For alkalis and some rare earth elements, high ionization efficiencies can be achieved using the surface ionization technique. This kind of 1^+ source has good efficiency and selectivity for the elements like Rb, Cs, Ba. For most of the other elements it is preferable to apply laser resonant photo-ionization. It uses the same hot cavity cell and it is a powerful method to achieve sufficiently selected ionization. This technique, now under development in the laboratory, has been implemented with the aim to produce beams as pure as possible (chemical selectivity) also for metal isotopes. Very recently laser ionization of Sn ions has been achieved at LNL. Laser ionization schemes for other elements are presently under development in the newly created laser laboratory at LNL.

Beam selection and transport: the ions after the extraction from the source undergo a first mass selection performed by a Wien Filter having a mass resolution of 1/100. The Wien filter is installed just after the first electrostatic quadrupole triplet inside the production bunker with the aim to confine the larger part of radioactivity inside this high shielded area. The transfer line toward the ALPI superconductive linac is equipped with several beam handling systems to purify the secondary ion beams. Purification is achieved by the combination of a Beam Cooler and an High Resolution Mass Separator (HRMS) designed to reach 1/20000 mass resolution and installed inside the SPES building. Special efforts have been dedicated to reach such high mass resolution taking advantage of the emittance reduction provided by the Beam Cooler and of a 260 keV high voltage platform on which the mass separator is mounted. The mass separator is a scaled-up version of the separator designed for the CARIBU facility[5] in ANL (USA). In the SPES configuration the calculated resolution is of the order of 1/20000 constrained by an emittance of 3 mm mrad and an energy spread of 1.3 eV. Such high selectivity results in a strong advantage also for the safety issue, reducing the contaminations along the beam transport area, the re-accelerator and the final target location. The injection of the secondary beams in the ALPI superconducting linac requires an increase of the ion charge state from 1^+ to n^+. This is performed by means of a Charge Breeder. The SPES Charge Breeder is based on ECR method[6] and aims to produce ions with A/q up to 7 for masses of the order of A 130. The design and construction of the Charge-Breeder has been performed in the framework of the SPES-SPIRAL2 collaboration at LPCS (Grenoble, France). The new design represents an up-graded version of the Phoenix booster in operation at LPSC. After the Charge Breeder a second mass separator with 1/1000 mass resolution is installed with the aim to clean the secondary beam from the contaminants introduced by the Charge Breeder itself (mainly contamination from stable gasses). The same basic configuration used for the HRMS is adopted. Finally an RFQ pre-accelerator is used to increase the secondary beam energy to match the ALPI acceptance.

Exotic beam re-acceleration: the secondary beams after mass selection and n^+ ionization are injected into the ALPI superconducting linac by a new RFQ that is operating in a CW mode (100% duty factor) at a resonant frequency of 80 MHz. This frequency is the same as that of the superconducting cavities used for the low-energy section of the ALPI linear accelerator. The injection energy of the ions is set to 5.7 keV/u. This choice is a compromise between the desire to reduce the ion energy to simplify the low energy injection and the RFQ bunching section design and the need to increase the beam rigidity in the 1^+ transport line to reduce space charge effects. The extraction energy of the RFQ is set to 727 keV/u to optimize the beam dynamics of the superconducting linac ALPI. In the last year the ALPI linac has gone through a continuous upgrade in the number and performance of its accelerating cavities, and consequently in the maximum achievable transport efficiency and beam energy. In the context of such upgrade all the electromagnetic elements have been realigned and transport efficiency has been maximized. The cavities of the medium energy QWR section have been upgraded using a new Nb sputtered coating in substitution to the original Pb sputtered layer. A further upgrade program has been launched to improve the accelerating fields of the present QWRs and adding more cavities in the Low Beta section. The final equivalent voltage will exceed 70 MV going well beyond the Coulomb barrier for nuclear reactions using medium and heavy mass projectiles and heavy targets.

Status of the SPES project: the SPES facility is presently in an advanced stage of construction at the Legnaro National Laboratories. The production building hosting the cyclotron and up to two Isol sources has been completed. The cyclotron, provided by the BEST company in Canada, after the low energy commissioning in the factory, has been delivered to the Legnaro National Laboratories and commissioned - Fig. 3. The authorizations to operate the cyclotron have been already obtained allowing the acceleration of the proton beam on non-fissile target at high power and testing the UCx target up to a current of 5 μA. Very recently the proton beam has been accelerated inside the cyclotron up to the maximum energy and intensity. The physics design of the radioactive ion beam transfer line (1^+ and n^+) is completed and the reshaping of the ALPI beam line in the experimental hall has been performed to allow the installation of the

Fig. 3. The SPES high intensity cyclotron. Inserted in his final location from the roof of the production building (top) and during the installation (bottom).

Charge Breeder and of the low and high resolution mass separators. The Isol front-end is completed and under operation in the off-line laboratory to test ion sources, assembly and controls[7].

References

1. www.eurisol.org
2. R. Surman *et al.*, arXiv:13090059v1 (2013).
3. D. Scarpa, L. Biasetto, S. Corradetti, M. Manzolaro, A. Andrighetto, S. Carturan, G. Prete, P. Zanonato and D.W. Stracener, *Eur. Phys. J.* A (2011) 47.
4. M. Lindroos, *CERN-AB-2004-086* (2004).
5. T. Lamy, J.L. Bouly, J.C. Curdy, P. Sole, P. Sortais, T. Thuillier, J.L. Vieux-Rochaz and D. Voulot, *Rev. Sci. Instrum* 75, 1624 (2004).
6. G. Savard, R. Pardo, S. Baker, C. Davids, A. Levand, D. Peterson, D. Phyllips, T. Sun, R. Vondrasek, B. Zabransky and G. Zinkann *Hyperfine Interactions*, Vol. 199 Issue 1-3, (2011) 301.
7. *http://web.infn.it/spes/*

Precision Mass Measurements at TRIUMF Using Highly Charged Ions

J. Dilling
TRIUMF, 4004 Wesbrook Mall, Vancouver, BC, Canada, and
University of British Columbia, Vancouver, BC, Canada

for the TITAN Collaboration

The atomic mass of nuclei is a fundamental property and an essential ingredient in understanding binding and the strong interaction. Recent progress in theory as well as experimental techniques and access to rare isotopes are at the very center in getting closer to answering these questions.

One of the premier facilities for rare isotopes is the ISAC complex at TRIUMF, Vancouver, Canada. The isotopes are produced, often only in minuscule quantities, and with half-lives as short as few milliseconds, hence the name rare. To overcome the research obstacles of rare isotopes and extract information about the atoms and their fundamental interactions dedicated instruments are required. We have developed very sensitive and fast methods using ion trap techniques at TITAN (TRIUMF's Ion Trap of Atomic and Nuclear science). We were able to do such experiments using a unique combination of traps, including a Paul trap, an electron beam ion trap (EBIT, to generate highly charged ions), and a set of Penning traps. What makes the TITAN system unique and special, is its ability to charge breed the isotopes to higher charge states prior to the mass determination with the EBIT. As the mass resolution essentially scales with the charge state, we are able to improve the mass uncertainty, or in many cases have unique access to isomeric states, in particular for n-rich systems. In this talk I will report on recent measurements and how they are carried out, and what the opportunities and limitations are.

Plans of Spectrometer Experiments at Super-FRS of FAIR*

Juha Äystö for the Super-FRS Collaboration
Helsinki Institute of Physics, FI-00014 University of Helsinki, Finland
Department of Physics, FI-40014 University of Jyväskylä, Finland

Super-FRS will be a powerful superconducting in-flight separator and also a versatile high-resolution spectrometer system for exotic nuclei over a large energy range up to 1500 MeV/u $^{238}U^{92+}$ of ions equivalent to a maximum magnetic rigidity 20 Tm [1]. In this contribution we present the planned physics program of the Super-FRS Collaboration, a sub-collaboration within the NUSTAR, which is partially based on the previous experimental results obtained with the present FRS but will also extend the research to new fields [2].

Search for new isotopes and precise measurements of the production cross sections belong to topics carried out by the Super-FRS collaboration. These together with basic atomic collision measurements will be beneficial for all other planned NUSTAR experiments at the Super-FRS. Unique high-resolution spectrometer experiments will also include the studies of hyper-nuclei, the delta-resonances in exotic nuclei and the spectroscopy of atoms characterized by bound mesons or antiprotons. These latter three experimental fields are based on the high energy of the ions which exceeds

*In collaboration with K.-H. Behr, J. Benlliure, A. Bracco, P. Egelhof, A. Fomichev, S. Galès, H. Geissel, T. Grahn, L.V. Grigorenko, M.N. Harakeh, R. Hayano, S. Heinz, K. Itahashi, A. Jokinen, N. Kalantar-Nayestanaki, R. Kanungo, H. Lenske, I. Mukha, G. Münzenberg, C. Nociforo, H.J. Ong, S. Pietri, M. Pfützner, W. Plaß, A. Prochazka, S. Purushothaman, T. Saito, C. Scheidenberger, H. Simon, I. Tanihata, S. Terashima, H. Toki, L. Trache, H. Weick, J.S. Winfield, M. Winkler, V. Zamfir and the Super-FRS Collaboration.

the range of all other planned or existing RIB facilities. Besides the science case also the detector developments will be briefly discussed.

References

1. Scientific Program of the Super-FRS Collaboration, GSI Report 2014-4.
2. J. Äystö *et al.*, *Nucl. Instr. Meth B* 376, 111 (2016).

Nuclear Structure II

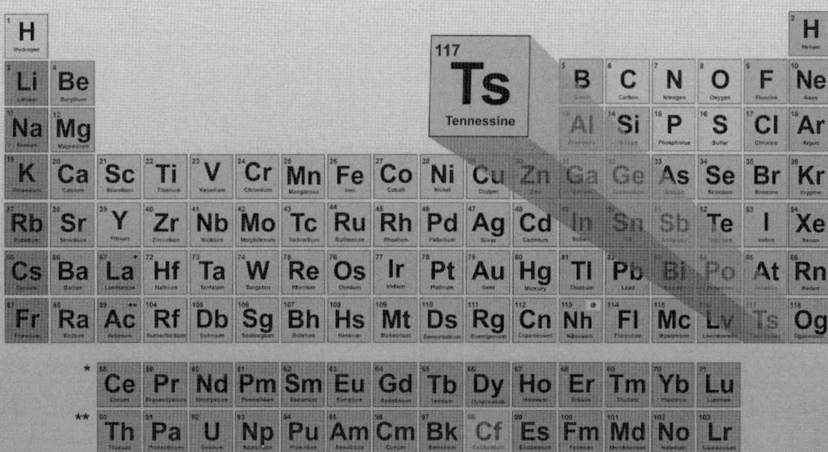

Measurements of the Dipole Polarizability of Neutron-Rich Unstable Nuclei*

Thomas Aumann[1,2], Philipp Schrock[1], Andrea Horvat[1], Fabia Schindler[1], Dominic Rossi[1] and Konstanze Boretzky[2], for the S412 Collaboration

[1]*Technische Universität Darmstadt, 64289 Darmstadt, Germany*
[2]*GSI Helmholtzzentrum für Schwerionenforschung, 64291 Darmstadt, Germany*

Large efforts are presently being undertaken aiming at providing experimental constraints to the equation of state of neutron-rich matter. Several nuclear observables have been identified which can provide potentially stringent constraints on the symmetry energy and its density dependence around saturation density. Among them are, for instance, the neutron-skin thickness in heavy neutron-rich nuclei [1] and the dipole polarizability of neutron-rich nuclei [2, 3]. The dipole polarizability α_D is experimentally even for stable nuclei not well measured and an accurate value for ^{208}Pb was only determined recently at RCNP [4]. A first measurement of α_D for a neutron-rich unstable nucleus was recently performed at GSI using Coulomb excitation at relativistic energies in conjunction with invariant-mass spectroscopy [5]. The same method we have applied to determine α_D for neutron-rich unstable Sn isotopes up to ^{132}Sn. A precise measurement along a long isotopic chain including as neutron-rich as possible isotopes should provide better constraints on the symmetry energy. I will discuss the experimental method and first preliminary results for ^{132}Sn in comparison with previously obtained values for other nuclei and the related theoretical analysis [6].

*This work is supported in parts by the DFG through the CRC1245, by the BMBF Verbundforschung under contract 05P15RDFN1, and the TU Darmstadt-GSI Cooperation Agreement.

References

1. S. Typel and B.A. Brown, *Phys. Rev. C* **64**, 027302 (2001).
2. P.-G. Reinhard and W. Nazarewicz, *Phys. Rev. C* **81**, 051303 (2010).
3. J. Piekarewicz, *Phys. Rev. C* **83**, 034319 (2011).
4. A. Tamii *et al.*, *Phys. Rev. Lett.* **107**, 062502 (2011).
5. D.M. Rossi, *Phys. Rev. Lett.* **111**, 242503 (2013).
6. X. Rocca Maza *et al.*, *Phys. Rev. C* **92**, 064304 (2015).

Direct Measurement of Octupole Deformation in ^{146}Ba

B. Bucher[1,2,*], S. Zhu[3], C. Y. Wu[1], R. V. F. Janssens[3], D. Cline[4], A. B. Hayes[4], A. D. Ayangeakaa[3], M. Q. Buckner[1], C. M. Campbell[5], M. P. Carpenter[3], J. A. Clark[3], H. L. Crawford[5], H. M. David[3], C. Dickerson[3], J. Harker[3], C. R. Hoffman[3], B. P. Kay[3], F. G. Kondev[3], T. Lauritsen[3], A. O. Macchiavelli[5], R. C. Pardo[3], G. Savard[3], D. Seweryniak[3] and R. Vondrasek[3]

[1]*Lawrence Livermore National Laboratory, Livermore, California 94550, USA*
[2]*Idaho National Laboratory, Idaho Falls, Idaho 83415, USA*
[3]*Argonne National Laboratory, Argonne, Illinois 60439, USA*
[4]*University of Rochester, Rochester, New York 14627, USA*
[5]*Lawrence Berkeley National Laboratory, Berkeley, California 94720, USA*
[*]*E-mail: brian.bucher@inl.gov*

The octupole strength present in the neutron-rich, radiocative nucleus ^{146}Ba has been experimentally determined for the first time using Coulomb excitation. To achieve this, A=146 fission fragments from CARIBU were post-accelerated by the Argonne Tandem Linac Accelerator System (ATLAS) and impinged on a thin ^{208}Pb target. Using the GRETINA γ-ray spectrometer and the CHICO2 heavy-ion counter, the reduced transition probability $B(E3; 3^- \to 0^+)$ was determined as $48(^{+21}_{-29})$ W.u. The new result provides further experimental evidence for the presence of a region of octupole deformation surrounding the neutron-rich barium isotopes.

Keywords: Octupole, Coulomb Excitation, Radioactive Beam, Barium.

1. Introduction

The first experimental evidence indicating that the atomic nuclei of certain isotopes exhibit octupole-deformed shapes was provided several decades ago through the observations of low-lying 1^- states populated in α decay [1]. Since then, there has been a wealth of experimental and theoretical investigations attempting to understand and characterize the importance of octupole correlations in nuclei (see Ref. 2 and references therein). Indeed, there is now ample evidence to suggest that the most octupole-deformed nuclei will generally display a low-lying negative-parity band of states that interleaves with the ground-state 0^+ band with relatively fast $E1$ transitions linking the two. The strongest octupole correlations have been observed in isotopes having certain combinations of protons and neutrons, namely

$Z,N\sim34$, 56, 88, 134,[3] which is where valence nucleons occupy pairs of single-particle orbitals near the Fermi surface that differ in orbital (ℓ) and total (j) angular momenta by 3 units of $\hbar$.[2] However, the magnitude of octupole strength present in what is expected to be the two most octupole-enhanced regions, has only recently been quantified in the neutron-deficient radium isotopes[4,5] and the neutron-rich barium isotopes[6] providing the only direct experimental evidence for octupole-deformed nuclear shapes. Such kind of measurements are difficult since they involve Coulomb excitation of radioactive beams or targets. In particular, the radioactive barium isotopes relevant for these measurements are especially difficult to measure since the lifetimes are on the order of seconds only. Those experiments have only been made possible within the past few years through the development of re-accelerated radioactive beam facilities such as CARIBU[7,8] at Argonne National Lab or REX-ISOLDE[9] at CERN. The following report highlights the newest of these studies where octupole measurements have been extended to the neutron-rich isotope ^{146}Ba with N=90.

2. Experiment

As stated above, in order to investigate the octupole strength in ^{146}Ba, a Coulomb excitation experiment was performed similar to the one done for ^{144}Ba.[6] The experimental setup was identical utilizing the CHICO2 heavy ion counter[10] and the GRETINA γ-ray tracking array[11], while the radioactive ^{146}Ba ($T_{1/2}$=2.2 sec.[12]) beam was produced by the CARIBU ^{252}Cf fission source[7,8] and accelerated through the Argonne Tandem Linac Accelerator System (ATLAS). In addition to the beam of ^{146}Ba ions, CARIBU also produced the isobaric contaminants ^{146}La and ^{146}Ce which could not be separated. The A=146 beam was injected into an electron cyclotron resonance (ECR) ion source and charge-bred to q=28$^+$. It was then accelerated through the ATLAS radio-frequency quadrupole (RFQ) and linac to 659 MeV and focused onto a 1.1 mg/cm^2 ^{208}Pb target (99.86% enriched). The average ^{146}Ba beam intensity was 3×10^3 ions per second over 12 days. There were additional stable contaminants (having the same A/q) introduced into the beam from the ECR source, including ^{94}Mo^{18+}, ^{94}Zr^{18+}, ^{120}Sn^{23+}, ^{193}Ir^{37+}, and ^{198}Hg^{38+}, however all of them were easily separated from the A=146 beam ions based on time-of-flight (TOF) and scattering angle (θ) data recorded in CHICO2 (Fig. 1). This allowed for a clean A=146 γ-ray spectrum in GRETINA recorded from the Coulomb-excited beam ions (Fig. 2).

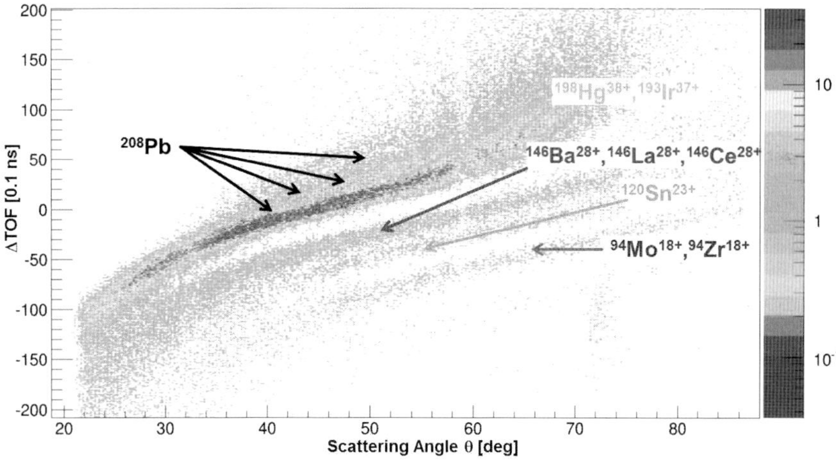

Fig. 1. TOF vs. scattering angle recorded by CHICO2 in coincidence with a γ ray recorded in GRETINA. The A=146 group is easily distinguished from the stable beam contaminants.

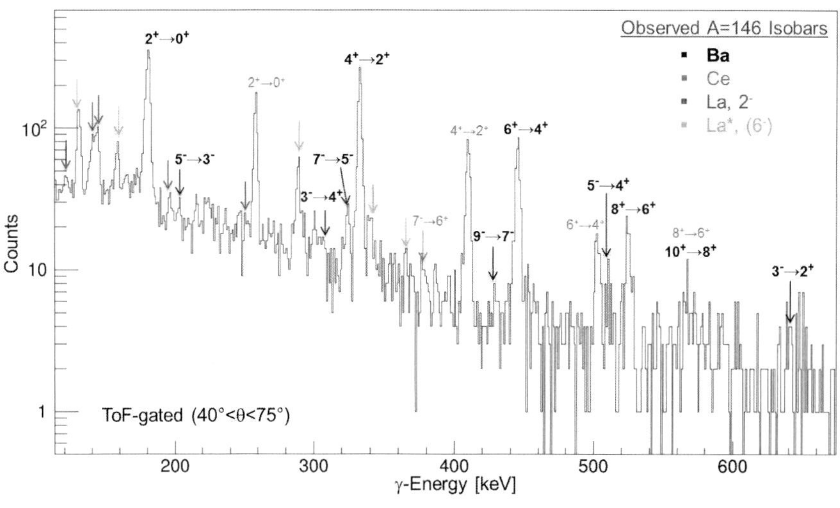

Fig. 2. The coincident γ-ray spectrum by gating on the A=146 group in the CHICO2 TOF spectrum (Fig. 1). Many ^{146}Ba lines are seen along with lines from the radioactive isobaric contaminants that were also produced in CARIBU.

3. Analysis

In the γ-ray spectrum of Fig. 2, several peaks from ^{146}Ba are apparent, especially ones belonging to positive-parity levels in the ground-state band which are excited (and decay) by $E2$ transitions. The negative-parity levels are excited much less frequently, but the excitation occurs predominantly through $E3$ transitions and their decay yields provide a measurement of the corresponding $E3$ matrix elements.

The decay yields were extracted for two separate scattering angle ranges: 30°–40° and 40°–75°. At lower angles, it is difficult to isolate the A=146 ions from the other beam contaminants in the TOF spectrum, while at higher angles there is simply not enough statistics for a useful analysis as the cross section falls off with the well-known Rutherford angular dependence $1/\sin^4(\frac{\theta}{2})$. The yields for each set of angles were analyzed using the Coulomb excitation least-squares search code, GOSIA[13,14]. Decay yields were observed for levels up to $10\hbar$ in the ground-state band and $9\hbar$ in the negative-parity band. The energies of all observed γ rays, along with several branching ratios and lifetimes, were known in advance based on previous experiments[12,15,16]. The latter information was useful for constraining the GOSIA fit. The sets of $E1$, $E2$, and $E3$ matrix elements between levels with no previously known lifetimes, however, were coupled according to the rigid-rotor prescription[17] governed by the individual intrinsic $E\lambda$ moments (see also Refs. 4,6,18). This was done to help search for the χ^2 minimum while excluding minima in unphysical regions of the parameter space which often arise from matrix elements that are relatively insensitive to the observed yields and other available experimental data (so-called nuisance parameters, see Ref. 14). Once the χ^2 minimum was found, the rigid-rotor constraint was removed to properly determine the full uncertainties, including correlations between matrix elements[19].

4. Results and Discussion

The final results from the GOSIA analysis for the $E1$ matrix elements did not display much sensitivity to the experimental data. This is not surprising since the $E1$ strength is known to be relatively weak[15] with the only observed $E1$ decays coming from the 3^- and 5^- levels. Moreover, the relative sign between the intrinsic $E1$ and $E3$ moments was found to be completely insensitive to the data. On the other hand, a number of new $E2$ and $E3$ matrix elements were well-determined from the new Coulomb excitation yields; the full set of which are reported in Ref. 19. The most significant result is the ground-state $E3$ matrix element $\langle 3_1^- \| \mathcal{M}(E3) \| 0_1^+ \rangle$ determined to be $0.65(^{+14}_{-20})$ $eb^{3/2}$ which reflects the amplitude of octupole deformation present in the ground state[2]. This corresponds to a $B(E3; 3^- \to 0^+)$ strength of $48(^{+21}_{-29})$ W.u. which is very close to the value recently reported[6] for ^{144}Ba (Fig. 3).

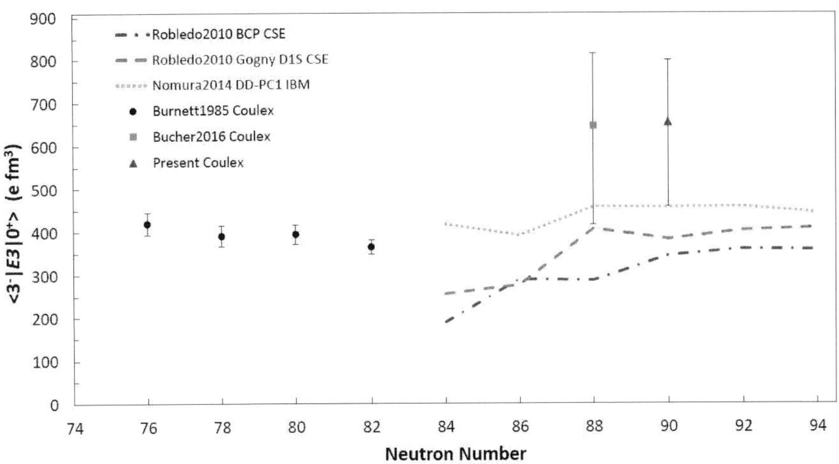

Fig. 3. Values of the ground-state $E3$ matrix elements for neutron-rich barium (Z=56) isotopes are shown based on experimental work by Burnett *et al.*[20] and Bucher *et al.*[6] as well as theoretical work by Robledo *et al.*[21] and Nomura *et al.*[22].

The new measurement lends strong support to the idea that octupole collectivity is indeed enhanced in these barium isotopes near N=90, even exceeding the magnitude predicted in past beyond-mean-field calculations[21,22]. This is also apparent when compared to Coulomb excitation measurements performed on the stable barium isotopes[20], where no octupole enhancement is expected to come from the valence neutrons. The observed enhancement at N=90 is also consistent with the interpretation that the strong octupole correlations are driven by the presence of the high-j, unique-parity $\nu i_{13/2}$ orbital near the Fermi surface (Fig. 4) and its interaction with a nearly-filled $\nu f_{7/2}$ orbital[23]. Indeed, recent symmetry-conserving configuration-mixing (SCCM) calculations have provided quantitative support of this interpretation[19,24].

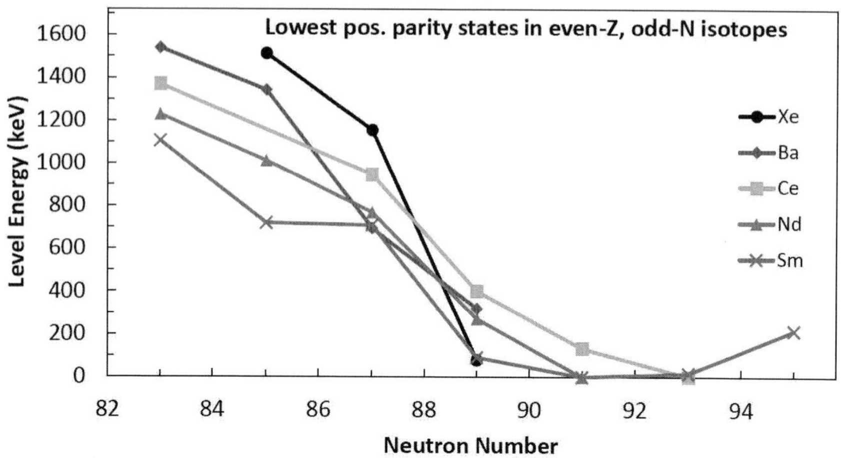

Fig. 4. Energies of the first-excited positive-parity states in nearby even-Z, odd-N isotopes. These low-energy states must originate from the unique-parity spherical $\nu i_{13/2}$ single-particle orbitals. Clearly they approach the Fermi surface as neutrons are added towards N=90. (Data taken from the NNDC nuclear database[25].)

5. Conclusions

With the present measurement, the low-energy octupole strength in ^{146}Ba has been quantified. Such a measurement was only made possible by recent advances in radioactive beam acceleration (CARIBU) and state-of-the-art detector arrays (GRETINA and CHICO2). The results confirm past theoretical predictions of enhanced $E3$ transitions in this nucleus and are consistent with an octupole-deformed ground state, similar to observations in neighboring ^{144}Ba. The strong octupole correlations exhibited by these neutron-rich barium isotopes can be linked to the presence of $\nu i_{13/2}$-originating orbitals near the Fermi surface at N=90 in conjuction with a nearly-full set of $f_{7/2}$ valence neutrons populated there. As a result, this region of isotopes surrounding ^{146}Ba, other than the neutron-deficient actinides around ^{226}Ra, may represent the most suitable combinations of protons and neutrons for nuclei to form stable octupole-deformed shapes.

Acknowledgements

This material is based upon work supported by the U.S. Department of Energy, Office of Science, Office of Nuclear Physics under Contract No. DE-AC02-06CH11357 (ANL). Work at LLNL and INL is supported by the

U.S. DOE under respective Contracts No. DE-AC52-07NA27344 and No. DE-AC07-05ID14517. GRETINA was funded by the U.S. DOE-Office of Science, Office of Nuclear Physics by the ANL contract No. above and by Contract No. DE-AC02-05CH11231 (LBNL). This research used resources of ANL's ATLAS facility, which is a DOE Office of Science User Facility.

References

1. F. S. Stephens, F. Asaro and I. Perlman, *Phys. Rev.* **100**, 1543 (1955).
2. P. A. Butler and W. Nazarewicz, *Rev. Mod. Phys.* **68**, p. 349 (1996).
3. R. H. Spear and W. N. Catford, *Phys. Rev. C* **41**, p. R1351 (1990).
4. H. J. Wollersheim *et al.*, *Nucl. Phys. A* **556**, 261 (1993).
5. L. P. Gaffney *et al.*, *Nature (London)* **497**, 199 (2013).
6. B. Bucher *et al.*, *Phys. Rev. Lett.* **116**, p. 112503 (2016).
7. G. Savard, S. Baker, C. Davids, A. F. Levand, E. F. Moore, R. C. Pardo, R. Vondrasek, B. J. Zabransky and G. Zinkann, *Nucl. Instrum. Methods B* **266**, 4086 (2008).
8. G. Savard, A. Levand, R. Pardo, R. Vondrasek and B. Zabransky, *JPS Conf. Proc.* **6**, p. 010008 (2015).
9. D. Voulot, F. Wenander, E. Piselli, R. Scrivens, M. Lindroos, H. B. Jeppesen, L. M. Fraile, S. Sturm and P. Delahaye, *Nucl. Instrum. Methods B* **266**, 4103 (2008).
10. C. Y. Wu, D. Cline, A. Hayes, R. S. Flight, A. M. Melchionna, C. Zhou, I. Y. Lee, D. Swan, R. Fox and J. T. Anderson, *Nucl. Instrum. Methods A* **814**, 6 (2016).
11. S. Paschalis *et al.*, *Nucl. Instrum. Methods A* **709**, 44 (2013).
12. L. K. Peker and J. K. Tuli, *Nucl. Data Sheets* **82**, p. 187 (1997).
13. T. Czosnyka, D. Cline and C. Y. Wu, *Bull. Am. Phys. Soc.* **28**, p. 745 (1983).
14. GOSIA user manual: `http://www.pas.rochester.edu/~cline/Gosia/Gosia_Manual_20120510.pdf`, accessed March 2014.
15. W. Urban *et al.*, *Nucl. Phys. A* **613**, 107 (1997).
16. H. Mach, W. Nazarewicz, D. Kusnezov, M. Moszyński, B. Fogelberg, M. Hellstrom, L. Spanier, R. L. Gill, R. F. Casten and A. Wolf, *Phys. Rev. C* **41**, p. R2469 (1990).
17. A. Bohr and B. R. Mottelson, *Nuclear Structure* (W. A. Benjamin, Inc., Reading, MA, 1975).
18. W. R. Phillips, I. Ahmad, H. Emling, R. Holzmann, R. V. F. Janssens, T.-L. Khoo and M. W. Drigert, *Phys. Rev. Lett.* **57**, 3257 (1986).

19. B. Bucher *et al.*, Direct Evidence for Octupole Deformation in ^{146}Ba and the Origin of Large $E1$ Moment Variations in Reflection-Asymmetric Nuclei, submitted to *Phys. Rev. Lett.*, (2017).
20. S. M. Burnett, A. M. Baxter, S. Hinds, F. Pribac, R. H. Spear and W. J. Vermeer, *Nucl. Phys. A* **432**, 514 (1985).
21. L. M. Robledo, M. Baldo, P. Schuck and X. Viñas, *Phys. Rev. C* **81**, p. 034315 (2010).
22. K. Nomura, D. Vretenar, T. Nikšić and B.-N. Lu, *Phys. Rev. C* **89**, p. 024312 (2014).
23. I. Ahmad and P. A. Butler, *Annu. Rev. Nucl. Part. Sci.* **43**, 71 (1993).
24. T. R. Rodríguez and L. M. Robledo, private communication.
25. Brookhaven National Laboratory, NNDC Nuclear Database `http://www.nndc.bnl.gov/`, accessed May 2016.

Octupole Deformation in ^{144}Ba and ^{148}Ce

N.T. Brewer[1], E.H. Wang[1], W.A. Yzaguirre[1], J.H. Hamilton[1], A.V. Ramayya[1], S.H. Liu[1], J.K. Hwang[1], Y.X. Luo[1,2], C.J. Zachary[1], J.O. Rasmussen[2], S.J. Zhu[1,3], C. Goodin[1], G.M. Ter-Akopian[4,5], A.V. Daniel[4,5] and Yu. Ts. Oganessian[4]

[1]*Physics Department, Vanderbilt University, Nashville, TN 37235*
[2]*Lawrence Berkeley National Laboratory, Berkeley, CA 94720*
[3]*Physics Department, Tsinghua University, Beijing 100084, People's Republic of China*
[4]*Joint Institute for Nuclear Research, Dubna 141980, Russia*
[5]*Joint Institute for Heavy Ion Research, Oak Ridge, TN, 37830*

Gamma rays emitted by fission fragments of ^{252}Cf and measured using the Gammasphere array continue to give insight into neutron rich nuclei. Using our high statistics data, we have reexamined high-spin states and linking transitions associated with octupole correlations in ^{144}Ba and ^{148}Ce. In an even-A system, it is possible that rotational band structures are explainable by simplex quantum numbers, s = +1 and s = -1. In ^{144}Ba and ^{148}Ce we have deduced spin values from angular correlations and assigned parities from mixing ratios for s = -1 characterized states. These are the first examples of both simplex bands in even-even nuclei. Extensions to higher spins of bands in ^{144}Ba and in ^{148}Ce are also reported. In addition to collective model analysis of energy displacement and rotational frequency, the intrinsic dipole moment via B(E2)/B(E1) ratios are also analyzed.

Keywords: Octupole Deformation, Angular Correlation, Barium, Cerium.

1. Introduction

Reflection asymmetry in the intrinsic frame was first predicted for nuclei near ^{145}Ba based on the single particle energy levels from the deformed shell model [1], which showed octupole deformed shell gaps for Z = 56 and N = 88. Reflection asymmetry should be evidenced [2] by two, alternating parity rotational bands with opposite parities distinguishing the two. A detailed review of octupole deformed nuclei and some of their properties and importance is given in ref. [3]. More recent calculations have been done that continue to support the assignment of octupole deformation for these nuclei [4, 5].

In experiment, octupole deformation has been extensively studied in this region [6–13]. However, until now firm assignments of spins and parities had not been given to members of both s = +1 and s = -1 in any even-even isotopes.

Both s = +i and s = -i bands have been clearly seen in nuclei with odd mass number [7, 8, 12]. While the s = +1 band is easily seen in even-A nuclei, the s = -1 band had only been seen in ^{148}Ce [14] with a measure of alternating parity, but not a measure of the spin of each level. We now firmly assign I^{π}, spins and parities, for the levels in the s = -1 bands in ^{148}Ce as well as ^{144}Ba through the use of angular correlation measurements. We find extensions of existing bands in ^{148}Ce built on previously known levels. We then carried out analyses in terms of a geometrical model [2, 15].

2. Experiment and Technique

This experiment was performed for four weeks using Gammasphere with 101 Compton suppressed gamma ray detectors and a 62 μCi ^{252}Cf source. This gives us access to high statistics, $5.7*10^{11}$ triple and higher fold events and $1.9*10^{11}$ quadruple and higher fold events, as well as high spin states. To determine the weakest transitions in these isotopes, they were examined with up to 4 gamma ray coincidences. For details of the experiment consider ref. [16]. In addition, Gammasphere also has excellent angular coverage (≈ 47%) and is well suited to angular correlation measurements [17], with possible mixing ratios δ extracted [18, 19].

3. Results

3.1. ^{144}*Ba*

The level scheme of ^{144}Ba shown in Fig. 1 now goes up to (20^+) and (21^-) in the previously assigned s = +1 band [10]. Now, via angular correlations, very strong evidence has been found to support the s = -1 assignment for the side band found in [20]. Four new levels extend the s = +1 band including four new linking E1 transitions. In addition, many gammas have been found that confirm the previous assignment of levels based on selection rules. Furthermore, we propose a level at 1769.0 keV as the 3^+ level for the s = -1 band. In addition, angular correlation measurements now reveal levels consistent only with spin values of 5 at 1892.0 keV, 6 at 1992.7 keV, and 7 at 2160.4 keV and with mixing ratios only consistent with the alternating parity assignment of a 5^+,6^-,7^+. In addition, the transitions now seen within the s = -1 and connecting the s = -1 band to the s = +1 band are sufficient to assign the spins and parities of the 8^-, 9^+, 10^-, and 11^+ levels as well.

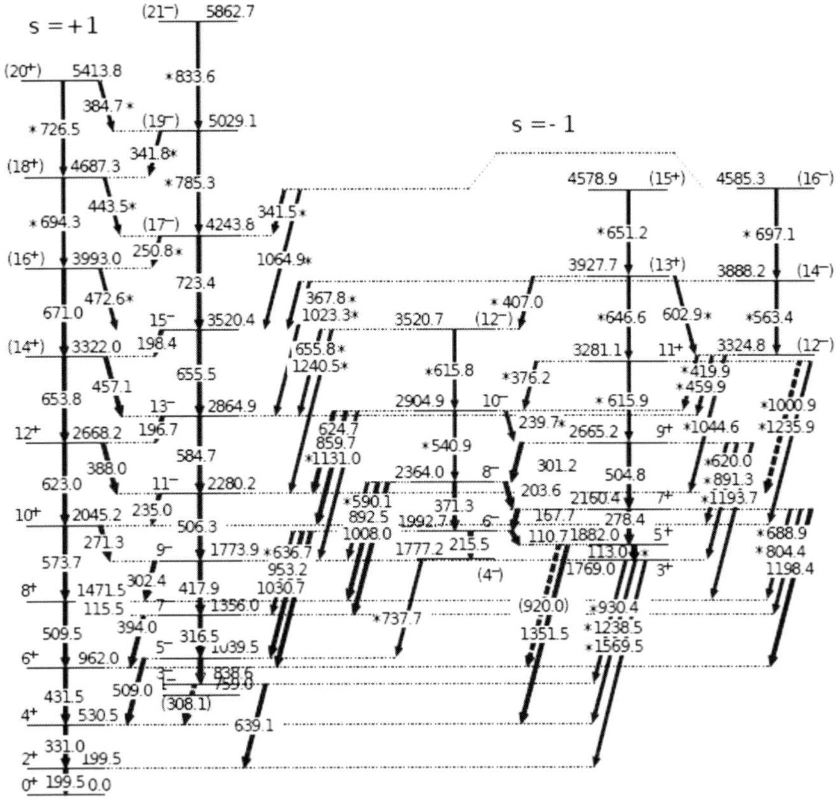

Fig. 1. Level scheme showing new levels and transitions (marked with *) in ^{144}Ba.

The new level scheme of ^{148}Ce over previous work [14] has 7 new levels, 22 new gamma rays and 9 new angular correlation measurements. The new levels and transitions are shown in Fig. 2. We have measured the angular correlations of gamma transitions, in order to assign spins and parities for the states at 1224.8 keV and 1585.1 keV to be 4^+ and 6^+ respectively. Additionally, we have been able to extend the s = +1 to high spin (22^+) and the correlation between the 463.5 and 451.1 keV gamma rays yields δ = 0.01(3) in the 463.5 keV gamma ray depopulating a spin 9 level at 1754.6 keV. The gamma ray is therefore assigned to be of an E1 character and the level it depopulates is assigned to be 9^-. We also confirmed that for the s = -1 band, two angular correlations firmly assign the spin parity values to the 3^+ and 5^+ levels, while two other cascades, the

353.0-167.0 keV and the 167.0-363.5 keV cascades both show that there is no dipole-quadrupole mixing in the 167.0 keV transition. This gamma is therefore a pure E1 and there is a parity shift between the states as was previously reported from conversion coefficient measurement [14]. This, with what has been seen in ^{144}Ba, confirms that even-even isotopes would have two simplex characterized rotational bands [2].

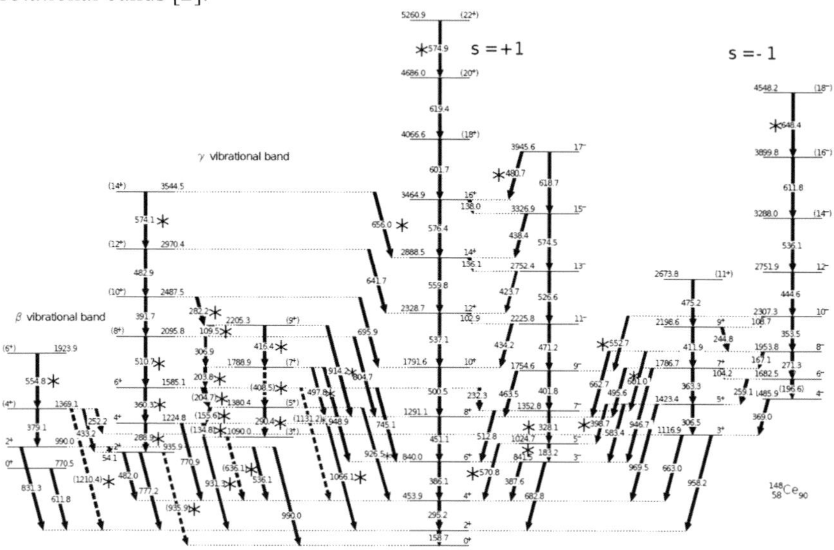

Fig. 2. Level scheme showing new levels and transitions (marked with *) in ^{148}Ce.

4. Discussion of Octupole Deformation in ^{144}Ba and ^{148}Ce

Energy displacement ≈ 0 indicates likely spin states where the nuclear angular momentum stabilizes the deformation of the nucleus. The low spin states, which tend to have higher values, likely have a vibrational component. In all cases we see values, at medium to high spin, near to $\delta E = 0$. The s = +1 band in ^{144}Ba continues to stay very near -100 keV. For the first time we present the values of these variables in this isotope for the s = -1 and we note that these values are all less than 100 keV below $\delta E = 0$. We also note that curve for ^{148}Ce continues to be higher at high spin and does not reach $\delta E = 0$ in the s = +1 band. Yet there is convincing evidence in its s = -1 band as well for near zero values. This supports our argument that these are simplex bands, because in ^{144}Ba and ^{148}Ce the $\pi = -$ and $\pi = +$ members show similar rotation properties.

The rotational frequency ratio of the positive and negative parity states is expected to be 1 for stabilized octupole deformation and the limit we expect to see for a rotational level built off an octupole phonon is, $\omega(I)^-/\omega(I)^+ = 2I-5/2I+1$, called the vibrational limit. The values for ^{144}Ba and ^{148}Ce all trend, at higher spin values, toward one, the stable octupole deformation limit.

Furthermore, the intrinsic dipole moment [15] is a way to analyze deformation and provides further support for octupole deformation. There are very different moments in each band in ^{144}Ba but not ^{148}Ce. This variety may guide the theoretical interpretation of these band structures.

5. Summary

In conclusion, we have now firmly assigned s = +1 and -1 band structures in two even-even isotopes near Z = 56, N = 88. We show remarkable consistency with simplex band structures and the geometrical model for observables δE, $\omega(I)^-/\omega(I)^+$, and D_0, which may be surprising for isotopes that are so neutron rich. We have now given the best evidence to date of the rotational consequences of octupole deformation in even-even isotopes. We can see the trends in these nuclei as rotation stabilizes the vibrational degrees of freedom. We also observe in ^{144}Ba and ^{148}Ce additional degrees of freedom at high spin which could breakdown rotational stabilization. It will be interesting to see if the deformation properties might still be more pronounced in ^{148,150}Ba as predicted by theory [4, 5].

Acknowledgements

The work at Vanderbilt University and Lawrence Berkeley National Laboratory was supported by the U.S. Department of Energy under Grant and Contract Nos. DE-FG05-88ER40407s and DE-AC03-76SF00098. The work at Tsinghua University was supported by the National Natural Science Foundation of China under Grant Nos. 11175095 and 10975082. The Joint Institute for Heavy Ion Research is supported by the University of Tennessee, Vanderbilt University, and the US DOE through Contract No. DE-FG05-87ER40311 with the University of Tennessee. The work at JINR was supported by the Russian Foundation for Basic Research Grant No. 08-02-00089 and by the INTAS Grant No. 03-51-4496.

References

1. G. A. Leander *et al.*, *Phys. Lett. B* **152**, 284 (1985).
2. W. Nazarewicz and P. Olanders, *Nucl. Phys. A* **441**, 420 (1985).
3. P. A. Butler and W. Nazarewicz, *Rev. Mod. Phys.* **68**, 349 (1996).
4. L. M. Robledo *et al.*, *Phys. Rev. C* **81**, 034315 (2010).

5. W. Zhang, Z. P. Li and S. Q. Zhang, *Chinese Phys. C* **34**, 1094 (2010).
6. S. J. Zhu *et al.*, *Phys. Lett. B* **357**, 273 (1995).
7. M. A. Jones *et al.*, *Nucl. Phys. A* **605**, 133 (1996).
8. S. J. Zhu *et al.*, *Phys. Rev. C* **60**, 051304 (1999).
9. R. L. Gill *et al.*, *Phys. Rev. C* **27**, 1732 (1983).
10. W. R. Phillips *et al.*, *Phys. Rev. Lett.* **57**, 3257 (1986).
11. S. J. Zhu *et al.*, *Phys. Rev. C* **85**, 014330 (2012).
12. Y. X. Luo *et al.*, *Phys. Rev. C* **66**, 014305 (2002).
13. M. Sakhaee *et al.*, *Phys. Rev. C* **60**, 067303 (1999).
14. Y. J. Chen *et al.*, *Phys. Rev. C* **73**, 054316 (2006).
15. P. A. Butler and W. Nazarewicz, *Nucl. Phys. A* **533**, 249 (1991).
16. J. H. Hamilton *et al.*, *Prog. Part. Nucl. Phys.* **35**, 635 (1995).
17. A. V. Daniel *et al.*, *Nucl. Instrum. Methods B* **262**, 399 (2007).
18. H. W. Taylor *et al.*, *Nucl. Data Tables A* **9**, 1 (1971).
19. P. Haustein *et al.*, *At. Data Nucl. Data Tables* **10**, 321 (1972).
20. W. Urban *et al.*, *Nucl. Phys. A* **613**, 107 (1997).

Doublet Octupole Bands in Neutron-Rich $^{140,141}Xe$

Y. Huang[1], S. J. Zhu[1], J. H. Hamilton[2], E. H. Wang[2], A. V. Ramayya[2], Y. J. Chen[3], Z. G. Xiao[1], H. J. Li[1], Y. X. Luo[2,4], J. O. Rasmussen[4], G. M. Ter-Akopian[5] and Yu. Ts. Oganessian[5]

[1]*Department of Physics, Tsinghua University, Beijing 100084, People's Republic of China*
[2]*Department of Physics, Vanderbilt University, Nashville, Tennessee 37235, USA*
[3]*Department of Nuclear Physics, China Institute of Atomic Energy, Beijing 102413, China*
[4]*Lawrence Berkeley National Laboratory, Berkeley, California 94720, USA*
[5]*Joint Institute for Nuclear Research, Ru-141980 Dubna, Russia Federation*

High-spin level structures of neutron-rich ^{140}Xe and ^{141}Xe nuclei have been reinvestigated by analyzing the triple and four-fold γ coincidence data obtained from the spontaneous fission of ^{252}Cf. Several new levels and transitions are identified in both nuclei. The doublet octupole bands are observed in ^{140}Xe and expanded in ^{141}Xe. The characteristics of octupole deformation and octupole correlations are discussed in both nuclei.

1. Introduction

Theoretical calculations in the deformed shell model predicted the existence of octupole deformation and octupole correlations in the Z = 56 and N = 88 neutron-rich nuclear region [1-3]. So far, octupole-deformed bands and octupole correlations were observed in many nuclei in this region, for examples, in Ba (Z = 56) [3, 4-9], La (Z = 57) [10] and Ce (Z = 58) [5, 11-18] isotopes. However, most of the observed octupole bands belong to single simplex with $s = +1$ in even-even nuclei, or $s = +i$ or $s = -i$ in odd-A nuclei. The two sets of parity doublet bands were only reported in a few nuclei, for example, in odd-A ^{143}Ba (with $s = \pm i$) [9] and in even-even ^{148}Ce (with $s = \pm 1$) [16]. The neutron-rich Xe isotopes with Z = 54 are located at the Z = 56, N = 88 octupole deformed region. In previous reports [5, 19-22], the high-spin-level structures have been studied in $^{140,141}Xe$. The $s = +1$ octupole band structure was observed in ^{140}Xe [21] and the $s = \pm i$ octupole bands were proposed in ^{141}Xe [22]. Here we reinvestigated the level structures of ^{140,141}Xe. The parity doublets with opposite simplex quantum numbers are established in ^{140}Xe and expanded in ^{141}Xe. The detail results of ^{140}Xe have been published in a recent paper [23].

2. Experiment and Results

The level structures of ^{140,141}Xe have been studied by measuring the prompt γ-rays from the spontaneous fission of ^{252}Cf with Gammasphere detector array. The experiment was carried out at the Lawrence Berkeley National Laboratory. A total of 5.7×10^{11} triple- and higher fold γ-coincidence events were collected. The coincidence data were analyzed with the Radware software package [30].

The new level schemes of ^{140,141}Xe obtained in the present work are shown in Figs. 1 and 2. In both nuclei, five collective band structures with $\Delta I = 2$ linking transitions inside each band are labeled on top of the bands with numbers (1)-(5). Most of the levels and transitions reported in Refs. [5, 19-22] are confirmed in this work. A total of 8 transitions and 4 levels in ^{140}Xe and 12 transitions and 6 levels in ^{141}Xe are newly observed. We assigned another set of doublets consisting of a positive parity band (3) and a negative parity band (4) as an octupole band structure with simplex quantum number s = -1 in ^{140}Xe. Our angular correlation measurements [31] give support for our assignments in both nuclei.

3. Discussion

Now we discuss the characteristics of the octupole deformation and octupole correlations in ^{140,141}Xe. Figure 3 presents the systematic level comparison of the observed octupole bands in ^{140,141}Xe with those in ^{143}Ba [9]. They exhibit very similar patterns. The assigned octupole band structures in ^{140,141}Xe agree with the systematics.

In an octupole band structure, the $B(E1)/B(E2)$ ratios can be calculated from the observed transition intensity ratios $I_\gamma(E1)/I_\gamma(E2)$ by using the equation given in Ref. [16]. The calculated average $B(E1)/B(E2)$ values for the doublet octupole structures in ^{140}Xe and ^{141}Xe is 0.132 and 0.055 (10^{-6}.fm^{-2}) respectively, while the average values are 0.78 and 0.52 (10^{-6}.fm^{-2}) in ^{143}Ba and ^{145}Ba [9], respectively. This shows that the observed octupole correlations in ^{140,141}Xe are rather weaker than that in ^{143,145}Ba.

The energy differences δE between the positive- and negative-parity bands in an octupole band can be used to discuss the stability of octupole deformation and octupole correlations with spin variation. Such δE values can be obtained from the experimental level energies by using the relation [16, 17]:

$$\delta E(I) = E(I)^{-} - [(I+1)E(I-1)^{+} + IE(I+1)^{+}] / (2I+1) \quad (1)$$

Here the superscripts indicate the parities of the levels. The $\delta E(I)$ values should be close to zero in the limit of stable octupole deformation. As seen in Fig. 4, the $\delta E(I)$ decreases with the spin increasing in each octupole band. It is close to the stable point at I ~ $12\hbar$ for s = −1 band in ^{140}Xe, $13\hbar$ for s = −i band in ^{141}Xe, and 9 and $11\hbar$ for s = ±i bands in ^{143}Ba, respectively. However, for s = +1 band in ^{140}Xe and s = +i band in ^{141}Xe, they do not reach the stable points until spin up to $17\hbar$ and $25/2\hbar$, respectively. These results show that the octupole deformation and octupole correlations in ^{140,141}Xe are more unstable than that in ^{143}Ba.

The kinematic moments of inertia J_1 against the rotation frequencies $\hbar\omega$ for the s = ±1 octupole bands in ^{140}Xe are plotted in Fig. 5. One can see the J_1 curves for the ΔI = 2 bands show rather different behaviors. Moreover, with the increasing of the frequency $\hbar\omega$, backbending occurs for both the positive and the negative parity bands of the s = +1 octupole band structure at $\hbar\omega \approx$ 0.25-0.3 MeV in ^{140}Xe.

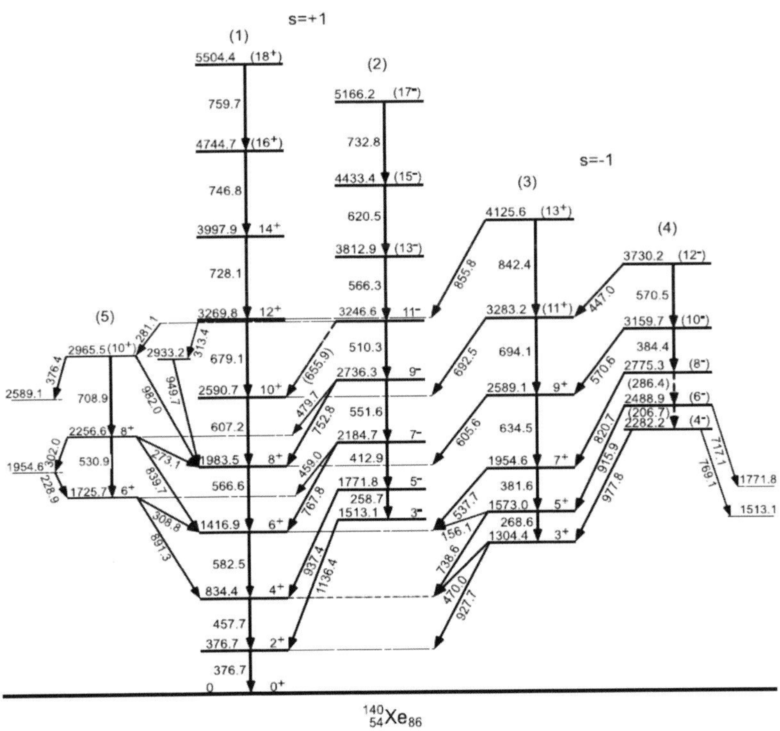

Fig. 1. The level scheme of ^{140}Xe.

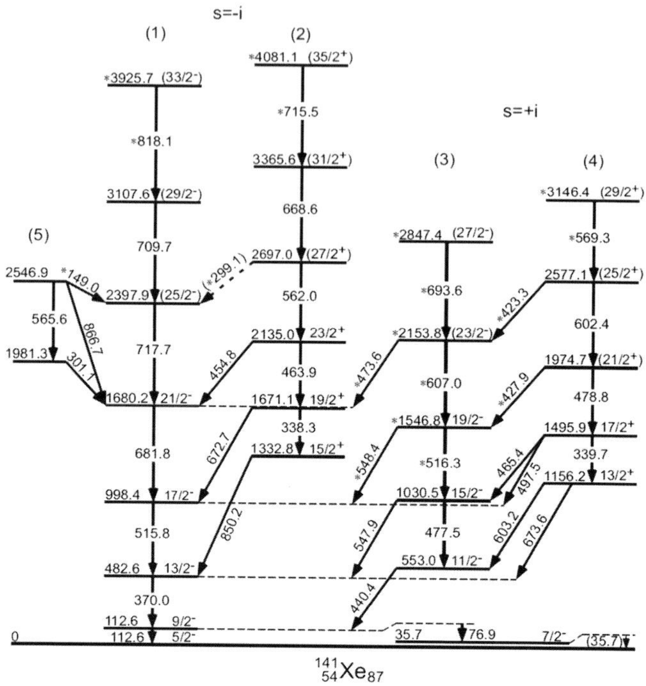

Fig. 2. The level scheme of ^{141}Xe.

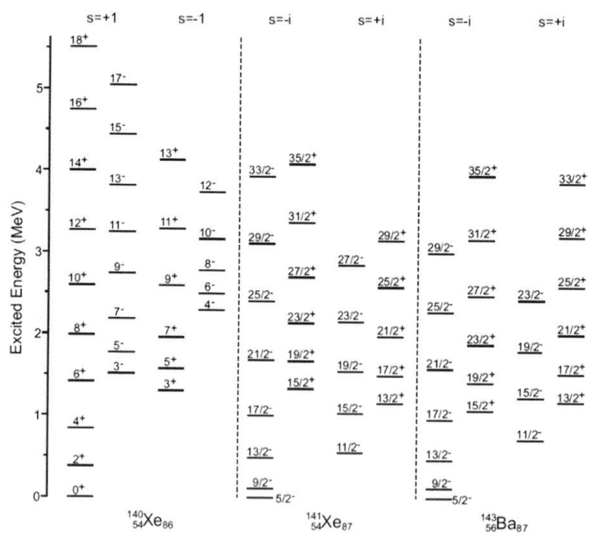

Fig. 3. The systematic comparison for the levels of the octupole bands in ^{140}Xe, ^{141}Xe and ^{143}Ba [9].

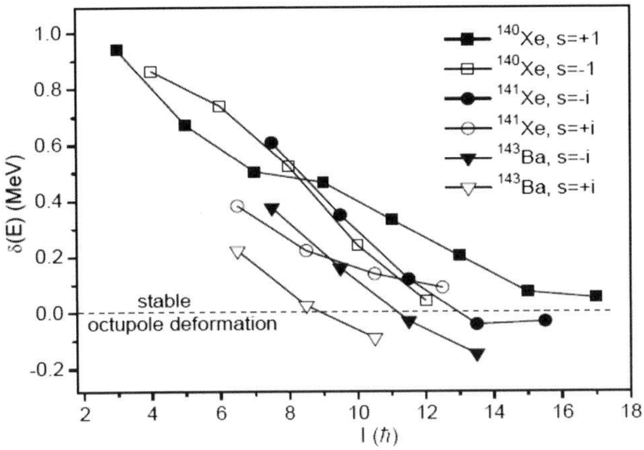

Fig. 4. The comparison for $\delta E(I)$ versus spin I for s = ±1 octupole bands in ^{140}Xe, s = ±i octupole bands in ^{141}Xe and ^{143}Ba.

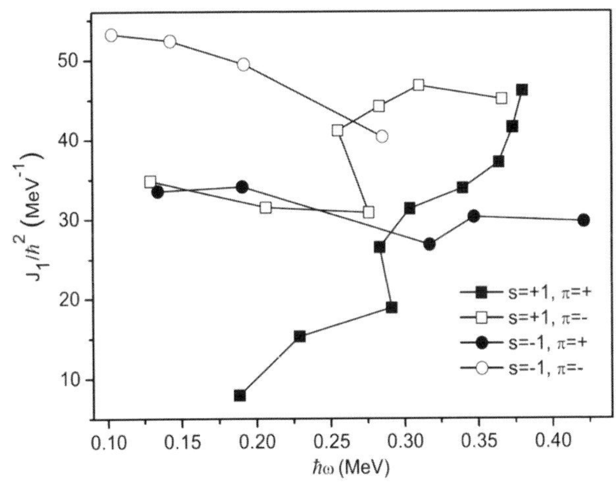

Fig. 5. Moments of inertia J_1 as a function of the rotation frequency $\hbar\omega$ for s = ±1 octupole bands in ^{140}Xe.

To understand the characteristics of the octupole deformation and octupole correlations in ^{141}Xe, we have carried out RASM calculations. The details about RASM can be found in Ref. [24, 25]. The calculated results and comparison with the experimental data are shown in Fig. 6. The calculations indicate that both the s = ±i octupole bands in ^{141}Xe originate from the $\nu h_{9/2}$ [530]1/2 (K = 1/2) mixed with $\nu f_{7/2}$ [532]3/2 (K = 3/2) configuration.

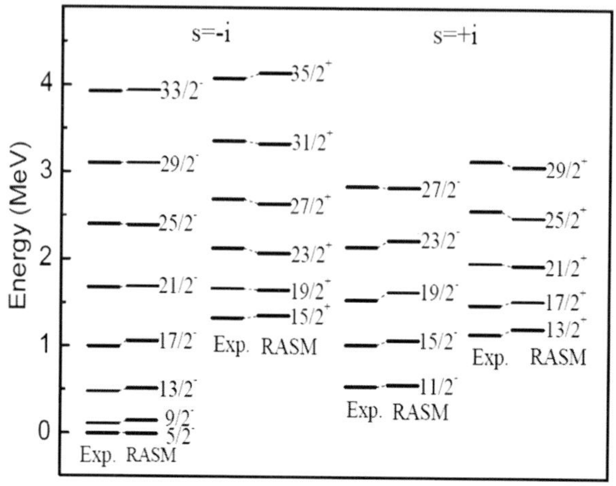

Fig. 6. Comparison of the level energies by the RASN calculations for s = ±i octupole bands in ^{141}Xe with experimental data.

4. Summary

The high-spin band structures of neutron-rich ^{140,141}Xe have been reinvestigated. A new s = −1 octupole band structure has been proposed and so the s = ±1 doublet octupole bands are completed in ^{140}Xe. The s = ±i octupole band structures in ^{141}Xe have been significantly expanded.

Acknowledgments

The work at Tsinghua University was supported by the National Natural Science Foundation of China under Grants No. 11175095, No. 11205112, and No. 11375094, and by Tsinghua University Initiative Scientific Research Program. The work at Vanderbilt University, Lawrence Berkeley National Laboratory was supported by the U.S. Department of Energy under Grant No. DE-FG05-88ER40407 and Contract No. DE-AC03-76SF0098.

References

1. W. Nazarewicz *et al.*, *Nucl. Phys.* **A429**, 269 (1984).
2. W. Nazarewicz and S. L. Tabor, *Phys. Rev.* **C45**, 2226 (1992).
3. P. A. Butler and W. Nazarewicz, *Rev. Mod. Phys.* **68**, 349 (1996).

4. W. R. Phillips *et al.*, *Phys. Rev. Lett.* **57**, 3257 (1986).
5. J. H. Hamilton *et al.*, *Prog. Part. Nucl. Phys.* **35**, 635 (1995).
6. S. J. Zhu *et al.*, *Phys. Lett.* **B357**, 273 (1995).
7. W. Urban *et al.*, *Nucl. Phys.* **A613**, 107 (1997).
8. S. J. Zhu *et al.*, *Chin. Phys. Lett.* **14**, 569 (1997).
9. S. J. Zhu *et al.*, *Phys. Rev.* **C60**, 051304 (1999).
10. S. J. Zhu *et al.*, *Phys. Rev.* **C59**, 1316 (1999).
11. W. R. Phillips *et al.*, *Phys. Lett.* **B212**, 402 (1988).
12. L. Y. Zhu *et al.*, *High Energy Phys. and Nucl. Phys.-Chinese Edition* **22**(10), 885 (1998).
13. Y. J. Chen *et al.*, *High Energy Phys. and Nucl. Phys-Chinese Edition.* **30**(8), 740 (2006).
14. Ts. Venkova *et al.*, *Eur. Phys. J.* **A26**, 315 (2005).
15. H. J. Li *et al.*, *Phys. Rev.* **C90**, 047303 (2014).
16. Y. J. Chen *et al.*, *Phys. Rev.* **C73**, 054316 (2006).
17. S. J. Zhu *et al.*, *Phys. Rev.* **C85**, 014330 (2012).
18. H. J. Li *et al.*, *Phys. Rev.* **C86**, 067302 (2012).
19. M. Bentaleb *et al.*, *Z. Phys.* **A354**, 143 (1996).
20. J. H. Hamilton *et al.*, *Prog. Part. Nucl. Phys.* **38**, 273 (1997).
21. W. Urban *et al.*, *Eur. Phys. J.* **A16**, 303 (2003).
22. W. Urban *et al.*, *Eur. Phys. J.* **A8**, 5 (2000).
23. Y. Huang *et al.*, *Phys. Rev.* **C93**, 064321 (2016).
24. Y. J. Chen *et al.*, *Int. J. Mod. Phys.* **E24**, 1550081 (2015).
25. Y. J. Chen *et al.*, *Phys. Rev.* **C91**, 014317 (2015).
26. M. A. Jones *et al.*, *Nucl. Phys.* **A605**, 133 (1996).
27. Y. X. Luo *et al.*, *Phys. Rev.* **C66**, 014305 (2002).
28. W. R. Phillips *et al.*, *Phys. Lett.* **B212**, 402 (1988).
29. S. J. Zhu *et al.*, *J. Phys. G: Nucl. Part. Phys.* **21,** L75 (1995)
30. D. C. Radford, *Nucl. Instrum. Methods Phys. Res.* **A361**, 297 (1995).
31. A .V. Daniel *et al.*, *Nucl. Instrum. Methods* **B262**, 399 (2007).
32. L. S. Geng *et al.*, *Chin. Phys. Lett.* **24**, 1865 (2007).
33. J. Y. Guo, P. Jiao, and X. Z. Fang, *Phys. Rev.* **C82**, 047301 (2010).
34. P. G. Reinhard *et al.*, *Z. Phys.* **A323**, 13 (1986).

Mass Measurements

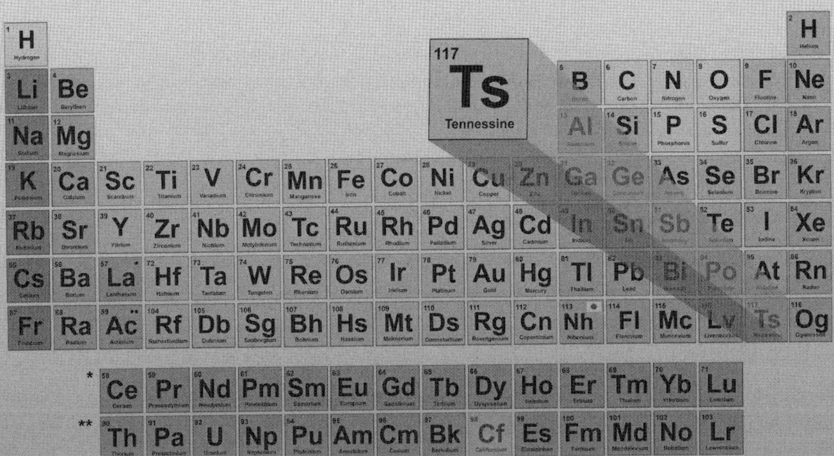

Recent Advances in High-Precision Nuclear Mass Measurements at ISOLTRAP at ISOLDE/CERN

K. Blaum

Max-Planck-Institut für Kernphysik, Saupfercheckweg 1, 69117 Heidelberg, Germany

The mass of the nucleus reflects the total energy of this many-body system and thus is a key property for a variety of nuclear structure and fundamental investigations. Modern experimental techniques, like storage-ring or Penning-trap mass spectrometry, have pushed in recent years the limits of sensitivity, resolution and accuracy. This has allowed to access exotic species very far from the valley of beta-stability.

The mass accuracy achieved even for very short-lived species in the ms regime and below allowed, e.g., to probe the shell structures and their evolution toward the neutron dripline or to perform in some regions fine examinations of the mass surface. This includes many exciting results like, for instance, the establishment of a new, prominent shell closure at neutron number $N = 32$, in excellent agreement with theoretical calculations.

In addition, with the nowadays achievable accuracy in Penning-trap mass spectrometry on short-lived exotic nuclides, precision fundamental tests can be performed, among them a test of the Standard Model, in particular with regard to the weak interaction and the unitarity of the Cabibbo-Kobayashi-Maskawa quark mixing matrix. Furthermore, accurate mass values of specific nuclides are important for nuclear astrophysics and neutrino physics as well as for the search of physics beyond the Standard Model.

In this contribution, recent results in the determination of nuclear masses at the high-precision Penning-trap mass spectrometer ISOLTRAP at the ISOLDE/CERN facility, their impact on nuclear structure, nuclear astrophysics and fundamental studies and the comparison to modern calculations will be presented.

Mass Measurements of the Heaviest Elements

M. Block
GSI Helmholtz Center for Heavy Ion Research,
64291 Darmstadt, Germany
Helmholtz Institute Mainz,
55099 Mainz, Germany
Johannes Gutenberg University,
55099 Mainz, Germany
E-mail: m.block@gsi.de

Mass spectrometry is a well-established technique for nuclear structure studies in exotic nuclei far off stability. The Penning-trap mass spectrometer SHIPTRAP at GSI has focused on mass measurements of the heaviest elements to map shell effects around $N = 152$. Several upgrades are presently ongoing to extend these measurements to heavier elements. A major step was the implementation of the phase-imaging technique (PI-ICR) that has been used for mass measurements of long-lived and stable nuclides in the context of neutrino physics. In this contribution recent achievements are briefly reviewed.

Keywords: Penning Trap Mass Measurements; Superheavy Elements; Neutrino Mass.

1. Introduction

Nuclear masses contain information on all the fundamental interactions between the constituents of the atomic nucleus. Mass measurements of radionuclides thus allow us to study the nature of the strong interaction that governs the behavior of nuclear matter from light nuclei to neutron stars. They also provide signatures of nuclear shell structure and other effects such as the onset of deformation[1]. Mass measurements in unknown territory often provide first signatures of changes in the nuclear suture and they are well suited to map the strength of shell effects and their evolution far off stability for extreme proton-to-neutron ratios.

Investigations of nuclei at the extremes may also feature new phenomena. Besides pushing towards the drip lines, in particular on the neutron-rich side of the valley of stability, the region of superheavy nuclei remains at the forefront of nuclear physics research. Superheavy nuclei owe their very existence to nuclear shell effects that provide additional binding and stabilizes them against spontaneous fission. In the last decades many nuclides in the region above Fermium (Fm, $Z = 100$) have been synthesized and new elements up to $Z = 118$ (Oganesson) have been discovered[2–4].

However, theoretical predictions concerning the center and the extension of the long-sought island of stability still disagree. Magic numbers of $Z = 114, 120$, or 126 are obtained for protons and $N = 172, 184$ for neutrons in different models[5–7]. In fact many models actually predict larger regions of enhanced stability rather than distinct magic numbers, see for example[8].

Nuclear models and their predictions are constantly challenged by more and increasingly accurate experimental data of nuclear properties, for example from nuclear decay spectroscopy[9] and mass measurements[10]. In the heaviest elements such experiments are very challenging due to low production rates on the order of one particle per second in the nobelium region with todays's accelerators. Future efforts will not only aim on the discovery of new elements beyond $Z = 118$, but also try to get more detailed information on known nuclides with sufficient statistics. This concerns for example the nuclides around the deformed shell closures at $N = 152, 162$.

2. Direct mass measurements with Penning traps

Direct mass measurements of radionuclides are nowadays mainly performed by Penning trap mass spectrometry (PTMS)[11,12], in storage rings[13], and by time-of-flight mass spectrometry[14]. They provide nuclear structure information without detailed knowledge of the nuclear level scheme. The most accurate mass measurements are presently performed with Penning traps[12,15]. In such a trap an ion's mass is obtained from the measurement of its cyclotron frequency with high precision on the order of $\delta m/m = 10^{-9}$[16].

The status of direct mass measurements of the heaviest elements has been reviewed recently[10]. In this mass region typical mass uncertainties of about 10–20 keV can be obtained as demonstrated by the first direct mass measurements of nobelium and lawrencium isotopes with SHIPTRAP[17–19]. The uncertainty was limited by statistics and can be further improved in the future. The most exotic nuclide reached by these measurements was ^{256}Lr. It was produced in the reaction ^{209}Bi(^{48}Ca,1n)^{256}Lr with a cross sections of about 60 nanobarns and a single mass measurement took about 93 hours[19]. The mass values obtained with SHIPTRAP serve as anchor points in this mass region. In addition, precise values of the empirical shell gap δ_n have been derived from the mass values to map the strength of the $N = 152$ subshell closure in nobelium and lawrencium[19].

Moreover, based on ground-state masses of ^{252}No ($N - Z = 48$) and ^{254}No ($N - Z = 50$) obtained with SHIPTRAP and precisely measured Q_α-values of even-even nuclei up to $Z = 108$ obtained at SHIP, the two neutron separation energies of even-even $N - Z = 50$ nuclei up to $Z = 108$ have been determined. The experimental values exhibited a slower increase towards the neutron shell at $N = 162$ than predicted[20]. This can be regarded as an indication that either the $N = 162$ is weaker than predicted or that the region of strong shell effects is more localized.

2.1. *Improvements to extend PTMS towards more exotic nuclei*

Several activities are pursued to extend direct mass measurements with SHIPTRAP to heavier nuclides despite ever-decreasing production rates. As in most mass spectrometers continuous efforts focus on pushing the sensitivity, the efficiency and the precision to the limits. The ultimate goal is to develop a method that enables a fast measurement to access short-lived nuclides without sacrificing precision and in which eventually one particle suffices for a mass determination.

An major step towards faster online mass measurements was taken with the recent development of the novel phase-imaging ion-cyclotron-resonance (PI-ICR) technique at SHIPTRAP[21]. In this method the ion motion in imaged with a position-sensitive channelplate detector. The frequencies of the circular ion motions are then obtained from the ion's phase[22]. The new method features a forty times higher mass resolving power and allows a gain in precision by a factor of five compared to the conventional method[21]. It is thus well suited for measurements on short-lived nuclides without sacrificing precision. For the moderate precision required in the context of nuclear structure studies the new method is about 25 times faster than the present standard technique. In experiments on stable and very long-lived nuclides the PI-ICR technique allows pushing towards the high-precision frontier.

A relative precision of 10^{-9} or below is required for mass differences of nuclides that are relevant in neutrino physics[23]. This comprises for example the masses of candidate nuclides for the search for neutrinoless double β transformations, processes that would prove the Majorana character of neutrinos if observed. Another key question in contemporary physics is related to the mass hierarchy and the absolute mass on neutrinos. In experiments that aim for the mass determination of neutrinos or antineutrino such as KATRIN[24] or ECHo[25] the endpoint of the β-decay spectrum should be measured independently with a very high precision on the order of electronvolts or even less[23]. To this end, dedicated Penning traps setups such as PENTATRAP have been designed[26]. However, already a mass uncertainty of tens of electronvolts as it can be reached utilizing the PI-ICR technique provide important information for the design of neutrino experiments like ECHo. This was recently demonstrated by the mass difference measurement of the isobar pair ^{163}Ho-^{163}Dy with SHIPTRAP in which an uncertainty of about 30 eV was achieved[27].

The approach followed to increase the sensitivity towards single-ion mass measurements capability is the implementation of the Fourier-transform ion-cyclotron-resonance (FT-ICR) technique[28]. In this method image currents induced by the trapped ion into the trap electrodes are amplified and Fourier-analyzed to obtain the motional frequency. This nondestructive method has already been used for mass measurements of stable particles successfully. Efforts for the use with radioactive ions are ongoing within the LEBIT project[29], at TRIGATRAP[30] and at SHIPTRAP where a dedicated superconducting solenoid with 4.7 T field strength has recently been installed to facilitate the FT-ICR developments without interfering with the ongoing mass measurement program.

The efficiency for the preparation of the ions of interest for a Penning trap mass measurement is usually limited by the slowing down process. In the previous SHIPTRAP experiments on nobelium and lawrencium isotopes the reaction products were slowed down in about 50 mbar helium gas at room temperature in a gas cell that featured an overall efficiency of 5-10%[31]. A higher efficiency will be achieved by a new cryogenic stopping cell[32]. It provides a larger stopping volume and the incoming ion beam is injected along the symmetry axis of the system. In addition, the new cell is operated at a temperature of about 40 K to improve the cleanliness by freezing out impurities. Extraction tests performed offline with a radioactive source have demonstrated a gain in efficiency by factor of two to three when cooling the system from room temperature to 40 K[32]. First online tests with nobelium ions have also shown promising results[33]. Thus, the SHIPTRAP setup was recently rearranged in line with the SHIP velocity filter to accommodate the new cryogenic stopping cell. This will pave the way for mass measurements of transactinides in the near future.

References

1. D. Lunney, J. Pearson and C. Thibault, Recent trends in the determination of nuclear masses, *Rev. Mod. Phys.* **75**, p. 1021 (2003).
2. Yu. T. Oganessian and V. K. Utyonkov, Superheavy nuclei from 48 Ca-induced reactions, *Nucl. Phys.* **A944**, 62 (2015).
3. K. Morita, SHE research at RIKEN/GARIS, *Nucl. Phys.* **A944**, 30 (2015).
4. G. Mnzenberg, From bohrium to copernicium and beyond SHE research at SHIP, *Nucl. Phys.* **A944**, 5 (2015).
5. K. Rutz, M. Bender, T. Bürvenich, T. Schilling, P.-G. Reinhard, J. Maruhn and W. Greiner, Superheavy nuclei in self-consistent nuclear calculations, *Phys. Rev. C* **56**, p. 238 (1997).
6. S. Cwiok, J. Dobaczewski, P. H. Heenen, P. Magierski and W. Nazarewicz, Shell structure of the superheavy elements, *Nucl. Phys.* **A611**, 211 (1996).
7. A. Sobiczewski and K. Pomorski, Description of structure and properties of superheavy nuclei, *Progress in Particle and Nuclear Physics* **58**, 292 (2007).
8. M. Bender, W. Nazarewicz and P.-G. Reinhard, Shell stabilization of super-and hyperheavy nuclei without magic gaps, *Phys. Lett. B* **515**, 42 (2001).
9. M. Asai, F. P. Heberger and A. Lopez-Martens, Nuclear structure of elements with 100 Z 109 from alpha spectroscopy, *Nucl. Phys.* **A944**, 308 (2015).
10. M. Block, Direct mass measurements of the heaviest elements with penning traps, *Nuclear Physics A* **944**, 471 (2015).
11. H.-J. Kluge, Penning trap mass spectrometry of radionuclides, *Int. J. Mass Spectrom.* **349**, 26 (2013).
12. K. Blaum, J. Dilling and W. Nörtershäuser, Precision atomic physics techniques for nuclear physics with radioactive beams, *Phys. Scr.* **2013**, p. 014017 (2013).
13. Y. H. Zhang, Y. A. Litvinov, T. Uesaka and H. S. Xu, Storage ring mass spectrometry for nuclear structure and astrophysics research, *Phys. Scripta* **91**, p. 073002 (2016).
14. H. Wollnik, History of mass measurements in time-of-flight mass analyzers, *Int. J. Mass Spectrom.* **349**, 38 (2013).
15. K. Blaum, High-accuracy mass spectrometry with stored ions, *Phys. Rep.* **425**, 1 (2006).

16. G. Bollen, D. Davies, M. Facina, J. Huikari, E. Kwan, P. Lofy, D. Morrissey, A. Prinke, R. Ringle, J. Savory *et al.*, Experiments with thermalized rare isotope beams from projectile fragmentation: A precision mass measurement of the superallowed β emitter ^{38}Ca, *Phys. Rev. Lett.* **96**, p. 152501 (2006).
17. M. Block, D. Ackermann, K. Blaum, C. Droese, M. Dworschak, S. Eliseev, T. Fleckenstein, E. Haettner, F. Herfurth, F. Heßberger *et al.*, Direct mass measurements above uranium bridge the gap to the island of stability, *Nature* **463**, 785 (2010).
18. M. Dworschak, M. Block, D. Ackermann, G. Audi, K. Blaum, C. Droese, S. Eliseev, T. Fleckenstein, E. Haettner, F. Herfurth *et al.*, Penning trap mass measurements on nobelium isotopes, *Phys. Rev. C* **81**, p. 064312 (2010).
19. E. M. Ramirez, D. Ackermann, K. Blaum, M. Block, C. Droese, C. E. Düllmann, M. Dworschak, M. Eibach, S. Eliseev, E. Haettner *et al.*, Direct mapping of nuclear shell effects in the heaviest elements, *Science* **337**, 1207 (2012).
20. F. P. Hessberger, Nuclear Structure of the Heaviest Elements - Investigated at SHIP-GSI, *EPJ Web Conf.* **66**, p. 02048 (2014).
21. S. Eliseev, K. Blaum, M. Block, C. Droese, M. Goncharov, E. M. Ramirez, D. Nesterenko, Y. N. Novikov and L. Schweikhard, Phase-imaging ion-cyclotron-resonance measurements for short-lived nuclides, *Phys. Rev. Lett.* **110**, p. 082501 (2013).
22. S. Eliseev, K. Blaum, M. Block, A. Dörr, C. Droese, T. Eronen, M. Goncharov, M. Höcker, J. Ketter, E. M. Ramirez *et al.*, A phase-imaging technique for cyclotron-frequency measurements, *Appl. Phys. B* **114**, 107 (2014).
23. S. Eliseev, Y. N. Novikov and K. Blaum, Penning-trap mass spectrometry and neutrino physics, *Annalen der Physik* **525**, 707 (2013).
24. F. Fraenkle, Status of the neutrino mass experiments KATRIN and Project 8, *PoS* **EPS-HEP2015**, p. 084 (2015).
25. L. Gastaldo, P.-O. Ranitzsch, F. von Seggern, J.-P. Porst, S. Schäfer, C. Pies, S. Kempf, T. Wolf, A. Fleischmann, C. Enss *et al.*, Characterization of low temperature metallic magnetic calorimeters having gold absorbers with implanted ^{163}Ho ions, *Nucl. Instrum. and Meth. A* **711**, 150 (2013).
26. J. Repp, C. Böhm, J. C. López-Urrutia, A. Dörr, S. Eliseev, S. George, M. Goncharov, Y. N. Novikov, C. Roux, S. Sturm *et al.*, PENTATRAP: a novel cryogenic multi-Penning-trap experiment for high-precision

mass measurements on highly charged ions, *Appl. Phys. B* **107**, 983 (2012).

27. S. Eliseev *et al.*, Direct measurement of the mass difference of ^{163}Ho and ^{163}Dy solves the Q-value puzzle for the neutrino mass determination, *Phys. Rev. Lett.* **115**, p. 062501 (2015).
28. R. Ferrer, K. Blaum, M. Block, F. Herfurth, J. Ketelaer, S. Nagy, D. Neidherr and C. Weber, Development of a fourier-transform ion-cyclotron-resonance detection for short-lived radionuclides at SHIP-TRAP, *The European Physical Journal-Special Topics* **150**, 347 (2007).
29. M. Redshaw, G. Bollen, S. Bustabad, A. Kwiatkowski, D. Lincoln, S. Novario, R. Ringle, S. Schwarz and A. Valverde, LEBIT II: Upgrades and developments for high precision Penning trap mass measurements with rare isotopes, *Nucl. Instrum. and Meth. B* **317**, 510 (2013).
30. J. Ketelaer, K. Blaum, M. Block, K. Eberhardt, M. Eibach, R. Ferrer, S. George, F. Herfurth, J. Ketter, S. Nagy *et al.*, Recent developments in ion detection techniques for Penning trap mass spectrometry at TRIGA-TRAP, *Eur. J. Phys. A* **42**, 311 (2009).
31. J. Neumayr, L. Beck, D. Habs, S. Heinz, J. Szerypo, P. Thirolf, V. Varentsov, F. Voit, D. Ackermann, D. Beck *et al.*, The ion-catcher device for SHIPTRAP, *Nucl. Instrum. and Meth. B* **244**, 489 (2006).
32. C. Droese, S. Eliseev, K. Blaum, M. Block, F. Herfurth, M. Laatiaoui, F. Lautenschläger, E. M. Ramirez, L. Schweikhard, V. Simon *et al.*, The cryogenic gas stopping cell of SHIPTRAP, *Nucl. Instrum. and Meth. B* **338**, 126 (2014).
33. O. Kaleja, Massenmessung von ^{197}Au mit SHIPTRAP und der erste on-line Test der neuen kryogenen Gaszelle an SHIP, Master's thesis, Technische Universitaet Darmstadt (2016).

Searching for Superheavy Elements in Terrestrial Matter with Accelerator Mass Spectrometry*

Walter Kutschera[1] and Gunther Korschinek[2]

[1]*Faculty of Physics, University of Vienna, A-1090 Vienna, Austria*

[2]*Department of Physics, Technical University of Munich, D-85748 Garching, Germany*

Ever since the concept of an island of increased stability around A = 300 (Z = 114, N = 186) was proposed in the late 1960s, numerous searches with various different techniques were performed to find traces of long-lived ($t_{1/2} > 10^8$ yr) nuclides of superheavy elements (SHEs) in nature. So far, no events have been observed which could unequivocally attributed to the presence of SHE nuclides, either in cosmic rays, extraterrestrial or terrestrial matter [1, 2].

One of the most sensitive methods to measure minute traces of long-lived nuclides throughout the nuclear chart is accelerator mass spectrometry [3, 4]. In some cases, isotopic abundance levels of 10^{-16} have been reached. In contrast to searches for SHE nuclides which depend on the observation of rare decay events (e.g. spontaneous fission [1]), the direct detection of SHE nuclides with AMS does not depend on the half-life [2], and therefore allows one to use much smaller samples. The major challenge in AMS searches is the correct tuning of the facility to a "white spot" on the nuclear chart, since finding no events can always mean that one missed the mark. A brief discussion on how to accomplish a valid search with AMS for a SHE nuclide will be presented.

*We acknowledge the contributions of F. Dellinger, J. Lachner and P. Ludwig on searches for SHEs in terrestrial materials at the Munich and Vienna AMS facilities.

Besides the technical aspect of pinpointing a particular SHE nuclide, there remains the question of its production in explosive stellar nucleosynthesis, of its actual half-life and the geochemical fate once it becomes incorporated into solar system material. Even though these loopholes remain open as long as one observes null results, it still seems meaningful to continue searching for "something which may exist", as long as one can improve the detection sensitivity of the expected very rare events.

We will summarize the current situation of searches for SHE nuclides with AMS, and also briefly mention searches with different methods.

References

1. G.M. Ter-Akopian and S.N. Dmitriev, *Nucl. Phys. A* **944**, 177 (2015).
2. G. Korschinek and W. Kutschera, *Nucl. Phys. A* **944**, 190 (2015).
3. H.-A. Synal, *Int. J. Mass Spectr.* **349-350**, 192 (2013).
4. W. Kutschera, *Int. J. Mass Spectr.* **349-350**, 203 (2013).

Nuclear Structure III

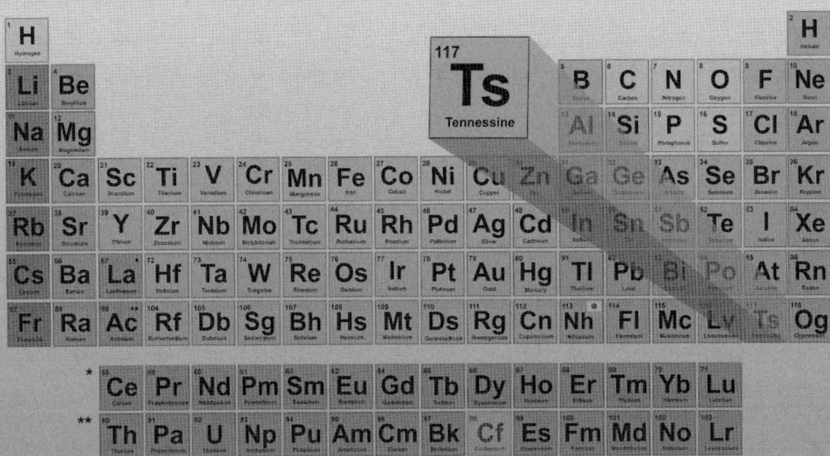

Production of $N = 126$ Nuclei and Beyond Using Multinucleon Transfer Reactions

Y.X. Watanabe*, Y. Hirayama, H.S. Jung, Y. Kakiguchi, H. Miyatake, M. Oyaizu, P. Schury and M. Wada

Wako Nuclear Science Center, Institute of Particle and Nuclear Studies, High Energy Accelerator Research Organization (KEK), Wako, Saitama 351-0198, Japan

Y.H. Kim, E. Clement, G. de France, A. Navin, M. Rejmund and C. Schmitt

Grand Accélérateur National d'Ions Lourds (GANIL), F-14076 Caen Cedex 5, France

S.C. Jeong, H. Ishiyama, J.Y. Moon and J.H. Park

Institute for Basic Science, Daejeon 34047, Korea

N. Imai

Center for Nuclear Study, University of Tokyo, Tokyo 113-0033, Japan

S. Kimura, M. Mukai and M. Ahmed

Department of Physics, University of Tsukuba, Tsukuba, Ibaraki 305-8571, Japan

S.H. Choi and J.S. Song

Department of Physics and Astronomy, Seoul National University, Seoul 08826, Korea

G. Pollarolo

Dipartimento di Fisica Teorica, Universitá di Torino, and Istituto Nazionale di Fisica Nucleare, I-10125 Torino, Italy

L. Corradi and E. Fioretto

Istituto Nazionale di Fisica Nucleare, Laboratori Nazionali di Legnaro, I-35020 Legnaro, Italy

D. Montanari

Dipartimento di Fisica, Universitá di Padova, and Istituto Nazionale di Fisica Nucleare, I-35131 Padova, Italy

M. Niikura and D. Suzuki
Institut de Physique Nucléaire (IPN), IN2P3-CNRS, F-91406 Orsay Cedex, France

H. Nishibata and J. Takatsu
Department of Physics, Osaka University, Osaka 560-0043, Japan

Production of neutron-rich nuclei around $N = 126$ by multinucleon transfer reactions has recently gained a renewal of interest both theoretically and experimentally. We are now advancing the KISS project at RIKEN to produce those nuclei by multinucleon transfer reactions and measure their lifetimes to investigate the astrophysical environments of the r-process. To investigate the relevance of the production mechanism, we performed measurements for the reaction system ^{136}Xe + ^{198}Pt, using the spectrometer VAMOS++ and the gamma-ray array EXOGAM at GANIL. Cross sections deduced from the measurements for neutron-rich nuclei around $N = 126$ show a modest enhancement when compared with the GRAZING calculations, indicating the advantage of the multinucleon transfer reactions as compared with fragmentation for production of $N = 126$ neutron-rich isotones. The possibility for production of neutron-rich nuclei around uranium beyond $N = 126$ by multinucleon transfer reactions is also discussed.

Keywords: Multinucleon Transfer Reaction, Deep Inelastic Collision, Gas Cell, Laser Resonance Ionization.

1. Introduction

Nuclear properties at the neutron closed shell $N = 126$ are very important for understanding the r-process in the astrophysical nucleosynthesis. The r-process goes through a very neutron-rich unstable region, where the neutron magic number $N = 126$ leads to waiting points on its path[1]. The corresponding nuclei are considered as progenitors of the peak around mass number $A = 195$ in the observed solar r-abundance distribution. The masses and lifetimes of those nuclei are significant to identify the astrophysical environments of the r-process. However the area of those nuclei remains unexplored, because it is difficult to access by fusion, fission and fragmentation with stable projectiles.

We are now advancing the KISS project to produce those neutron-rich nuclei and measure their lifetimes to investigate astrophysical environments of the r-process[2]. In this project, we selected multinucleon transfer (MNT) reactions between the ^{136}Xe projectile and the ^{198}Pt target to access neutron-rich nuclei around $N = 126$ by proton pick-up and neutron stripping transfer.

Recently, promising expectations have focused on the nuclear production in this region by MNT reactions theoretically and experimentally. Using a macroscopic approach based on the Langevin-type dynamical equation of motion, Zagrebaev *et al.* demonstrated that the MNT reaction between Xe and Pb provided cross sections large enough to produce those neutron-rich nuclei[3]. Measurements of MNT reactions in this reaction system were performed at ANL, and indicated a general agreement with macroscopic calculations[4]. However, they did not reach $N = 126$ for $Z < 80$, and experimental data are scarce to evaluate the nuclear production with $N = 126$.

2. Measurements of MNT reactions between ^{136}Xe and ^{198}Pt

To investigate the feasibility of our reaction system to produce neutron-rich target-like fragments (TLFs), we performed measurements using the large acceptance magnetic spectrometer VAMOS++[5] coupled to the EXOGAM gamma-ray array[6] at GANIL[7,8]. The ^{136}Xe beam impinged on the ^{198}Pt target with the thickness of 1.3 mg/cm^2 at the energy of 8 MeV/nucleon, 55% above the Coulomb barrier. The projectile-like fragments (PLFs) produced in the MNT reactions were detected by the spectrometer located at the angle of 30 degrees around grazing, and were isotopically identified, that is, their mass and atomic numbers were unambiguously determined.

Figure 1 shows the distribution of the detected PLFs. The large contribution of the proton pick-up and the neutron stripping channels was observed. It would be promoted by the use of the heavy projectile such as

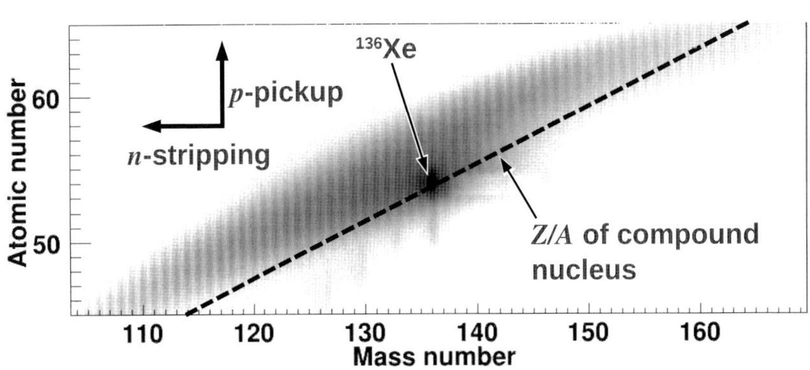

Fig. 1. Distribution of projectile-like fragments produced in ^{136}Xe (8 MeV/nucleon) + ^{198}Pt.

^{136}Xe. The dashed line indicates the loci of isotopes having the same A/Z as the compound nucleus. The distribution stretches along the dashed line, although it is shifted to the smaller mass side. It indicates that the isotopic distributions are mainly governed by the Z/A equilibration associated with neutron evaporation caused by strong energy damping in the deep inelastic collision at the reaction energy of 8 MeV/nucleon.

Since the PLFs were detected after particle evaporation by the spectrometer, they are not straightforwardly related to the neutron-rich isotopes of interest, which are produced as TLFs after particle evaporation. Nevertheless, the (not directly measured) TLF distributions can be deduced by evaluating particle evaporation from the primary PLFs and TLFs in the cascade calculations using the measured total kinetic energy loss (TKEL) in the reaction. This procedure to identify the TLFs was checked and found supported by the gamma-ray transitions detected in the EXOGAM array.

Solid circles in Fig. 2 show the so-deduced distribution of Os TLF isotopes, where the ordinate indicates the cross sections. The histogram corresponds to the GRAZING calculation[9,10], which is a semi-classical approach based on single-particle transfer probabilities. The mass number at the maximum of the experimental distribution shifts to the neutron-deficient side as compared with the GRAZING calculation, and the maximal experimental cross section is greater. It is interpreted as indication of strong energy damping with Z/A equilibration and the neutron evaporation. The discrepancies between the measured cross sections and the calculation are

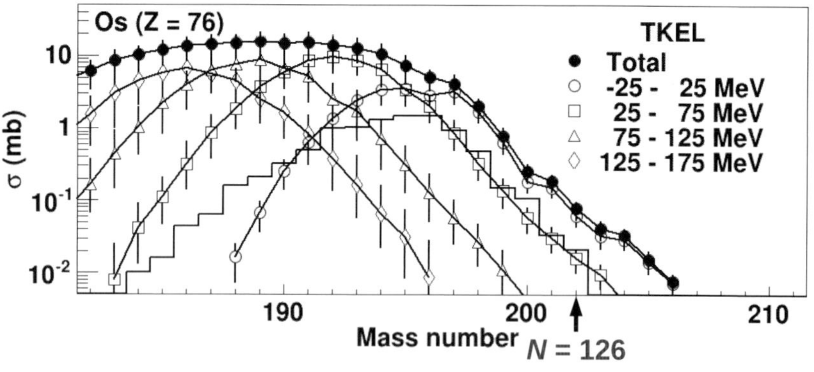

Fig. 2. Distributions of Os target-like fragments for different total kinetic energy losses in the reaction deduced from the measurements of projectile-like fragments for ^{136}Xe (8 MeV/nucleon) + ^{198}Pt.

modest at the neutron-rich side around the neutron magic number 126. This is understood by decomposing the isotopic distribution for different TKELs as shown by open symbols in Fig. 2. The neutron-rich side of the isotopic distribution is observed to mainly arise from the lowest TKEL values, less than 75 MeV. The high-TKEL components contribute to the production of neutron-deficient Os isotopes, however, such components are not included in the GRAZING calculations, showing the large discrepancies at the neutron-deficient side and the modest ones at the neutron-rich side. Relatively large cross sections for production of neutron-rich nuclei around $N = 126$ as compared with fragmentation[11] suggest the advantage of the MNT reactions.

3. KISS project

We have been developing the KEK Isotope Separation System (KISS) at RIKEN to produce neutron-rich nuclei around $N = 126$ and measure their lifetimes. Figure 3 shows a schematic view of KISS. It consists of a gas cell system, an extraction chamber, an Isotope Separator On-Line (ISOL) and a detection system. The ^{198}Pt target in the gas cell is bombarded by the ^{136}Xe beam (10.75 MeV/nucleon) from RIKEN ring cyclotron. The TLFs produced by MNT reactions are stopped by argon gas filled in the gas cell and neutralized. Those neutral atoms are transported by laminar gas flow in the gas cell, and are element-selectively ionized by laser resonance ionization just before the exit of the gas cell. The ions are extracted by application of a high voltage (20 kV), and are mass-separated with the ISOL method. Finally they are transported to the detection system for β-decay spectroscopy.

The measured extraction efficiency for $^{198}Pt^+$ ions, which are the elastic recoils, was found to amount to about 0.15%, where the extraction efficiency is defined as the ratio between the number of the detected $^{198}Pt^+$ ions and the ^{198}Pt atoms emitted from the target in a calculation. The measured extracted beam purity was larger than 99.5% for the $^{198}Pt^+$ ions, which is high enough for the β-decay lifetime measurements.

The laser-ionized $^{199}Pt^+$ and $^{196}Ir^+$, produced in the MNT reactions, were extracted and their lifetimes were measured as 31.3 ± 1.5 minutes and 67 ± 13 seconds, respectively. They are in good agreement with the reported values[12], indicating the feasibility to measure the lifetimes of radioactive nuclei around $N = 126$ with the present extraction efficiency and purity.

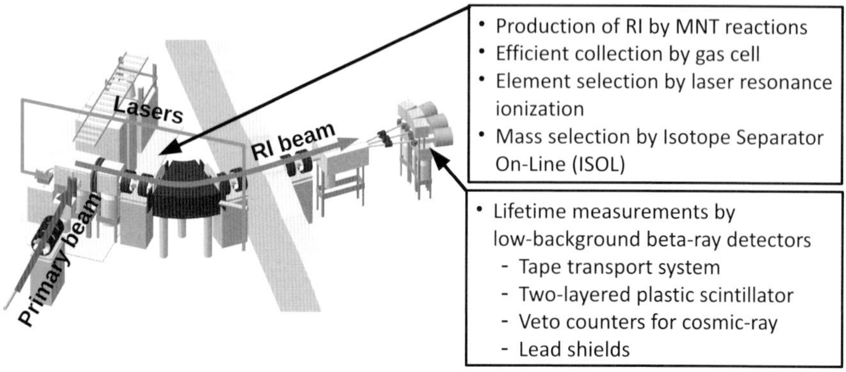

Fig. 3. Schematic view of KEK Isotope Separation System.

4. Nuclear production beyond $N = 126$

Beyond $N = 126$, neutron-rich nuclei around uranium would be accessible by using MNT reactions. Those nuclei are located at the terminal region of the r-process nucleosynthesis, where fission plays an important role, affecting the production of trans-uranium elements, especially possible super-heavy elements, the initial value of ^{232}Th/^{238}U cosmochronometer[13] and the production of heavy nuclei in the mass region between 100 to 160 as fission fragments[14].

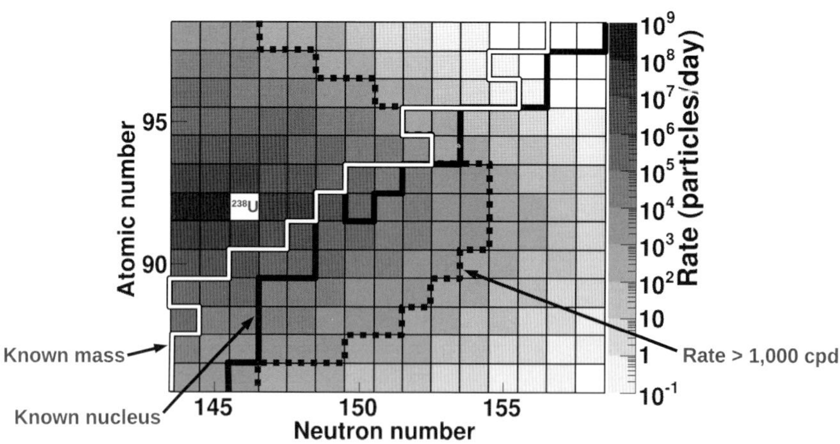

Fig. 4. Expected extraction rates for neutron-rich nuclei around ^{238}U produced by multinucleon transfer reactions between ^{238}U and ^{238}U at KISS.

β-delayed fission probabilities are one of important parameters, which strongly depend on evaluation of nuclear mass, fission barrier and β-decay strength function[15], although they are unknown experimentally in this region. Figure 4 shows expected extraction rates at KISS for isotopes around ^{238}U produced by the ^{238}U beam (9 MeV/nucleon, 500 pnA) and the ^{238}U target (10 mg/cm^2) in the GRAZING calculations, where an extraction efficiency of 0.01% was assumed. White and black thick lines indicate the boundaries of known masses and known nuclei. The measurements of masses and β-delayed fission probabilities for isotopes at the left of dotted lines, where the extraction rates are expected higher than 1,000 counts per day, will be considered as a future plan at KISS. Previous measurements for MNT reactions of ^{238}U and ^{238}U[16] shows smaller cross sections compared with the GRAZING calculations. In a recent theoretical study, these discrepancies were demonstrated to be due to fission of the reaction products[17]. Thus, more careful considerations for the extracted yields are necessary to investigate nuclear properties of those nuclei.

5. Summary

Recently, the production of neutron-rich nuclei around $N = 126$ by MNT reactions has attracted much attention both theoretically and experimentally. In our KISS project, we aim to measure lifetimes of them to investigate the astrophysical environments of the r-process. However experimental data are scarce to evaluate the production. To investigate the feasibility of the nuclear production by MNT in the selected system, ^{136}Xe + ^{198}Pt, we performed a measurement using the spectrometer VAMOS++ and the gamma-ray array EXOGAM. The detected PLFs show a large contribution of proton pick-up channels. However the isotopic distribution is mainly governed by Z/A equilibration. The TLFs were deduced and their cross sections around $N = 126$ were found to be connected with the lowest TKELs, indicating a modest enhancement as compared with GRAZING calculations. The result of the GANIL measurement indicates the advantage of MNT over fragmentation for producing $N = 126$ neutron-rich isotones. We are now advancing the KISS project to perform lifetime measurements of nuclei around $N = 126$. Measurements of mass and β-delayed fission probabilities for neutron-rich nuclei around uranium are considered in the future plan of KISS.

Acknowledgments

The experiments were performed at GANIL and RI Beam Factory operated by RIKEN Nishina Center and CNS, the University of Tokyo. The authors acknowledge the staff of the accelerators for their support. The works were supported by Grant-in-Aids for Scientific Research (A) (grant no. 23244060) from the Japan Society for the Promotion of Science (JSPS).

References

1. H. Grawe *et al.*, *Rept. Prog. Phys.* **70**, 1525 (2007).
2. Y. Hirayama, *Nucl. Instrum. and Meth. B* **353**, 4 (2015).
3. V. Zagrebaev and W. Griner, *Phys. Rev. Lett.* **101**, 122701 (2008).
4. J.S. Barrett *et al.*, *Phys. Rev. C* **91**, 064615 (2015).
5. M. Rejmund *et al.*, *Nucl. Instrum. Methods Phys. Res. Sect. A* **646**, 184 (2011).
6. J. Simson *et al.*, *Acta Phys. Hung. new Ser.: Heavy Ion Phys.* **11**, 159 (2000).
7. Y.H. Kim *et al.*, *EPJ Web of Conferences* **66**, 03044 (2014).
8. Y.X. Watanabe *et al.*, *Phys. Rev. Lett.* **115**, 172503 (2015).
9. A. Winther, *Nucl. Phys. A* **572**, 191 (1994).
10. A. Winther, *Nucl. Phys. A* **594**, 203 (1995).
11. T. Kurtukian-Nieto *et al.*, *Phys. Rev. C* **89**, 024616 (2014).
12. R.B. Firestone, *Table of Isotopes* (Wiley & Sons, New York) 1996.
13. J.J. Cowan *et al.*, *Astrophys. J.* **521**, 194 (1999).
14. M. Eichler *et al.*, *J. Phys.* **665**, 012054 (2016).
15. I.V. Panov *et al.*, *Phys. Atom. Nucl.* **76**, 88 (2013).
16. J.V. Kratz *et al.*, *Phys. Rev. C* **88**, 054615 (2013).
17. R. Yanez and W. Loveland, *Phys. Rev. C* **91**, 044608 (2015).

Using α-Transfer Reactions to Populate Radioactive Beam Species, Present and Future Uses to Study Neutron-Rich Nuclei

D. A. Torres

Departamento de Física, Universidad Nacional de Colombia, Bogotá, Colombia
E-mail: datorresg@unal.edu.co
www.unal.edu.co

The use of α-transfer reactions to populate states in nuclear species that cannot be produced in the present radioactive beam facilities is presented. Some examples of the use of these type of reactions to measure nuclear g factors and lifetimes will be discussed. The advantages and challenges for the use of α-transfer reactions with stable and radioactive beams will be discussed.

Keywords: Alpha Transfer, Coulomb Excitation, Magnetic Moments, Electric Moments.

1. Introduction

In recent years the use of α-transfer reactions has been presented as an alternative to the production of nuclear species that cannot be created in the present radioactive beam facilities. This has allowed the study of nuclear properties such as electric and magnetic moments in several nuclear mass regions such as $A \approx 40$[1], $A \approx 60$[2,3], $A \approx 80$[4,5] and $A \approx 100$[6,7]; contributing to the understanding of nuclear phenomena of transition from single-particle to collective behavior in the regions under study[8].

The use of α-transfer opens interesting options for the study and characterization of nuclear matter. Microscopic characterization of states larger than 2^+ permits comparisons between theoretical models, and provides an alternative to the use of Coulomb excitation reactions, which mainly populate 2_1^+ states, with enough intensity to allow the study of lifetimes and magnetic moment measurements with the Doppler-Shift Attenuation Method (DSAM) and the Transient Field Technique (TF) respectively[9]. A hindrance to the use of the α-transfer mechanism is that the alignment of the nuclear spin, required to observe a precession of the nuclear magnetic moment in an hyperfine magnetic field, is much reduced compared to that observed in experiments using Coulomb excitation[6].

In this contribution an overview to the α-transfer technique will be presented, along with some comments about the use of the reaction mechanism in future radioactive beam facilities.

2. Alpha-Transfer Reactions

Experiments using α-transfer reactions have some specific requirements to be successful, these are:

(1) use of a beam energy around the Coulomb barrier. Acceptable calculations of Coulomb barriers can be a challenge for different nuclear mass regions, and in some cases different energies should be scanned during the experiment to obtain a good reaction[7].

(2) a target that facilitates both, the pick up of α particles by the beam and the posterior use of new nuclear species produced after the reaction. Figure 1 shows an example of a multi-layered target commonly used in g factors and lifetime measurements, carbon is the favorite material to produce the target due to its cluster-like behavior.

(3) the use of inverse kinematic to produce the reaction. This type of reaction has proven to be the most convenient to study created isotopes. The recoil velocity of the products allows its use for the study of electric and magnetic moments, not to mention other type of reactions.

Figure 1 shows an example of a combination of a target and an α-transfer reaction, experimental details are described in Ref.[7]. The α transfer takes place in the first thin carbon layer, the Gd and Cu layers are utilized to measure g factors and magnetic moments. The detection in coincidence of the emitted α particles and γ rays ensures a clean analysis.

Figure 2 shows the γ-ray spectra produced in coincidence with the emitted charged particle of the reaction. The particle spectra is also presented. A clean distinction is observed between the signals produced by the emission of carbon particles (Coulomb excitation) and the 2α particles (alpha transfer).

The number of events obtained in the α-transfer process is less than in the Coulex process. The spin alignment for the recoil nuclei is far better in the Coulex process than in the α-transfer process. Also it is interesting to note that spin alignment for the α-transfer improves with spin of the state, as larger the spin better the alignment. No states with spins larger

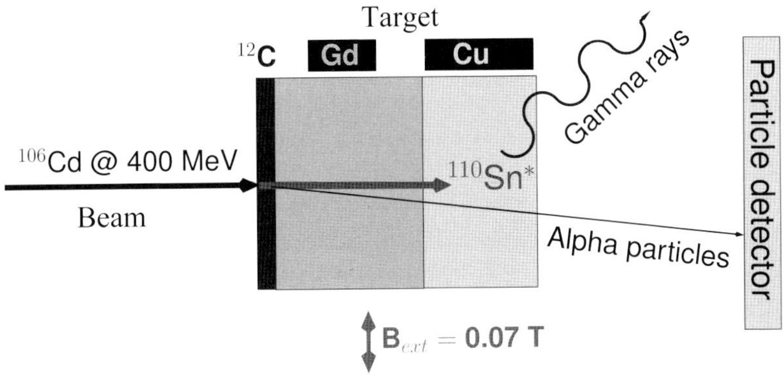

Fig. 1. Multi-layered target utilized for g factor and magnetic moment measurements with α-transfer reactions. The reaction ^{12}C(^{106}Cd,^{8}Be)^{110}Sn at 400 MeV beam energy is presented as an example. The ^{106}Cd beam captures one α particle from the carbon layer in the target, the radioactive ^{110}Sn ($T_{1/2}$ = 4.11 h) is produced and goes through the Gd layer where it slows down and precess due to the TF. Finally the ^{110}Sn stops in the Cu layer. The detection in coincidence of the emitted γ rays from the ^{110}Sn and the two α particles from the carbon, ensures clean results for the experiment. The in-flight γ-ray emission during the slowing down inside the target of the ^{110}Sn isotope permits the use of the DSAM to measure lifetime of the excited states, while the precession of the states inside the ferromagnetic Gd layer allows the use of the Transient Field technique to measure magnetic moments of the precessing states. See text for details.

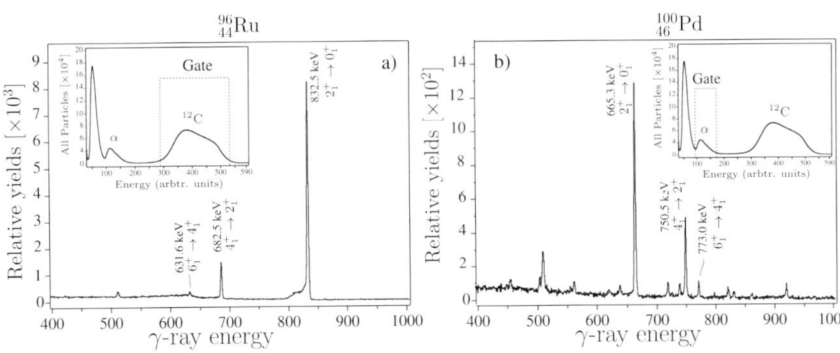

Fig. 2. Gamma-ray spectra in coincidence with particle signals from the charged-particle detectors. The reaction ^{12}C (^{96}Ru,^{8}Be)^{100}Pd at 350 MeV was utilized to study the magnetic moments of excited states in ^{100}Pd[6]. **a) Left**: the inner canvas shows the charged particle spectra, the rectangular region presents the gate made and the obtained γ-ray coincidence spectrum is shown below, a clean ^{96}Ru γ-ray spectrum produced by the Coulomb excitation channel is obtained. **b) Right**: the coincidence in the particles are made in the α-particle region. A clean γ-ray spectra from the radioactive ^{100}Pd is obtained.

than $6\hbar$ have been observed in the α-transfer experiments of Refs.[1,3,4,6,7,10]. The low-spin alignment is a hindrance for the use of α-transfer reactions in future radioactive beam facilities, and a radioactive beam with enough intensity will be needed to compensate this lack of alignment.

3. Some Results for Magnetic Moments and Lifetimes

Table 1 shows results for the radioactive ^{110}Sn produced in an α-transfer reaction[7], and also its comparison with the results from the initial beam ^{106}Cd[11]. These results were obtained simultaneously during the same experiment, which constitutes one of the advantages for α-transfer experiments, the possibility to use two reactions simultaneously, Coulomb excitation (Coulex) and α transfer, to study the nuclear structure of neighbor nuclei.

Table 1. Mean life and g-factor values obtained with the coulomb excitation reaction ^{12}C (^{106}Cd,^{12}C)^{106}Cd^{*a} and α-transfer reaction ^{12}C (^{106}Cd,^{8}Be)^{110}Sn^{*b}. See text for details.

Nucleus	State (I^{π})	Type of Reaction	Mean life (τ) (ps)	g-factor
^{106}Sn	2^+	α-transfer	0.81(10)	+0.29(11)
	4^+	α-transfer		+0.05(14)
^{106}Cd	2^+	Coulex	0.28(2)	+0.398(22)
	4^+	Coulex		+0.23(5)

Note: [a] See Ref.[11].
[b] See Ref.[7].

From Table 1 it is worth to noticing the errors for α-transfer reactions, the errors are larger for the α-transfer than for the Coulex reaction. The low-spin alignment and the low statistics are the responsible for these large errors in the *alpha*-transfer reactions. The goal for the next years is to improve the quality of the data in α-transfer process.

4. Populating States in Neutron-Rich Nuclei

Any reaction that can be made with stable beams, can also be made with radioactive beams, but every experiment using radioactive beams is unique, and the experimental setup will be defined by the lifetime of the beam and the sub-products of the reaction.

Despite to the fact that the statement may be true, in fact experiments using radioactive beams are unique, and three factors define the type of experiment that can be performed: lifetime of the nuclei in the radioactive beam, the intensity of the beam and the contamination produced by the decay products from the beam. The use of α-transfer reactions, to populate states in neutron-rich nuclei, will need neutron-rich beams with enough intensity and long lifetimes to ensure the success or the reaction.

5. Conclusions

The use of α-transfer reactions allows the production of nuclear states in radioactive species that are not yet available. The study of the nuclear spin alignments, obtained by the produced nuclei, is necessary to improve the precision of the experimental results for α-transfer reactions. Such studies should give a detailed characterization of the reaction and the angular particle correlations as a final result.

Future studies of g factors and lifetime for neutron-rich nuclei could take advantage of the use of α-transfer reaction using radioactive beams of some nuclear species with enough intensity and a reasonable long lifetime to populate new neutron-rich species. It is worth to point out that the main challenges will arise from the available beam intensities.

Acknowledgments

This work was partially supported by Colciencias under contract 110165842984 - 2015. For Paulo Gomes, in memoriam.

References

1. S. S. *et al.*, First measurement and shell model interpretation of the g factor of the 2_1^+ state in self-conjugate radioactive ^{44}Ti, *Physics Letters B* **567**, 153 (2003).
2. J. Leske, K.-H. Speidel, S. Schielke, O. Kenn, J. Gerber, P. Maier-Komor, S. J. Q. Robinson, A. Escuderos, Y. Y. Sharon and L. Zamick, Nuclear structure of the first 2+ state in radioactive ^{68}Ge based on g factor and lifetime measurements, *Phys. Rev. C* **71**, p. 044316 (Apr 2005).
3. K.-H. Speidel, J. Leske, S. Schielke, S. Bedi, O. Zell, P. Maier-Komor, S. Robinson, Y. Y. Sharon and L. Zamick, Low-level structure of ^{52}Ti based on g factor and lifetime measurements, *Physics Letters B* **633**, 219 (2006).

4. G. K. *et al.*, Structure of the Sr-Zr isotopes near and at the magic $N = 50$ shell from g-factor and lifetime measurements in $^{88}_{40}$Zr and $^{84,86,88}_{38}$Sr, *Phys. Rev. C* **85**, p. 044322 (Apr 2012).
5. G. Kumbartzki, Transition from collectivity to single-particle degrees of freedom from magnetic moment measurements on $^{82}_{38}Sr_{44}$ and $^{90}_{38}Sr_{52}$, *Phys. Rev. C* **89**, p. 064305 (Jun 2014).
6. D. T. *et al.*, First g-factor measurements of the 2^+_1 and the 4^+_1 states of radioactive ^{100}Pd, *Phys. Rev. C* **84**, p. 044327 (Oct 2011).
7. G. J. Kumbartzki, N. Benczer-Koller, K.-H. Speidel, D. A. Torres, J. M. Allmond, P. Fallon, I. Abramovic, L. A. Bernstein, J. E. Bevins, H. L. Crawford, Z. E. Guevara, G. Gürdal, A. M. Hurst, L. Kirsch, T. A. Laplace, A. Lo, E. F. Matthews, I. Mayers, L. W. Phair, F. Ramirez, S. J. Q. Robinson, Y. Y. Sharon and A. Wiens, $z = 50$ core stability in ^{110}Sn from magnetic-moment and lifetime measurements, *Phys. Rev. C* **93**, p. 044316 (Apr 2016).
8. D. Torres and F. Ramirez, Nuclear structure aspects via g-factor measurements: pushing the frontiers, *PoS Proceedings of Science, 10th Latin American Symposium on Nuclear Physics and Applications (2013)* (2013).
9. N. Benczer-Koller and G. Kumbartzki, Magnetic moments of short-lived excited nuclear states: measurements and challenges, *Journal of Physics G: Nuclear and Particle Physics* **34**, p. R321 (2007).
10. K.-H. S. *et al.*, Experimental g factors and B(E2) values in Ar isotopes: Crossing the N=20 semi-magic divide, *Physics Letters B* **632**, 207 (2006).
11. N. Benczer-Koller, G. J. Kumbartzki, K.-H. Speidel, D. A. Torres, S. J. Q. Robinson, Y. Y. Sharon, J. M. Allmond, P. Fallon, I. Abramovic, L. A. Bernstein, J. E. Bevins, H. L. Crawford, Z. E. Guevara, A. M. Hurst, L. Kirsch, T. A. Laplace, A. Lo, E. F. Matthews, I. Mayers, L. W. Phair, F. Ramirez and A. Wiens, Magnetic moment and lifetime measurements of coulomb-excited states in ^{106}Cd, *Phys. Rev. C* **94**, p. 034303 (Sep 2016).

Odd Features of Some Odd-Odd Prolate Spheroidal Nuclei

J. O. Rasmussen[1,†], Y. X. Luo[1,2], Y. Liu[3], E. Wang[2], J. H. Hamilton[2], A. V. Ramayya[2], W. Y. Liang[4], F. R. Xu[4] and Y. Sun[5]

[1]*Lawrence Berkeley National Laboratory, Berkeley, CA/94720, USA*
[†]*E-mail: oxras1@gmail.com*
[2]*Physics Dept. Vanderbilt University, Nashville, TN 37235 USA*
[3]*Physics Dept. Huzhou University, Huzhou, China*
[4]*School of Physics, Peking University, Beijing 100871, China*
[5]*Dept. of Physics, Shanghai Jiao Tong University, Shanghai 200240, China*

We address the theoretical interpretation of rotational bands and low-lying states in odd-odd prolate fission-products Nb and La nuclei, which are characteristic of shape transition and coexistence. Our Projected Shell Model (PSM), Total Routhian Surface (TRS) and Potential Energy Surface (PES) calculations were applied to the odd-odd ^{104,106}Nb. Results for ^{142}La show theoretically a high-spin state as band-head in the case of both odd neutron and proton in K = 1/2 states.

1. Theories

The Projected Shell Model (PSM)[1] uses an axially-deformed basis and selects low-lying states based on Nilsson + BCS representation to form the basis. Shell model diagonalization is carried out within the space spanned by the angular-momentum-projected quasiparticle (qp) vacuum, 2-qp and 4-qp states for even-even nuclei[1]. One expects that the collectivity of low-lying states may be affected by mixing many 2-qp and 4-qp states. Indeed, a multi-qp admixture can cause significant effects in band crossing regions. However, in the low-spin region before any band crossings ($I \leq 10$), the admixture is weak and the calculated γ band exhibits mainly the characteristics of an axially symmetric rotor.

The PSM uses a Nilsson potential having axial deformation to build a model basis. In the present article the axial-deformation parameters are adjusted around the value taken from Ref. [2] to obtain suitable deformation bases. Pairing correlations are treated by the usual BCS approximation to establish the Nilsson+BCS basis. The angular momentum projection is then carried out on the

Nilsson+BCS qp states to obtain the many-body basis, and the Hamiltonian is diagonalized in this projected basis. For even-even system, the multi-qp configurations consisting of 0-, 2- and 4-qp states are as follows:

$$\begin{aligned}\{ &\hat{P}^I_{MK}|0\rangle, \hat{P}^I_{MK}a^+_{\pi 1}a^+_{\pi 2}|0\rangle, \hat{P}^I_{MK}a^+_{\nu 1}a^+_{\nu 2}|0\rangle, \\ &\hat{P}^I_{MK}a^+_{\pi 1}a^+_{\pi 2}a^+_{\nu 1}a^+_{\nu 2}|0\rangle \}\end{aligned} \tag{1}$$

where |Φ> denote the qp vacuum and $a^+\pi$ and $a^+\nu$ the qp creation operators, with the index π (ν) being the proton (neutron) quantum numbers. The angular-momentum-projection operator is

$$\hat{P}^I_{MK} = \frac{2I+1}{8\pi^2}\int d\Omega D^I_{MK}(\Omega)\hat{R}(\Omega) \tag{2}$$

As in the usual PSM calculations [1], we use the pairing-plus-quadrupole-quadrupole Hamiltonian

$$\begin{aligned}\hat{H} = \hat{H}_0 - \frac{1}{2}\chi_{QQ}\sum_{\mu}\hat{Q}^+_{2\mu}\hat{Q}_{2\mu} - G_M\hat{P}^+\hat{P} \\ - G_Q\sum_{\mu}\hat{P}^+_{2\mu}P_{2\mu}\end{aligned} \tag{3}$$

and the corresponding Nilsson Hamiltonian is given by

$$\hat{H}_N = \hat{H}_0 - \frac{2}{3}\hbar\omega\varepsilon Q_0 \tag{4}$$

Here $\hat{H}_0$ is the spherical single-particle Hamiltonian with inclusion of appropriate spin-orbit forces parametrized in Ref. [3]. The QQ-force strength is adjusted such that the physical quadrupole deformation is obtained as a result of the self-consistent mean-field [Hartree-Fock-Bogoliubov (HFB)] calculation[1]. The quadrupole-pairing strength GQ is assumed to be proportional to GM, with the proportionality constant being fixed to be 0.20. The PSM calculation has been successfully applied for nuclei far away from the stability[4-6], the coexistence of normal, super-, and hyper-deformation in nuclei[7,8] and isomeric states in nuclei[9].

The PSM, Total Routhian Surface (TRS) and Potential Energy Surface (PES) theories were applied to the odd-odd nuclei ^{104,106}Nb[10,11] and ^{142}La[12], where we have fission-γ data.

Theory of odd-odd spheroidal nuclei is difficult to formulate, since there are three angular momentum vectors to be coupled, the j values of the odd nucleons and the rotational angular momentum of the spheroidal core. In the case of both odd nucleons having z-axis projection $\Omega = 1/2$, the three angular momentum vectors lie in the same equatorial plane. We postulate (guess) that the TRS case (and possibly our fission-γ data case) have the odd proton and odd neutron in $h_{11/2}$ [550] orbitals. Both odd nucleons can be in nearly the same plane along longitude lines going through the poles, but not necessarily so. The attractive n-p force will favor their proximity. The spheroidal collective rotational momentum vector will also be in the equatorial plane.

As to Prof. Goennenwein's observation at the 2012 Sanibel conference that odd-odd neutron fission of ^{239}Pu showed the highest kinetic energy, higher than even-even products, we would suggest that the neck snapping might occur sooner for odd-odd. Also there may be a bending mode near scission that brings the centers-of-charge closer together at scission.

2. Interpretations of the shapes and structure of the odd-odd Nb and La nuclei

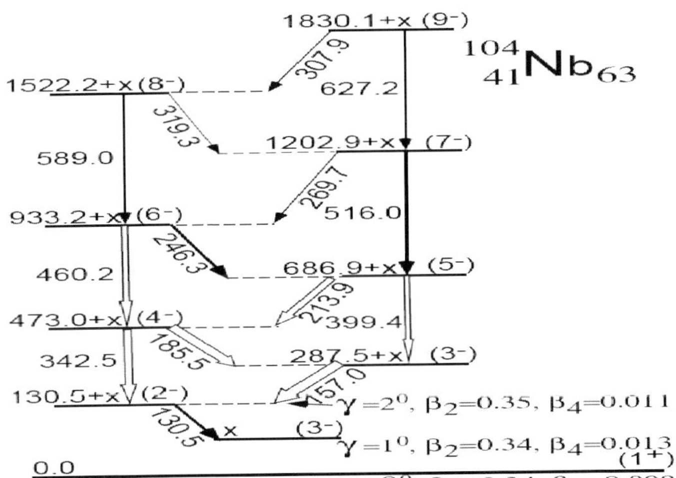

Fig. 1. High-spin level scheme proposed in ^{104}Nb[10,11]. An analogous band was identified in ^{106}Nb, which, however, has different decay-out pattern.

Based on our fission-γ data, very weakly-populated bands were identified and extended in the odd-odd ^{104,106}Nb[10,11] and ^{142}La[12]. Fig. 1 shows the level scheme of ^{104}Nb with deduced shapes. Analogous band structure and shapes but different decay-out patterns in ^{104,106}Nb were well reproduced by PSM and PES calculations. A $K^{\pi} = 2^{-}$, $\pi 1/2^{+}[431]$ (intruder) $\times$ $\nu 5/2^{-}[532]$ configuration was assigned to the band-heads of ^{106,104}Nb. 3^{+}, $\pi 1/2^{+}[431] \times \nu 5/2^{+}[413]$ was assigned to the 204.8 keV isomer in ^{106}Nb, and the M2 character of the 204.8 keV transition explained the 204.8 keV isomeric decay from the 0.66 μs isomer[10]. However, in ^{104}Nb the 3^+ state is bypassed by the very low excitation of the band-head. The proton intruder $\pi 1/2^+[431]$ plays an important role in the shape evolution with regard to triaxiality.

The PSM calculations were performed to reproduce the level excitations E_{exc} and moments of inertia $J^{(1)}$ of the bands of ^{104,106}Nb[10,11]. The PSM calculations of E_{exc} on 2^-, and 3^-, $\pi 1/2^+[431]\times\nu 5/2^-[532]$ configurations of ^{106}Nb are shown in Fig. 2. One can see that the calculations based on $K^\pi = 2^-$, $\pi 1/2^+[431]\times\nu 5/2^-[532]$, which was also deduced by the PES calculations, well described the band of ^{106}Nb. All the PSM calculations with spin/parity/configuration other than $K^\pi = 2^-$, $\pi 1/2^+[431]\times\nu 5/2^-[532]$ yielded large discrepancies in comparison to the experiment.

Successful reproductions for E_{exc} and $J^{(1)}$ of the analogous band in ^{104}Nb by PSM calculations also based on $K^\pi = 2^-$, $\pi 1/2^+[431]\times\nu 5/2^-[532]$ were achieved as well. A near axialy-symmetric shape with large quadrupole deformation $\gamma = 2^0$, $\beta_2 = 0.35$, $\beta_4 = 0.011$ was also deduced for the band of ^{104}Nb[10,11].

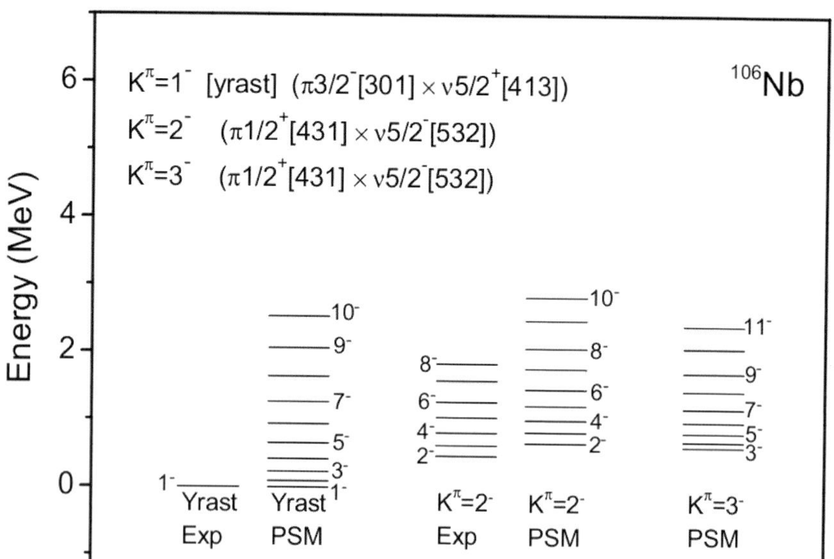

Fig. 2. PSM calculations for E_{exc} based on 2^-, and 3^-, $\pi 1/2^+[431]\times\nu 5/2^-[532]$ for ^{106}Nb[10,11]. Calculations based on 2^-, $\pi 1/2^+[431]\times\nu 5/2^-[532]$ well reproduced the level excitations. A near axially-symmetric shape with large quadrupole deformation $\gamma = 2^0$, $\beta_2 = 0.35$, $\beta_4 = -0.005$ was deduced for the band of ^{106}Nb.

One intriguing result on ^{142}La by TRS calculations comes for the case of $1/2^-$ spin-parity and negative signature for both the odd proton and the odd neutron. The uncranked solution shows the minimum-energy shape to be weakly prolate with a quadrupole β_2 of 0.14 and hexadecapole β_4 of 0.046. The triaxiality is nearly zero at 5^0. The surprising feature is that the total spin is 5 with 4 units from the protons and 1 unit from the neutrons. With modest cranking (ω = 0.4) the total spin rises to 14. The shape is hardly changed. Our experimental paper on fission-γ spectra of ^{142}La[12] is now being refereed, in which there is an excited band unconnected to lower levels, that may have these features of the TRS results.

Acknowledgments

Work partially supported by US DOE Grants and Contracts DE-FG-05-88ER40407 at VU and LBNL. Work at HUTC, China was supported by the National Natural Science Foundation (NNSF) of China under Grants No. 11305059, 11275063 and 11275068. The work at SJTU was supported by the NNSF of China under Grant No. 1113500511075103 and by the 973 Program of China, with Grant No. 2013CB83440. The work at Peking University was supported by the NNSF of China under Grants No. 11235001 and 11320101004.

References

1. K. Hara and Y. Sun, *Int. J Mod. Phys.*, E**4**, 637(1995).
2. P. Möller *et al.*, At Data Nucl Data Tables, **98**, 149(2012).
3. T. Bengtsson and I. Ragnarsson, *Nucl. Phys.* A**436**, 14(1985).
4. Y.C. Yang *et al.*, *J Phys G: Nucl. Part. Phys.*, **37**, 085110(2010).
5. Y.X. Liu, Y. Sun, X.H. Zhou *et al.*, *Nucl. Phys.* A**858**, 11(2011).
6. Y. Sun, Y.C. Yang, H.L. Liu, K. Kaneko, M. Hasegawa and T. Mizusaki, *Phys. Rev.* C**80**, 054306(2009).
7. Y. Sun, J.Y. Zhang, G.L. Long *et al.*, *Chinese Sci. Bull*, **54(3)**, 358(2009).
8. Z.Y. Zhou, Y.X. Liu, Y.C. Yang *et al.*, *Chinese Sci. Bull*, **59**, 3853(2014).
9. Y. Sun, X.R. Zhou, G.L. Long, E.G. Zhao, P.M. Walker, *Phys. Lett.* B**589**, 83(2004).
10. Y.X. Luo *et al.*, *Phys. Rev.* C**89**, 044326(2014).
11. Y.X. Luo *et al.*, *Chinese Nucl. Phys. Rev.*, Vol. **32**, No. 1, 1(2015).
12. Y.X. Luo *et al.*, submitted to *Phys. Rev. C.*

How Well Do We Know Nuclear Electric Quadrupole Moments?

N. J. Stone

Dept. of Physics and Astronomy, University of Tennessee,
Knoxville, TN 37996, USA
E-mail: nick.stone@physics.ox.ac.uk

A new Table of Recommended Nuclear Electric Quadrupole Moments was published recently [1]. The main feature of the table is the adoption, wherever possible, of the best available calculation of the electric field gradient (efg), required to extract the moment from the measured quantity in most measurement methods. This paper presents a short discussion of the accuracy to which quadrupole moments of nuclear states of all elements are known, highlighting that fact that few, if any, are known to better than a few percent. For some 19 elements no reliable efg has been adopted and up-to-date calculation for these elements remains an outstanding need.

Keywords: Nuclear Electric Quadrupole Moments.

1. Introduction

The electric quadrupole moment is a feature of atomic nuclei which finds relevance in a wide range of fields of research and more general measurement from chemical analysis to explosive and narcotics detection. The quadrupole moment is most obviously of value in the study of nuclei themselves, that is in nuclear structure physics, being a measure of the deviation of the nuclear charge distribution from a spherical shape. However, in addition, measurement of nuclear quadrupole interaction, by aiding the elucidation of complex electronic structures and their functions, forms a vital investigatory tool in other areas extending into chemical, biological and medical science.

Establishing an accepted value of the quadrupole moment of any nuclear state faces a fundamental problem. The difficulty lies in the fact that what can actually be measured is the electric quadrupole interaction, usually written as $h\nu_Q = e^2qQ$. The interaction energy involves the product of the moment, eQ, with the electric field gradient (efg), eq, acting at the nucleus. The situation can be contrasted to the challenge of measuring the nuclear magnetic dipole moment, μ, which similarly involves its product with the magnetic field B acting at the nucleus producing the magnetic dipole interaction energy, $h\nu_M = \mu B$. In the latter case it is a relatively simple matter to apply a known magnetic field, generated by a suitable arrangement of coils carrying a precisely measured current, which can be varied at the will of the experimenter. The product $h\nu_M$ can be measured by a wide range of techniques and the nuclear magnetic moment extracted. However it is not possible to generate an efg in the laboratory which is sufficiently large to produce measurable energy differences in the product $h\nu_Q$. Given the small scale of nuclear quadrupole moments, ~one barn [100 fm^2], to provide an energy splitting between levels of 10 MHz requires an efg ~10^{17} Vcm^{-2}. Sufficiently large gradients are, however, to be found at nuclei when in atoms, ions or molecules and also in a wide range of non-cubic solids. This does not solve the problem of extracting the quadrupole moment since the efg has to be calculated from first principles in a multi-electron system. The difficulty of such calculations has been a serious barrier to separation of the nuclear quadrupole moment from the measured interaction energy.

Recent compilations of measurements of nuclear electric quadrupole moments contain a full record of all published results on ~1050 nuclear states of which over 600 refer to stable isotopes, long-lived ground states or isomers and about 450 to short lived (< ms) excited states [1,2]. Many results, published some time ago, utilized efg values which, although the best available at the time, have been more recently improved upon. Earlier this year a new table, of recommended or 'best' moment values, was published in which, as far as possible, all results have been brought up-to-date taking advantage of our improved ability to calculate efg's in many systems [3]. The present short paper describes the current state of our knowledge of these moments, with reference to their general precision, or the lack of it, detailing where there have been significant improvements and for which elements large uncertainties still remain. For all specific information readers are referred to ref [3] for the present recommended 'best' moment values and their source, whilst those who wish to be aware of all original work should consult the more complete compilation [2].

2. Adopted primary standard electric field gradients

This is not the place to enter into a history of multi-electron electric field distribution theory, however until the recent advent of powerful computers only the simplest atoms or ions, with few electrons, were open to accurate calculation. For this reason the study of quadrupole hyperfine interactions in muonic atoms, in which the muon wavefunction [and hence charge distribution and efg] at the nucleus can be accurately predicted, was early seen as a fruitful way to establish reliable values of quadrupole moments, especially in heavier elements where the interaction can be readily resolved. Until the past two decades or so, for multi-electron atoms and ions, the best efg estimates involved approximations and corrections related to the distortion of inner closed electron shells, collectively associated with the name Sternheimer shielding [or anti-shielding] factors, which could be uncertain to 10% or more. Thus, although experimentally the nuclear quadrupole interaction could be measured to very much better precision, the nuclear quadrupole moments extracted from the measurements were hard to pin down.

In recent years multi-electron system computation has advanced to the extent that it is now possible to make calculations of efgs in most atoms and ions and also in many molecules. The results show admirable self-consistency and are providing reliable efg's of accuracy estimated to be comparable to that of the best muonic atom results. First in 1992 and again in 2001 and 2008 Pyykko, one of the pioneers of such calculations, published a listing of recommended values for one or two quadrupole moments of selected nuclear states, usually stable isotopic ground states, in the majority of elements, including those most commonly used in applications beyond the realm of nuclear structure physics [4, 5, 6]. These moments use the best available efg calculation for the system in which the interaction was measured. Adopting a standard moment value for a state in one isotope of an element allows calibration of efgs in other environments. Hence moments of ground and excited states in all isotopes of the same element, can, in principle, be related to the single reference moment. [To date there has been no consideration of possible small effects corresponding to the hyperfine anomaly for magnetic interactions.] The modern efg calculations include all effects previously grouped under the name of shielding.

There exist in the literature many excellent experimental results giving rise to values of nuclear quadrupole moments. Each measurement required the authors to specify the efg at the nucleus in their particular experimental system. The published moments depend upon the individual choice of efg and can thus be difficult to compare consistently without considerable effort. This is especially true if the experiments were made many years apart, during which time the best efg estimates may have changed. Existing tabulations do not always specify adopted standards and where they do, the standard values can also change over the years. Only recently has it been possible, with the aid of the much improved efg calculations, to prepare a tabulation in which the adopted standards are likely to prove stable over at least a period of years.

The new table [3] presents a listing of measured quadrupole moments normalised, as far as possible, to the set of identified 'reference' moments, one or more for each element, which have been derived from measurement of $h\nu_Q$ in a situation in which the efg has been calculated either by a modern computation or in a simpler system such as a muonic atom. For the majority of elements the reference values chosen by Pyykko in his most recent listing [6] have been adopted. There are still some 19 'problem' elements, in the sequence to einsteinium at Z = 99 for which no basic reference standard is adopted.

3. Adopted secondary standard electric field gradients

It is frequently not possible to normalize all reported quadrupole interaction measurements on every nuclear state in all isotopes of any given element to the adopted primary reference moment for that element. This arises not only when there is no adopted primary standard, as described above, but also frequently in connection with excited state $h\nu_Q$ measurements when an equivalent measurement on the reference (usually stable) isotope in the same system cannot be made. In such cases single or multiple secondary standard efg's have been adopted. These are briefly specified in the table in the heading for each element and full reference is given. The most commonly used secondary standards utilise efg's calculated at regular sites in non-cubic metals such as Zn, Ga, Os and hcp Co. Other quadrupole moments are extracted from a measured ratio with the moment of a second state for which the moments may be reliably estimated from nuclear theory, or from the transition rate (B(E2)) of the decay of the state in question.

4. Methods needing no efg calculation

In contrast to the foregoing discussion there do exist methods of measurement of the quadrupole moment which are free of uncertainty of estimation of a complex, multi-electron, efg but involve calculation of field gradients provided by free electric charges. They include electron scattering [ES] and re-orientation of the angular distribution during coulomb excitation [CER]. Both of these have been extensively used in the study of the electric quadrupole interactions of short lived excited nuclear states. Calculation of the electromagnetic interaction with the nucleus is straightforward in these systems and the quadrupole moment can be directly extracted, although the experiments themselves have limited accuracy. Results based on these methods are therefore not directly related to, or dependent upon, the value of the adopted standard quadrupole moment for that element.

5. Experimental methods

The method used for each measurement is indicated briefly in the table. For most stable nuclear ground states and for very many radioactive ground states and longer-lived isomers the method used involves an atomic, ionic or molecular beam. Application of many sophisticated methods of laser spectroscopy has allowed accurate measurement of magnetic dipole and electric quadrupole hyperfine interactions with increasing sensitivity. The second prevalent source of accurate measurements, particularly in the heavier elements, has been resolved hyperfine interactions in muonic and pionic atoms. Low temperature nuclear orientation, extended to resolved NMR and the MAPON technique, has also made contributions. For shorter-lived nuclear excited states common methods include, for lighter elements, β-NMR and β-NQR, with perturbed angular correlation and distribution, time independent and time dependent, for heavier elements. Coulomb excitation reorientation has been the predominant method to study collective excited states in even-even nuclei.

6. Signs and estimates of uncertainty

The sign of the quadrupole moment is given when the sign can be determined from experimental data. Where the sign is not given by the measurement, no sign is offered in the published Table [1], although it can sometimes be inferred either from systematics or from the magnitude of the result. Estimates of uncertainty are not straightforward as in many other measurements since the efg is the product of a calculation. Authors of these calculations usually provide uncertainty estimates, which are combined with those of the measurements to give the moment error estimate. This has been done where updated efgs have been used to recalculate moments from older experimental results.

One independent check on the correctness of calculated efg's, and of their suggested accuracy, is the degree to which calculations in different systems lead to a consistent quadrupole moment of a single nuclear state from measurements in those systems. Pyykko has highlighted some cases of agreement and some of disagreement. Table 1 gives examples of successes and failures at different levels of claimed accuracy. They demonstrate that there may exist systematic differences between calculations on multi-electron atomic, ionic and molecular systems and those in solid state systems (see e.g. Cl and Au entries) and all three show some differences with the older mesonic atom estimates (see Lu entry, for a discussion on Na and Al see Pyykko [5]). Where there have been several atomic calculations these also can show differences greater than the authors' stated uncertainties.

Table 1. Variation between latest ground state Q moment values (in barns) from different types of measurement and analysis. Where two isotopes are indicated the entry is their moment ratio.

Isotope	Atomic	Molecular	Solid state	Mesonic	Variation (%)
^{23}Na	0.104(1)	0.104(1)			0(1)
^{35}Cl	−0.0817(8)		0.0850(11)		4.0(16)
^{69}Ga	0.174(3)	0.171(2)			1.8(22)
^{79}Br	0.302(5)	0.313(3)			3.6(19)
^{85,87}Rb	0.48050(16)	0.48383(2)			0.69(3)
^{115}In	0.772(5)	0.770(8)			0.3(8)
^{127}I	−0.680(10)			0.722(21)	6(3)
^{175}Lu		3.415(34)		+3.49(2)	2.2(11)
^{197}Au	0.521(7)	0.510(15)	0.560(30)		9.8(54)

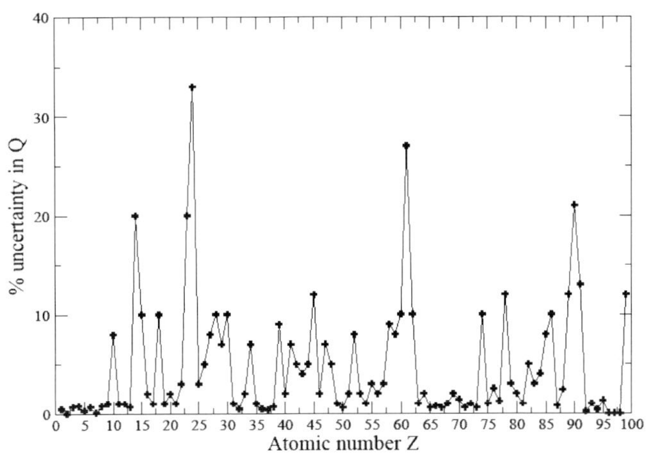

Fig. 1. Uncertainty (%) in reference Q values vs atomic number.

Fig. 1 shows the estimated uncertainty in the adopted standard 'best' quadrupole moment measurement for each element, plotted against the elemental atomic number. This plot shows the present state of the art demonstrating that even the best calculations are considered to have uncertainties of order 1 %, with many others at the 5-10 % level. The one exception to this conclusion is the result of recent calculation for the quadrupole moment of the deuteron, 0.0028578(3) b [7]. The reported uncertainty is reduced by a factor of 50 as the result of new efg calculation technique applied to HD and D_2 molecules.

7. Notable changes to previous moment values

Preparation of the new table required updating many moments from well determined hν_Q values using improved efg calculations. Not only did this change the extracted value of the moment, but also the uncertainty of the result was frequently reduced.

Comments in this section refer to entries in the published Table [1] as compared with the comprehensive listing of all measurements which appeared in 2005 [8]. Many entries are unchanged as they were already based on the adopted efg's and for most others the changes are small, of the order of a few percent or less. However in a several notable cases recent efg calculations using up-to-date methods as reviewed by Pyykko [5, 6] have led to major change in the extracted quadrupole moment. Details are given in Table 2.A. The cases of fluorine, calcium, germanium, silicon and tin involve more recent efg calculations being applied to older measurements. In the case of the 14.4 keV state in iron the large increase was close to a factor of two. Table 2.B lists cases for which the improved efg calculations have led to major reduction in the quoted uncertainty in the extracted quadrupole moment.

8. Problem elements

There remain a total of 19 elements for which no efg calculation has been adopted as reliable and up-to-date. In this the author has been guided by the work of Pyykko, who is the acknowledged authority, especially in the area of multi-electron atomic, ionic and molecular calculations. The problem elements are: He (no quadrupole interaction data exist for this element), Si, P, Cd, Sn, Te, Ce, Tm, W, Pt, Tl, Po, At, Ac, Th, Pa, Cm, Bk and Cf. For several of these the best estimates of quadrupole moments have larger uncertainties than discussed above (see Table 2.C) caused by the large uncertainties in older efg calculations – examples are Nd (10%) and Pm (27%) where Sternheimer corrections are still part of the estimates. It is clearly desirable that improved efg calculations be made for these elements.

9. Conclusion

The determination of nuclear electric quadrupole moments has undergone a fundamental change consequent upon recent developments in multi-electron configuration computation. Nevertheless, whilst for many elements we can now take the best-measured quadrupole moment to be accurate to a few percent, there

Table 2. A. Elements for which new efgs have produced major change in extracted quadrupole moments.
B. Elements for which new efgs reduced uncertainty by a substantial factor.
C. Elements for which efg inaccuracy causes large (> 9%) quadrupole moment uncertainty.

A	Element	Change in Q (%)	Element	Change in Q (%)
	F	−22	Sn	+25.7
	Ca	−26	Sb	+52
	Ge	+15.3	Cs	−6.0
	Se	−30.9	Ba	+9.6
	Sr	−7.6	Gd	−5.0
	In	−5.1		

B	Element	Reduction factor	Element	Reduction factor
	C	0.10	Sr	0.1
	F	0.25	In	0.4
	Ge	0.03	Sn	0.7
	Se	0.14		

C	Element	Unc. in efg (%)	Element	Unc. in efg (%)
	Cr	33	Rn	10
	Zn	10	Ac	12
	Nd	10	Th	21
	Pm	27	Es	12
	Sm			

remain a significant number for which the uncertainty in the absolute value is considerably larger. Recent evidence has also shown that new, improved, field gradient calculations not only lead to smaller uncertainties in extracted quadrupole moments, but also to large adjustments in their magnitude – and these adjustments can be significant both for theoretical nuclear physics interpretation and for the use of the quadrupole interaction in its many applications.

Acknowledgements

The author gratefully acknowledges help and advice from Pekka Pyykko and assistance in obtaining NSR references and obscure publications from the staff of the National Nuclear Data Centre, Brookhaven National Laboratory, in particular Joann B. Totans, Boris Pritichenko and Jagdish Tuli. Financial support for this work has come in part from DoE under grant DE-FG02-96ER40983 and in part from the Nuclear Data Section, IAEA, Vienna.

References

1. N. J. Stone, *Atomic Data and Nuclear Data Tables* **111-112**, 1 (2016).
2. N. J. Stone, *IAEA Vienna Report INDC(NDS)-0658* (2014).
3. N. J. Stone, *IAEA Vienna Report INDC(NDS)-0594* (2011) and *Atomic Data and Nuclear Data Tables* **90,** 75 (2005).
4. P. Pyykko, *Z. Naturforsch.* **47a** 189, (1992).
5. P. Pyykko, *Molecular Physics* **99**, 1617 (2001).
6. P. Pyykko, *Molecular Physics* **106**, 1965 (2008) and private communication.
7. M. Pavanello, W.-C. Tung and L. Adamowicz, *Phys. Rev* **A81,** 042526 (2010).

Chiral Vibrations and Collective Bands in ^{104}Mo

B. Musangu[1,2], E. H. Wang[1], J. H. Hamilton[1], S. Frauendorf[3], C. J. Zachary[1], J. M. Eldridge[1], A. V. Ramayya[1], W. H. Brantley[2], J. O. Rasmussen[4], Y. X. Luo[1,4], G. M. Ter-Akopian[5], Yu. Ts. Oganessian[5] and S. J. Zhu[6]

[1]*Department of Physics, Vanderbilt University, Nashville, TN 37235, USA*
[2]*Department of Physics, Furman University, Greenville, SC 29613, USA*
[3]*Department of Physics, University of Notre Dame, Notre Dame, IN 46556, USA*
[4]*Lawrence Berkeley National Laboratory, Berkeley, CA 94720, USA*
[5]*Joint Institute for Nuclear Research, RU-141980 Dubna, Russian Federation*
[6]*Department of Physics, Tsinghua University, Beijing 100084, China*

High spin states of the neutron-rich ^{104}Mo nucleus have been reinvestigated by analyzing the γ-rays in the spontaneous fission of ^{252}Cf with Gammasphere. Both γ-γ-γ and γ-γ-γ-γ coincidence data were analyzed. A new $\Delta I = 1$ band has been discovered with a tentative 5^- band-head, and is proposed to form a class of chiral doublets with another 4^- band previously found. Angular correlation measurements have been performed to determine the spins and parities. The origin of the chiral doublet bands in ^{104}Mo is interpreted as neutron $h_{11/2}$ particle and mixed $d_{5/2}$, $g_{7/2}$ hole coupled to the short and long axis, respectively.

1. Introduction

In nuclear structure research, chirality, or handedness is the study of left- and right- handed symmetry in a nucleus which yields in a pair of nearly degenerate $\Delta I = 1$ rotational bands. Such phenomenon was predicted theoretically in 1997 [1]. Chiral doublet occur when the angular momentum has components along all the three principle axes. Since then, chiral nuclei have been suggested experimentally in A~80 [2, 3], A~100 [4-16], A~130 [17-20], A~190 [21, 22] mass regions. For Mo nucleus, the soft triaxial ^{106}Mo was suggested to have chiral doublet bands [9]. In ^{106}Mo, chirality is generated by neutron $h_{11/2}$ particle and mixed $d_{5/2}$, $g_{7/2}$ hole coupled to the short and long axis, respectively.

The ^{104}Mo nucleus, as well as ^{106}Mo, has one-and two-phonon gamma vibrational bands which indicate the softness with respect to triaxial deformations [23, 24]. In these soft nuclei, nuclear shapes may be driven to stable triaxial due to the excitation of quasi-particles [25, 26]. In the present work, we find the candidates for chiral doublet bands in ^{104}Mo with more degenerate energy (~60 keV) than ^{106}Mo (~100 to 140 keV). Close agreement is expected theoretically for chiral bands.

2. Experimental Technique and Result

The experiment was carried out at the Lawrence Berkeley National Laboratory (LBNL). A 62 μCi ^{252}Cf source was sandwiched between two 10mg/cm^2 Fe foils and was mounted in a 7.62 cm diameter plastic (CH) ball. By using 101 Ge detectors of Gammasphere, high statistics of prompt γ-rays emitted in spontaneous fission of ^{252}Cf were detected. The data were sorted into 5.7×10^{11} γ-γ-γ and higher fold and 1.9×10^{11} γ-γ-γ-γ and higher fold γ coincident events. These data were analyzed by the RADWARE software package [27]. The angular resolution of Gammasphere is used to obtain γ-γ angular correlations [28]. More details of this experiment can be found in Ref. [29].

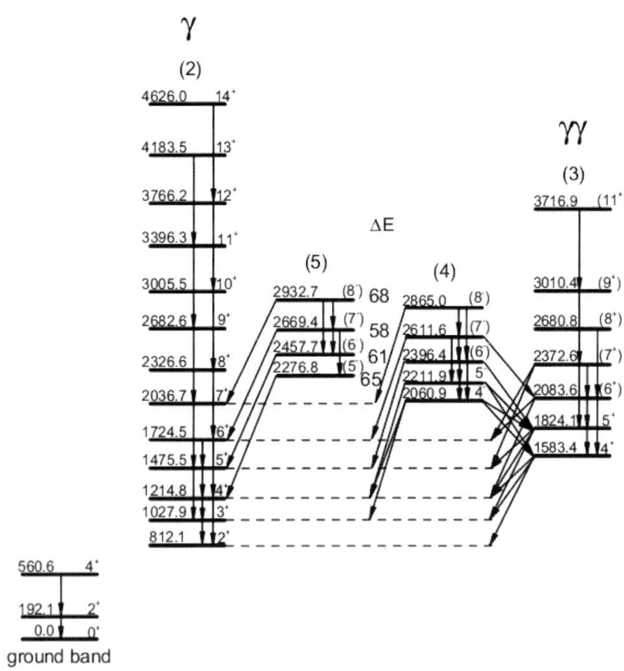

Fig. 1. Partial decay patterns of γ and γγ bands in ^{104}Mo. Energy separations at the same spins between bands (4) and (5) are given in between. Transitions depopulating the γ band (2) are not plotted.

The partial level scheme of ^{104}Mo obtained in the present work is shown in Fig. 1 with 5 band structures. A total of 13 new transitions and 5 new levels are identified compared to the previous high spin study of this nucleus [24, 30, 31]. The previously observed band (4) and the newly identified band (5) are proposed to form a chiral doublet. Bands (2) and (3) are suggested as one and two phonon γ vibrational bands, respectively [24]. Some transitions in band (1) and populating band (1) are omitted in the figure. Fig. 2 provides the evidence of the new band (5). The small part on top denotes a double gate on 368 (560 to 192) keV transition and 915 (1475 to 560) keV transition, which illustrate the existence of the 982 keV (2457 to 1475) one. The main part represents an "&" of two double gates, 982 keV transition with 192 keV one, and 915 with 982 keV one, respectively. In such spectrum, the M1, E2 transitions in band (5) as well as the $^{144,146}Ba$ fission partner transitions can be seen.

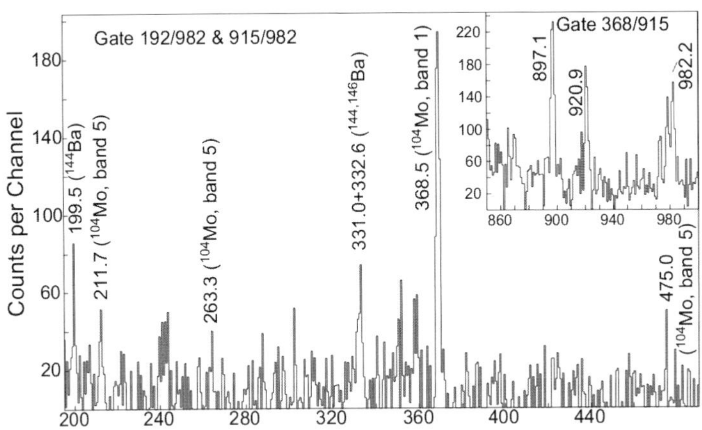

Fig. 2. Partial γ-ray coincidence spectra by gating on 192.5 keV $2\to0$ transition and 982.2 keV one depopulating 6^- state in band (5) & 914.9 keV depopulating 5^+ in band (2) and 982.2 keV one in ^{104}Mo. The right-top part denotes a double gate on the 368.5 keV $4\to2$ transition and 914.9 keV one.

Angular correlation measurements have been made to determine the spins and parities in ^{104}Mo. The results are $A_2 = 0.178(9)$, $A_4 = -0.027(14)$ for the 477.4 (band (4) $4^- \to$ band (3) 4^+) - 771.3 (band (3) $4^+ \to$ band (2) 2^+) cascade, and $A_2 = 0.251(46)$, $A_4 = 0.009(69)$ for the 387.8 (band (4) $5^- \to$ band (3) 5^+) - 796.2 (band (3) $5^+ \to$band (2) 3^+) cascade. The two results agree with theoretical 4(D)4(Q)2 and 5(D)5(Q)3 values, which are $A_2 = 0.196$, $A_4 = 0$, and $A_2 = 0.186$, $A_4 = 0$, respectively. This measurement confirms the assignment of the 4^- and 5^- states in band (4). The band-head of band (5) is tentatively argued as 5^- according to the decay pattern and level energy differences.

3. Discussion

The kinetic moments of inertia of bands (4) and (5) in the proposed triaxial ^{104}Mo remain almost constant and equal with increasing spin. Both these bands exhibit small signature splitting, as shown by the very small staggering of kinetic moments of inertia indicated as $J^1 = I/(E(I)-E(I-1))$ in Fig. 3. Such behavior is very analogous to the case in ^{106}Mo in Fig. 3. The energy differences of these doublet bands are quite small and almost constant with increasing spin, as listed in Fig. 1, in particular, about half of the differences in ^{106}Mo. These are the characteristics of very soft chiral vibrations.

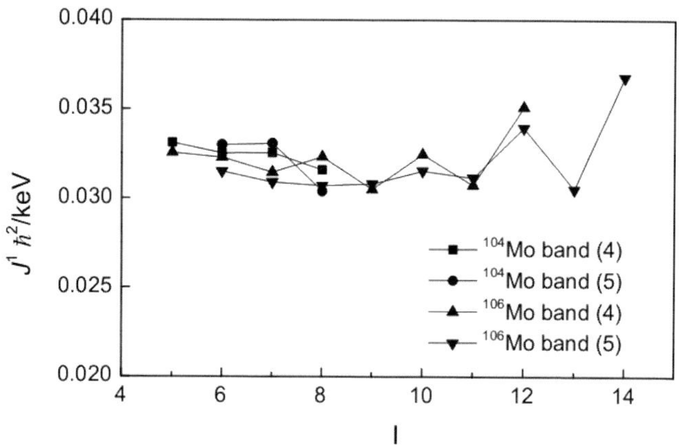

Fig. 3. Plots of moments of inertia (J^1) vs I for ^{104,106}Mo.

4. Conclusion

In summary, new levels and transitions have been identified in neutron-rich ^{104}Mo. Bands (4) and (5) in this nucleus are proposed as chiral vibrational doublet bands. These doublet rotational bands in ^{104}Mo show similar behavior to those in ^{104}Mo but exhibit closer separation energies as expected theoretically.

Acknowledgments

The work at Vanderbilt University and Lawrence Berkeley National Laboratory are supported by the US Department of Energy under Grant No. DE-FG05-88ER40407 and Contract No. DE-AC03-76SF00098. The work at Tsinghua University was supported by the National Natural Science Foundation of China under Grant No. 11175095. The work at JINR was supported by the Russian

Foundation for Basic Research Grant No. 08-02-00089 and by the INTAS Grant No. 03-51-4496. In the summer of 2016 Brooks Musangu at Vanderbilt University was also supported by the Furman University through Furman Advantage.

References

1. S. Frauendorf and J. Meng, *Nucl. Phys. A* **617**, 131 (1997).
2. S. Y. Wang *et al., Phys. Lett. B* **703**, 40 (2011).
3. T. Koike *et al., Int. J. Mod. Phys. E* **20**, 520 (2011).
4. C. Vaman *et al., Phys. Rev. Lett.* **92**, 032501 (2004).
5. P. Joshi *et al., Phys. Lett. B* **595**, 135 (2004).
6. J. Timár *et al., Phys. Lett. B* **598**, 178 (2004).
7. J. A. Alcántara-Núñez *et al., Phys. Rev. C* **69**, 024317 (2004).
8. P. Joshi *et al., Eur. Phys. J. A* **24**, 23 (2005).
9. S. J. Zhu *et al., Eur. Phys. J. A* **25**, 459 (2005).
10. P. Joshi *et al., Phys. Rev. Lett.* **98**, 102501 (2007).
11. T. Suzuki *et al., Phys. Rev. C* **78**, 031302R (2008).
12. Y. X. Luo *et al., Phys. Lett. B* **670**, 307 (2009).
13. J. Sethi *et al., Phys. Lett. B* **725**, 85 (2013).
14. D. Tonev *et al., Phys. Rev. Lett.* **112**, 052501 (2014)
15. E. O. Lieder *et al., Phys. Rev. Lett.* **112**, 202502 (2014).
16. N. Rather *et al., Phys. Rev. Lett.* **112**, 202503 (2014).
17. K. Starosta *et al., Phys. Rev. Lett.* **86**, 971 (2001).
18. S. Zhu *et al., Phys. Rev. Lett.* 91132501 (2003).
19. E. Grodner *et al., Phys. Rev. Lett.* **97**, 172501 (2006).
20. S. Mukhopadhyay *et al., Phys. Rev. Lett.* **99**, 172501 (2007).
21. D. L. Balabanski *et al., Phys. Rev. C* **70**, 044305 (2004).
22. P. L. Masiteng *et al., Phys. Lett. B* **719**, 83 (2013).
23. A. Guessous *et al., Phys. Rev. Lett.* **75**, 2280 (1995).
24. A. Guessous *et al., Phys. Rev. C* **53**, 1191 (1996).
25. S. Frauendorf *et al., Rev. Mod. Phys.* **73**, 463 (2001).
26. V. Dimitrov *et al., AIP Conf. Proc.* **656**, 151 (2003).
27. D. C. Radford, *Nucl. Instrum. Methods Phys. Res. A* **361**, 297 (1995).
28. A. V. Daniel *et al., Nucl. Instrum. Methods B* **262**, 399 (2007).
29. J. H. Hamilton *et al., Prog. Part. Nucl. Phys.* **35**, 635 (1995).
30. H. Hua *et al., Phys. Rev. C* **69**, 014317 (2004).
31. E. F. Jones *et al., Phys. Atom. Nuclei* **69**, 1198 (2006).

A Revisit to The Nuclear Structure of ^{154}Gd in DPPQ Model

J.B. Gupta[1] and J.H. Hamilton[2]

[1]*Ramjas College, University of Delhi, Delhi-110007 India*
[2]*Physics Department, Vanderbilt University, Nashville, TN, U.S.A*

In a previous study of ^{154}Gd spectrum, using the dynamic pairing plus quadrupole model, data for eight *K*-bands were analyzed. Here we present a revised analysis of the new data and new concepts. A discussion is given on the view of the shape coexistence of β-band with ground band and on the nature of the $K^\pi = 0^+$ excited bands. Comparison is made with the X(5) symmetry. The validity of the multi-phonon view of the $K^\pi = 4^+$ and 2_2^+ bands is also illustrated.

1. Introduction

The nucleus ^{154}Gd is one of the four $N = 90$ isotones ($Z = 60$-66), associated with the X(5) critical point symmetry [1, 2]. Earlier in (1979) [3], its band structure comprising eight *K*-bands (ground, β-, γ-), $K^\pi = 0_3^+$, and ($K^\pi = 0_4^+, 2_2^+, 4_1^+$) and 0_5^+ bands were studied in the dynamic pairing plus quadrupole model [4]. New data [5, 6] in the last four decades mainly concerns the revision of the levels in K^π - $0_{3,4,5}^+$ bands. The 1295 keV 0_3^+ is replaced by 1182 keV and 4^+ at 1698 keV by 1701 keV one. In the $K^\pi = 0_4^+$ band, 1574 keV, 1716 keV and 2080.23 keV are assigned. In the $K^\pi = 0_5^+$ band, 1650.3 keV band head and 1775.4 keV 2^+ are new assignments. So, a revisit to the DPPQ model predictions is warranted. Here we discuss the different views on the multiphonon nature of the excited 0^+ bands and on the $K^\pi = 2_2^+$ βγ band at 1531 keV, and the $K^\pi = 4_1^+$ band at 1646 keV.

In the DPPQ model, the parameters of the collective Hamiltonian

$$H_{coll} = V(\beta,\gamma) + T_{vib}(\beta,\gamma) + T_{rot}(\beta,\gamma), \tag{1}$$

are derived from the microscopic Hamiltonian [4]

$$H_{PPQ} = H_{sph} + H_Q + H_P. \tag{2}$$

The first term represents the average potential of all the nucleons in the n = 4,5 for protons (n = 5,6 for neutrons) harmonic oscillator shells, and the term H_Q represents the quadrupole part and H_P the monopole pairing part of the residual nuclear interaction. The H_{coll} is determined at 92 points of a (β,γ) mesh on a triangle space with β = 0-0.5, and γ = 0°-60°. Summation over the full (β,γ) space provides the dynamics of the Hamiltonian. The wave functions and the transition data are derived for the nucleus. Results from Interacting boson model [7] calculation are also included.

2. Results from theory

2.1. *Energy level spectrum*

The seven *K*-bands spectrum is given in Fig. 1 and the DPPQM spectrum is given in Fig. 2. The vibrational bands lie slightly higher in DPPQM spectrum, but the overall pattern is well reproduced. The IBM-1 calculation is limited to 4 bands.

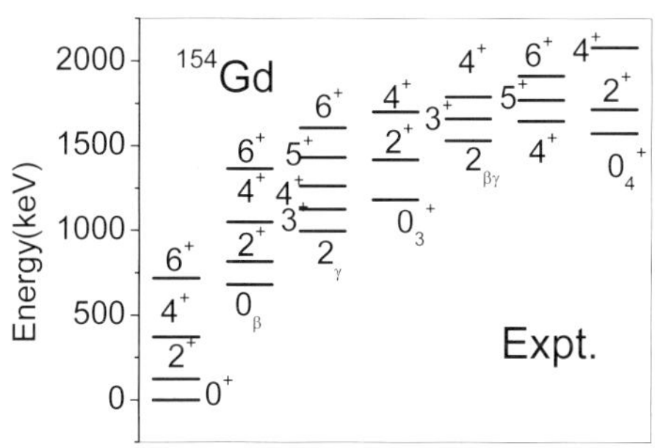

Fig. 1. The partial energy level spectrum in ^{154}Gd [5].

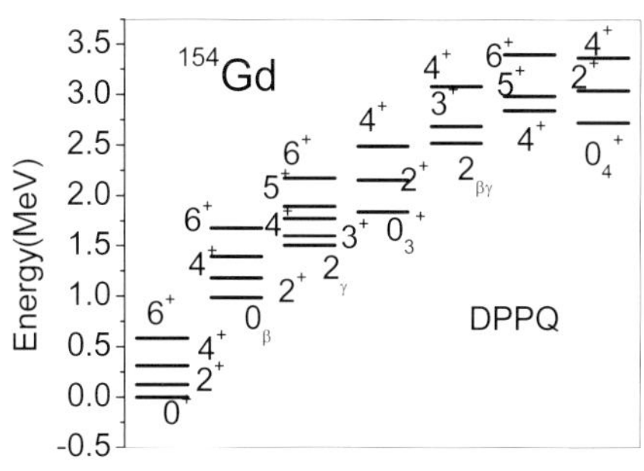

Fig. 2. The partial energy spectrum of ^{154}Gd from DPPQM.

2.2. *K-band structure from DPPQ model*

In g-, β-bands $K = 0$, and in γ- $K = 2$, exceed 98%, signifying pure K–bands. The $K = 0$, 2 components in (1418, 1531, 1716, 1775) keV 2^+ states are (81.4+ 18.6)%, (21+79)%, (95.7+4.3)%, (44+56)%. The 4^+ states at (1701, 1646, and 1789) keV are (70+22+8)%, (22+13+65)% and (9+63+28)%.respectively. While the lower 3 bands lie in the prolate part of the V(β,γ) potential energy surface (PES), the higher bands lie above the spherical barrier, representing the mixing of the prolate, oblate and spherical shapes.

2.3. *B(E2) values and B(E2) ratios*

2.3.1. Ground, β- and γ-band

Table 1. Absolute B(E2) values (e^2b^2) in ^{154}Gd.

I_i	I_f	Expt.[2,6]		DPPQ	IBM-1
0_g	2_g	3.85	15	3.86	3.82
	2_β	022	2	0.018	0.0025
	2_γ	140	5	0.139	0.135
	$2_{02}{}^+$			0.0065	0.0035
	$2_{5=\beta\gamma}$	0.0002		0.0002	

Table 2. Absolute B(E2) values (×100) (e^2b^2) for β-band.

I_i	I_f	Expt.[3]	Expt.[2]	DPPQ	X(5)	IBM-1
0_β	2_g	26 4	22 2	22	49^a	4^b
2_β	0_β	49 16	27 2	76	61	51
	0_g	0.48 4	0.45 5	0.37	1.5	0.05^b
	2_g	4.0 4	3.3 2	3.3	6.8	1.3
4_β	2_β	122 35	93 7	120	93	75

a) Normalized to B(E2, $2_g{}^+ \rightarrow 0_g{}^+$).
b) IBM-1 β-g transitions are weak.

Theory reproduces well the varying excitation strength from ground state (Table 1). The rotational intraband transitions of 2_β-0_β, 4_β-2_β (Table 2) reproduced in theory and other evidences indicate $0_2{}^+$ band as a rotational band. The previous assumption [8] of β-band as vibrational band co-existing with deformed ground band, based on weak B(E2, $2_\gamma \rightarrow 0_2{}^+$) (Table 3) is not supported. Thus a shape coexistence of spherical β-band and deformed g-band has to be abandoned. The γ-g interband $E2$ transitions are well reproduced in

DPPQM and IBM-1 (Table 3). Large intraband B(E2)s reproduced in DPPQM, also support the rotational nature of the γ-band.

Table 3. Absolute B(E2) $(e^2b^2)\times 100$ for γ-band.

Ii	If	Expt.[6]	DPPQ	IBM-1	X(5)
2_γ	0g	2.86 22	2.8	2.7	2.86[b]
	2_g	6.1 5	4.9	4.3	4.3
	4_g	0.86 7	0.44	0.65	0.22
	0_β	0.08[a]	0.07	0.10	0.25
3_γ	$2_\gamma/2_g$	19 1	26	23	
4_γ	$2_\gamma/2_g$	43 9	29	33	

a) Ref. [8]
b) Normalized. Ref. [9].

2.3.2. Nature of $K^\pi=0_3^+$ band at 1182 keV

Ratio $E(0_3^+)/E(0_2^+) = 1.74$ implies anharmonicity. Large B(E2, 0_3^+-$2_\beta/2_g$) supports 2β character. Large intraband transition from 1418 keV 2^+ to 0_3^+ and 1701 keV 4^+ to 1418 keV supports the *K*-band formation. Cross section σ(t,p) is large, but σ(p,t) g.s to g.s. is = 2.37(4), and to 0_2^+ = 20% of g.s. For 0_3^+ it is only 0.005%. Kulp *et al.* [10] called 0_3^+ state a pairing isomer. Clark *et al.* [11] explained large σ(t, p) on account of shape difference of target and product nucleus ground states. Fossion *et al.* [12] obtained in IBM peaks in σ(t,p) due to shape difference. Pairing isomer view is based on assumption of 0_3^+ being spherical.

Table 4. B(E2) ratios for $K^\pi = 0_3^+$ (2β)-band at 1182.1 keV.

Transitions		B(E2)ratios		
I_i	I_f/I_f	Expt.[3]	Expt.[5]	DPPQ
0_3^+	$2_\beta/2_g$		42 18	1910[a]
2_{03}	$2_\beta/2_\gamma$	>4.8	3.6 28	4.4
1418	$2_\gamma/2_g$	>96	32 24	560
	$0_3/0_\beta$	b)	225 50	810
4_4 -	$2_{\beta\beta}/2_\gamma$		742	144
1701	$2_\gamma/2_g$		13	5.2

a) In DPPQ model, B(E2, 0_3^+-2_g) is too small.
b) 2_{03}^+-0_3^+ 125 keV transition is replaced by 236 keV.

2.3.3. $K^{\pi} = 4^+$ band at 1646 keV

Table 5. B(E2) ratios for $K^{\pi}=4_1{}^+ I=4$ (1645.8 keV) band.

Transitions	Expt.[3]	Expt.[5]	DPPQ
$4_{K=4}$ $2_{\gamma}/2_{\beta}$	57 8	48 5	28
$2_g/2_{\gamma}$	0.004 1	0.0014 3	0.021
$4_{\gamma}/4_{\beta}$	6.6 10	8.5 16	22
$4_g/4_{\gamma}$	0.0013	0.0004 4	0.026
$3_{\gamma}/2_{\gamma}$	1.72 17	1.77 9	1.29
$3_{\gamma}/4_{\gamma}$	1.30 22	1.15 8	1.26

The B(E2) ratios for transitions from the $I = 4$ 1646 keV state (and 5^+, 6^+ members) to γ-, β- and g-band, indicating preferential decay to γ-band are fairly reproduced in DPPQM (Table 5). The absolute B(E2, $4_{\gamma\gamma} \rightarrow 2_{\gamma}$) in DPPQM of 0.047 e^2b^2 or~9 W.u, compared to B(E2, $2_{\gamma} \rightarrow 0_g$) = 0.029 e^2b^2 or ~6 W.u. indicates good collectivity. The energy ratio $E(4_{K=4})/E(2_{\gamma})$~1.65 indicates anharmonicity, a general feature of γγ (K = 0,2,4) bands. The possible admixture of *E*4 hexadecapole or 2 quasiparticle component, requires further considerations.

2.3.4. Nature of 1531 keV $K^{\pi} = 2_2$ band

Table 6. B(E2) ratios for $K^{\pi}=2^+$ (βγ), $I^{\pi}=2_5{}^+$ 1531.3 keV band.

I_i	I_f/I_f'	Expt.[3]	Expt.[5]	DPPQ
$2_{\beta\gamma}$	$0_3/0_{\beta}$		2.54 10	2.0
	$0_{\beta}/0_g$	713 51	716 30	550
$3_{\beta\gamma}$	$2_{\beta}/2_g$	219 11	197 8	368
	$2_{\beta\gamma}/2_{\gamma}$		216 36	188
	$2_{\beta\beta}/2_{\gamma}$		26 3	12
$4_{\beta\gamma}$	$4_{\beta}/4_g$		1	83

Kulp *et al.* [13] assigned the 349.2 keV γ-ray to the newly assigned 1182.2 keV $0_3{}^+$ level from the $2_{\beta\gamma}$ state. The B(E2) ratio relative to $0_{\beta}{}^+$ state agrees with the DPPQ model value. Fair reproduction of most B(E2) ratios from $K^{\pi} = 2_2{}^+$ band in DPPQM (Table 6) supports its collective nature.

2.3.5. Nature of $0_4{}^+$ and $0_5{}^+$ bands

The B(E2, 0_4 - $2_{\gamma}/2_{\beta}$) = 3.3 supports $K = 0$ γγ character. Same is the trend for 1716 keV 2^+ state. The 364.3 keV 4_{04} to 2_{04} decay supports band assignment.

The DPPQ model calculation yields reasonable B(E2) ratios corresponding to data (Table 7), with some ratios far off too, due to very weak transitions to g.s.

The 0_5^+ state decays predominantly to 2_β level, compared to 2_g level, with B(E2) ratio ~ 5.5 (Table 8). The 2^+ at 1775 keV decays to β-, and g-band levels. The preferential decay of 2_{05}^+ to 0_β^+ is given in DPPQM, though with larger ratio. Thus it supports the 3β character.

Table 7. B(E2) ratios for $K^\pi=0_4^+$ band.

I_i	I_f/I_f'	B(E2)ratios	DPPQ
0_4^+	$2_\gamma/2_\beta$	3.3 2	308
2_{04}^+	$2_\gamma/2_g$	32 18	200
1716	$3_\gamma/2_\gamma$	4.5 9	5.0
4_{04}^+	$3_{\beta\gamma}/2_{04}$		1.2
2080	$2_{04}/2_\gamma$	10 4	5380

Table 8. B(E2) ratios for 1650.34 keV, $K^\pi=0_5^+$-band.

I_i	I_f/I_f'	B(E2)ratios	DPPQ
0_5^+	$2_\beta/2_g$	5.5 19	0.1
2^+	$0_\beta/0_g$	5.1 13	34
1775	$0_g/2_g$	0.32 6	0.10
keV	$4_g/2_g$	0.90 19	0.07
	$2_\beta/4_\beta$	0.15 6	0.40

3. Summary

In the present study, data for all the newly assigned states in ^{154}Gd, in the last 4 decades, are compared with the predictions of the DPPQ model. Their *K*-band assignments and *E*2 transition rates are fairly well explained. The anomaly of shape coexistence of vibrational β-band with deformed ground band is discussed in detail. The possibility of 0_3^+ band being a pairing isomer band, as interpreted in literature, from the 2n transfer (t,p) and (p,t) reaction data is in conflict with the *E*2 transition data in the well formed band. The model predicts collective nature in view of the fair reproduction of many B(E2) ratios, and the calculated absolute B(E2) values.

The anharmonicities observed for the higher γγ (K = 0_4, $2_2$and 4) bands lead to varied views on their character. The fair prediction from DPPQ model supports their collective multi-vibrational interpretation, as discussed here and in detail in a Paper to be published.

References

1. F. Iachello, *Phys. Rev. Lett.* **87**, 052502 (2001).
2. D. Tonev *et al.*, *Phys. Rev.* **C69**, 034334 (2004).
3. K. Kumar, J.B. Gupta and J.H. Hamilton, *Aust. J. Phys.* **32**, 307 (1978).
4. K. Kumar and M. Baranger, *Nucl. Phys.* **A110**, 529 (1968).
5. http:\\www.nndc.bnl.gov/ENSDF.
6. C.W. Reich, Nuclear Data Sheets **110**, 2257 (2009).
7. F. Iachello and A. Arima, The Interacting Boson Model (Cambridge University Press, Cambridge, 1987).
8. N.V. Zamfir *et al.*, *Phys. Rev.* **C60**, 054312 (1988).
9. R. Bijker *et al.*, *Phys. Rev.* **C68**, 064310 (2003).
10. W.D. Kulp *et al.*, *Phys. Rev. Lett.* **91**, 102501 (2003).
11. R.M. Clark *et al.*, *Phys. Rev.* **C80**, 011303R (2009).
12. R. Fossion *et al.*, *Phys. Rev.* **C76**, 01 4316 (2007).
13. W.D. Kulp *et al.*, *Phys. Rev.* **C69**, 064309 (2004).

Transfer Reactions with ^{134}Xe

S. D. Pain*, K. A. Chipps, M. Febbraro, I. Marsh, M. S. Smith and R. L. Varner

Physics Division, Oak Ridge National Laboratory, Oak Ridge, TN 37831, USA
**E-mail: painsd@ornl.gov*

T. Baugher, R. Blanchard, J. A. Cizewski, H. Garland, A. Lepailleur, A. Ratkiewicz, H. Sims, P. L. Tai and D. Walter

Department of Physics and Astronomy, Rutgers University, New Brunswick, NJ 08903, USA

S. Burcher, K. L. Jones, K. Smith, P. Thompson and C. Thornsberry

Department of Physics and Astronomy, University of Tennessee, Knoxville, TN 37996, USA

A. Engelhardt, R. L. Kozub and S. Shadrick

Department of Physics, Tennessee Technological University, Cookeville, TN 38505, USA

J. Anderson, M. Carpenter, D. Seweryniak and S. Zhu

Physics Division, Argonne National Laboratory, Argonne, IL 60439, USA

J. Allen, D. W. Bardayan, M. R. Hall, O. Hall, J. Hu and P. D. O'Malley

Department of Physics, University of Notre Dame, Notre Dame, Indiana 46556, USA

S. M. Cha, K. Y. Chae and E. J. Lee

Department of Physics, Sungkyunkwan University, Suwon 16419, South Korea

G. L. Wilson

Department of Physics and Applied Physics, University of Massachusetts Lowell, Lowell, MA 01854, USA

The GODDESS coupling of charged-particle and photon spectrometers has been commissioned using a beam of ^{134}Xe at 10 MeV/A. This measurement demonstrates the capability of the device for measuring inelastic scattering and (d,pγ) reactions in inverse kinematics under radioactive beam conditions, and provides the first data on the single-particle structure of states in ^{135}Xe, including previously unobserved states based on the orbitals above the $N = 82$ shell closure.

Keywords: Transfer Reactions, Nuclear Structure, (d,pγ), ^{134}Xe, $N = 82$.

1. Introduction

The single-particle structure of atomic nuclei provides a cornerstone of the nuclear shell model by constraining the location of single-particle states and their mixing. Furthermore, many astrophysical reaction rates are determined by the location and properties of excited states with single-proton or single-neutron character that lie close to particle thresholds. Transfer reactions, in particular those involving hydrogen isotopes, are well-characterized probes of single-particle states, having been employed for decades using light-ion beams.

The locations and properties of low-spin states with single-neutron character in heavy neutron-rich nuclei are of interest for understanding r-process nucleosynthesis, impacting the neutron-capture cross sections in the astrophysical environment, and hence final abundances in r-process nucleosynthesis models[1]. The (d,p) reaction preferentially populates these important states, and is one of the best-characterized reactions in terms of quantitative reaction theory. It is therefore well-suited to extracting neutron spectroscopic information on excited states, which fold directly into calculations of neutron-capture cross sections[1,2].

However, in order to measure these reactions on the unstable nuclei of interest for astrophysics, and to extend nuclear-structure studies away from the valley of stability, transfer reactions must be measured in inverse kinematics with a beam of heavy nuclei incident on a light-ion target. The experimental issues and challenges of inverse-kinematics measurements are well documented[3,4]. When performing such reaction measurements, the coincident measurement of γ rays brings numerous advantages[4–7]. For separating closely-spaced states populated in the reaction, in addition to the improved energy resolution of γ-ray detection compared to particle-derived measurements, in many cases the decays from neighboring levels proceed via decay to different states, leading to better separation in γ-ray energy than simply the difference in excitation energy. Beyond energy-resolution arguments, γ rays carry additional information on the populated states and decay modes, and can provide information on states not populated directly in the transfer reaction. The detection of γ rays is also critical to the surrogate technique for constraining statistical neutron capture[8,9], where low-lying collecting transitions are used to constrain the γ branches of collections of nuclear states in the continuum, populated just above the neutron separation energy (as determined by detection of the ejectile).

2. GODDESS

In order to facilitate such transfer reaction measurements in inverse kinematics, the charged particle detector ORRUBA[10] (Oak Ridge Rutgers University Barrel Array) has been coupled to the Gammasphere[11] array of HPGe detectors, to form GODDESS (Gammasphere ORRUBA: Dual Detectors for Experimental Structure Studies)[7]. GODDESS is designed for high resolution particle-γ measurement of reactions with stable and radioactive beams, including beams of fission fragments from CARIBU[12], and lighter ($A \lesssim 30$) beams produced in-flight at Argonne National Laboratory (from the present in-flight facility and the AIRIS upgrade currently under development[13]).

The silicon array is mounted in a thin-walled spherical vacuum chamber (14-inch ⌀, 1/16-inch wall thickness) inside Gammasphere, with the signals brought out via a conical chamber in the downstream direction, located in the volume vacated by the typically unused 17° ring of Gammasphere detectors. No other impacts were made on Gammasphere. The ORRUBA barrel detectors (which cover ~45° to ~135° with typically 1° resolution in angle) have been augmented with new endcap detectors, which have a graduated strip pitch providing finer angular resolution at larger angles from the beam axis where the kinematics are steepest. This detector system provides polar angular coverage from ~15° to ~165°, with ~1° polar angular resolution and 70% azimuthal coverage throughout. At backward angles, a single layer of 1000-μm Si is used throughout the array in its standard configuration. At forward angles, 65-1000μm telescopes are used in the barrel, and 100-1000-1000μm telescopes in the forward-angle endcap. Preamplifiers for ~720 channels of silicon are used, in versatile configurable 72-channel preamplifier boxes mounted on the downstream conical chamber. Within this chamber, a compact tilted-grid fast ionization chamber can be mounted to provide identification and counting of beam particles at rates up to a few 10^5 ions per second, or a coupling can be made to the first quadrupole of the Fragment Mass Analyzer.

Due to the large number of silicon electronics channels, the silicon detector array was split (beam-left and beam-right) across two separate data acquisition systems - an analog system of CAEN V785 peak sensing ADCs (ORPHAS), and the digital FMA (DFMA) system of 100-MHz sampling digitizers[14,15]. γ rays were measured using Digital Gammasphere (i.e. Gammasphere read out with 100-MHz sampling digitizers; DGS), which was operated as a slave to the silicon detectors, receiving triggers from both ORPHAS and DFMA, and distributing a clock to the silicon data acquisition systems for time-stamping. Particle singles and particle-γ coincidences were stored; events from all three data acquisition systems were merged and sorted off-line. For further details on the GODDESS setup, see Ref. 16.

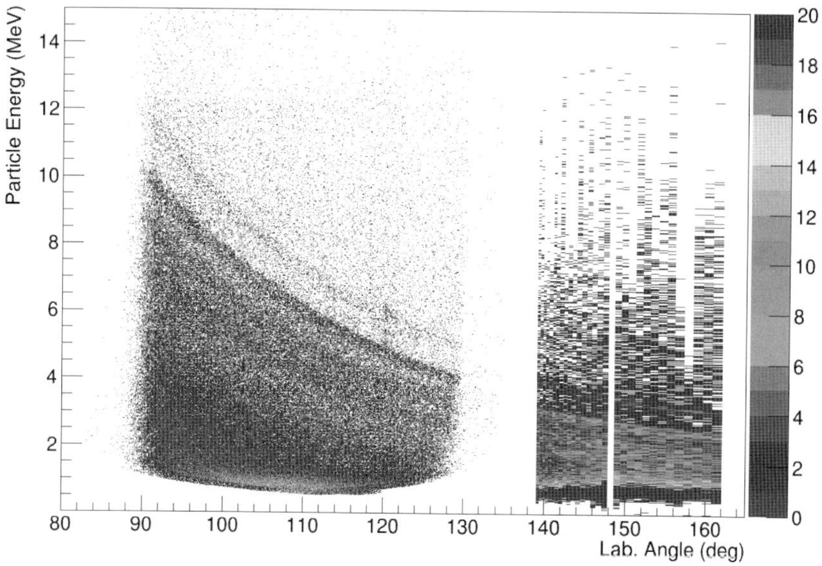

Fig. 1. Preliminary plot of particle energy vs laboratory angle, for a subset of upstream silicon detectors, for a beam of ^{134}Xe incident at 10 MeV/A on a 765-μg/cm^2 CD_2 target. Kinematic loci corresponding to the ^{134}Xe(d,p)^{135}Xe reaction are evident. Data for $90^\circ \lesssim \theta_{lab} \lesssim 130^\circ$ correspond to the barrel detectors, and data for $140^\circ \lesssim \theta_{lab} \lesssim 165^\circ$ to the upstream endcap detectors.

3. ^{134}Xe(d,p)^{135}Xe commissioning experiment

A beam of ^{134}Xe at 10 MeV/A, delivered from ATLAS, was used to commission GODDESS. ^{134}Xe is the only stable isotope of xenon on which the (d,p) reaction has not been measured. Correspondingly, the single-neutron states above the $N = 82$ shell closure (in particular those based on the f

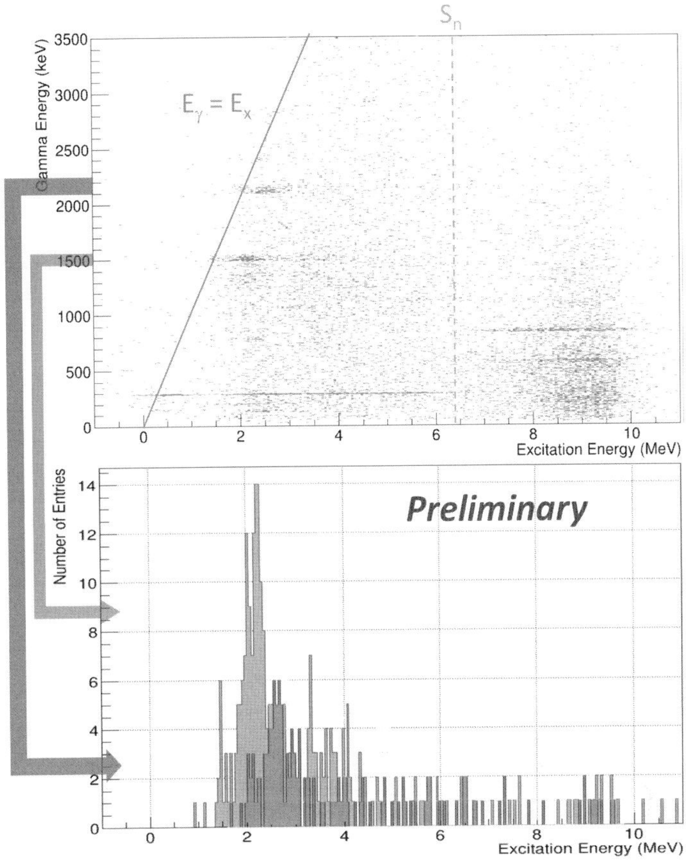

Fig. 2. Preliminary plot (top panel) of γ-ray energy vs excitation energy in ^{135}Xe as derived from the Q value determined from proton-ejectiles measured in the backward-angle endcap detector ($135^\circ \lesssim \theta_{lab} \lesssim 165^\circ, 10^\circ \lesssim \theta_{CM} \lesssim 35^\circ$) using a 765 μg/cm^2 CD_2 target. States that decay via single γ-ray emission to the ground state lie on the line of $E_\gamma = E_x$. The approximate location of the neutron separation energy in ^{135}Xe, at 6.364 MeV, is indicated by the vertical dashed line. Events to the right of this line correspond to neutron-unbound states in ^{135}Xe, which may decay to excited states in ^{134}Xe. The lower panel shows projections of the particle spectrum gated on γ rays at $\sim$1500 keV and $\sim$2122 keV.

and p orbitals) have not been previously observed in ^{135}Xe. Target thicknesses spanning the range typically used for inverse-kinematic transfer reactions with radioactive beam ($\sim$200 to $\sim$1000 μg/cm^2 CD_2) were employed, with beam intensities ranging from $\sim 10^4$ to $\sim 10^8$ ions/second, in order to simulate performance under radioactive beam conditions. Data shown herein come exclusively from runs using a 765-μg/cm^2 CD_2 target.

Fig. 1 (preliminary analysis) shows particle energy vs laboratory angle from the backward-angle silicon detectors, for a subset of runs (particle singles). The highest energy visible locus corresponds to population of states below the $N = 82$ shell closure ($\ell_n = 2, 0$ transfer to the $3/2^+$ ground state and $1/2^+$ first-excited state at 288 keV, respectively). About 2 MeV below are loci corresponding predominantly to population of previously unobserved levels above the $N = 82$ shell closure, exhibiting a comparable shell gap as inferred in the nearby $N = 81$ isotone ^{131}Sn[2].

A preliminary particle-γ matrix is displayed in Fig. 2 for events where a proton was detected in the upstream endcap detectors. The 288-keV γ ray can be seen from direct population of the $1/2^+$ state, but also as an effective collecting transition for many higher-lying states above the $N = 82$ shell closure, with possible application as a surrogate reaction for ^{134}Xe(n,γ). γ rays at ~1500 keV and ~ 2122 keV correspond to the deexcitation of states above the $N = 82$ shell closure. It can also be seen that the ^{135}Xe gamma rays diminish in intensity above the neutron separation energy, as the neutron widths dominate over the gamma widths for more unbound states, and γ rays corresponding to collecting ^{134}Xe transitions (notably the 847-keV $2_1^+ \rightarrow 0_{gs}^+$ transition) become apparent, potentially providing a surrogate for ^{134}Xe$(n, n'\gamma)$. A projection of particle-derived excitation energy, gated on ~1500 keV and ~ 2122 keV γ rays, is shown in the lower panel of Fig. 2. The data shown herein are preliminary, and are currently under more detailed analysis.

Acknowledgments

This material is based upon work supported by the U.S. Department of Energy, Office of Science, Office of Nuclear Physics, National Nuclear Security Administration (NNSA) Stewardship Science Academic Alliance Programs, the National Science Foundation, and a National Research Foundation of Korea (NRF) grant funded by the Korean government. This research used resources of the Argonne Tandem Linear Accelerator System, which is a DOE Office of Science User Facility operated by Argonne National Laboratory.

References

1. M. R. Mumpower, R. Surman, G. C. McLaughlin and A. Aprahamian, *Prog. in Part. and Nucl. Phys.* **86**, p. 86 (2016).

2. R. L. Kozub, G. Arbanas *et al.*, *Physical Review Letters* **109**, p. 172501 (2012).
3. W. Catford, Nucleon transfer studies with radioactive beams, *Nuclear Physics A* **701**, 1 (2002), 5th International Conference on Radioactive Nuclear Beams.
4. S. D. Pain, *AIP Advances* **4**, p. 041015 (2014).
5. M. Labiche, W. Catford, R. Lemmon, C. Timis, R. Chapman, N. Orr, Fernández-Domínguez, G. Moores *et al.*, *Nucl. Instr. and Meth. in Phys. Res. A* **614**, p. 439 (2010).
6. C. A. Diget, S. P. Fox, A. Smith, S. Williams, M. Porter-Peden *et al.*, *Journal of Instrumentation* **6**, p. P02005 (2011).
7. A. Ratkiewicz, S. D. Pain, J. A. Cizewski, D. W. Bardayan, J. C. Blackmon, K. A. Chipps, S. Hardy, K. L. Jones, R. L. Kozub, C. J. Lister, B. Manning, M. Matos, W. A. Peters, D. Seweryniak and C. Shand, *AIP Conf. Proc.* **1525**, p. 487 (2013).
8. J. E. Escher, J. T. Burke, F. S. Dietrich, N. D. Scielzo, I. J. Thompson and W. Younes, Compound-nuclear reaction cross sections from surrogate measurements, *Rev. Mod. Phys.* **84**, 353 (Mar 2012).
9. A. Ratkiewicz, J. A. Cizewski, S. D. Pain *et al.*, *EPJ Web of Conferences* **93**, p. 02012 (2015).
10. S. D. Pain, J. A. Cizewski, R. Hatarik, K. L. Jones, J. S. Thomas, D. W. Bardayan, J. C. Blackmon, C. D. Nesaraja, M. Smith, R. Kozub and M. Johnson, *Nucl. Instr. and Meth. in Phys. Res. B* **261**, 1122 (2007).
11. I.-Y. Lee, *Nuclear Physics A* **520**, c641 (1990).
12. G. Savard *et al.*, *JPS Conf. Proc.* **6**, p. 010008 (2015).
13. https://www.phy.anl.gov/airis/airis.pdf.
14. J. T. Anderson *et al.*, A digital data acquisition system for the detectors at gammasphere, *Nuclear Science Symposium and Medical Imaging Conference (NSS/MIC), IEEE* (2012).
15. M. P. Carpenter *et al.*, The new digital data acquisition system for Gammasphere, in *APS Division of Nuclear Physics Meeting Abstracts*, 2015.
16. S. D. Pain *et al.*, Direct reactions using GODDESS, *Physics Procedia* (2017), Proceedings of the 24th International Conference on the Application of Accelerators in Research and Industry. In preparation.

Pseudo Spin Doublet Bands and Gallagher Moszkowski Doublet Bands in ^{100}Y

E. H. Wang[1], J. H. Hamilton[1], A. V. Ramayya[1], C. J. Zachary[1], A. Lemasson[2], A. Navin[2], M. Rejmund[2], S. Bhattacharyya[3], Q. B. Chen[4], S. Q. Zhang[4], J. M. Eldridge[1], J. K. Hwang[1], N. T. Brewer[1], Y. X. Luo[1,5], J. O. Rasmussen[5], S. J. Zhu[6], G. M. Ter-Akopian[7], Yu. Ts. Oganessian[7], M. Caamaño[8], E. Clément[2], O. Delaune[2], F. Farget[2], G. de France[2] and B. Jacquot[2]

[1]*Department of Physics, Vanderbilt University, Nashville, TN 37235, USA*
[2]*GANIL, Bd Henri Becquerel, BP 55027, F-14076 Caen Cedex 5, France*
[3]*Variable Energy Cyclotron Centre, 1/AF Bidhan Nagar, Kolkata 700064, India*
[4]*School of Physics, Peking University, Beijing 100871, China*
[5]*Lawrence Berkeley National Laboratory, Berkeley, CA 94720, USA*
[6]*Department of Physics, Tsinghua University, Beijing 100084, China*
[7]*Joint Institute for Nuclear Research, RU-141980 Dubna, Russian Federation*
[8]*USC, Universidad de Santiago de Compostela, E-15706 Santiago de Compostela, Spain*

New transitions in the deformed neutron-rich ^{100}Y uniquely identified using the VAMOS++ spectrometer (A, Z) in conjunction with EXOGAM, produced in fission reactions of ^{238}U on a ^{9}Be target at energies around the barrier are reported. High spin states were obtained from an analysis of the large fold γ coincidence data from the spontaneous fission of ^{252}Cf using the Gammasphere detector array. The new $K^{\pi} = 4^{+}$ band decaying to an isomeric state (1s) is assigned to be the high-K Gallagher-Moszkowski (GM) partner of the known $K^{\pi} = 1^{+}$ band, and is also proposed to be the pseudo spin (PS) partner of another new $K^{\pi} = 5^{+}$ band. Constrained triaxial covariant density functional and quantal particle rotor model calculations for the band structure and electromagnetic transition probabilities are in good agreement with the measurements. The work shows for the first time a coexistance of GM and PS bands in the same nucleus and could point towards a common source of these bands despite their very different origins.

Keywords: Pseudo Spin; GM Partner; ^{100}Y.

1. Introduction

In the mass region A~100, the shape transition and shape coexistence of the neutron-rich nuclei have long been of interest [1, 2]. In this region, there exists a shell closure effect with Z = 40 and N = 56 spherical subshells and a sudden onset of large ground state axially symmetric deformation for N~60 occurs in the Sr (Z = 38), Y (Z = 39) and Zr (Z = 40) isotopes [2]. Thus, it is interesting to investigate the odd-odd ^{100}Y with Z = 39 and N = 61.

In this paper, we report the new high spin structure of ^{100}Y with identification of three new bands and the extension of a previously reported band [3, 4]. Pseudo spin doublet and GM doublet bands are proposed in this nucleus. The constrained triaxial covariant density functional theory and quantal particle rotor model calculations were performed to compare with such assignments.

2. Experimental Technique

Two complementary methods have been used to investigate the level structure of ^{100}Y. The first experimental measurements were performed at GANIL using a ^{238}U beam at 6.2 MeV/u, with a intensity of 0.2 pnA, impinging on a 10-μm thick ^{9}Be target. The large-acceptance VAMOS++ spectrometer [5] was used to identify the fission fragments. The prompt γ rays were measured with the EXOGAM array [6]. The velocity of the fragment along with the angle of the segment of the relevant clover detector were used to obtain the γ-ray energy in the rest frame of the emitting fragment [7].

Second, the spontaneous fission (SF) of ^{252}Cf experimental work was done by examining the prompt γ rays detected with Gammasphere array at the Lawrence Berkeley National Laboratory (LBNL). A 62 μCi ^{252}Cf source was sandwiched between two iron foils of 10 mg/cm^2. A plastic (CH) ball of 7.62 cm in diameter was placed surrounding the source. About 5.7×10^{11} γ-γ-γ and higher fold γ-events, and 1.9×10^{11} γ-γ-γ-γ and higher fold γ coincident events were recorded. These γ coincident data were analyzed by the RADWARE software package [8]. More details can be found in Ref. [2].

3. Experimental Result

The level scheme of ^{100}Y is shown in Fig. 1. The ground state of ^{100}Y was assigned as 1^- or 2^- in β-decay work [3]. A 1^+ $\pi5/2[422] \otimes \nu3/2[411]$ band was reported in Refs. [3, 4] with transitions identified Refs. [3, 4, 9]. In the current work, this band (1) has been extended up to 10^+ with three new transitions and levels. The new bands (2) and (3) identified in the present work are proposed to decay to the 4^+ isomeric state (0.94s) at 145(15) keV reported in Refs. [10-12].

Band (1) was assigned to have a 1^+ $\pi 5/2[422] \otimes \nu 3/2[411]$ configuration [3, 4]. Such configuration is adopted in the current work. Note that the M1 transition energies of the same spins in bands (1) - (3) are very close. Therefore, the moments of inertias in these two bands are also very close to rigid body.

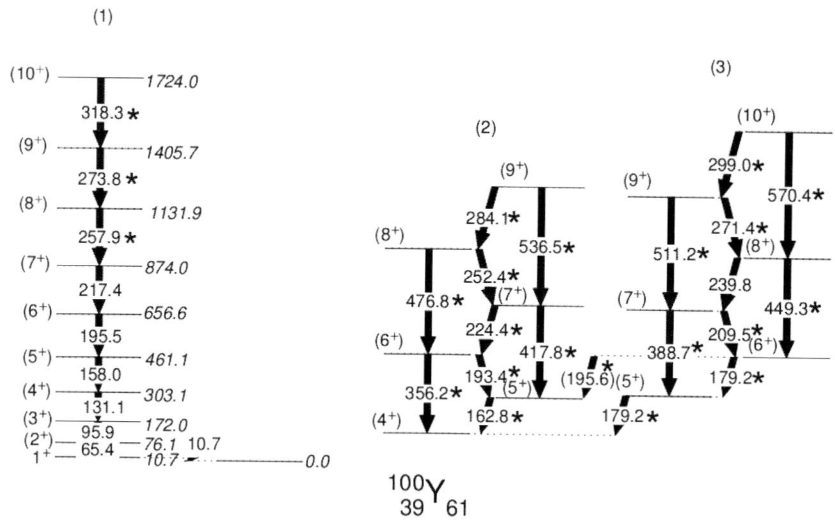

Fig. 1. The new level scheme of ^{100}Y. New transitions are labeled with an asterisk.

4. Theoretical Calculation and Discussion

Constrained triaxial covariant density functional theory (CDFT) calculations [13-19] were performed to obtain the potential energy surface (PES). The calculated ground state of ^{100}Y locates at $\beta_2 = 0.45$, $\gamma = 0°$. The unpaired nucleon configuration of the ground state is found to be $\pi(1g_{9/2})^1 \otimes \nu(2d_{5/2})^1$ or $\pi(1g_{9/2})^1 \otimes \nu(1g_{7/2})^{-1}$. The former one corresponds to the configuration $\pi 5/2[422] \otimes \nu 3/2[411]$ and will be assigned to bands (1) and (2), and the latter one $\pi 5/2[422] \otimes \nu 5/2[413]$ to band (3).

Then the many-particle-many-hole PRM [20-28] calculations are carried out for bands (1) and (2) with the configurations $\pi(1g_{9/2})^5 \otimes \nu(2d_{5/2})^3$, and for band (3) with $\pi(1g_{9/2})^5 \otimes \nu(1g_{7/2})^5$. The moments of inertia are chosen to be $\mathcal{J}$=20.0ħ^2/MeV.

In Fig. 2 (a) and (b), the energy spectra for bands (1), (2), and (3) calculated by PRM excellently reproduce the data. These three bands are separated by a nearly constant value at the whole spin region. This can be understood by the parallel and antiparallel coupling of intrinsic spins of proton and neutron, respectively, for bands (1) and (2), and a pair of nearly degenerated pseudo spin partner configurations for bands (2) and (3).

The calculated B(E2, ΔI = 2) increases smoothly with spin. It becomes smaller with larger K. Contrary to B(E2, ΔI = 2), the B(E2, ΔI = 1), decreases with spin. Moreover, it shows strong competitions with B(E2, ΔI = 2). Large B(E2, ΔI = 1) is accompanied by small B(E2, ΔI = 2), vice versa.

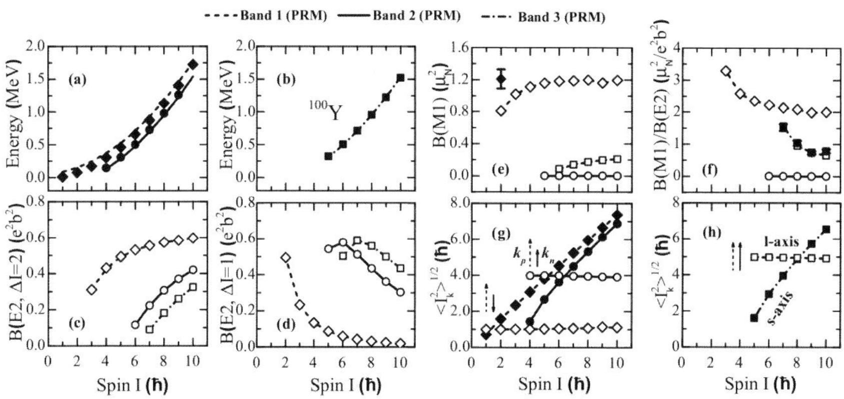

Fig. 2. PRM calculations for level energies, B(M1), B(E2) and short and long axis of spin.

In Fig. 2(e), the calculated B(M1) of band (1) shows a large value, which is in good agreement with the experiment value of 2^+ state that obtained from β decay measurement [29]. However, its partner band (2) presents almost vanished B(M1) values. The reason can be understood from the value of the g-factor and the antiparallel and parallel coupling of proton and neutron in bands (1) and (2), respectively. Similarly, as $g_\nu(g_{7/2})$-g_R = 0.31 is small, the B(M1) values of band (3) are small.

For the B(M1)/B(E2) ratio, as shown in Fig. 2(f), it decreases with spin for bands (1) and (3), and almost vanishes for band (2). The former one is caused by the increased B(E2), and the latter one by the vanished B(M1). The available experimental values of band (3) are excellently reproduced by the PRM, further in favor of the configuration assignment for this band.

The coupling modes in the three bands are further examined. In Fig. 2 (g) and (h), the root mean square components along the short (s-, circles) and long (l-, triangles) axes of the total spin (***I***) are illustrated. The s-components of ***I*** linearly increase with spin. Meanwhile, the l-components of ***I*** keep almost constant values, i.e, K=1, 4, and 5ħ in the bands 1, 2, and 3, respectively. This strongly supports the GM doublet interpretation for bands (1) and (2), and the pseudo spin doublet interpretation for bands (2) and (3) and provides direct experimental evidence for the coexistence of the GM doublet and pseudo spin doublet bands in the same nucleus.

5. Summary

In summary, the new transitions in neutron rich ^{100}Y have been identified by ^{9}Be+^{238}U experiment with mass- and Z- gate with full fragment identification and from the spontaneous fission of ^{252}Cf at Gammasphere detector array with high statistics. Constrained triaxial CDFT calculations and quantal PRM calculations have been performed to explain the observed band structures. The calculation confirmed that the K = 4^+ new band is a high-K GM partner of the known K = 1^+ band, with the $\pi 5/2[522] \otimes \nu 3/2[411]$ configuration, and is also proposed to be the pseudo spin partner of the new K = 5^+ band with a $\pi 5/2[422]$ $\otimes \nu 5/2[413]$ configuration, to form a neutron pseudospin doublet. The observation provides solid evidence for the coexistence of both the GM doublet and pseudo spin doublet bands in a same nucleus for the first time.

Acknowledgments

The work at Vanderbilt University and Lawrence Berkeley National Laboratory are supported by the US Department of Energy under Grant No. DE-FG05-88ER40407 and Contract No. DE-AC03-76SF00098. The work at Tsinghua University was supported by the National Natural Science Foundation of China under Grant No. 11175095. The work at JINR was supported by the Russian Foundation for Basic Research Grant No. 08-02-00089 and by the INTAS Grant No. 03-51-4496. The work at Peking University are supported by the Chinese Major State 973 Program No. 2013CB834400, the National Natural Science Foundation of China (Grants No. 11335002, No. 11375015, and No. 11461141002), and the China Postdoctoral Science Foundation under Grants No. 2015M580007 and No. 2016T90007. One of us (S.B.) acknowledges partial financial support through the LIA France-India agreement. We would like to thank J. Goupil, G. Fremont, L. Ménager, J. Ropert, C. Spitaels, and the GANIL accelerator staff for their technical contributions.

References

1. J. Skalski, S. Mizutori and W. Nazarewicz, *Nucl. Phys. A* **617**, 282 (1997).
2. J. H. Hamilton *et al.*, *Prog. Part. Nucl. Phys.* **35**, 635(1995).
3. F. K. Wohn *et al.*, *Phys. Rev. C* **36**, 1118 (1987).
4. J. K. Hwang *et al.*, *Phys. Rev. C* **58**, 3252 (1998).
5. M. Rejmund *et al.*, *Nucl. Inst. Meth. Phys. Res. A* **646**,184 (2011).
6. J. Simpson *et al.*, Acta Physica Hungarica, New Series, Heavy Ion Physics **11**, 159 (2000).
7. A. Navin *et al.*, *Phys. Lett. B* **728**, 136 (2014).
8. D. C. Radford, *Nucl. Instrum. Methods Phys. Res. A* **361**, 297(1995).
9. S. Bottoni *et al.*, *Phys. Rev. C* **92**, 024322 (2015).
10. T. A. Khan *et al.*, *Z. Phys. A* **283**, 105 (1977).
11. U. Hager *et al.*, *Nuclear Physics A* **793**, 20 (2007).
12. K. Baczynska *et al.*, *J. Phys. G: Nucl. Part. Phys.* **37**, 105103 (2010).
13. J. Meng *et al.*, *Phys. Rev. C* **73**, 037303 (2006).
14. P. Ring, *Prog. Part. Nucl. Phys.* **37**, 193 (1996).
15. D. Vretenar *et al.*, *Phys. Rep.* **409**, 101 (2005).
16. J. Meng *et al.*, *Prog. Part. Nucl. Phys.* **57**, 470 (2006).
17. J. Meng *et al.*, *Prog. in Phys.* **31**, 199 (2011).
18. J. Meng ed., Relativistic density functional for nuclear structure (World Scientific, Singapore, 2016).
19. P. W. Zhao *et al.*, *Phys. Rev. C* **82**, 054319 (2010).
20. S. Frauendorf and J. Meng, *Nucl. Phys. A* **617**, 131 (1997).
21. J. Peng, J. Meng and S. Q. Zhang, *Phys. Rev. C* **68**, 044324 (2003).
22. S. Q. Zhang *et al.*, *Phys. Rev.C* **75**, 044307 (2007).
23. B. Qi *et al.*, *Phys. Lett. B* 675, **175** (2009).
24. Q. B. Chen *et al.*, *Phys. Rev.* C82, 067302 (2010).
25. B. Qi *et al.*, *Phys. Rev.* C83, 034303 (2011).
26. A. D. Ayangeakaa *et al.*, *Phys. Rev. Lett.* **110**, 172504 (2013).
27. E. O. Lieder *et al.*, *Phys. Rev. Lett.* **112**, 202502 (2014).
28. I. Kuti *et al.*, *Phys. Rev. Lett.* **113**, 032501 (2014).
29. H. Mach *et al.*, *Phys. Rev. C* **41**, 1141 (1990).

Triaxiality in Neutron-Rich Nuclei with A ~ 100-126, Z Beyond and Below Ru (Z=44)

Y.X. Luo[1,2], J.H. Hamilton[1], J.O. Rasmussen[2,3], A.V. Ramayya[1], S. Frauendorf[4,5], E. Wang[1], J.K. Hwang[1], J.G. Wang[6], H.J. Li[6], E.Y. Yeoh[6], S.J. Zhu[1,6], Y.X. Liu[7], C.F. Jiao[8], W.Y. Liang[8], Yue Shi[8], F.R. Xu[8], Y. Sun[9,10], S.H. Liu[1,11], N.T. Brewer[1,12], I.Y. Lee[2], G.M. Ter-Akopian[13], A.V. Daniel[13], Yu. Oganessian[13], M.A. Stoyer[14], R. Donangelo[15] and W. C. Ma[16]

[1]Physics Department, Vanderbilt University, Nashville, TN 37235, USA
[2]Lawrence Berkeley National Laboratory, Berkeley, CA 94720, USA
[3]Department of Chemistry, U.C. Berkeley, Berkeley, CA 94720, USA
[4]Department of Physics, University of Notre Dame, Notre Dame IN 46556, USA
[5]Institut für Strahlenphysik, FZD-Rossendorf, Postfach, D-01314 Dresden, Germany
[6]Department of Physics, Tsinghua University, Beijing, 100084, China
[7]School of Science, Huzhou Teachers College, Huzhou, 31300, China
[8]School of Physics, Peking University, Beijing 100871, China
[9]Department of Physics and Astronomy, Shanghai Jiao Tong University, Shanghai, 200240, China
[10]Institute of Modern Physics, Chinese Academy of Sciences, Lanzhou, 730000, China
[11]Chemistry Department, University of Kentucky, Lexington, KY 40505, USA
[12]Physics Division, Oak Ridge National Laboratory, Oak Ridge, TN 37831, USA
[13]Flerov Laboratory for Nuclear Reactions, JINR, Dubna, Russia
[14]Lawrence Livermore National Laboratory, Livermore, CA 94550, USA
[15]Facultad de Ingenieria, C.C.30, 11300 Montevideo, Uruguay
[16]Physics Department, Mississippi State University, Mississippi State, MS 39762, USA

This paper reviews the systematic investigations of the shape transitions and coexistence with regard to triaxial deformations in neutron-rich nuclei with A ~ 100-126, Z from Zr (Z = 40) to Cd (Z = 48). Shape changes with Z, N and spins are addressed. γ vibrations, chiral symmetry breaking and wobbling motions identified in the region are discussed.

Keywords: Triaxial Deformation; Shape Transition and Coexistence; γ Vibrations; Chiral Symmetry Breaking; Wobbling Motion; Neutron-Rich Nuclei; Fission γ Data.

1. Introduction

The neutron-rich nuclei with A~100-126 over long chains in Zr (Z = 40), Nb (Z = 41), Mo (Z = 42), Tc (Z = 43), Ru (Z = 44), Rh (Z = 45), Pd (Z = 46),

Ag (Z = 47) and Cd (Z = 48) are intermediate between the spherical doubly magic ^{132}Sn and the strongly deformed axially symmetric Sr (Z = 38) and Y (Z = 39) isotopes, exhibiting richness of nuclear structure and shapes[1-4]. In this region, the Fermi level relative to the high-j sub-shells, $\nu h_{11/2}$ and $\pi g_{9/2}$, ranges from the bottom of the shells, through the near middle (half filled) shells to upper half of the shells, favoring prolate (or triaxial prolate), through large triaxial deformations to oblate shapes, respectively. This nuclear region is characteristic of shape transitions and coexistence and provides good opportunities for investigation of shape changes and new excitations with regard to triaxial deformations[5-12].

Level schemes established for the first time and new ones significantly extended based on our fission γ data accumulated at Gammasphere provide rich information for the systematic studies[13]. The "gold mine" for exploring in neutron-rich nuclei is still productive. The triple-gated 4D data have been found to be powerful for the identifications of the very weakly populated bands.

2. An overview of shapes with changing Z, N and spins in the region[9]

Figure 1 shows the systematics of shapes evolving with changing odd-Z[4], except for those of Ag / Cd exhibiting complexity in the vicinity of Z = 50 shell.

In the region at or beyond Ru (Z ≥ 44), Ru (Z = 44) and Rh (Z = 45) isotopes were found to be rigid triaxial rotors with maximum/near maximum triaxiality, $\gamma \sim -30^0$, -28^0, respectively[7,12,14,15], with medium quadruple deformations $\varepsilon_2 = 0.27\sim0.28$ calculated. For Pd (Z = 46), a smaller energy gain

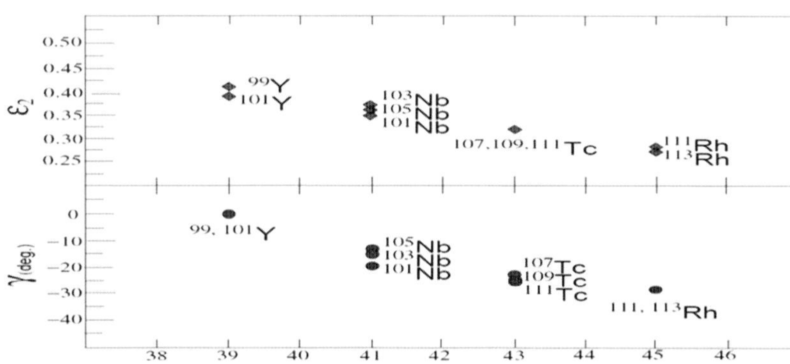

Fig. 1. The interpretations for the levels of odd-Z isotopes in terms of RTRP model revealed a correlation of quadruple deformation and triaxiality in the region[4]. The results for even-Z nuclei in terms of other models follow the same trend, triaxiality getting larger with Z approaching Ru.

0.32 MeV of ground state with regard to triaxiality was calculated, in contrast to 0.67 MeV in Ru, and thus less pronounced triaxiality, $\gamma \sim -41^0$ in ^{114}Pd for example, was deduced in Pd isotopes[16]. An overall shape evolution from triaxial prolate via triaxial oblate to oblate, and then back to prolate were identified in ^{110,112,114,116,118}Pd isotopes (Fig. 2). $(h_{11/2})^2$ neutron crossings (first band crossing) are seen to correlate to dramatic up-slopping γs, and $(g_{9/2})^2$ proton crossings (second crossing) to down-slopping γs (also see Section 5). Studies of $^{104-117}$Ag (e.g. [17, 18]) revealed rich shape information such as softness towards triaxiality, magnetic rotation, triaxial band structure, possible chiral structure and so on. For Cd isotopes, spherical vibrations, quasi-particle couplings, quadruple-octupole couplings, quasi-rotations, and soft-triaxiality with $\gamma \sim -10^0$ were proposed. Onset of collectivity was identified in the vicinity of the Z = 50, N = 82 closed shells[19,20].

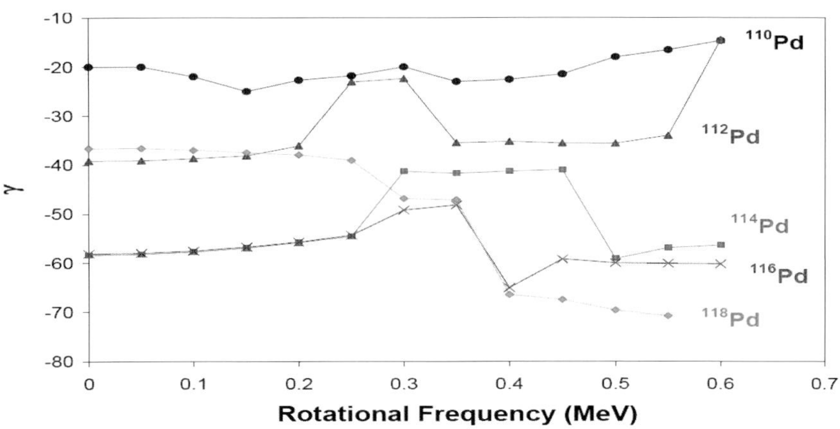

Fig. 2. Shape evolutions with changing N and rotational frequency, a more complete but complex picture than earlier predictions for $^{110-118}$Pd. The triaxial deformation parameter γs correspond to energy minima in the contour plots by TRS calculations, which well fit the band crossings[16].

In the region below Ru (Z < 44), super-deformations with $\varepsilon_2 = 0.39 \sim 0.40$ were deduced in Y (Z = 39) and Zr (Z = 40) isotopes. Axially-symmetric shapes were identified in Y and in light Zr isotopes with A<104 (e.g. [1]). Onset of triaxiality was seen in ^{104,106}Zr[21]. In Nb (Z=41) with increasing N, small to medium soft triaxia deformations, $\gamma \sim 2^0$–15^0, and large quadruple deformations, $\varepsilon_2 = 0.35 \sim 0.36$, were deduced[4,9,22](e.g. see Fig. 3). A triaxial shape coexistence in the two signature partners of the $\pi 5/2^+[422]$ bands in ^{101,103}Nb were proposed[9]. Mo (Z = 42) isotopes were found to be rigid rotors[1,8] with $\gamma \sim 20^0$. Even larger

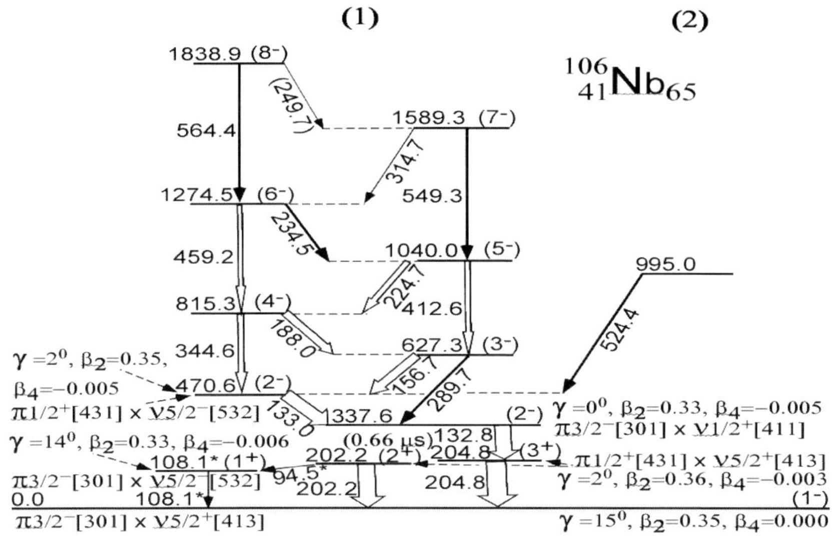

Fig. 3. Level scheme of ^{106}Nb, configurations and shapes deduced by PES and PSM calculations[22].

and an increase of γs with increasing N, -22.5^0–-26^0, were proposed in Tc isotopes[11].

3. One- and two-phonon γ vibrations in triaxial nuclei in the region

The observations and theoretical interpretations of the one- and two- phonon γ vibration bands in Ru, Tc, Mo, Nb, Rh, and Pd isotopes are the earlier studies of excitation modes with regard to triaxial deformations in the region[3,23-25]. New bands observed in ^{103,105}Nb are the new progress, and interpreted as one- and two-phonon γ vibrations by PES / TPSM calculations[25].

4. Chirality identified in the triaxial nuclei in the region[9,27]

Chiral symmetry breaking was predicted to occur in the nuclei with large and stable triaxial deformations[26]. In the nuclear region, chiral bands have been identified in ^{106,108}Mo[8,27], ^{100}Tc[28], ^{110,112}Ru[14,12], $^{103\text{-}106}$Rh[15], ^{104,105}Ag[17], and disturbed chiral structure in ^{108}Ru[14]and ^{112,114,116}Pd[16] by softness or less pronounced triaxiality. The very weakly populated odd-parity doublet bands in ^{112,114,116}Pd (see Fig. 4 for ^{114}Pd) are very similar to the chiral doublets in ^{110,112}Ru and ^{106}Mo. However, the signature splittings of the doublets in Pd

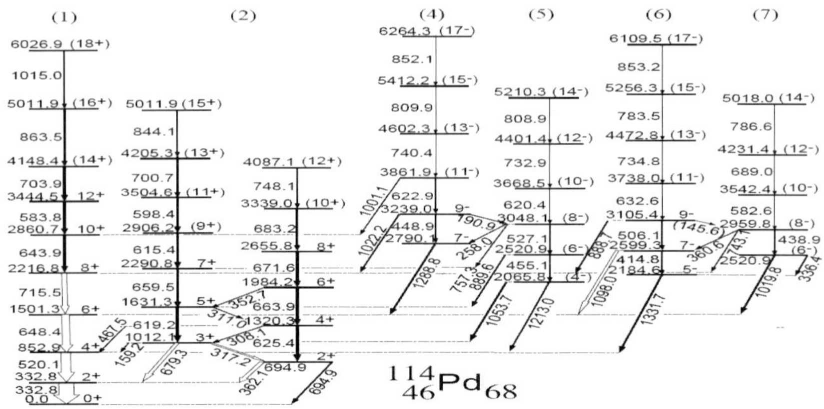

Fig. 4. Partial level scheme of ^{114}Pd identified using our fission data[16]. Note that the very weakly populated doublet bands 4 and 6 extend to excitations even higher than that of the ground band.

exhibit large level staggerings, not fulfilling the criterion of chiral doubling, and were explained by disturbance by the less pronounced triaxiaity in Pd[16].

5. Wobbling motions in triaxial nuclei in the region

The wobbling motions constitute a revolving motion of J about an axis of a triaxial nucleus[29,30], which are expected to occur at moderate spins if the predicted triaxial shapes lead to different moments of inertia for the three principal axes. In this region, onset of wobbling motions has been identified in ^{112}Ru and ^{114}Pd[30,16], the N = 68 isotones, likely also in ^{114}Ru[16]. The evolution of wobbling motions and triaxiality with changing N and spins in Ru and Pd isotopes were explained by TRS calculations well reproducing the band-crossings. In Fig. 5 one can see the triaxiality, $\gamma \sim -41^0$, over a wide spin range[16].

Acknowledgments

The work on the present review at Vanderbilt University and LBNL was supported by the U.S. DOE Grants and Contracts No. DE-FG-05-88ER40407, DE-FG02-95ER40934 and DE-FG02-95EER4093.

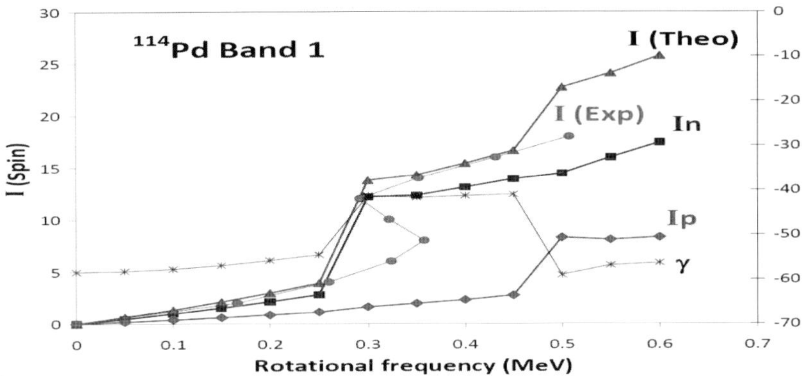

Fig. 5. TRS calculations for band 1 of ^{114}Pd[16]. The first band crossing caused by $(\nu h_{11/2})^2$ alignment correlates to a rapid uprising γ, and the second band crossing caused by $(\pi g_{9/2})^2$ alignment to a dramatic down-sloping γ. It is the driving effects of both alignments and the softness towards triaxiality that created a stable triaxial oblate deformation $\gamma \sim -41^0$ over a wide spin range.

References

1. J. Skalski, S. Mizutori and W. Nazarewicz, *Nucl. Phys. A* **617**, (1997) 282.
2. H. Hamilton, *Prog. in Particle and Nucl. Phys.* **15**, (1985) 107.
3. S.J. Zhu *et al.*, *Inter. Jour. of Modern Phys. E***18(8)**, 1717(2009).
4. Y.X. Luo *et al.*, *J. Phys. G***31**, 1303(2005).
5. F.R. Xu *et al.*, *Phys. Rev. C***65**, 021303(R)(2002).
6. W. Nazarewicz *et al.*, *Acta. Phys. Pol. B***32**, 2349(2001).
7. P. Möller *et al.*, *Phys. Rev Lett,* **97**, 162502 (2006).
8. S.J. Zhu *et al.*, *Eur. Phys. J, A***25**, Supl, **1** 459(2005).
9. Y.X. Luo *et al.*, *Chin. Journal of Nucl. Phys. Rev. Vol.* **32**, 1, 1(2015).
10. Y.X. Luo *et al.*, *Phys. Rev. C***69**, 024315(2004).
11. Y.X. Luo *et al.*, *Phys. Rev. C***74**, 024308(2006).
12. I. Stefanescu *et al.*, *Nucl. Phys. A***789**, 125(2007).
13. J.H. Hamilton *et al.*, *Prog. Part. Nucl. Phys.* **35,** 635(1995).
14. Y.X. Luo *et al.*, *Phys. Lett. B***670**, 307(2009).
15. J. Timmar *et al.*, *Phys. Rev. C***73**, 011301(R)(2006).
16. Y.X. Luo *et al.*, *Nucl. Phys. A***919**, 67(2013).
17. Z.G. Wang *et al.*, *Phys. Rev. C***88**, 024306(2013).
18. J.Sethi *et al.*, *Phys. Lett. B***725**, 85(2013).
19. Y.X. Luo *et al.*, *Nucl. Phys. A***874**, 32(2012).
20. S. LLieva *et al.*, *Phys. Rev. C***89**, 014313(2014).

21. A. Navin *et al.*, *Phys. Lett. B***728**, 136(2014).
22. Y.X. Luo *et al.*, *Phys. Rev. C***89**, 044326(2014).
23. J.H. Hamilton *et al.*, *Eur. Phys. J, A***15**, 175(2002).
24. Jian-Guo Wang *et al.*, *Phys. Lett. B***675**, 420(2009).
25. H.J. Li *et al.*, *Phys. Rev. C***88**, 054311(2013).
26. S. Frauendorf and J. Meng, *Nucl. Phys. A***617**, 131(1997).
27. Y.X. Luo *et al.*, *Chin. Journal of Nucl. Phys. Rev. Vol.* **27**, 3, 229(2010).
28. P. Joshi *et al.*, *Eur. Phys. J, A***24**, 23(2005).
29. M. Caprio, *Phys. Rev. C***83**, 064309(2011).
30. J.H. Hamilton *et al.*, *Nucl. Phys. A***834**, 28c(2010).

Probing the Low-Energy Structure of A = 109 Ru – Pd

M.K. Smith[1,2,3], B. Bucher[1,4], H. Mach[1,5], A. Aprahamian[1], G.S. Simpson[6], J. Rissanen[7], D. G. Ghiţă[8], B. Olaizola[9], W. Kurcewicz[10], J. Äystö[7], T. Eronen[7], L. M. Fraile[9], A. Jokinen[7], P. Karvonen[7], I. D. Moore[7], H. Penttilä[7], M. Reponen[7], E. Ruchowska[5], A. Saastomoinen[7] and C. Weber[7]

[1]*Department of Physics, University of Notre Dame, Notre Dame, Indiana 46556, USA*
[2]*National Superconducting Cyclotron Laboratory, Michigan State University, East Lansing 48824, USA*
[3]*Joint Institute for Nuclear Astrophysics, Michigan State University, East Lansing, Michigan 48824, USA*
[4]*Lawrence Livermore National Laboratory, Livermore, California 94550, USA*
[5]*Division of Nuclear Physics, BPI, National Centre for Nuclear Research, ul. Hoża 69, 00-681, Warsaw, Poland*
[6]*LPSC, Université Joseph Fourier Grenoble 1, CNRS / IN2P3, Institut National Polytechnique de Grenoble, F-38026 Grenoble Cedex, France*
[7]*Department of Physics, University of Jyväskylä, P. O. Box 35, F1-40014 Jyväskylä, Finland*
[8]*National Institute for Physics and Nuclear Engineering, R-77125 Bucharest-Magurele, Romania*
[9]*Grupo de Física Nuclear, Facultad de Físicas, Universidad Complutense - CEI Moncloa, E-28040 Madrid, Spain*
[10]*Faculty of Physics, University of Warsaw, Pasteura 5, PL 02-093 Warsaw, Poland*

Far from stability, nuclear structure and the onset of deformation evolves rapidly. Reliable structure information is crucial in understanding this shape evolution across isotopic and isobaric chains. In particular the A$\sim$110 nuclei are known to exhibit rapid shape changes. The onset of deformation in the A = 110 region remains under discussion, and signatures of shape-phase transitions, triaxial deformations, shape coexistence and oblate configurations have been reported within a small window of the nuclear landscape. We have investigated the A = 109 β-decay chain from the fission of ^{238}U at the Univ. of Jyväskylä IGISOL facility. Level lifetimes and gamma-ray transitions were measured with a multi-detector array consisting of 2 HPGe, 2 LaBr and 1 plastic scintillator. Triple coincidence β-γ-γ events were recorded and used to construct/check level schemes, and extract picosecond-range level lifetimes via the fast-timing method pioneered by H. Mach. Results will be presented on the low energy structure of ^{109}Tc, ^{109}Ru and ^{109}Pd in the context of the A = 109 decay chain and the region.

Acknowledgments

This project was funded by the National Science Foundation through Grants No. PHY0822648 and No. PHY0758100.

Results on the Influence of Neutron-Richness on Quasifission in Intermediate Mass Reactions

K. M. Hammerton*, D. J. Morrissey, Z. Kohley, A. Wakhle and K. Stiefel

National Superconducting Cyclotron Laboratory
Michigan State University, East Lansing, Michigan 48824, USA
Department of Chemistry, Michigan State University
East Lansing, Michigan 48824, USA
**E-mail: hammerto@nscl.msu.edu*

D. J. Hinde, M. Dasgupta, E. Williams, I. P. Carter, K. J. Cook, D. Y. Jeong, D. H. Luong, S. D. McNeil, C. Palshetkar, D. C. Rafferty and C. Simenel

Department of Nuclear Physics, Research School of Physics and Engineering,
Australian National University
Canberra, Australian Capital Territory 2601, Australia

J. Greene

Physics Division, Argonne National Laboratory
Lemont, Illinois 60473, USA

Superheavy elements are primarily formed through heavy ion fusion reactions[1–3]. Formation of a fully equilibrated compound nucleus is a critical step in this reaction mechanism but can be hindered by orders of magnitude by quasifission, a process in which the dinuclear system breaks apart prior to full equilibration[1,4–8]. To provide a complete description of heavy-ion fusion it is important to characterize the quasifission process. The interplay between the fusion-fission and quasifission reaction channels was explored by measuring fission mass distributions in eight different combinations of Cr+W reactions, with varying neutron-richness, at the Australian National University. The reactions were measured in two energy regimes: one at 13% above the Bass fusion barrier[9] and one at 52.0 MeV of excitation energy in the compound nucleus, E^*_{CN}. For the systems measured at $E_{c.m.}/V_{Bass} = 1.13$ the dependence on the neutron-richness is clear[10]. However, for the reactions at $E^*_{CN} = 52.0$ MeV, the dependence is less clear and additional factors are shown to play a vital role, especially the influence of deformation on the effective fusion barrier. The present work demonstrates that quasifission is an important process in competition with heavy-ion fusion in reactions with intermediate mass projectiles, particularly with more neutron-rich systems.

Keywords: Heavy-Ion Reactions, Quasifission.

1. Introduction

The production of each new superheavy elements (SHE) extends the boundaries of the periodic table and chart of the nuclides[1]. Recently, new SHEs have been produced in heavy-ion fusion reactions[1–3]. Theory predicts that the region around $N = 184$ will contain the next spherical shell closure[11]. Further exploration of this region and synthesis of new neutron-rich heavy or superheavy nuclei will require increasingly neutron-rich projectiles and targets[1,2,11–13]. Neutron-rich beams from next-generation radioactive ion beam facilities will offer the first opportunities to explore their use for the production of new neutron-rich "light" superheavy isotopes[14–17].

The production cross section for the superheavy evaporation residues (σ_{ER})[14] can be determined from the product of the capture cross section (σ_{cap}), the probability that the dinuclear system fuses (P_{CN}) rather than separates in the quasifission process[4,5], and the survival probability against fission (W_{sur}). It is imperative to understand the dependence of each component of σ_{ER} on the neutron-richness to properly assess the opportunities neutron-rich beams may offer. The largest uncertainty in the predictions for SHE production rates is P_{CN}, estimated to have an uncertainty of $1-2$ orders of magnitude[14], owing to the complex nature of the quasifission process[1,4–8]. Experimental studies have indicated that quasifission is dependent on the neutron-richness of the compound system, $(N/Z)_{\mathrm{CN}}$. However, the previous work led to contradictory conclusions as to the nature of this dependence[18–24]. If quasifission is reduced in more neutron-rich systems then fusion probabilities could be enhanced and help compensate for the decreased beam intensities of neutron-rich radioactive ion beams.

2. Experimental Details

This work explored the effect of varying the neutron-richness on the competition between fusion-fission and quasifission. To minimize the sensitivity to other factors known to influence the strength of the quasifission reaction channel, a set of Cr + W reactions was chosen (Table 1). Between the most neutron-deficient (^{50}Cr +^{180}W) and neutron-rich (^{54}Cr +^{186}W) systems there is a change of 10 neutrons. The Cr + W reactions have the same charge product, Z_pZ_t, of 1776 and only ^{52}Cr has a closed shell with $N = 28$. The W targets are prolate deformed with β_2 from 0.225 to 0.254[25].

Beams of ^{50,52,54}Cr were accelerated by the Heavy Ion Accelerator Facility at the Australian National University and bombarded targets of ^{180,182,184,186}W[26] at $E_{\mathrm{c.m.}}/V_{\mathrm{B}} = 1.13$[10] and $E^*_{\mathrm{CN}} = 52.0$ MeV at the energies listed in Table 1. Fragments resulting from fusion-fission and quasi-fission (termed fission-like) were detected in coincidence using the ANU CUBE detector system[27,28]. From the measured velocity vectors of the fragments, the mass ratios, $M_R = m_1/(m_1 + m_2)$, where m_1 and m_2 are the masses of the fragments, were determined[28].

Table 1. Cr + W systems, compound nuclei, relative change in neutrons (ΔN), Bass interaction barrier (V_B)[9], $E_{c.m.}$ for both energy regimes.

			$E_{c.m.}/V_B = 1.13$		$E^*_{CN} = 52.0$ MeV	
Reaction	ΔN	V_B (MeV)	$E_{c.m.}$ (MeV)	E^*_{CN} (MeV)	$E_{c.m.}$ (MeV)	$E_{c.m.}/V_B$
$^{50}Cr + ^{180}W \rightarrow ^{230}Cf$	0	196.95	222.6	64.6	210.0	1.07
$^{50}Cr + ^{186}W \rightarrow ^{236}Cf$	6	195.59	221.0	71.7	201.4	1.03
$^{52}Cr + ^{180}W \rightarrow ^{232}Cf$	2	195.75	221.2	59.1	214.1	1.09
$^{52}Cr + ^{184}W \rightarrow ^{236}Cf$	6	194.80	220.1	62.4	209.7	1.08
$^{54}Cr + ^{180}W \rightarrow ^{234}Cf$	4	194.56	219.8	56.7	215.4	1.11
$^{54}Cr + ^{182}W \rightarrow ^{236}Cf$	6	194.12	219.3	57.6	213.8	1.10
$^{54}Cr + ^{184}W \rightarrow ^{238}Cf$	8	193.67	218.9	59.0	211.8	1.09
$^{54}Cr + ^{186}W \rightarrow ^{240}Cf$	10	193.22	218.3	73.1	207.5	1.08

3. Results and Discussion

The mass distributions determined for the Cr + W systems are shown in Figure 1. For the systems shown in Panels B, C, E-G, the mass distribution has a maximum at $M_R = 0.5$. However, the mass distributions from $^{50}Cr + ^{180}W \rightarrow ^{230}Cf$ and $^{50}Cr + ^{186}W \rightarrow ^{236}Cf$ (Panels A and D in Figure 1, respectively) are flat or have a slight minimum at $M_R = 0.5$. This is different from the systems measured at $E_{c.m.}/V_B = 1.13$ where all of the mass distributions had a maximum at a mass ratio of 0.5[10].

Each mass distribution was fitted across the fission-like region with a second degree polynomial function. The results of these fits are represented by the lines in Figure 1. The second derivative was used to characterize the shape of the mass distributions and termed the curvature parameter. A more negative curvature parameter indicates that the distribution has a sharper peak at $M_R = 0.5$ and relatively less quasifission than a system with a larger curvature parameter.

The curvature parameters determined from the mass distributions of the Cr + W systems are shown in Panel A of Figure 2 as a function of $(N/Z)_{CN}$. While the curvature parameters generally decrease with

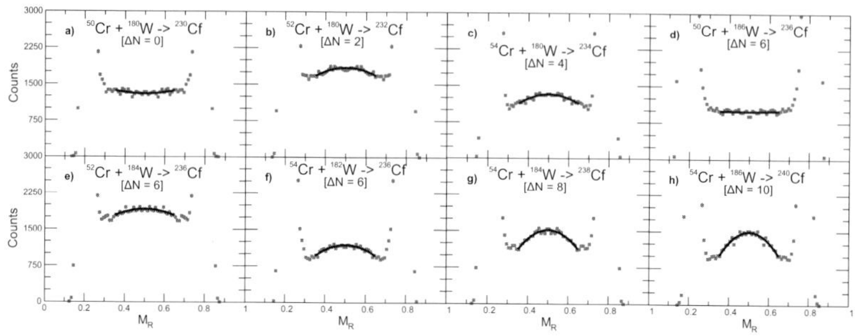

Fig. 1. Mass distributions for Cr + W systems presented in this work at $E^*_{CN} = 52.0$ MeV. The solid line represents the second degree polynomial function fitted to the data.

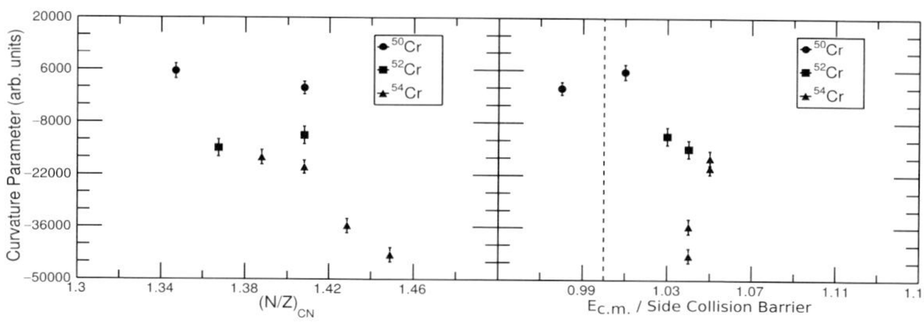

Fig. 2. A) Curvature parameter in arbitrary units deduced from the mass distribution for each system measured at $E^*_{CN} = 52.0$ MeV as a function $(N/Z)_{CN}$. B) Curvature parameters as a function of $E_{c.m.}/V_B$ (side collision).

increasing neutron-richness the correlation is not conclusive. Two systems of note are the two where ^{50}Cr was the projectile which have a positive curvature indicating that the quasifission reaction channel was enhanced in these systems compared to the others. This effect can be explained by considering the importance of the orientation of the deformed tungsten targets on the reaction dynamics.

^{180,182,184,186}W are all prolate deformed and in reactions of deformed nuclei the interaction barrier is dependent on the orientation of the nucleus. To reach E^*_{CN} of 52.0 MeV, the center-of-mass energies of the reactions are closer to the interaction barrier[9] than those required to reach $E_{c.m.}/V_B = 1.13$. As shown by previous works[27,30–32], deformation has a larger impact near the interaction barrier where the quasifission probability is suggested to increase[27,31–36].

It is useful to consider two extreme collision types. In one type the Cr collides with the tip of the prolate W, which forms an elongated compound nucleus, has a relatively low interaction barrier, and enhances the quasifission reaction channel[27,31,32]. The other extreme type occurs when the Cr interacts with the elongated side of the prolate W, which forms a compact dinuclear system, has a relatively high interaction barrier, and has been suggested to enhance the probability of forming a compound nucleus[27,31,32]. In Panel B of Figure 2, the curvature is shown as a function of $E_{c.m.}$ / V_B for the side collision barriers. For the two reactions of ^{50}Cr, this collision type is strongly hindered since $E_{c.m.}$ / V_B(sidecollision) is 0.98 and 1.01. The systems with minimums in yield at $M_R = 0.5$ are below or at the barrier for side collisions.

In conclusion, eight Cr + W systems were measured at $E^*_{CN} = 52.0$ MeV to characterize the effect of neutron-richness on the competition between fusion-fission and quasifission. A curvature parameter was determined for the mass distribution from each system and used to compare the prominence of the quasifission reaction channel. The curvature parameters generally decrease from the most neutron-deficient to the most neutron-rich system, however there is not a strong correlation. The orientation of the prolate deformed tungsten targets played a role in the enhancement of the quasifission reaction channel in the two systems where ^{50}Cr was the projectile. Overall, quasifission was shown to play an important role in the reaction dynamics in the Cr + W systems measured in this work.

Acknowledgments

The authors are grateful for the staff at the ANU accelerator facility. This work is supported by the National Science Foundation under Grants No. PHY-1102511 and No. IIA-1341088, and the Australian Research Council Grants DP160101254, DP140101337, FL110100098, DP130101569, FT120100760, and DE140100784. This material is based upon work supported by the Department of Energy National Nuclear Security Administration under Award Number DE-NA0000979. This material is based upon work supported by the U.S. Department of Energy, Office of Science, Office of Nuclear Physics, under Contract No. DE-AC02-06CH11357. This research used resources of ANL's ATLAS facility, which is a DOE Office of Science User Facility.

References

1. J. Hamilton, S. Hofmann and Y. Oganessian, *Annu. Rev. Nucl. Part. S.*, **63**, 383, (2013).

2. Y. T. Oganessian *et al.*, *Phys. Rev. Lett*, **109**, 162501, (2012).
3. Y. T. Oganessian, *J. Phys. G*, **34**, R165, (2007).
4. J. Tok, *et al.*, *Nucl. Phys. A*, **440**, 327, (1985).
5. B. B. Back, *Phys. Rev. C*, **31**, 2104, (1985).
6. V. Zagrebaev *et al.*, *Phys. Rev. C*, **65**, 014607, (2001).
7. A. C. Berriman *et al.*, *Nature*, **413**, 144, (2001).
8. R. Yanez *et al.*, *Phys. Rev. C*, **88**, 014606, (2013).
9. R. Bass, *Phys. Rev. Lett.*, **39**, 265, (1977).
10. K. Hammerton *et al.*, *Phys. Rev. C*, **91**, 041602(R) (2015).
11. M. Bender *et al.*, *Phys. Rev. C*, **60**, 034304 (1999).
12. M. Bender, W. Nazarewicz and P.-G. Reinhard, *Phys. Lett. B*, **515**, 42, (2001).
13. W. Nazarewicz *et al.*, *Nucl. Phys. A*, **701**, 165, (2002).
14. W. Loveland, *Phys. Rev. C*, **76**, 014612, (2007).
15. R. Smolanczuk, *Phys. Rev. C*, **81**, 067602, (2010).
16. S. Hofmann, *Prog. Part. Nucl. Phys.*, **46**, 293, (2001).
17. V. V. Sargsyan *et al.*, *Phys. Rev. C*, **88**, 054609, (2013).
18. J. F. Liang *et al.*, *Phys. Rev. C*, **85**, 031601, (2012).
19. C. C. Sahm *et al.*, *Nucl. Phys. A*, **441**, 316, (1985).
20. K. T. Lesko *et al.*, *Phys. Rev. C*, **34**, 2155, (1986).
21. A. Vinodkumar *et al.*, *Phys. Rev. C*, **78**, 054608, (2008).
22. C. Wang *et al.*, *Phys. Rev. C*, **82**, 054605, (2010).
23. G. G. Adamian, N. V. Antonenko and W. Scheid, *Phys. Rev. C*, **68**, 034601, (2003).
24. C. Simenel *et al.*, *Phys. Lett. B*, **710**, 607, (2012).
25. P. Möller *et al.*, *At. Data Nucl. Data Tables*, **59**, 185, (1995).
26. J. P. Greene and Z. Kohley, *J. Radioanal. Nucl. Chem.*, **305**, 743, (2015).
27. D. J. Hinde *et al.*, *Phys. Rev. C*, **53**, 1290, (1996).
28. R. du Rietz *et al.*, *Phys. Rev. C*, **88**, 054618, (2013).
29. C. J. Lin *et al.*, *Phys. Rev. C*, **85**, 014611, (2012).
30. A. Wakhle *et al.*, *Phys. Rev. Lett.*, **113**, 182502, (2014).
31. D. J. Hinde *et al.*, *Phys. Rev. Lett.*, **101**, 092701, (2008).
32. D. J. Hinde *et al.*, *Phys. Rev. Lett.*, **4**, 1295, (1995).
33. D. J. Hinde *et al.*, *Phys. Rev. Lett.*, **89**, 272701, (2002).
34. R. Rafiei *et al.*, *Phys. Rev. C*, **77**, 024606, (2008).
35. S. Mitsuoka *et al.*, *Phys. Rev. C*, **65**, 054608, (2002).
36. K. Nishio *et al.*, *Phys. Rev. C*, **63**, 044610, (2001).

Beta-Delayed Neutron Studies of Fission Fragments Using VANDLE

S. V. Paulauskas*, M. Madurga, R. Grzywacz, S. Go, A. Keeler, T. T. King and S. Z. Taylor

Department of Physics & Astronomy,
The University of Tennessee at Knoxville,
Knoxville, TN 37996-1200, The United States of America
**E-mail: spaulaus@utk.edu*
https://www.utk.edu/

C. J. Zachary

Department of Physics and Astronomy,
Vanderbilt University
Nashville, TN 37235, The United States of America

for the VANDLE Collaboration

Beta-delayed neutrons emitted from the decay of fission fragments play an important role in the r-process and nuclear reactor physics. Neutron energy spectra provide information necessary to accurately reconstruct a level scheme that reflects nuclear structure. Decay cascades where neutron emission populate excited states in the (N-1) daughter complicate level scheme construction. These excited states in the (N-1) daughter often emit gamma-rays. These situations necessitate measurement of three-fold coincidence between the beta, neutron and gamma-ray. However, there is a scarcity of measurements providing these data.

To address this need, we designed the Versatile Array of Neutron Detectors at Low Energy (VANDLE)[1], which measures neutron energies via the time-of-flight technique. VANDLE is instrumented with a digital data acquisition system. The digital system provides a neutron detection threshold as low as 70 keV and time resolutions better than 1 ns. A high-efficiency gamma-ray detection system enables complete calorimetry of the decay by detecting gamma-rays that follow neutron emission. High purity Ge clovers present isotope identification, but suffer from poor geometric efficiency. A high efficiency array of NaI and $LaBr_3$ detectors boost neutron-gamma coincidences.

Keywords: Plastic Scintillators; Digitizers; Time-of-Flight; Timing.

1. Introduction

Beta-delayed neutron spectroscopy provides deep insight into nuclear structure above the neutron separation energy. The nuclear structure may change as nuclei become more neutron rich. The imbalance of the protons and neutrons in their orbitals cause a shift in the energetics of the nucleus, which causes a decrease in the neutron separation energy. This decrease allow competition between de-excitations of the beta-decay daughter via gamma-ray neutron emission, see Figure 1.

Discrete lines in the neutron energy spectrum were originally understood as transitions from distinct states in the daughter nucleus emitting neutrons[2,3]. Hardy challenged this interpretation by proposing the Pandemonium effect[4,5]. Pandemonium is a fictional nucleus meant to demonstrate that the structure seen in particle emission spectra (gamma or neutron) is due to statistical fluctuations arising from high level densities. The neutron emission spectra for this nucleus behaves similarly to those studied by Kratz *et al.*[3]. While the spectra do not agree with complete accuracy, the overall gross structure and behavior of the spectra agree quite well. This leads Hardy et al. to state that one should exercise caution when interpreting beta-delayed neutron spectra and that any interpretation of these data must consider pandemonium.

These arguments lead one to immediately recognize the need for new measurements of the available data. New measurements require a complete beta, neutron and gamma-ray detection system. To achieve high

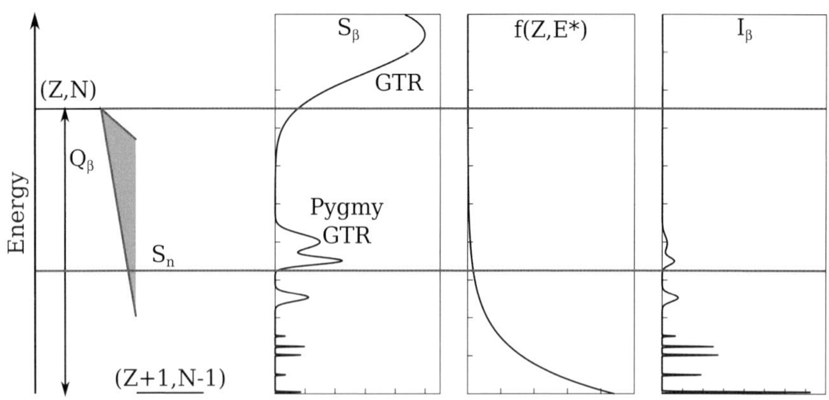

Fig. 1. A schematic of the energetics of the beta decay. S_β is the β strength distribution, f(Z,E*) is the Fermi integral, and I_β is the measured intensity spectrum.

coincidence rates each of these systems must have both high intrinsic and geometric efficiencies. The Versatile Array of Neutron Detectors at Low Energy (VANDLE) boasts both high intrinsic efficiency and sufficient energy resolution. VANDLE couples with a variety gamma-ray detector arrays. These proceedings report on the measurement of beta-delayed neutrons following the decay of ^{97}Rb and ^{72}Co as measured using VANDLE.

2. The Versatile Array of Neutron Detectors at Low Energy

The Versatile Array of Neutron Detectors at Low Energy (VANDLE)[1] is versatile by design, meaning that it can achieve a variety of scientific goals with many different experimental conditions. VANDLE consists of three different sizes of plastic scintillators. For example, the medium bars used in the work presented has dimensions of 3x6x120 cm^3. The variety of sizes allows VANDLE to maximize its geometric efficiency. Each VANDLE module has a PMT attached to each end of the bar, which allows the longitudinal interaction point to be determined. VANDLE detects neutrons and measures their energy via the time-of-flight (ToF) technique. This technique uses the time difference between two detectors in order to measure the time required for a particle to cover a known distance. We calculate the non-relativistic kinematic energy using this information. Even though this is a conceptually simple technique it is one of the easiest and most powerful methods for measuring neutron energies[6–10].

A fully digital data acquisition system provides VANDLE with several integral features. First is low energy neutron detection thresholds. This is accomplished using a coincidence triggering scheme that is fully implemented on the acquisition board. The low threshold triggering means that it is possible to detect neutrons with energies of 70 keV or lower and run the system in the photomultiplier's thermal noise. Second, the timing for the system is better than 1 ns for photomultiplier (PMT) signals[11].

VANDLE excels when coupled with gamma-ray detection systems. Previously, VANDLE was coupled with only a high purity Ge clover[12]. Due to the geometry of the experiments these detectors had poor total efficiency (1%). The experiments presented in these proceedings explore the use of both $LaBr_3$ and NaI arrays to boost the total gamma-ray detection efficiency of the system.

3. Preliminary Results from ^{97}Rb

This study surveyed mid-shell nuclei of interest to decay heat, the reactor anti-neutrino spectrum anomaly and nuclear structure. Of particular interest, is the amount of energy released via neutrons from fission products. Brady studied the fission energy release[13] and used these data to create a theoretical model to produce neutron spectra for nuclei with no measured spectroscopy data. However, the data used in this evaluation did not contain gamma-neutron coincidences. This means that the calculated energy release from gammas and neutrons is incorrect.

We performed this experiment at the On-line Test Facility (OLTF) using the Oak Ridge National Laboratory tandem at Oak Ridge National Laboratory (ORNL). The ORNL tandem accelerates protons that impinge on an UCx target in the OLTF. The fission fragments are mass and isotope separated and delivered a moving tape collector. An electrostatic beam deflector allowed the collection of samples of the isotope of interest and their subsequent decays measured by the detector suite. The detector suite consisted of VANDLE coupled with a high purity Ge clover, an array of NaI

Fig. 2. The experimental setup for the measurement of ^{97}Rb. This setup has a total neutron detection efficiency of approximately 10%.

detectors, an array of $LaBr_3$ detectors and an in vacuum beta-particle detector, see Figure 2. The setup boasted a total neutron detection efficiency of 10% at 1 MeV. These data are under analysis.

4. Preliminary Results from ^{72}Co

This experiment aimed to study decays of cobalt isotopes that populate excited states in proton-closed-shell Z=28 nickel isotopes. The beta decay of Co isotopes has recently attracted experimental and theoretical interest due to its unusual properties: large B(GT) and large gamma emission from neutron unbound states[14]. The proton shell gap crossing for neutron-rich cobalts (Z=27) results in Q_β energies exceeding 10 MeV and lifetimes below 100 ms. In the extreme-single particle picture, the decays of neutron rich Co isotopes should be dominated by the GT transformation of an $f_{5/2}$ neutron into an $f_{7/2}$ proton. Such transitions are observed in previous measurements using gamma spectroscopy[15]. Calculations using shell model with effective interactions and a ^{48}Ca core reveal that transformations involving $\nu f_{5/2}$, $\nu p_{1/2}$ and $\nu p_{3/2}$ orbitals dominate the Gamow-Teller strength at high excitation energies. These transformations are separated from the $f_{5/2} \rightarrow f_{7/2}$ transitions by the Z=28 shell gap. The increased energy means that these states will only be observed via beta-delayed neutron emission.

Fig. 3. The experimental setup for the measurement of ^{72}Co.

We performed this experiment at the National Superconducting Cyclotron Laboratory (NSCL) at Michigan State University. The coupled cyclotrons and the A1900 provided a cocktail beam consisting of isotope masses between 70 and 78 to the experimental end station. The detector suite was nearly identical to that described in Section 3. A pixelated implantation detector made from YAP replaced the beta particle detectors and provided better than 2 mm position resolution. Si PIN detectors provided ΔE information for reconstruction of the particle identification spectrum. Figure 3 displays the partial experimental setup. This setup had a neutron detection efficiency of 10%. The analysis of these data is ongoing.

5. Conclusions

Measurements of beta-delayed neutrons from fission fragments provide insight into nuclear structure above the neutron separation energy. These spectra are critical to improve our understanding of the structure of neutron-rich nuclei since as neutron emission begins to become a major decay path. Accurate neutron energy spectra are critical for decay heat measurements because they are inputs into reactor modeling simulations. These spectra also shed light on the reactor anti-neutrino anomaly. Experiments conducted at OLTF and NSCL provided excellent data on mid-shell and closed shell nuclei. The preliminary data show that beta-delayed neutrons are observed. The neutron-gamma coincidences demonstrate that the neutron energy spectra alone are insufficient to determine underlying nuclear structure.

Acknowledgments

The material is based upon work supported by the U.S. Department of Energy (DOE), Office of Science and the Office of Nuclear Physics. The National Nuclear Security Administration under the Stewardship Science Academic Alliances through DOE Award No. DE-FG52-08NA28552 sponsored part of this work. This research was also sponsored by the Office of Nuclear Physics and the U. S. Department of Energy under contracts DE-FG02-96ER40983 (UTK), DE-AC05-00OR22725 (ORNL).

References

1. W. Peters, S. Ilyushkin, M. Madurga, C. Matei, S. Paulauskas, R. Grzywacz, D. Bardayan, C. Brune, J. Allen, J. Allen, Z. Bergstrom,

J. Blackmon, N. Brewer, J. Cizewski, P. Copp, M. Howard, R. Ikeyama, R. Kozub, B. Manning, T. Massey, M. Matos, E. Merino, P. O'Malley, F. Raiola, C. Reingold, F. Sarazin, I. Spassova, S. Taylor and D. Walter, Performance of the versatile array of neutron detectors at low energy VANDLE, *NIMA* **836**, 122 (2016).

2. G. Rudstam, S. Shalev and O. Jonsson, Delayed neutron emission from separated fission products, *NIM* **120**, 333 (1974).
3. K. L. Kratz, W. Rudolph, H. Ohm, H. Franz, M. Zendel, G. Herrmann and H. V. Kalapdor, Investigation of Beta strength functions by neutron and gamma-ray spectroscopy, *Nucl. Phys. A* **317**, 335 (1979).
4. J. Hardy, L. Carraz, B. Jonson and P. Hansen, The essential decay of pandemonium: A demonstration of errors in complex beta-decay schemes, *Phys. Lett. B* **71**, 307 (1977).
5. J. C. Hardy, B. Jonson and P. G. Hansen, The essential decay of pandemonium: β-delayed neutrons, *Nucl. Phys. A* **305**, 15 (1978).
6. T. Blaich, T. Elze, H. Emling, H. Freiesleben, K. Grimm, W. Henning, R. Holzmann, G. Ickert, J. Keller, H. Klingler, W. Kneissl, R. König, R. Kulessa, J. Kratz, D. Lambrecht, J. Lange, Y. Leifels, E. Lubkiewicz, M. Proft, W. Prokopowicz, C. Schütter, R. Schmidt, H. Spies, K. Stelzer, J. Stroth, W. Walus, E. Wajda, H. Wollersheim, M. Zinser and E. Zude, A large area detector for high-energy neutrons, *NIMA* **314**, 136 (1992).
7. P. Zecher, A. Galonsky, J. Kruse, S. Gaff, J. Ottarson, J. Wang, F. Deák, A. Horváth, A. Kiss, Z. Seres, K. Ieki, Y. Iwata and H. Schelin, A large-area, position-sensitive neutron detector with neutron/γ-ray discrimination capabilities, *NIMA* **401**, 329 (1997).
8. A. Buţă, T. Martin, C. Timiş, N. Achouri, J. C. Angélique, C. Borecea, I. Cruceru, A. Genoux-Lubain, S. Grévy, M. Lewitowicz, E. Liénard, F. M. Marqués, F. Negoiţă, F. de Oliveria, N. A. Orr, J. Péter and M. Sandu, TONNERRE: an array for delayed-neutron decay spectroscopy, *NIMA* **455**, 412 (2000).
9. B. Luther, T. Baumann, M. Thoennessen, J. Brown, P. DeYoung, J. Finck, J. Hinnefeld, R. Howes, K. Kemper, P. Pancella, G. Peaslee, W. Rogers and S. Tabor, MoNA — The Modular Neutron Array, *NIMA* **505**, 33 (2003).
10. G. Perdikakis, M. Sasano, S. M. Austin, D. Bazin, C. Caesar, S. Cannon, J. Deaven, H. Doster, C. Guess, G. Hitt, J. Marks, R. Meharchand, D. Nguyen, D. Peterman, A. Prinke, M. Scott, Y. Shimbara, K. Thorne, L. Valdez and R. Zegers, LENDA: A low

energy neutron detector array for experiments with radioactive beams in inverse kinematics, *NIMA* **686**, 117 (2012).

11. S. V. Paulauskas, M. Madurga, R. Grzywacz, D. Miller, S. Padgett and H. Tan, A digital data acquisition framework for the Versatile Array of Neutron Detectors at Low Energy (VANDLE), *NIMA* **737**, 22 (2014).
12. M. Madurga, S. V. Paulauskas, R. Grzywacz, D. Miller, D. W. Bardayan, J. C. Batchelder, N. T. Brewer, J. A. Cizewski, A. Fijałkowska, C. J. Gross, M. E. Howard, S. V. Ilyushkin, B. Manning, M. Matoš, A. J. Mendez, K. Miernik, S. W. Padgett, W. A. Peters, B. C. Rasco, A. Ratkiewicz, K. P. Rykaczewski, D. W. Stracener, E. H. Wang, M. Wolińska Cichocka and E. F. Zganjar, Evidence for gamow-teller decay of ^{78}Ni core from beta-delayed neutron emission studies, *Phys. Rev. Lett.* **117**, p. 092502 (Aug 2016).
13. M. C. Brady, Evaluation and Application of Delayed Neutron Precursor Data, PhD thesis, Texas A&M University, April 1989.
14. A. Spyrou, S. N. Liddick, F. Naqvi, B. P. Crider, A. C. Dombos, D. L. Bleuel, B. A. Brown, A. Couture, L. Crespo Campo, M. Guttormsen, A. C. Larsen, R. Lewis, P. Möller, S. Mosby, M. R. Mumpower, G. Perdikakis, C. J. Prokop, T. Renstrøm, S. Siem, S. J. Quinn and S. Valenta, Strong neutron-γ competition above the neutron threshold in the decay of ^{70}Co, *Phys. Rev. Lett.* **117**, p. 142701 (Sep 2016).
15. M. M. Rajabali, R. Grzywacz, S. N. Liddick, C. Mazzocchi, J. C. Batchelder, T. Baumann, C. R. Bingham, I. G. Darby, T. N. Ginter, S. V. Ilyushkin, M. Karny, W. Królas, P. F. Mantica, K. Miernik, M. Pfützner, K. P. Rykaczewski, D. Weisshaar and J. A. Winger, β decay of $^{71;73}$Co: Probing single-particle states approaching doubly magic ^{78}Ni, *Phys. Rev. C* **85**, p. 034326 (Mar 2012).

The Analysis of ^{89}Rb β-Decay Using the Modular Total Absorption Spectrometer

A. Fijałkowska* on behalf of MTAS Collaboration

Department of Physics and Astronomy, Rutgers University, New Brunswick, NJ 08903, USA

Faculty of Physics, University of Warsaw, PL-02-093 Warsaw, Poland

Department of Physics and Astronomy, University of Tennessee, Knoxville, TN 37966, USA

**E-mail: aleksandra.kuzniak@gmail.com*

Total Absorption Spectrometers are characterized by high efficiency detection of γ-ray radiation, which is due to their large volume and nearly 4π solid angle coverage. This high efficiency, used in the study of β-decay of unstable nuclei, allows for the total detection of the deexcitation path of daughter nuclei. This makes total absorption spectroscopy an ideal technique to establish true β-decay feeding patterns, which helps determine β-strength distributions as well as to understanding decay heat and the anti-neutrino spectrum emitted from nuclear reactors. The Modular Total Absorption Spectrometer (MTAS) constructed at the Holifield Radioactive Ion Beam Facility has been used to study the β decay of over 70 fission products. The measurements were focused on nuclei abundantly produced in the reactor core. In this contribution we present aspects of the analysis of the β-decay of ^{89}Rb measured with the MTAS detector.

Keywords: Total Absorption Spectroscopy; MTAS; β-Decay.

1. Introduction

Total absorption spectroscopy using sodium-iodine NaI(Tl) crystals has been around for over 40 years. The use of this method for the β-decay study of unstable nuclei helps establish the true β-decay feeding patterns, especially for the decays of nuclei suffering from the Pandemonium Effect [1]. Recent studies show that the measurements of fission products using total absorption spectroscopy are extremely important in determining β-strength distributions as well as to understanding the decay heat and anti-neutrino spectra emitted from nuclear reactors [2–5]. A disadvantage of this technique is a limited energy resolution as well as the unique analysis procedure that requires precise knowledge of the detector response to γ and β radiation, and in the case of β-delayed neutron emitters also neutrons[6]. This contribution presents aspects of the analysis of the β-decay of ^{89}Rb using the Modular Total Absorption Spectrometer (MTAS). ^{89}Rb is a nuclide abundantly produced in the reactor core. The cumulative ^{235}U fission yields for ^{89}Rb is 0.0469(6)[7]. Due to the low Q_β value (4496(5) keV), long half-life ($T_{1/2}$=15.3(1) min), and relatively simple decay scheme of ^{89}Rb, this measurement allows for testing of the complex analysis procedure for the MTAS detector. Moreover the β-decay of ^{89}Rb was previously studied by Total Absorption Gamma Spectrometer (TAGS)[8], which enables a comparison of the results obtained by the same technique, but different detectors.

2. Experimental set-up

The experiment was performed at the Holifeld Radioactive Ion Beam Facility (HRIBF) at Oak Ridge National Laboratory (ORNL). The 40 MeV proton beam impinged on a $^{238}UC_x$ target placed inside the ion source. Fission products were mass selected by means of magnetic separation and implanted into a moving tape that transported the radioactive samples into the center of MTAS where they were measured, and then moved them away to a shielded box. The details of the experimental system are shown in the M. Karny *et al.*[9] The main component of the detection system is built of 19 NaI(Tl) hexagonal-shaped crystals Modular Total Absorption Spectrometer. The single crystals are arranged in a honeycomb structure. Depending on the distance from the center of the detector, crystals are separated in 4 parts – central module, inner, middle, and outer rings. Each of the crystals, with the exception of the central, is provided with two photomultiplier tubes disposed on opposite sides of the crystal. The central detector has a cylindrical cavity surrounded by six smaller photomultipliers, in which two 7-strip silicon β detectors are placed. Silicon detectors surround the radioactive source implanted on the movable transport tape. Implantation point was located in front of MTAS. Schematic drawing of the set-up is presented in Fig. 1.

The tape operating cycle covered the implantation of a radioactive sample, possibly waiting to reduce the activity of the sample, sliding sample into the detector, measurement, and then sliding the radioactive sample into the shielded box.

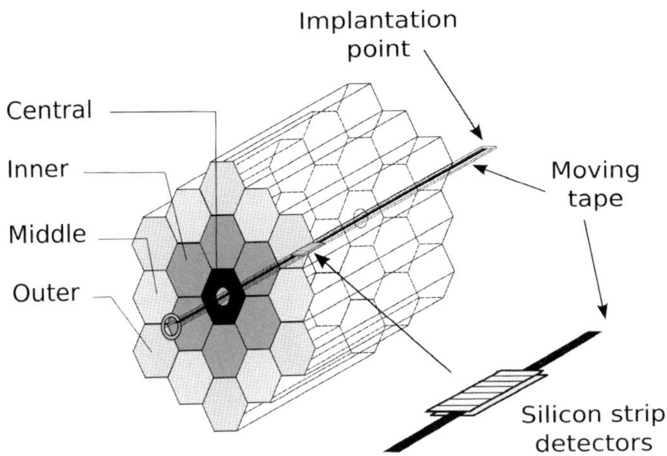

Fig. 1. Modular Total Absorption Spectrometer coupled with silicon strip detector.

3. Preparation of the experimental spectra

3.1. *Walk correction*

The signals from the 48 photomultipliers and 14 strips of silicon detectors were joined in the event by the time of their detection and the length of coincidence gate. Selecting the length of time window requires to take into account the risk of accidental coincidence with the background, in case of adoption to long gate, and loss of the efficiency, in case of establishing to short coincidence window. To improve the efficiency, while minimizing the risk of accidental coincidence with the background, detection time was corrected on the walk.

In order to determine the scale of the walk occurrence, the time difference between the signals from the tested detector (t_{std}) and the sufficiently high energy signal from arbitrarily chosen detector (t_{ref}) has been plotted. Sample result is shown at the left part of the Fig. 2. The x-axis shows the difference in time detection of test and reference signal determined according to the equation: $\Delta T = t_{ref} - t_{std} + 1\mu\text{s}$, y-axis shows the energy of the test signal. In order to find walk correction, the spectrum was divided along the y-axis of the 20 keV energy bands and then projected onto the time-line. For each of the resulting histograms the center of the distribution was found, and used to plot the detection time difference as a function of test signal energy ($\Delta T(E)$). The curve described by the Eq. 1 was fitted to received time-energy distribution and used to apply walk correction.

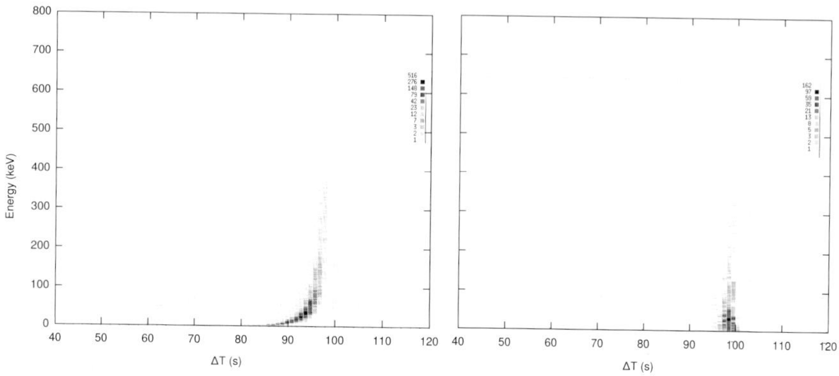

Fig. 2. The difference of the detection time of test and reference signal (x-axis) as a function of test signal energy (y-axis). The left panel shows the time difference before walk correction, while right panel after the walk correction.

$$\Delta T = a + b \cdot E^c \tag{1}$$

Sets of three different curves were used, one for silicon detectors, another for the central detector and the third for the others modules. The time difference between the signals from the tested and reference detector after walk correction completion was presented in the right part of the Fig. 2. The correction allowed to reduce the length of coincidence gate from 1 μs to 150 ns without the efficiency loss.

3.2. *Spectra construction*

The primary spectrum used in the analysis of MTAS data are the sum of all the signals recorded in the event (the so-called TAS spectra). Individual count in such prepared histogram is the sum of energy deposits in all summed detectors. Peaks in the TAS spectra correspond to the levels in the daughter nucleus, while their height to β-feeding of these levels. Analysis of the so prepared spectrum requires to find the detector response on the basis of the available decay scheme. In this case, data published in ENSDF database were used[10]. For this purpose a Monte Carlo simulation code by means of the Geant4 toolkit in version 4.9.4.p02[11] was developed. Based on the comparison of experimental and simulated spectra decay schemes have been modified to make them agree. Maximum entropy method was preformed as a fitting procedure with β-intensities as a fitting parameters[12]. Missing high energy levels were restored by adding bins every 100 keV starting from determined individually for each decay minimum energy

E_{min}, ending with energy equal $Q_\beta - 300$ keV. The details of this part of the analysis are described in the M. Karny *et al.*[9]

4. Measurement of the β-decay of ^{89}Rb

Available literature data shows the ^{89}Rb β-decay scheme containing 19 levels and 62 γ-rays[10,13]. The data contained in the ENSDF database come from high-resolution measurements, however, data from the TAGS measurement are also available[8]. The main difference between high resolution and TAGS decay scheme concerns the β-feeding of the first-excited state ($1/2^+$, 1031.98 keV), which was defined as $< 4\%$ and 0% in the ENSDF and TAGS data respectively.

The spectrum obtained from MTAS was compared with the simulated detector response assuming decay scheme published in the ENSDF database. The β-feeding of $1/2^+$ state was taken as 4%. The comparison is shown in Fig. 3. The biggest differences observed are in the area of 1031 keV energy peak, which the β-feeding was, as suspected, overestimated.

The analysis of data obtained from the MTAS requires only slight modifications of intensities of known β transitions. In addition, two bins of

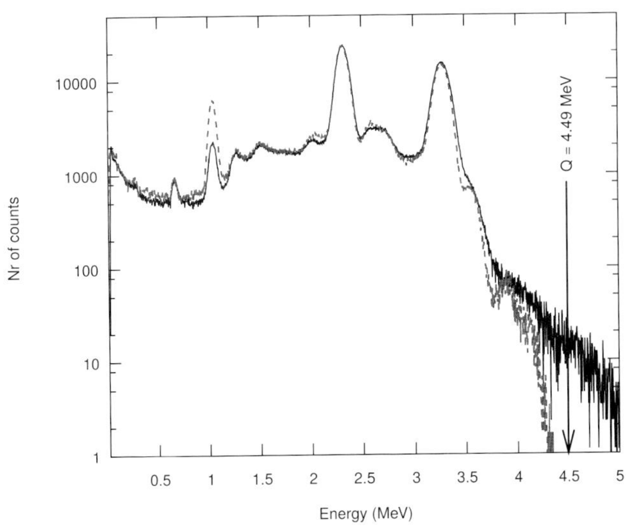

Fig. 3. Experimental total energy deposit in MTAS detector emitted in the decay of ^{89}Rb (solid) compared with the simulated detector response based on ENSDF entry (dashed).

energy 4200 and 4300 keV, with a small, less than 3.5%, feeding probability were added. The final result of the fit is shown in Fig. 4. The figure shows a comparison of the experimental spectrum (top dark, solid line), simulated spectrum based on the new decay scheme (dashed line) and detector reponse function of individual levels. Counts above energy Q_β have been identified as a pile-up and random coincidence with background, which also has been calculated, but is not shown in the Fig. 4.

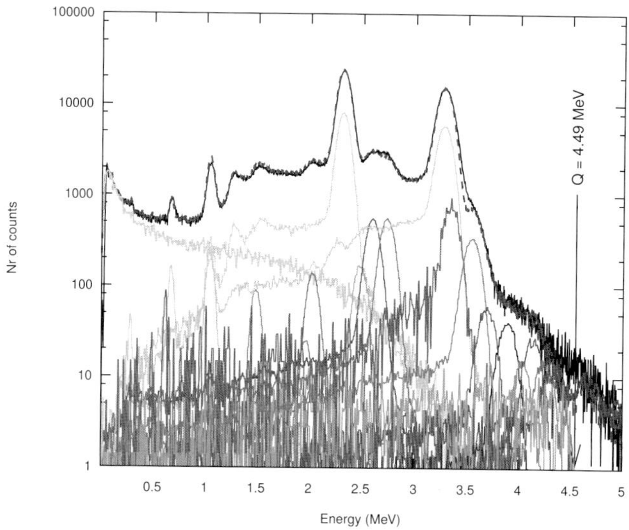

Fig. 4. Experimental total energy deposit in MTAS detector emitted in the decay of ^{89}Rb (top dark, solid) compared with the simulated detector response based on new decay scheme (dashed). Other curves represent detector reponse function of individual levels.

Fig. 5 presents the changes in the intensities of β transitions. The solid line shows the cumulative feeding contained in the ENSDF, the dashed curve shows the result of the presented analysis, while dotted β intensities obtained from TAGS[8]. The data received from total absorption spectroscopy (TAGS and MTAS) are in very good agreement and only slightly differ from the results of high-resolution measurements. The results MTAS measurement confirm that the 1031 keV level is not populated in the β decay.

A small modification in decay scheme resulted in a slight change of average γ and β energy per decay. Average γ-ray energy changed from

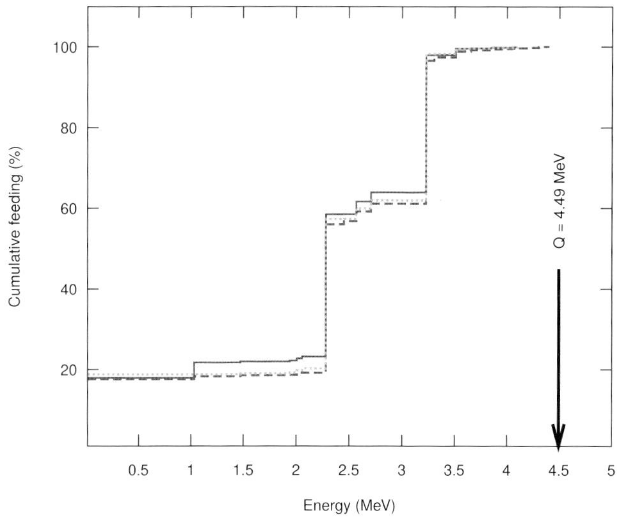

Fig. 5. Cumulative feeding of ^{89}Sr states after ^{89}Rb β-decay. Solid line follows ENSDF data, upper dashed line corresponds to the presented MTAS data evaluation, dotted line shows β intensities obtained from TAGS measurement.

2.187(86) to 2.264(69) MeV (+4%), while average β energy decreased form 0.956(39) to 0.939(32) MeV (-2%) compared to ENSDF.

5. Summary

This contribution presents some aspects concerning the analysis of data from Modular Total Absorption Spectrometer based on the analysis of the β-decay measurement of ^{89}Rb, in particular the details of walk correction procedure. The obtained ^{89}Rb decay scheme is slightly different than published in ENSDF database, while being in a very good agreement with the TAGS results. Data from MTAS confirmed that the first-excited state in the ^{89}Sr nucleus is not populated in the β-decay. Analysis of ^{89}Rb shows that in the case of well-known decay schemes the result of total absorption spectroscopy study are consistent with the previous measurements.

Acknowledgments

We would like to thank the HRIBF operations staff for the production of exceptional proton beams and for assisting with the experiments. This work is supported in part by the U.S. Department of Energy (DOE), National Nuclear Security Administration, Stewardship Science Academic Alliances, DE-NA0002132 (Rutgers, UTK), DE-AC05-00OR22725 (ORNL), and by the Polish National Science Center under contracts UMO-2013/08/T/ST2/00624.

References

1. J. C. Hardy, L. C. Carraz, B. Jonson and P. G. Hansen, *Phys. Lett. B* **71**, p. 307 (1977).
2. T. Yoshida and A. L. Nichols, *Assessment of Fission Product Decay Data for Decay Heat Calculations: A report by the Working Party on International Evaluation Co-operation of the Nuclear Energy Agency Nuclear Science Committee* (Nuclear Energy Agency, Organization for Economic Co-operation and Development, Paris, France, 2007).
3. A. Algora, D. Jordan, J. L. Tain, B. Rubio, J. Agramunt, A. B. Perez-Cerdan, F. Molina, L. Caballero, E. Nácher, A. Krasznahorkay, M. D. Hunyadi, J. Gulyás, A. Vitéz, M. Csatlós, L. Csige, J. Äysto, H. Penttilä, I. D. Moore, T. Eronen, A. Jokinen, A. Nieminen, J. Hakala, P. Karvonen, A. Kankainen, A. Saastamoinen, J. Rissanen, T. Kessler, C. Weber, J. Ronkainen, S. Rahaman, V. Elomaa, S. Rinta-Antila, U. Hager, T. Sonoda, K. Burkard, W. Hüller, L. Batist, W. Gelletly, A. L. Nichols, T. Yoshida, A. A. Sonzogni and K. Peräjärvi, Reactor Decay Heat in ^{239}Pu: Solving the γ Discrepancy in the 4-3000-s Cooling Period, *Phys. Rev. Lett.* **105**, p. 202501 (Nov 2010).
4. M. Fallot, S. Cormon, M. Estienne, A. Algora, V. M. Bui, A. Cucoanes, M. Elnimr, L. Giot, D. Jordan, J. Martino, A. Onillon, A. Porta, G. Pronost, A. Remoto, J. L. Taín, F. Yermia and A.-A. Zakari-Issoufou, New antineutrino energy spectra predictions from the summation of beta decay branches of the fission products, *Phys. Rev. Lett.* **109**, p. 202504 (Nov 2012).
5. B. C. Rasco, M. Wolińska-Cichocka, A. Fijałkowska, K. P. Rykaczewski, M. Karny, R. K. Grzywacz, K. C. Goetz, C. J. Gross, D. W. Stracener, E. F. Zganjar, J. C. Batchelder, J. C. Blackmon, N. T. Brewer, S. Go, B. Heffron, T. King, J. T. Matta, K. Miernik, C. D. Nesaraja, S. V. Paulauskas, M. M. Rajabali, E. H. Wang, J. A. Winger, Y. Xiao and C. J. Zachary, Decays of the three top contributors to the reactor $\overline{\nu}_e$ high-energy spectrum, ^{92}Rb, ^{96gs}Y, and ^{142}Cs, studied with total absorption spectroscopy, *Phys. Rev. Lett.* **117**, p. 092501 (Aug 2016).

6. B. Rasco, Complete β-decay pattern for the high priority decay heat isotopes ^{137}i and ^{137}xe determined using total absorption spectroscopy, *Submitted to Physical Review C* (2017).
7. JEFF-3.1 evaluated data library.
8. R. Greenwood, R. Helmer, M. Putnam and K. Watts, Measurement of β-decay intensity distributions of several fission-product isotopes using a total absorption γ-ray spectrometer, *Nuclear Instruments and Methods in Physics Research, Section A: Accelerators, Spectrometers, Detectors and Associated Equipment* **390**, 95 (1997).
9. M. Karny, K. Rykaczewski, A. Fijałkowska, B. Rasco, M. Wolińska-Cichocka, R. Grzywacz, K. Goetz, D. Miller and E. Zganjar, Modular total absorption spectrometer, *Nuclear Instruments and Methods in Physics Research Section A: Accelerators, Spectrometers, Detectors and Associated Equipment* **836**, 83 (2016).
10. Evaluated Nuclear Structure Data File.
11. J. Allison, K. Amako, J. Apostolakis, H. Araujo, P. Dubois, M. Asai, G. Barrand, R. Capra, S. Chauvie, R. Chytracek, G. Cirrone, G. Cooperman, G. Cosmo, G. Cuttone, G. Daquino, M. Donszelmann, M. Dressel, G. Folger, F. Foppiano, J. Generowicz, V. Grichine, S. Guatelli, P. Gumplinger, A. Heikkinen, I. Hrivnacova, A. Howard, S. Incerti, V. Ivanchenko, T. Johnson, F. Jones, T. Koi, R. Kokoulin, M. Kossov, H. Kurashige, V. Lara, S. Larsson, F. Lei, O. Link, F. Longo, M. Maire, A. Mantero, B. Mascialino, I. McLaren, P. Lorenzo, K. Minamimoto, K. Murakami, P. Nieminen, L. Pandola, S. Parlati, L. Peralta, J. Perl, A. Pfeiffer, M. Pia, A. Ribon, P. Rodrigues, G. Russo, S. Sadilov, G. Santin, T. Sasaki, D. Smith, N. Starkov, S. Tanaka, E. Tcherniaev, B. Tome, A. Trindade, P. Truscott, L. Urban, M. Verderi, A. Walkden, J. Wellisch, D. Williams, D. Wright and H. Yoshida, Geant4 developments and applications, *Nuclear Science, IEEE Transactions on* **53**, 270 (Feb. 2006).
12. J. L. Taín and D. Cano-Ott, Algorithms for the analysis of β-decay total absorption spectra, *Nuclear Instruments and Methods in Physics Research Section A: Accelerators, Spectrometers, Detectors and Associated Equipment* **571**, 728 (2007).
13. E. A. Henry, W. L. Talbert and J. R. McConnell, Gamma-ray decay schemes for ^{89}Kr and ^{89}Rb, *Phys. Rev. C* **7**, 222 (Jan 1973).

Microscopic CI Plus EDF Models for E0 Gamma Decay Matrix Elements

Alex Brown

Department of Physics and Astronomy, Michigan State University, East Lansing, MI

I present a microscopic model for electric monopole (E0) transition matrix elements. It is obtained by combining the configuration interaction model for orbital occupations with an energy-density functional model for the single-particle potential and radial wave-functions. The configuration interaction model is used to constrain the orbital occupations for the diagonal and off-diagonal matrix elements. These are used in an energy-density functional calculation to obtain a self-consistent transition density. This density contains the valence contribution, as well as the polarization of the protons by the valence protons and neutrons. We show connections between E0 matrix elements and isomer and isotope shifts of the charge radius. The spin-orbit correction to the charge density is important in some cases. This model accounts for a large part of the data over a wide region of the nuclear chart. It also accounts for the shape of the observed electron scattering form factors. The results depend on the Skyrme parameters used for energy-density functional and might be used to provide new constraints for them.

Exotic Cluster Structures in the Mean-Field Theory

J. A. Maruhn
Institut für Theoretische Physik, Goethe-Universität Frankfurt
Max-von-Laue-Straße 1, 60325 Frankfurt am Main, Germany
E-mail: maruhn@th.physik.uni-frankfurt.de

This talk contains results from several applications of the mean-field model for nuclei. A brief discussion is give of a comparison of nuclear shapes in heavy-ion collisions from time-dependent Hartree-Fock (TDHF) and the asymmetric two-center oscillator shell model (TCSM). The remainder of the talk is devoted to the structure of light nuclei: analysis in terms of α-clusters, the exotic 4-α chain state in ^{16}O, and a rotating torus configuration for ^{40}Ca.

Keywords: Nuclear Structure, α-Clusters, Toroidal Nuclei.

1. TDHF and TCSM

Before addressing cluster structures, a brief description of new results of relevance to topics in this conference will be given. At present, among many other models, two microscopic quantal descriptions of heavy-ion collisions coexist and it is interesting to compare them in practical applications. Time-dependent Hartree-Fock (TDHF) — here represented by the Sky3D[1] code — can allow almost arbitrary shapes and density distributions based on single-particle wave functions, but requires intensive computation, while the asymmetric two-center shell model[2] assumes a single-particle potential and nuclear shape with just five parameters and , moreover, the parametrization based on numerical efficiency. While many other properties of the two models are also quite different, most noticeably the treatment of dissipation, here we only try to check whether shapes attained during a collision with TDHF can be represented also in the TCSM parametrization.

Collisions in TDHF of ^{68}Ni + ^{208}Pb at 390 MeV in the center of mass and for various impact parameters were used to generate isodensity lines at $\rho = 0.08\,\mathrm{fm}^{-3}$ in the reaction plane and rotated to have the axis of smallest moment of inertia along the coordinate axis, allowing the axisymmetric shapes of the TCSM to be fitted to these curves. Aside from remaining non-axial effects for the larger impact parameters, such as on the lower right, after clutching this worked remarkably well until the rupture of the neck set in, which is not well described by a leptodermous model. Sample fits are shown in Fig. 1.

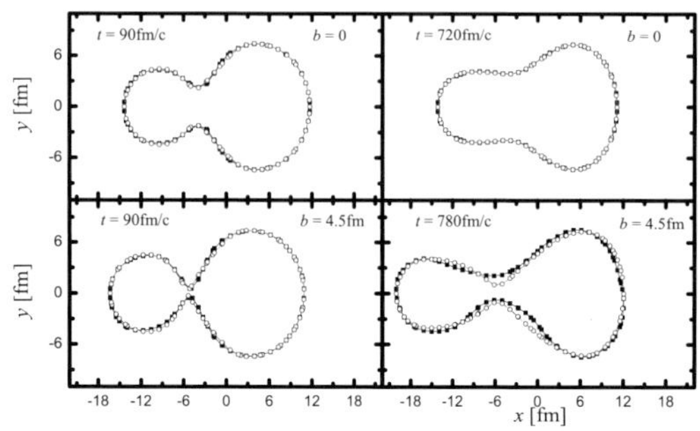

Fig. 1. Sample shapes obtained from a TDHF calculation (filled squares) fitted by the TCSM surface parametrization (open circles). The two upper figures are for a central collision and the lower ones for b=4.5 fm.

2. Cluster Structures in Light Nuclei

In a study of light nuclei[3] it was found that the mean-field approximation also allows the presence of alpha clusters. For the initial investigations, the ground-state Slater determinant of $n\alpha$ nuclei was obtained in the Sky3D code and then related to a cluster configuration in the following way: a second Slater determinant was constructed from spin and isospin saturated clusters of Gaussian wave functions located at various positions in space. Then these positions and the width of the Gaussians were varies to obtain maximal overlap with the mean-field wave function. The main results can briefly be characterized as follows:

(1) The nucleus ^{8}Be has a molecular structure formed by two α-particles. Depending on the Skyrme force used, the probability for this is around 80% and can even be increased to 98% by using the cluster distance as a GCM coordinate.
(2) For ^{12}C and ^{16}O the optimal placement of the clusters is very close to the origin, so that they are practically on top of each other. This does not correspond to a cluster structure, but antisymmetrization produces p-state wave functions to imitate the regular shell model states. The overlap probability for this structure is still only of the order of 1%.

Only a Skyrme force with weaker spin-orbit coupling leads to a larger cluster component. Spin-orbit coupling and the Coulomb force are the main mechanisms reducing cluster structure for the heavier nuclei.

(3) for ^{20}Ne the optimization leads to the ^{12}C structure in the center with two well-separated α-clusters on both sides with a probability of the order of 50%. The same quality can also be achieved with a ^{16}O nucleus in the center and a single α symmetrized between the two side positions or in a GCM wave function.

(4) For the heavier nuclei the cluster structure slowly fades away.

3. LocalizatIon Analysis

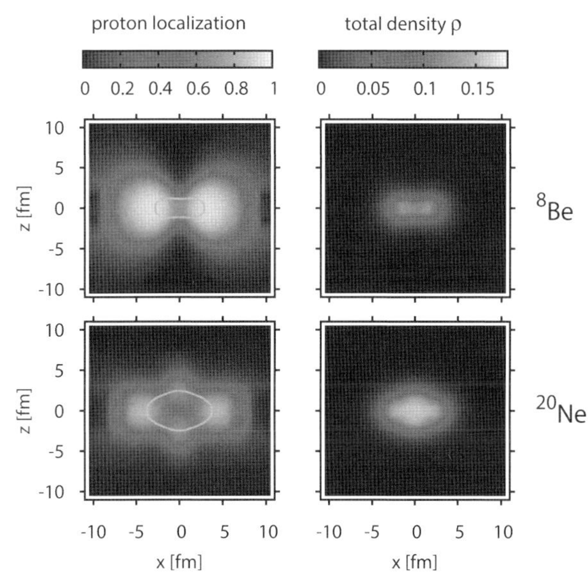

Fig. 2. Localization for ^{8}Be and ^{20}Ne nuclei (left) to be viewed in conjunction with the density distribution (right). Bright areas correspond to regions of strong localization. On the left, an isoline from the density distribution is added to guide the eye.

An alternative way to look for cluster correlations was proposed as *localization analysis* first developed for molecular structure[4,5] and then applied to nuclear physics[6]. The principal idea is to examine a particle at some point in space and then calculate the probability for having another particle *with the same spin and isospin* nearby. This probability is given through the one- and two-body densities

$$P_{q,\sigma}(\vec{r}.\vec{r}') = \rho_{q,\sigma}(\vec{r})\,\rho_{q,\sigma}(\vec{r}') - |\rho_{q,q,\sigma,\sigma}(\vec{r}, vecr')|^2$$

with q denoting the isospin and σ the spin. Averaged over a spherical shell of radius δ this can be rewritten in terms of the current and kinetic energy density and is finally inverted to look at the absence of such a particle to finally yield

$$R_{q,\sigma}(\vec{r},\delta) \approx \left(1 + \left[\frac{\tau_{q,\sigma}\rho_{q,\sigma} - \frac{1}{4}|\nabla\rho_{q,\sigma}|^2 - \vec{j}^{\,2}_{q,\sigma}}{\rho_{q,\sigma}\tau^{TF}_{q,\sigma}}\right]\right)^{-1}$$

with the Thomas-Fermi version of the kinetic energy

$$\tau^{TF}_{q,\sigma} = \frac{3}{5}\left(6\pi^2\right)^{2/3}\rho^{5/3}_{q,\sigma}.$$

and is normalized to the range $0\ldots 1$. This expression clearly gives information not involving single-particle wave functions in an unobservable way. Simple limiting cases are $R_{q,\sigma} \approx 1$ for the Fermi gas, while for the α-particle $R_{q,\sigma} \approx 1$ there is perfect localization, trivially reflecting that there is only one particle in the system.

Two-dimensional plots of the localization and density in two light nuclei are shown in Fig. 2. We find that, e. g., in ^{8}Be there is strong localization in both clusters, whereas in ^{20}Ne localization is large only in the regions associated with the clusters at the sides, agreeing with the interpretation of a ^{12}C core with two clusters attached.

4. Chain States

Exotic states in light nuclei like ^{12}C or ^{16}O made out of a chain of α-particles have been investigated for a long time. In the static TDHF code we also examined these configurations. It turned out that a linear arrangement of α-clusters as the initial configuration tends to converge to such a chain only if it is exactly linear and therefore cylindrically symmetric. A deviation for axial symmetry will lead to a folding over into the ground state. That the chain state appears as a very strong fixed point — convergence happening for a wide range of initial cluster distances — therefore is somewhat deceiving. It corresponds to a saddle point. Perturbing the chain in a bending deformation dynamically also leads to a folding, albeit only after some oscillations.

Later work[7,8], however, found out that the centrifugal force can stabilize the 4α chain, but not the 3α one (although there are arguments for a 3α

chain as an intermediate state in reactions[9]. Apparently the fact that the central particle is not subject to a centrifugal force makes it less effective in this case.

In our calculations[8] the static iterations were started with four α-particles along a line but slightly displaced to sides in alternating directions to avoid getting stuck in an axially symmetric saddle-point configuration. The outcome depended on the rotational frequency used in the cranking constraint. Fig. 3 shows the change of the rotational energy factor $\hbar^2/\Theta$ during the iterations. This should be viewed together with Fig. 4. The latter also shows the initial configuration in panel (a). For $\omega = 0$ the system actually settles into the highly deformed shape (c) and for $\hbar\omega$ increasing up to 1.5 MeV it folds over into the ground state (b). For $\hbar\omega = 2.0$ MeV we get the desired chain state (d) stabilized by the centrifugal force, while increasing ω even higher leads to breakup.

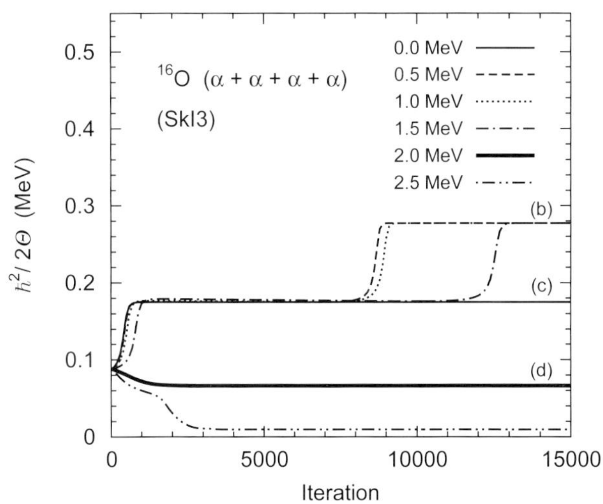

Fig. 3. HF convergence for the rotating α chain.

Fig. 5 shows the relation between angular momentum and ω for a few Skyrme forces in the range where the chain state is observed. The quantal angular momentum $J_{\rm crs}$ is also compared to the one calculated with the rigid-body moment. Both agree quite well, which is expected for such a strongly deformed shape. A slight upward bend shows the effect of centrifugal stretching.

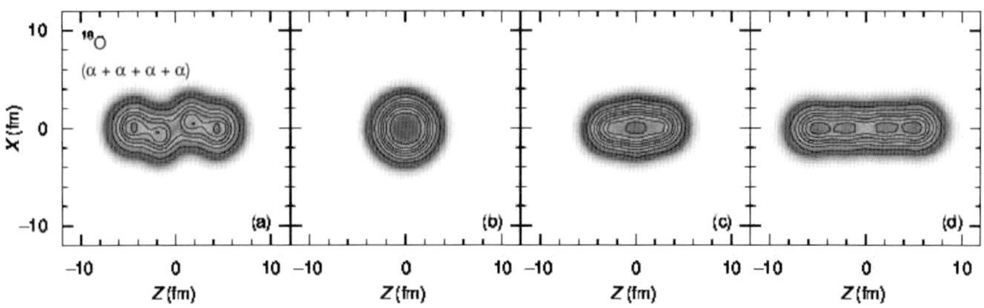

Fig. 4. Initial state (a) and three final configurations for the cranked ^{16}O system.

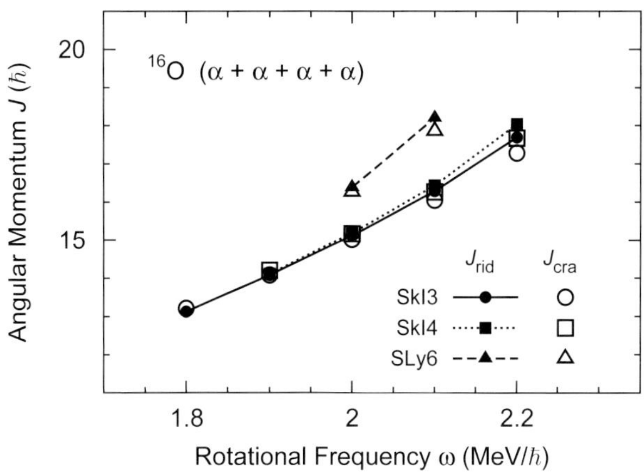

Fig. 5. Angular momentum as function of the rotational frequency for the rotating α chain.

5. Toroidal Nuclei

The idea that nuclei could have metastable states of toroidal type was first proposed by Wong[10] for heavy nuclei where the Coulomb force could stabilize the configuration. We have now investigated whether lighter nuclei like ^{40}Ca could be stabilized through rotation[11]. For this purpose a ring-like initial configuration of 10 α-particles was used and iterated with a cranking constraint. To check stability with respect to symmetry violations, the initial configuration sometimes was perfectly spherical, but perturbed configurations were also used but led to no difference in the final state.

For cranking parameters 1.12 MeV $< \hbar\omega \leq$ 1.8 MeV the static HF calculations converged to an axially symmetric torus with a major radius of about 6.5 fm and a minor radius roughly equivalent to that of an

α-particle. The single-particle wave functions in cylindrical coordinates can be well approximated by Gaussians around the major radius in the radial and z-directions and by a pure $\exp(im\phi)$ in the azimuthal angle.

The orbital angular momentum around the symmetry axis was precisely $60\hbar$, which is produced by an asymmetric occupation in the m quantum number: if all states fulfilling $-3 \leq m \leq 6$ are occupied with spin and isospin degeneracy, this results exactly in $L = 60\hbar$. Changing the range of m by one up or down yields alternatives of $L = 20\hbar$ or $L = 100\hbar$, of which slight indications were also seen but did not appear to be stable.

That these states really correspond to a classically rotating system was seen in two ways: giving the torus a slight "kick" off its axis produced a beautiful precession motion[12]. Also, to test stability with respect to the "sausage" deformation[10] we used an external repulsive potential of Gaussian shape at some point on its median to cut up the torus. For small strength of this potential the damage was repaired, but for 50 MeV the torus was cut open at that point and clearly continued rotating.

Acknowledgments

The results presented in this talk were produced with a large number of collaborators who are gratefully acknowledged here: H. Horiuchi, T. Ichikawa, N. Itagaki, M. Kimura, S. Ohkubo, P.-G. Reinhard, A. Tohsaki, S. Umar, Nan Wang.

References

1. J. A. Maruhn, P.-G. Reinhard, P. D. Stevenson and A. S. Umar, *Comp. Phys. Comm.* **185**, 2195 (2014).
2. J. Maruhn and W. Greiner, *Z. Physik* **251**, 431 (1972).
3. J. A. Maruhn, M. Kimura, S. Schramm, P.-G. Reinhard, H. Horiuchi and A. Tohsaki, *Phys. Rev. C***74**, 044311 (2006).
4. A. D. Becke and K. E. Edgecombe, *J. Chem. Phys.* **92**, 5397 (1990).
5. T. Burnus, M. A. L. Marques and E. K. U. Gross, *Phys. Rev. A* **71**, 010501 (2005).
6. P. G. Reinhard, J A. Maruhn, A. S. Umar and V. E. Oberacker, *Phys. Rev. C* **83**, 034312 (2011).
7. H. Flocard, P. H. Heenen, S. J. Krieger and M. S. Weiss, *Prog. Theor. Phys.* **72**, 1000 (1984).
8. T. Ichikawa, J. A. Maruhn, N. Itagaki and S. Ohkubo, *Phys. Rev. Lett.* **107**, 112501 (2011).

9. A. S. Umar, J. A. Maruhn, N. Itagaki and V. E. Oberacker, *Phys. Rev. Lett.* **104**, 212503 (2010).
10. C. Y. Wong, *Phys. Lett. B*41, 446 (1972).
11. T. Ichikawa, J. A. Maruhn, N. Itagaki, K. Matsuyanagi, P.-G. Reinhard and S. Ohkubo, *Phys. Rev. Lett.* **109**, 232503 (2012).
12. T. Ichikawa, K. Matsuyanagi, J. A. Maruhn and N. Itagaki, *Phys. Rev. C* **89**, 011305 (2014); *Phys. Rev. C* **90**, 034314 (2014).

Simple Patterns in Nuclei

A. C. Dreyfuss[1], J. P. Draayer[1,*], K. D. Launey[1], T. Dytrych[1,2] and R. B. Baker[1]

[1] *Dept. of Physics & Astronomy, Louisiana State University, Baton Rouge, LA 70803, USA*

[2] *Nuclear Physics Institute, 250 68 Řež, Czech Republic*

* *Email: draayer@sura.org*

We review state-of-the-art no-core shell model (NCSM) work that exploits symmetries which bare the emergence of simple patterns from Quantum Chromodynamics (QCD)-inspired effective interactions. Our theory of choice for this is the Symplectic Model (SpM) and a recent implementation that includes mixing of configurations of the no-core shell-model type, called NCSpM. The SpM is a fully microscopic theory that respects exact and exploits partial symmetries and hence, can uncover simple patterns hidden within the complexities of nuclei. Making use of the algebraic features of the SpM, the NCSpM complements the NCSM by "vertically extending" a physically relevant subspace into a model space that is needed to fully capture the collectivity of nuclei.

1. Introduction

Recent advances in the development of realistic inter-nucleon interactions[1–4] and the availability of massively parallel high performance computing (HPC) resources have placed *ab initio* nuclear structure models that have the capability to make use of these resources at the forefront of nuclear structure explorations. Building on the success of the independent-particle model of Mayer and Jensen[5,6], particle-driven models, such as the no-core shell model (NCSM), have been very successful in describing low-lying states of s- and p-shell nuclei[7,8]. The NCSM is a complete hamiltonian-based microscopic theory[7,9], ignoring all symmetries except for time reversal, parity, and translational invariance in its simplest implementation. It makes use of the harmonic oscillator (HO) single-particle basis, and organizes the complete space into "horizontal" HO shells, separated in energy by the $\hbar\Omega$ oscillator strength. The model space is limited by the $N_{\max}$ cutoff, which describes the cutoff in total oscillator quanta above the lowest HO configuration for the system. Additional techniques, such as the NCSM/RGM[10], Importance Truncation NCSM[11] and Monte Carlo NCSM[12] have augmented this approach, further.

Predating the NCSM is a more collective symplectic shell model[15,16], developed by Rowe and Rosensteel. Through partitioning the complete space into physically relevant subspaces that respect rotational invariance, this model exposes important dynamical symmetries in nuclei – namely, it reduces to the Bohr-Mottelson theory[17] in the macroscopic limit, and to the microscopic Elliott model[14,18] in the limit of a single shell. The symplectic $Sp(3,\mathbb{R})$ symmetry applied in a microscopic framework is directly related to the particle position and momentum coordinates, and naturally describes *rotations* and *vibrations* of an *equilibrium deformation*[19]. The existence of this symmetry and its slight symmetry breaking emerges naturally from first principles, as has been seen in recent *ab initio* symmetry-adapted no-core shell model (SA-NCSM) studies[20,21]. The SA-NCSM is an *ab initio* model that makes use of exact and partial symmetries, and which lays bare simple symmetry patterns that ultimately underpin nuclear structure. Indeed, the seminal work by Elliott[14,18] identifies the importance of the SU(3) symmetry (Fig. 1). In heavier nuclei, e.g., the upper *pf* and lower *sdg* shells, and in particular, in strongly deformed nuclei of the rare-earth and actinide regions, the SU(3) symmetry enters through the importance of the pseudo-spin symmetry and its pseudo-SU(3) complement in describing deformation[22,23].

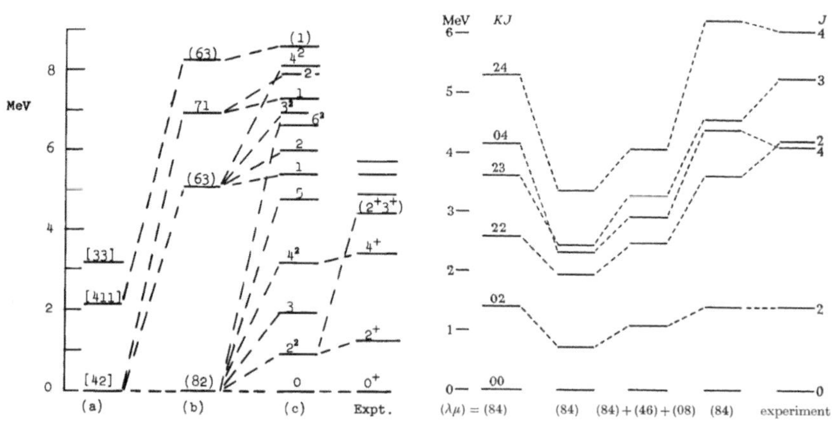

Fig. 1. Elliott's SU(3) model applied to *sd*-shell nuclei; both vertical axes represent energies in MeV. Left: energy spectrum of ^{20}Ne with (a) the Majorana potential, (b) the addition of the second-order SU(3) Casimir invariant C_2^{su3}, and (c) the further addition of the L^2 operator[13]. Right: energy spectrum of ^{24}Mg with a Gaussian central force. In both cases, the most deformed SU(3) configurations [(8 2) in ^{22}Ne and (8 4) in ^{24}Mg] play a key role in describing the low-energy structure.[14]

Informed by the emergence of the symplectic symmetry from *ab initio* studies, the fully microscopic no-core symplectic shell model (NCSpM)[24] makes use of the symplectic symmetry and a schematic interaction, and can approach model spaces beyond the reach of *ab initio* theories. Using this tool, we can probe the physics of the Hoyle state in ^{12}C, including the associated rotational band, plus the nature of other positive- and negative-parity states in ^{12}C and its giant resonance modes. This can be extended o the properties of other highly deformed intermediate-mass nuclei. Through reorganizing the $(LS)J$-coupled complete model space into irreducible representations (irreps) of SU(3) and then into irreps of the embedding $\mathrm{Sp}(3,\mathbb{R})$ – effectively, through reorganizing the space into physically relevant "vertical cones" – the NCSpM is able to describe bound states within a subspace that is a small fraction of the complete space.

Guided by insights into the underpinning nuclear structure laid bare by the schematic NCSpM, the particle-driven NCSM, and its symmetry-informed complement – the SA-NCSM – which incorporates both particle-driven and collective concepts, provide us with strong tools for new investigations into nuclear structure. These studies support and complement results of other first-principle approaches, such as Green's function Monte Carlo (GFMC)[25], Coupled-cluster (CC) method[26], In-medium SRG[27], Lattice Effective Field Theory (EFT)[28], and many other new developments.

2. Hidden symmetries in nuclei

Symmetry emerges from first principles – The *ab initio* SA-NCSM[20] is a no-core shell model with an SU(3)-coupled basis; namely, the many-particle basis states are organized with respect to the physically relevant, deformation-related $\mathrm{SU}(3)_{(\lambda\,\mu)} \overset{\kappa}{\supset} \mathrm{SO}(3)_L$ subgroup chain (see[21,29]). In a given complete $N_{\max}$ space, results obtained using the same interaction in the SA-NCSM are exactly equivalent to NCSM results. The symmetry-guided organization scheme used in the SA-NCSM allows for the consideration of only the most physically-relevant subspace of a complete $N_{\max}$ model space.

Results for p-shell nuclei computed in the *ab initio* SA-NCSM show the emergence, from first principles, of a simple pattern that favors large deformation and low spin. For example, the $N_{\max} = 10$ SA-NCSM wave function for the 1^+ ground state of ^{6}Li computed with the bare JISP16 interaction shows a dominance of the deformed $0\hbar\Omega$ $(2\,0)$ irrep and its symplectic excitations (e.g., $2\hbar\Omega$ $(4\,0)$, $4\hbar\Omega$ $(6\,0)$, etc.). This pattern is

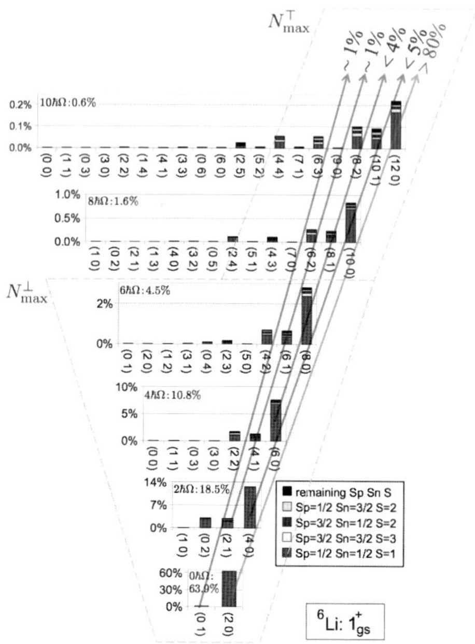

Fig. 2. Probability distributions for proton, neutron, and total intrinsic spin components $(S_p S_n S)$ across the Pauli-allowed deformation-related $(\lambda\,\mu)$ values for the 1^+ ground state of ^{6}Li, calculated in 12 HO shells with the JISP16 bare interaction ($\hbar\Omega = 20$ MeV). The most deformed configurations $(\lambda\,\mu)$ are at the right of each HO shell subspace. Note that the strengths are concentrated on that highly-deformed edge of the figure, indicating the dominance of collectivity; additionally, the lowest-spin configurations clearly dominate. A symmetry-guided model-space selection takes advantage of this emergent property by including the full space up through $N_{\rm max}^{\perp}$, but then selecting a subset of configurations with high deformation and low spin up through $N_{\rm max}^{\top}$. A model space constructed in this way is labeled $\langle N_{\rm max}^{\perp}\rangle N_{\rm max}^{\top}$. The projection onto symplectic vertical slices (with probability $\geq 1\%$) is schematically illustrated for ^{6}Li by arrows and clearly reveals the preponderance of a single symplectic irrep (vertical cone).

specific neither to this nucleus nor this interaction, as it has been seen in studies of other p-shell nuclei, including ^{6}He, ^{8}Be, and ^{12}C as well as in studies using the bare N^{3}LO interaction[3]. This emergent feature confirms the importance of the SU(3) and Sp(3, $\mathbb{R}$) symmetries in describing nuclear structure. Capitalizing on the use of these symmetries, new techniques are being employed to extend *ab initio* studies to larger model spaces (e.g., including up to 20 HO shells in ^{12}C), and extend the reach of *ab initio* methods to heavier nuclei[30].

Clustering and collectivity in ^{12}***C*** – The NCSpM is a fully microscopic no-core shell model based on the physically relevant symplectic Sp(3, $\mathbb{R}$) group[15,16] and its embedded SU(3) subgroup[14,18,31]. In the same complete $N_{\rm max}$ model space, NCSM results and results found using the NCSpM are identical. The NCSpM, however, organizes the space into Sp(3, $\mathbb{R}$) irreps, or "vertical slices," each of which describes a single $(\lambda\,\mu)$ equilibrium deformation and its rotations and vibrations. Selecting only the most physically relevant vertical slices in a given $N_{\max}$ space greatly reduces the model space, and allows for the NCSpM to extend to $N_{\max}$ model spaces far beyond the current NCSM limits.

We use a many-body Hamiltonian comprised of two main components: a single-particle piece, consisting of the HO potential and a spin-orbit term, and a collective piece that enters through the quadrupole-quadrupole interaction:

$$H_\gamma = \sum_{i=1}^{A} \left(\frac{\mathbf{p}_i^2}{2m} + \frac{m\Omega^2 \mathbf{r}_i^2}{2} \right) + \frac{\chi}{2} \frac{\left(e^{-\gamma Q\cdot Q} - 1\right)}{\gamma} + V_{NN}^{\mathrm{SB}}, \tag{1}$$

The HO Hamiltonian H_0, and the important quadrupole-quadrupole interaction, $\frac{1}{2} Q \cdot Q = \frac{1}{2} \sum_i q_i \cdot (\sum_j q_j)$ – which introduces the interaction of each particle with the total quadrupole moment of the system – preserve the symplectic $\mathrm{Sp}(3,\mathbb{R})$ symmetry.[a] The symmetry-breaking enters in the V_{NN}^{SB} term, applied to the symplectic bandheads. This can be either a realistic interaction, such as the chiral N^{3}LO interaction, or, e.g., the $-\sum_{i=1}^{A} \kappa l_i \cdot s_i$ spin-orbit term. The value of χ is fixed using self-consistent arguments[32] by the estimate used[33] in an $\mathrm{Sp}(3,\mathbb{R})$-based study of cluster-like states of ^{16}O, and the strength of the HO potential is fixed using an empirical estimate, $\hbar\Omega \approx 41/A^{1/3}$. The only adjustable parameter in the model is γ, which controls the presence of many-body interactions. Specifically, H_γ introduces many-body interactions in a hierarchical way, controlled by γ, such that, for sufficiently small γ, higher-order terms in the exponential of $Q \cdot Q$ become negligible. For example, we find that for ^{12}C, all terms in the expansion beyond $(Q \cdot Q)^2$ contribute negligibly to describing the ground state and its rotational band; for the Hoyle state band, however, it is necessary to include terms up through $(Q \cdot Q)^4$ (or third order in γ)[24].

Results for the energy spectrum for ^{12}C, computed in the NCSpM with $N_{\max} = 20$, agree remarkably well with experiment (Fig. 3). Results are shown for $N_{\max} = 20$, ($\hbar\Omega = 18$), down-selected to include only 5 symplectic irreps. We find the lowest 0^+, 2^+, and 4^+ states of the two 0p-0h irreps [0p-0h (4 0) and 0p-0h (1 2)] reproduce the ground state rotational band (the first 0^+, 2^+, and 4^+ of ^{12}C), the lowest 0^+ state of the 4p-4h (12 0) irrep coincides with the experimental Hoyle state, the lowest 0^+ state of the 2p-2h (6 2) irrep with the third 0^+ in ^{12}C, and the low-lying 3^-

[a]Note: The average value of $Q \cdot Q$ within an oscillator shell introduces a major renormalization of the HO shell structure, so as is normally done, this average is removed; that is, $Q \cdot Q \rightarrow (Q \cdot Q - < Q \cdot Q >)$ where the average $< Q \cdot Q >$ has a simple universal operator form that applies to HO shell.

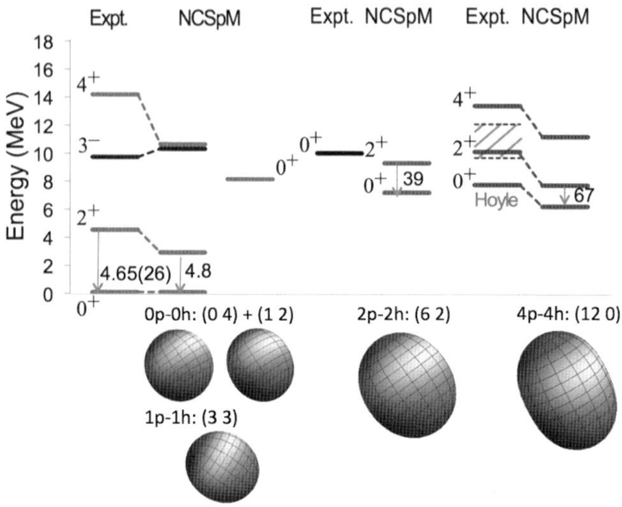

Fig. 3. Top: Energy spectrum for ^{12}C calculated using the NCSpM with the schematic interaction (1), with the bare JISP16 NN introduced as the symmetry-breaking term, and using 5 $\mathrm{Sp}(3,\mathbb{R})$ irreps (the average deformation of each is depicted at bottom) extended to $N_{\max} = 20$ ($\hbar\Omega = 18$), and compared to experiment[34]. The $B(E2)$ transition strengths are in W.u.

state is reproduced using the 1p-1h (3 3) irrep. In addition to the energy spectrum, the NCSpM reproduces observables such as $B(E2)$ transition strengths (shown in Fig. 3), matter rms radii, and electric quadrupole moments, and has been used to investigate the nature of the giant monopole and quadrupole resonances in select light- and intermediate-mass nuclei[35]. This model has also been applied to studies of other nuclei, including ^{8}Be, as well as various sd-shell nuclei without the need to adjust the γ strength parameter[36,37]. This is an indication that the NCSpM captures important components of the underlying nuclear physics.

Of distinct importance is the ability of the NCSpM to achieve higher $N_{\max}$ model spaces than those that are currently within the reach of other NCSM studies. While a smaller $N_{\max}$ model space is sufficient for convergence of the ground state rotational band, the wave function for the 0_2^+ state of ^{12}C has significant contributions from highly deformed configurations [e.g., (12 0), (14 0), (16 0), etc.] and requires a much larger $N_{\max}$ model space in order for the collectivity of the state to fully develop and for the energy to reach convergence (left side, Fig. 4). This large deformation is also seen in comparison of the one-body (matter) densities of the ground

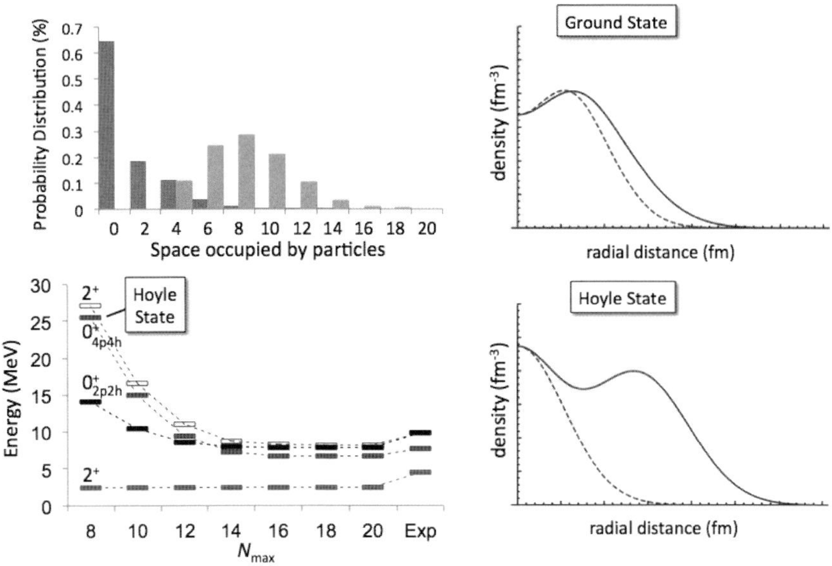

Fig. 4. **Top left**: probability distribution shown across total HO quanta excitations for the 0^+_{gs} state (dark) and 0^+_2 state (light) of ^{12}C, computed with the NCSpM $N_{\rm max} = 20$ ($\hbar\Omega = 18$); **Bottom left**: Dependence on $N_{\rm max}$ of the energies of the first three 0^+ states in ^{12}C (in MeV) computed in the NCSpM; **Right**: densities, shown along the x-axis (dashed) and z-axis (solid) of the intrinsic frame for (top) the ground state, and (bottom) the 0^+_2 – components of the wave function with probability $> 3\%$ are included in the calculation, comprising 95% of the ground state wave function, and 91% of the wave function for the 0^+_2 state.

state and 0^+_2 state. The ground state has an essentially symmetric probability density function (Fig. 4, top right), while the 0^+_2 state shows peaks in the probability density aligned along the z-axis (Fig. 4, bottom right), indicating overlapping clusters aligning along the z-axis in a spatially expansive fashion.

Moving beyond the lightest nuclei – Guided by the simple, emergent pattern that points to the importance of the symplectic Sp(3, $\mathbb{R}$) group and its embedded SU(3) subgroup, SA-NCSM studies are pushing toward heavier nuclei from a fully *ab initio* NCSM perspective, as well as to odd-A nuclei, and negative parity states. Studies of such *sd*-shell nuclei as ^{22}Mg, ^{24}Si, and ^{24}Ne were first undertaken in a NCSpM framework[36], which yielded insights into the collective nature of these nuclei and the basis needed for that collectivity to fully develop. Subsequent SA-NCSM studies of these

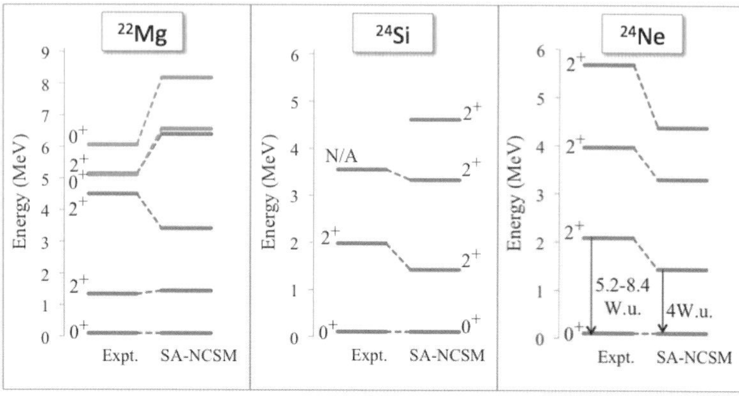

Fig. 5. Energy spectra for ^{22}Mg (left), ^{24}Si (center), and ^{24}Ne (right) computed using the SA-NCSM with SRG-N^{3}LO (2 fm^{-1}), $\hbar\Omega$ = 15MeV, in the $\langle N^{\perp}_{\rm max}\rangle N^{\top}_{\rm max} = \langle 2\rangle 6$ model space, compared to experimental spectra.

same nuclei with realistic chiral interactions (see Fig. 5), using model spaces reduced to include only the most important configurations, show impressive agreement with experiment.

In short, we have demonstrated the emergence of simple patterns in nuclei from first principles, and shown how the exploitation of these symmetries (in the implementation of the NCSpM and SA-NCSM) can gain us incredible insights into the underlying nuclear physics. In particular, we illustrate how the NCSpM can be used to gain insights into what types of particle excitations are most important for describing highly-deformed structures, and how that can guide *ab initio* NCSM studies, using the SA-NCSM, into more computationally challenging regimes.

Acknowledgements

We thank D. Rowe, G. Rosensteel, and C. Bahri, together with the PetaApps Collaboration, J. P. Vary, P. Maris, U. Catalyurek, and M. Sosonkina, for useful discussions. This work was supported by the U.S. NSF (OCI-0904874, ACI -1516338), the U.S. DOE (DE-SC0005248), SURA (Southeastern Universities Research Association), Louisiana State University (through its direct support of J.P.D.'s work with SURA), and the Czech Science Foundation under Grant No. 16-16772S. This work benefitted from HPC resources provided by Blue Waters, the Louisiana Optical Network Initiative, as well as LSU (`www.hpc.lsu.edu`).

References

1. R. B. Wiringa, V. G. J. Stoks and R. Schiavilla, *Phys. Rev. C* **51**, p. 38 (1995).
2. R. Machleidt, *Phys. Rev. C* **63**, p. 024001 (2001).
3. D. R. Entem and R. Machleidt, *Phys. Rev. C* **68**, p. 041001 (2003).
4. E. Epelbaum, *Prog. Part. Nucl. Phys.* **57**, p. 654 (2006).
5. M. G. Mayer, *Phys. Rev.* **75**, p. 1969 (1949).
6. O. Haxel, J. H. D. Jensen and H. E. Suess, *Phys. Rev.* **75**, p. 1766 (1949).
7. P. Navrátil, J. P. Vary and B. R. Barrett, *Phys. Rev. Lett.* **84**, p. 5728 (2000).
8. P. Maris, A. M. Shirokov and J. P. Vary, *Phys. Rev. C* **81**, p. 021301(R) (2010).
9. B. Barrett, P. Navrátil and J. Vary, *Prog. Part. Nucl. Phys.* **69**, p. 131 (2013).
10. P. Navrátil, S. Quaglioni, I. Stetcu and B. R. Barrett, *J. Phys. G: Nucl. Part.* **36**, p. 083101 (2009).
11. R. Roth and P. Navrátil, *Phys. Rev. Lett.* **99**, p. 092501 (2007).
12. T. Abe, P. Maris, T. Otsuka, N. Shimizu, Y. Utsuno and J. Vary, *Phys. Rev. C* **86**, p. 054301 (2012).
13. M. Harvey, *Adv. Nucl. Phys.* **1**, p. 67 (1968).
14. J. P. Elliott and M. Harvey, *Proc. Roy. Soc. A* **272**, p. 557 (1962).
15. G. Rosensteel and D. J. Rowe, Nuclear Sp(3,R) Model, *Phys. Rev. Lett.* **38**, p. 10 (1977).
16. D. J. Rowe, *Rep. Progr. Phys.* **48**, p. 1419 (1985).
17. A. Bohr and B. R. Mottelson, *Nuclear Structure* (Benjamin, New York, 1969).
18. J. P. Elliott, *Proc. Roy. Soc. A* **245**, p. 128 (1958).
19. D. J. Rowe, (World Scientific, 2017).
20. T. Dytrych, K. D. Launey, J. P. Draayer, P. Maris, J. P. Vary, E. Saule, U. Catalyurek, M. Sosonkina, D. Langr and M. A. Caprio, *Phys. Rev. Lett.* **111**, p. 252501 (2013).
21. K. D. Launey, T. Dytrych and J. P. Draayer, *Prog. Part. Nucl. Phys.* **89**, 101 (2016).
22. C. Bahri, J. Draayer and S. Moszkowski, *Phys. Rev. Lett.* **68**, p. 2133 (1992).
23. A. Zuker, J. Retamosa, A. Poves and E. Caurier, *Phys. Rev. C* **52**, p. R1741 (1995).

24. A. C. Dreyfuss, K. D. Launey, T. Dytrych, J. P. Draayer and C. Bahri, *Phys. Lett. B* **727**, p. 511 (2013).
25. S. C. Pieper, K. Varga and R. B. Wiringa, *Phys. Rev. C* **66**, p. 044310 (2002).
26. M. Wloch, D. J. Dean, J. R. Gour, M. Hjorth-Jensen, K. Kowalski, T. Papenbrock and P. Piecuch, *Phys. Rev. Lett.* **94**, p. 212501 (2005).
27. K. Tsukiyama, S. K. Bogner and A. Schwenk, *Phys. Rev. Lett.* **106**, p. 222502 (2011).
28. E. Epelbaum, H. Krebs, D. Lee and U.-G. Meißner, *Phys. Rev. Lett.* **106**, p. 192501 (2011).
29. T. Dytrych, K. D. Sviratcheva, J. P. Draayer, C. Bahri and J. P. Vary, *J. Phys. G: Nucl. Part. Phys.* **35**, p. 123101 (2008).
30. T. Dytrych, K. D. Launey and J. P. Draayer, in *McGraw-Hill Yearbook of Science and Technology*, (McGraw-Hill, 2014).
31. J. P. Elliott, *Proc. Roy. Soc. A* **245**, p. 562 (1958).
32. D. J. Rowe, *Phys. Rev.* **162**, p. 866 (1967).
33. D. J. Rowe, G. Thiamova and J. L. Wood, *Phys. Rev. Lett.* **97**, p. 202501 (2006).
34. K. D. Launey, A. C. Dreyfuss, J. P. Draayer, T. Dytrych and R. B. Baker, *J. Phys.: Conf. Ser.* **569**, p. 012061 (2014).
35. A. C. Dreyfuss, K. D. Launey, T. Dytrych, J. P. Draayer, R. B. Baker, C. Deibel and C. Bahri, *sumbitted to Phys. Rev. C* (2017).
36. G. K. Tobin, M. C. Ferriss, K. D. Launey, T. Dytrych, J. P. Draayer, A. C. Dreyfuss and C. Bahri, *Phys. Rev. C* **89**, p. 034312 (2014).
37. K. D. Launey, T. Dytrych, J. P. Draayer, G. K. Tobin, M. C. Ferriss, D. Langr, A. C. Dreyfuss, P. Maris, J. P. Vary and C. Bahri, in *Proceedings of the 5th International Conference on "Fission and properties of neutron-rich nuclei", ICFN5, November 4–10, 2012, Sanibel Island, Florida*, eds. J. H. Hamilton and A. V. Ramayya (World Scientific, Singapore, 2013).

Fission I

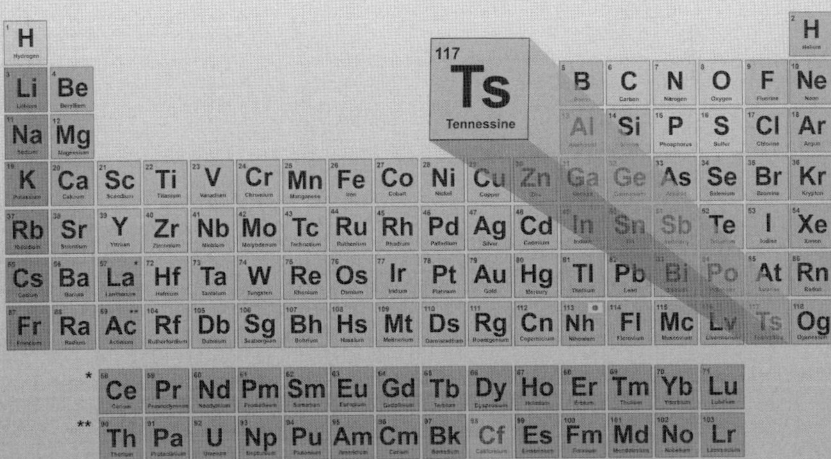

Fission Product Studies with the SPIDER Instrument at LANSCE

F. Tovesson*, D. Mayorov and K. Meierbachtol
Los Alamos National Laboratory,
Los Alamos, NM 87545, USA
**E-mail: tovesson@lanl.gov*

D. Shields
Colorado School of Mines,
Golden, CO 80401, USA

In order to advance our understanding of the fission process we need to measure changes in fission product yields as a function of excitation energy of the fissioning system, as well as study the correlations between kinetic energy, mass and charge of the products. In addition, fission product yields are used for diagnostics in nuclear technology and there is therefore interest in reducing the uncertainties in their yield.

The SPIDER instrument is based on the 2v-2E technique for measuring the mass of fission products, and has been used to study spontaneous fission of ^{252}Cf, as well as thermal neutron-induced fission of ^{233}U, ^{235}U, and ^{239}Pu. The current instrument has two spectrometer "arms", which provides sufficient detection efficiency for studying fission at thermal neutron energies. In order to study the change in fission yields at fast energies higher detection efficiency is needed, and a new version of SPIDER is therefore under construction. The new instrument will have a total of 16 individual arms, bringing the total solid angular coverage to about 1%.

Keywords: Nuclear Fission, Fission Product Yields, Pu-239.

1. Introduction

As our understanding of the fission process advances we need high quality experimental data to validate and guide new models. One major challenge in fission research is to model mass yields, and in particular how those yields change with the excitation energy of the compound nucleus. By measuring the relative mass yield changes as a function of excitation energy for multiple actinides we will have a much better data base that can be compared to model predictions. Fission product yields are also important for nuclear technology as they can be used to calculate burn-up in reactor fuel.

An instrument using the 2v-2E method has been developed to measure independent mass yields at the Los Alamos Neutron Science Center (LANSCE) for thermal neutrons, and a more efficient detector array is being developed to extend that capability to the fast neutron region. Thermal neutron-induced fission product yields have been measured for ^{233}U, ^{235}U, and ^{239}Pu, and there are plans to extend these measurements to higher incident neutron energies.

2. Experimental Method

The measurements were performed at the Los Alamos Neutron Science Center (LANSCE)[1] using the SPIDER instrument. The actinide samples are loaded into the detector and irradiated with neutrons. The fragments resulting from fission are characterized in order to determine their mass.

2.1. *Neutron Source*

The experiments took place at flight path 12 of the Lujan Center at LANSCE. An 800-MeV linear proton accelerator delivers beam to a proton storage ring, where several pulses are stacked on top of each other, and then delivered to the tungsten spallation target with a pulse width of 250 ns and repetition rate of 20 Hz. Flight path 12 is approximately 25 meters from the spallation target, and views a cold moderator. The usable part of the neutron spectrum accessible through time-of-flight ranges from about 1 meV to 100 keV.

2.2. *Samples*

The actinide samples were prepared at Oregon State University, and were deposited on carbon films with a thickness of 100 μg/cm^2. The actinide deposits also have an areal density of approximately 100 μg/cm^2, and at this areal density the fission fragments escape the targets with only minimal energy loss.

2.3. *Detector*

The fission detection system uses the 2v-2E method for inferring fragment masses. A mechanical drawing of the system is shown in Fig. 1. The

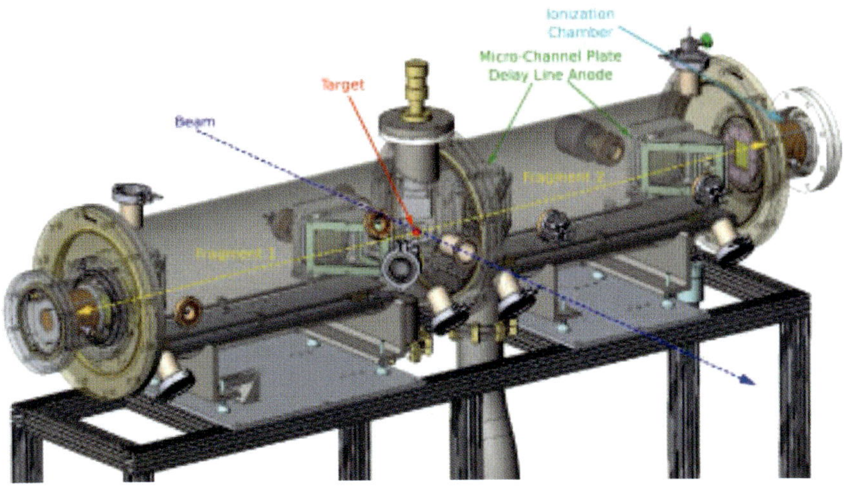

Fig. 1. Mechanical drawing of the SPIDER instrument.

samples are installed in the center of the vacuum chamber, which also holds timing pick-off detectors that are used to measure fragment time-of-flight. Two Ionization chamber are mounted to the main vacuum chamber, one on each end. The vacuum and ionization chamber volumes are separated by 200 nm thick silicon nitride windows. When fission occurs the two fragments are emitted collinearly, and for a certain range of emission angles they will be detected by SPIDER. Each fragment passes through two timing detectors, separated by 60-70 cm, in order to measure their time-of-flight. The fragment then enters the ionization chamber by going through the silicon nitride window, where it deposits its full energy.

The timing detectors consists of a carbon film, electrostatic mirror, and a microchannel detector (MCP). The carbon film has a thickness of 20 μg/cm^2, and when a fission fragment passes thought it secondary electrons are emitted. Some of those electrons are accelerated and bent 90 degrees perpendicular to the fragment path by the electrostatic mirror. The electrons are then detected by a 75 mm diameter MCP which provides a fast timing signal. A delay-line anode provides position information which can be used to deduce the location of the fission fragment hit on the carbon film. The timing detector assembly and its performance is described in Ref.[2].

The ionization chambers were chosen for their high energy resolution (0.5 - 1%) for typical fission fragments. The drift region is 15 cm long, and guard rings are used to create a highly uniform drift field. The anode is shielded by a Frisch-grid, which ensures that the detector signal response is proportional to the energy deposited in the active volume, independently of the track orientation. This type of ionization chamber has been used to identify nuclear charge of ions[3], but the feature has yet not been implemented in the SPIDER instrument.

The SPIDER detector system was commissioned and benchmarked using spontaneous fission from a ^{252}Cf source. By comparing the measured mass yield curve with the England and Rider evaluation[5] the mass resolution was estimated to be about 1.5 amu on average, with higher resolution for the light fragment group, and lower for the heavy fragment group. The full results from the ^{252}Cf study can be found in Ref.[4].

3. Result and Conclusions

The independent mass yields for thermal-neutron induced fission was measured for ^{233}U, ^{235}U, and ^{239}Pu. The preliminary result for ^{235}U is shown in Fig. 2, and compared an evaluation[5]. Measurements is in close agreement with the evaluation, although the effect of finite mass resolution is evident

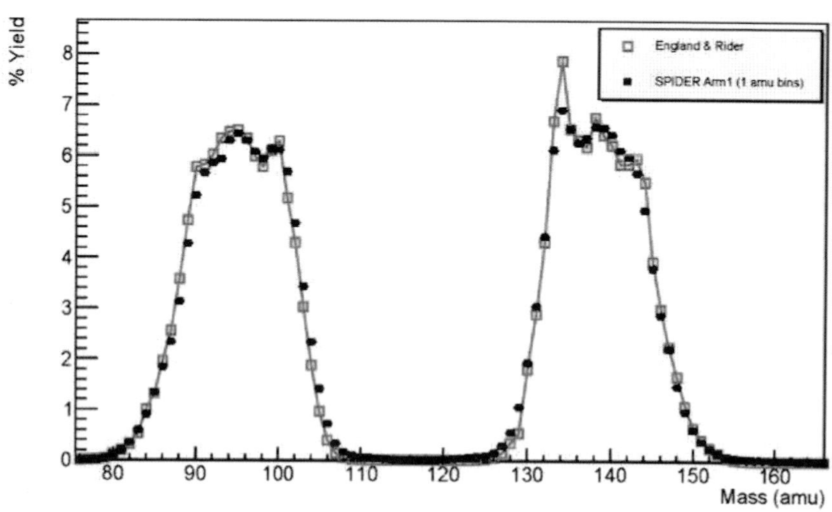

Fig. 2. Independent mass yields for thermal-neutron induced fission of ^{235}U. Also shown is the England and Rider evaluation.

of masses 103 and 104, where the yield changes rapidly from one mass to another. Excluding the resolution effect the same general structures are observe in the the experiment as in the evaluation.

A larger detector array using the same principles as SPIDER is currently being developed in order to increase the detection efficiency. The new instrument, MegaSPIDER, will take advantage of the higher efficiency to study fast neutron-induced fission, which has lower cross sections than thermal neutron-induced fission. The end goal is to obtain the relative change in mass yields over a wide range of incident neutron energies.

References

1. P. W. Lisowski and K. F. Schoenberg, *Nucl. Instr. and Meth.* **A 562**, 910 (2006).
2. C. W. Arnold, F. Tovesson, K. Meierbachtol *et al.*, *Nucl. Instr. and Meth.* **A 764**, 53 (2014).
3. N. Boucheneb, P. Geltenbort, M. Asghar *et al.*, *Nucl. Phys.* **A 502**, 261 (1989).
4. K. Meierbachtol, F. Tovesson, D. Shields *et al.*, *Nucl. Instr. and Meth.* **A 788**, 59 (2015).
5. T. R. England and B. F. Rider, Los Alamos National Laboratory Technical Report no. **LA-UR-94-3106** (1994).

High Precision Neutron-Induced Fission Cross Sections Using a Time Projection Chamber

B. Manning, for the NIFFTE Collaboration

Los Alamos National Laboratory, Los Alamos, NM 87545

Neutron-induced fission cross sections for actinides have long been of great interest for nuclear energy and stockpile stewardship. More recently, there has been increased effort to both minimize and fully quantify uncertainties in nuclear data used in these applications. Furthermore, precise data provides an excellent test of modern nuclear theory which continues to strive for a complete model of nuclear fission. Traditionally, measurements were performed using fission chambers which provided limited information about the detected fission events. The Neutron Induced Fission Fragment Tracking Experiment (NIFFTE) has developed the fission Time Projection Chamber (fissionTPC) to measure fission cross sections to unprecedented precision by providing 3D tracking of charged particles. Analyzing track parameters provides a wealth of information not previously available with standard fission chambers. The presentation will highlight capabilities of the NIFFTE fissionTPC that provide a direct measurement of systematic uncertainties that could not be well studied with a traditional fission chamber. Data from recent experiments at the Weapons Neutron Research facility at the Los Alamos Neutron Science Center will be presented.

Studies on Fission with Aladin

Precise and simultaneous measurement of fission total, yields kinetic energy and total prompt neutron multiplicity at GSI.

Julien Taieb[1], Lucie Grente[1], Audrey Chatillon[1], Julie-Fiona Martin[1], Gilbert Belier[1], Guillaume Boutoux[1], Adeline Ebran[1], Thomas Gorbinet[1], Benoit Laurent[1], Eric Pellereau[1], Hector Alvarez Pol[3], Laurent Audouin[2], Yassid Ayyad[3], Jose Benlliure[3], Enrique Casajeros[4], Dolores Cortina Gil[3], Manuel Caamano[3], Fanny Farget[7], Beatriz Fernandez Dominguez[3], Andreas Heinz[5], Beatriz Jurado[8], Aleksandra Kelic-Heil[6], Nikolaus Kurz[6], Chiara Nociforo[6], Carlos Paradela[3], Stephane Pietri[6], Diego Ramos[3], Jose-Luis Rodriguez Sanchez[3], Carme Rodriguez Tajes[7], Dominic Rossi[6], Karl-Heinz Schmidt[6], Haik Simon[6], Laurent Tassan-Got[2], Jositt Vargas[3], Bernd Voss[6] and Helmut Weick[6]

[1]*CEA DAM Bruyeres-le-Chatel, F-91297 Arpajon, France*
[2]*CNRS/IN2P3, IPNO, F-91406 Orsay, France*
[3]*Universidad de Santiago de Compostela, E-15782, Santiago de Compostela, Spain*
[4]*Universidad de Vigo, E-36200 Vigo, Spain*
[5]*Chalmers University of Technology, SE-41296 Gothenburg, Sweden*
[6]*GSI Helmholtzzentrum fuer Schwerionienforschung GmbH, D-64291 Darmstadt, Germany*
[7]*CNRS/IN2P3, GANIL, F-14076 Caen, France*
[8]*CNRS/IN2P3, CENBG, F-33175 Gradignan, France*

A novel technique for Fission studies, based on the inverse kinematics approach, is presented. Following pioneering work in the nineties, the SOFIA collaboration has designed and built an experimental set-up dedicated to the simultaneous measurement of isotopic yields, total kinetic energies and total prompt neutron multiplicities, by fully identifying both Fission fragments in coincidence, for the very first time. This experiment, performed at GSI, permits to study the fission of a wide variety of fissioning systems, ranging from mercury to neptunium, possibly far from the valley of stability. Two experiments, performed in 2012 and 2014, have provided a large array of unprecedented data regarding the nuclear fission process. An excerpt of the results is presented. With this solid starter, further improvements of the experimental set-up are considered, consistent with the expected developments at the GSI facility, in order to measure more fission observables in coincidence. The completeness, which is reached in the SOFIA data, permits to scrutinize the correlations between the interesting features of fission, offering a very detailed insight into this, still unraveled, mechanism.

Fission Dynamics with Microscopic Level Densities

D. Ward[1], B.G. Carlsson[1], Th. Døssing[2], P. Möller[3], J. Randrup[4*] and S. Åberg[1]

[1] *Tekniska Högskolan, Universitet i Lund, S-221 00 Lund, Sverige*
[2] *Niels Bohr Institutet, Københavns Universitet, DK-2100 København Ø, Danmark*
[3] *Los Alamos National Laboratory, Los Alamos, New Mexico 87545, USA*
[4] *Lawrence Berkeley National Laboratory, Berkeley, California 94720, USA*
[*] *E-mail: JRandrup@LBL.gov*

We present a consistent framework for treating the energy and angular-momentum dependence of the nuclear shape evolution in the fission process. It combines microscopically calculated level densities with the Metropolis walk method and it contains no new parameters. The treatment can elucidate the influence on the shape dynamics of warm nuclei of pairing and shell effects as a function of the excitation energy.

Keywords: Nuclear Fission, Shape Dynamics, Brownian Motion, Level Densities.

1. Introduction

Soon after the discovery of nuclear fission in 1938[1], it was recognized that the process can be viewed qualitatively as an evolution of the nuclear shape from that of a single compound nucleus to two receding fragments[2,3] and that Langevin transport theory provides an appropriate model framework[3,4]. A number of Langevin treatments of fission have been successfully developed and applied for excitations high enough to render the dynamics macroscopic, see for example Refs.[5–7].

If the collective shape dynamics is idealized as being highly dissipative, then the Langevin equation reduces to the Smoluchowski equation in which the evolution depends on the balance between the driving force provided by the potential energy of deformation and the dissipative force resulting from the coupling of the collective degrees of freedom to the remaining system. In this limit, it has proven possible to describe the Brownian shape motion as a Metropolis walk on the associated multi-dimensional deformation-energy surface[8]. This calculationally simple method has yielded remarkably good results for the fission-fragment mass distributions[9] and it has made it possible to predict fission-fragment mass distributions for poorly explored regions of the nuclear chart [10,11].

2. Shape evolution

The description of nuclear fission as a generalized Brownian motion builds on the assumption that the evolving nucleus can be characterized by its shape degrees of freedom χ. The associated shape parameters are treated as classical variables that are coupled dissipatively to the remaining microscopic degrees of freedom. The resulting large-amplitude collective motion then exhibits a strongly damped diffusive evolution that can be described as a random walk on the associated multi-dimensional potential-energy surface $U(\chi)$. When the values of the potential are available on a (five-dimensional) lattice of shapes [12], the Smoluchowski equation can be approximately simulated by means of a random walk on this lattice [8].

Because the random walk must satisfy detailed balance, the following relation must hold between the rates of transition between one lattice site χ and another χ',

$$\nu(\chi \to \chi')/\nu(\chi' \to \chi) \; = \; \rho(\chi')/\rho(\chi) \; , \tag{1}$$

where $\rho(\chi)$ is the level density at the lattice site having the shape χ. This condition can be satisfied in many ways and the Metropolis procedure is merely one particularly simple realization: A proposed shape change from χ to χ' is accepted unconditionally if $\rho(\chi') > \rho(\chi)$, whereas it it accepted only with the probabilty $\rho(\chi')/\rho(\chi)$ otherwise; it is readily verified that this procedure satisfies detailed balance.

In Ref. [8] and all subsequent applications until recently [13], smooth Fermi-gas level-density expressions were used. The ratio between the level densities for neighboring lattice sites could therefore be approximated as

$$\rho(\chi')/\rho(\chi) \; \approx \; \exp(-\Delta U/T) \; , \tag{2}$$

where $\Delta U = U(\chi') - U(\chi)$ is the change in the potential energy associated with the proposed shape change and $T = 1/[\partial \ln \rho(E^*)/\partial E^*]$ is the local nuclear temperature. In the new development [13], we wish to account for pairing correlations and shell effects and we therefore use the microscopic level densities (see Sect. 3) to evaluatee the required ratios $\rho(\chi')/\rho(\chi)$.

The Metropolis walks were performed on a discrete lattice of more than five million shapes given in the three-quadratic-surface parametrization [14]. The five independent parameters characterizing a shape are the overall elongation of the nucleus (in terms of the reduced quadrupole moment q_2), the constriction of the central part (in terms of the neck radius c), the spheroidal deformations ε_{f1} and ε_{f2} of the two nascent fragments, and the reflection (mass) asymmetry α_g.

3. Combinatorial level density

In Ref.[15] a method was developed for microscopic calculations of level densities for deformed nuclei and it has been adapted to the fission process[13]. For a specified shape $\boldsymbol{\chi}$, the single-particle levels for protons and neutrons needed for the combinatorial calculation of the level density are obtained by solving the Schrödinger equation in the associated folded-Yukawa potential. These are the same levels as those previously used in Ref.[11] to calculate the microscopic shell and pairing energies in the construction of the deformation energy surface, thus guaranteeing consistency of the treatment. The corresponding local many-body vacuum state $|0;\boldsymbol{\chi}\rangle$ has N neutrons and Z protons filling the lowest single-particle states and the uncorrelated excited states consist of all multiple particle-hole excitations,

$$|i;\boldsymbol{\chi}\rangle = \prod_{n\geq 1} a^{\dagger}_{\nu_n^{(i)}} a_{\mu_n^{(i)}} |0;\boldsymbol{\chi}\rangle, \tag{3}$$

For each many-body state $|i;\boldsymbol{\chi}\rangle$, blocked BCS calculations for neutrons and protons separately provide the state-dependent pairing gaps, $\Delta_i^{\rm n}(\boldsymbol{\chi})$ and $\Delta_i^{\rm p}(\boldsymbol{\chi})$, respectively, and the energy of the correlated intrinsic many-body state, $E_i(\boldsymbol{\chi}) = E_i^{\rm n}(\boldsymbol{\chi}) + E_i^{\rm p}(\boldsymbol{\chi})$.

For each such intrinsic statem the angular momentum along the nuclear symmetry axis is denoted by K_i and it is assumed that it forms a rotational band head. The resulting total angular momenta I may then take on the values $I = K_i, (K_i+1), (K_i+2), \ldots$ and the corresponding rotational energies of the band members are

$$E_i^{\rm rot}(I;\boldsymbol{\chi}) = \frac{I(I+1) - K_i(\boldsymbol{\chi})^2}{2\mathcal{J}_\perp(\boldsymbol{\chi}, \Delta_i^{\rm n}(\boldsymbol{\chi}), \Delta_i^{\rm p}(\boldsymbol{\chi}))} \,. \tag{4}$$

The moment of inertia $\mathcal{J}_\perp$ is approximated by the moment of inertia of a rigid body with the shape $\boldsymbol{\chi}$, modified by the calculated pairing gaps for the state.

In Ref.[15] it was found that the enhancements arising from collective vibrational modes are unimportant and they are therefore ignored in the present fission applications. The total energy of a state is then given by

$$E_i(I,\boldsymbol{\chi}) = E_i^{\rm n}(\boldsymbol{\chi}) + E_i^{\rm p}(\boldsymbol{\chi}) + E_i^{\rm rot}(I;\boldsymbol{\chi}) \,. \tag{5}$$

For each shape χ, the resulting states are binned according to their energy E_i and their total angular momentum I; the bin width was taken as $\Delta E = 200\,\mathrm{keV}$. The sensitivity of our results to the bin size has been tested and it was found that a doubling or tripling of ΔE produces negligible changes in the calculated mass distributions.

4. Analytical extrapolation

Because the computational effort required by the combinatorial method grows exponentially with excitation, it is practically important to develop a simple way of extending the results to high energy. We do that by switching to an analytical expression at $E^*(\chi) \approx 6\,\mathrm{MeV}$, below which most of the specific structure effects, such as spectral gaps or non-monotonic behavior, have washed out. In this way, important structures in the level density are maintained, while the numerical calculations are kept to a manageable level (about 3 CPU seconds per shape). For each particular shape χ, the extrapolation makes use of its specific shell and pairing energies, $E_{\mathrm{sh}}(\chi)$ and $E_{\mathrm{pc}}(\chi)$. Their influence diminishes with increasing energy and the level density approaches the analytical expression for a Fermi gas.

We employ the following simple analytical Fermi-gas expression[16], which is suitable for deformed nuclei with a fixed small angular momentum,

$$\rho(N, Z, E^*(\chi), I) = C(\chi)\, E_{\mathrm{intr}}^{-3/2} \exp\left(2\sqrt{a_0 E_{\mathrm{intr}}}\right), \tag{6}$$

where $E_{\mathrm{intr}} = E^*(\chi) - I(I+1)\hbar^2/2\mathcal{J}_\perp(\chi)$ is the approximate intrinsic energy. The single-particle level-density parameter is $a_0 = A/e_0$ for a nucleus having mass number A and the constant e_0 is determined below.

Accounting for the different energy scales of shell effects and pairing effects, we introduce a backshifted intrinsic excitation energy $\tilde{E}_{\mathrm{intr}}$, which is similar to the effective excitation E^*_{eff} of Ref.[9] and can be considered as a generalization of the precription originally employed by Ignatyuk *et al.*[17],

$$\begin{aligned}\tilde{E}_{\mathrm{intr}}(\chi) = E_{\mathrm{intr}}(\chi) &+ (1 - e^{-E_{\mathrm{intr}}(\chi)/E_{\mathrm{d,sh}}})E_{\mathrm{sh}} \\ &+ (1 - e^{-E_{\mathrm{intr}}(\chi)/E_{\mathrm{d,pc}}})E_{\mathrm{pc}}\,.\end{aligned} \tag{7}$$

Here $E_{\mathrm{sh}}(\chi)$ is the shell-correction energy and $E_{\mathrm{pc}}(\chi)$ is the pairing condensation energy, both calculated for the lowest state at the given shape χ. The full backshifted energy, $\tilde{E}_{\mathrm{intr}} = E_{\mathrm{intr}} + E_{\mathrm{sh}} + E_{\mathrm{pc}}$, emerges when E_{intr} is much larger than both damping scales.

The parameter $E_{\mathrm{d,sh}}$ sets the energy scale for the melting of shell structure and the parameter $E_{\mathrm{d,pc}}$ is the corresponding energy scale for the melting of pairing correlations. For simplicity, both quantities are here assumed to be common to all shapes. By fitting the analytic expression for a number of typical shapes, optimal values of $E_{\rm d,sh}$, $E_{\rm d,pc}$, and e_0 are determined and they are then applied for all shapes. The normalization constant $C(\chi)$ in (6) is determined by continuity with the corresponding microscopic value at the matching energy.

The validity of the analytical extrapolation is illustrated in Fig. 1:

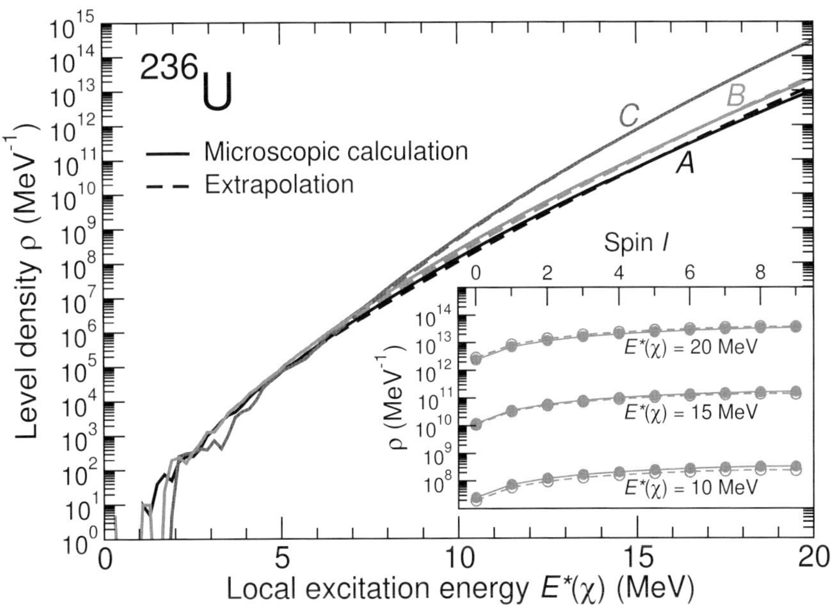

Fig. 1. Microscopic level densities (solid lines) compared to extrapolated values (dashed lines). Three different deformations for ^{236}U are considered: the second minimum (A), the asymmetric second saddle (B), and an elongated symmetric shape close to the outer barrier (C). Extrapolated values are shown for local excitation energies $E^*(\chi) \geq 6$ MeV. The inset shows the angular momentum dependence of the level density at the asymmetric saddle (B) for different excitation energies.

5. Energy dependence of the fragment yield

The use of microscopic level densities is particularly well-suited for studying the dependence of the shape evolution on the total excitation energy of the fissioning system because the microscopic pairing and shell effects automatically wash out as the energy is raised.

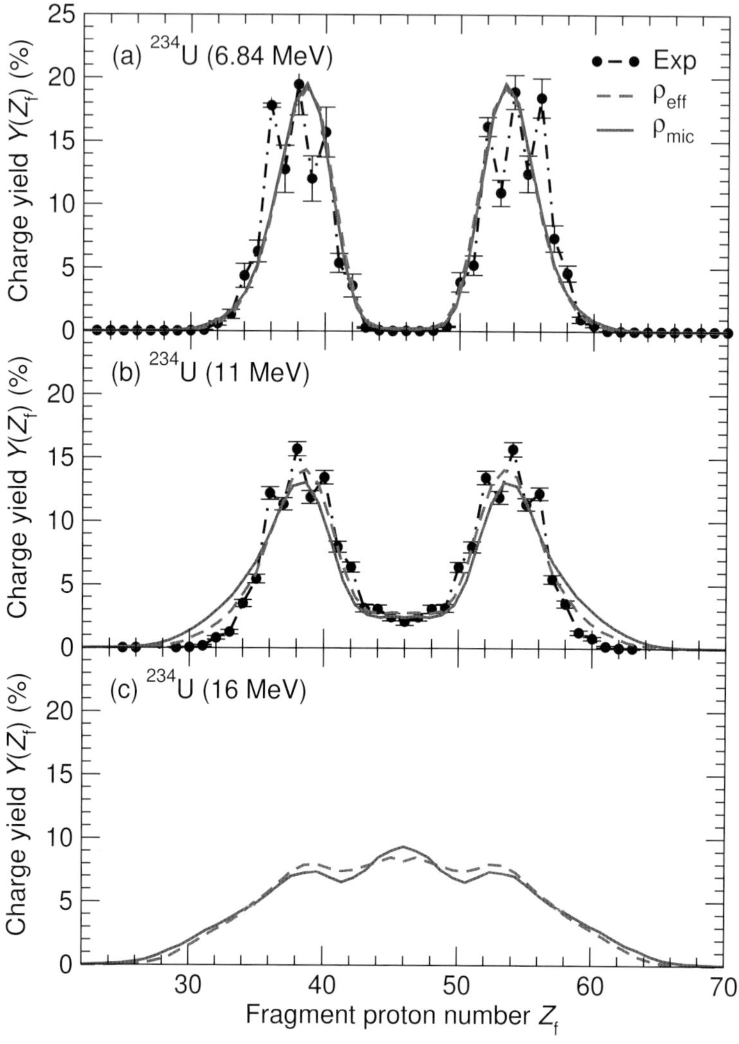

Fig. 2. Fission-fragment charge distributions for ^{234}U at three different excitation energies E^* (MeV): 6.84 (a), 11 (b), and 16 (c). The solid curves have been obtained with the microscopic level densities, while the dashed curves were calculated with the effective level density $\rho_{\rm eff}$ introduced in Ref.[9]. The results for $E^* = 6.84$ MeV are compared to ($\rm n_{th}$, f) data[18], while those for $E^* = 11$ MeV are compared to (γ, f) data[19]. The calculated results for $E^* = 6.84$ MeV are practically identical and both are in good agreement with the experimental data. For $E^* = 11$ MeV the current approach reproduces the symmetric yield of around 2%, while the very asymmetric yields are too large. The calculated yields for $E^* = 16$ MeV are also quite similar.

The preliminary treatment in Ref. 9 used an energy-dependent effective potential, $U_E(\chi) = U_{\rm macro}(\chi) + \mathcal{S}(E^*(\chi))E_{\rm micro}(\chi)$, in conjunction with a simplified Fermi-gas level density, $\rho(\chi) = \rho_{\rm TF}(E_{\rm total} - U_E(\chi))$, with a global suppression function of the form $\mathcal{S}(E^*) = (1+c)/[\exp(E^*/E_{\rm damp}) + c]$.

5.1. *The symmetric yield*

The importance of taking account of nuclear structure effects in the level density is well illustrated by the energy dependence of the fission-fragment yield for symmetric splits, see Fig. 3. The standard calculation leads to a local maximum at $E^* = 7\,\text{MeV}$, followed by a local minimum at 8 MeV; this bump is followed by a second bump ending with an inflection at around 10 MeV. The experimental data exhibit a qualitatively similar behavior, in particular a local maximum at $E^* = 7\,\text{MeV}$. Without pairing, the absence of a gap produces a larger level density, especially for shapes with positive

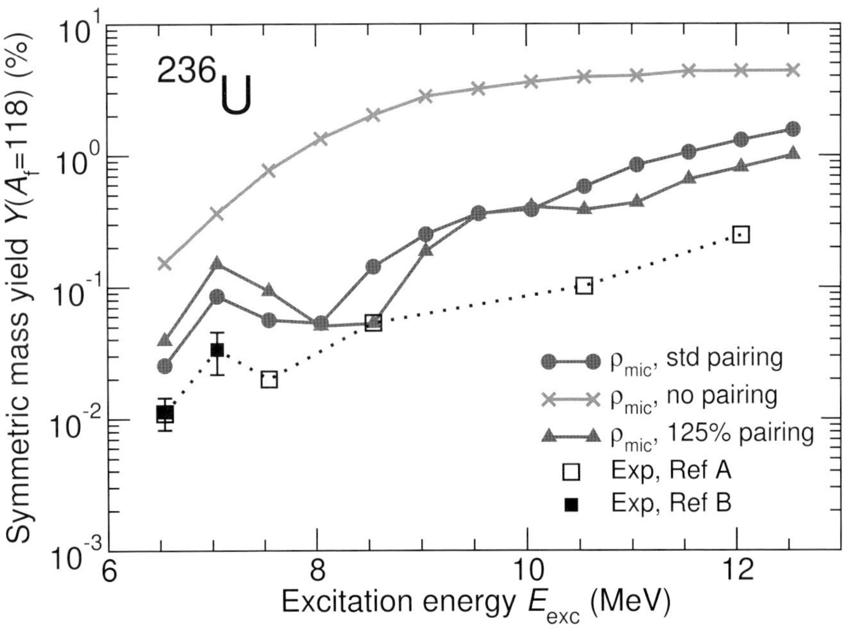

Fig. 3. The dependence of the symmetric mass yield for ^{236}U on the excitation energy. In addition to the standard calculation (filled circles) are shown two additional calculations: one where the level densities are obtained without pairing (crosses) and one where the pairing strength is increased by 25% (filled triangles). (The potential-energy surface used in the three calculations is the same, consistent with standard pairing.) Also shown are experimental data from Ref.[20] (open squares) and Ref.[18] (filled squares).

shell corrections, where both the gap and the single-particle level density are larger. Thus the symmetric yield is highly sensitive to the pairing properties of the level density.

6. Concluding remarks

We have refined the Metropolis-walk approximation to the Brownian-motion treatment of fission dynamics[8] by employing microscopic local level densities obtained by the recently developed combinatorial method[15]. Because the single-particle levels used are the same as those employed for the macroscopic-microscopic calculation of the potential-energy surfaces, the approach is consistent and no new parameters need be introduced. Moreover, the combinatorial procedure provides access to the level density for a fixed value of angular momentum.

Using these parameter-free microscopic level densities with the Metropolis-walk method, we calculate fission fragment charge distributions for ^{226}Th, ^{234,236}U, and ^{240}Pu. The agreement with experimental data was on par with or better than the yields obtained previously with a phenomenological level-density parametrization[9]. The angular-momentum dependence of the fission yields was found to be relatively small and decreasing with increasing excitation energy.

Because the microscopic level densities automatically contain the diminishing effects of pairing correlations and shell structure, the present refined model makes it possible to make more detailed predictions for the energy dependence of the fission yields. The gradual transition from asymmetric to symmetric fission and the detailed energy dependence of the symmetric yield were studied.

Particularly interesting is the finding that the symmetric fragment yield is not monotonically increasing with excitation energy. This perhaps counterintuitive effect appears to be a result of the large pairing correlations for shapes with positive shell-correction energies separating the asymmetric-fission path from symmetric shapes. It is in this connection intriguing that a recent experiment has reported a non-monotonic energy dependence of the asymmetric peak shape in the fragment mass yields for fission of ^{240}Pu[21]. We plan to investigate these phenomena in more detail.

The present refined model provides a consistent and computationally manageable theoretical framework for studying large-scale collective motion of warm nuclear systems far from equilibrium and, in particular, it provides a unique tool for calculating energy-dependent fission fragment mass distributions. The present work extends the use of microscopic level densities from nuclei in shape equilibrium to arbitrary shapes and our present studies have revealed the intriguing possibility that the pairing interaction in shapes far from equilibrium.

Acknowledgments

We are grateful to N. Schunck, A. Tonchev, and R. Vogt for discussions, helpful comments, and valuable suggestions. This work was supported by the Swedish Natural Science Research Council (BGC and SÅ), by the National Nuclear Security Administration of the U.S. Department of Energy at Los Alamos National Laboratory under Contract No. DE-AC52-06NA25396 (PM), and by the Office of Nuclear Physics in the U.S. Department of Energy's Office of Science under Contract No. DE-AC02-05CH11231 (JR).

References

1. O. Hahn and F. Straßmann, *Naturwiss.* **27**, 11 (1939).
2. L. Meitner and O.R. Frisch, *Nature* **143**, 239 (1939).
3. N. Bohr and J.A. Wheeler, *Phys. Rev.* **56**, 426 (1939).
4. H.A. Kramers, *Physica* **VII**, 284 (1940).
5. A.V. Karpov *et al.*, *Phys. Rev. C* **63**, 054610 (2001).
6. P.N. Nadtochy *et al.*, *Phys. Rev. C* **65**, 064615 (2002).
7. E.G. Ryabov *et al.*, *Phys. Rev. C* **78**, 044614 (2008).
8. J. Randrup and P. Möller, *Phys. Rev. Lett.* **106**, 132503 (2011).
9. J. Randrup and P. Möller, *Phys. Rev. C* **88**, 064606 (2013).
10. P. Möller, J. Randrup and A.J. Sierk, *Phys. Rev. C* **85**, 024306 (2012).
11. P. Möller and J. Randrup, *Phys. Rev. C* **91**, 044316 (2015).
12. P. Möller, A.J. Sierk, T. Ichikawa, A. Iwamoto, R. Bengtsson, H. Uhrenholt and S. Åberg, *Phys. Rev. C* **79**, 064304 (2009).
13. D. Ward, G. Carlsson, Th. Døssing, P. Möller, J. Randrup and S. Åberg, *Phys. Rev. C* **95**, 024618 (2017).
14. J.R. Nix, *Nucl. Phys. A* **130**, 241 (1969).
15. H. Uhrenholt, S. Åberg, A. Dobrowolski, Th. Døssing, T. Ichikawa and P. Möller, *Nucl. Phys. A* **913**, 127 (2013).
16. T. Ericsson, *Adv. Phys.* **9**, 425 (1960).
17. A.V. Ignatyuk, K.K. Istekov and G.N. Smirenkin, *Sov. J. Nucl. Phys.* **29**, 450 (1979).
18. M.B. Chadwick *et al.*, *Nucl. Data Sheets* **112**, 2887 (2011).

19. K.-H. Schmidt, S. Steinhäuser, C. Bäckstiegel, A. Grewe, A. Heinz, A.R. Junghans, J. Benlliure, H.-G. Clerc, M. de Jong, J. Müller, M. Pfützner and B. Voss, *Nucl. Phys. A* **665**, 221 (2000).
20. L.E. Glendenin, J.E. Gindler, D.J. Henderson and J.W. Meadows, *Phys. Rev. C* **24**, 2600 (1981).
21. M.E. Gooden *et al.*, *Nucl. Data Sheets* **131**, 319 (2016).

A Microscopic Theory of Fission

W. Younes
Lawrence Livermore Natl. Lab., Livermore, CA 94551

J. A. Brown and E. F. Matthews
Department of Nuclear Engineering, University of California, Berkeley, CA 94720

We examine the role of non-collective degrees of freedom in fission. we focus in particular on the use of the quasiparticle random-phase approximation (QRPA) to study the population mechanisms of the fissioning nucleus, and the Schrödinger collective intrinsic model (SCIM) to allow for quasiparticle excitations during fission. The QRPA calculations are performed with the D1S finite-range effective interaction for ^{240}Pu while the SCIM is applied to a level-density calculation for the schematic multi-$O(4)$ model.

1. Introduction

One the interesting aspects of fission theory concerns the role played by single-particle degrees of freedom. Fission is understood to be a primarily collective phenomenon, but non-collective (single-particle) states are expected to play an important role both in the initial excitation mechanism of the fissioning nucleus[1] and throughout its subsequent evolution to scission[3,4]. In this paper we explore two formalisms that can be used to include non-collective modes in the theoretical description of fission: the quasiparticle random phase approximation (QRPA) and the Schrödinger collective-intrinsic model (SCIM). This work builds on earlier studies performed using the generator-coordinate method within the Gaussian-overlap approximation (GCM+GOA)[5]. In order to provide a context for the new results presented here, we first recall the earlier GCM+GOA work.

A comprehensive review of microscopic approaches to fission in general, and with the GCM+GOA in particular, can be found in the recent publication by Schunck and Robledo[6]. Starting from constrained Hartree-Fock-Bogoliubov (HFB) calculations, performed in this case with the D1S effective interaction[7], the GCM constructs a nuclear wave function as a linear superposition of the HFB solutions that can be static,

$|\Psi\rangle = \int dq\, f(q)\, |\Phi(q)\rangle$, or time-dependent, $|\Psi(t)\rangle = \int dq\, f(q,t)\, |\Phi(q)\rangle$, where $|\Phi(q)\rangle$ is the HFB state calculated with a constraint (or with multiple constraints) value(s) q. The unknown coefficients $f(q)$ or $f(q,t)$ then have to be determined. A variational principle with respect to these unknown coefficients is applied to the energy $E = \langle\Psi|H|\Psi\rangle$ (or to $\langle\Psi(t)|H - \hbar i\partial/\partial t|\Psi(t)\rangle$ in the time-dependent case), where H is the HFB Hamiltonian. This produces the well-known Hill-Wheeler equation[8]. In practice, this integro-differential equation is very difficult to solve with more than one constraint[9,10]. Therefore, in previous work we adopted a standard set of approximations by assuming a Gaussian form for the overlap of HFB states, $\langle\Phi(q)|\Phi(q')\rangle$, and expanding the Hamiltonian overlaps $\langle\Phi(q)|H|\Phi(q')\rangle$ to 2nd order in $q - q'$. The result is a Schrödinger-like equation in the collective parameter(s) q where the collective Hamiltonian is constructed from the underlying single-particle degrees of freedom[7,11]. Thus the inertia tensor, the potential energy, and the zero-point energy (ZPE) correction are all calculated from the HFB wave functions. The static collective Schrödinger equation produces a spectrum of initial fissioning states, and the time-dependent equation evolves those initial states to a set of scission configurations (scission line) where we extract fission-fragment properties such as their mass distributions, total kinetic energies (TKE's), and total excitation energies (TXE's).

The calculations presented in Younes *et al.*[5] differed from previous work[7,11] in several respects. First, instead of the usual quadrupole (Q_{20}) and octupole (Q_{30}) moments, the separation distance and mass difference between pre-fragments were used as constraints in the HFB calculations in order to avoid a persistent problem with missing yield for certain mass ranges in the fragment distribution[12]. Second, quantum localization was used[13] to: 1) progressively recognize the wave functions of the pre-fragments within the wave function of the parent nucleus, 2) impose a quantum-mechanical criterion for scission, and 3) calculate the static contributions to the TKE and TXE at scission. Finally, the dynamical contribution to the TKE and TXE (i.e., the pre-scission energies) were estimated from an analysis of the flux in the fission direction and by using the WKB approximation to relate that flux to energy[5]. In this previous work we calculated the fragment mass distribution for the $^{239}\text{Pu}(n,f)$ reaction from thermal to 5-MeV incident neutron energies. These mass distributions were found to agree with data at thermal energies, and with a phenomenological model at higher energies, where data are not available. We also calculated the TKE and TXE for $^{239}\mathrm{Pu}(n_{th}, f)$ and compared them to data. Although the agreement between theory and experiment was good, the TKE were systematically overestimated by the theory, while the TXE were underestimated. These deviations could be caused by missing degrees of freedom, and in particular by quasiparticle states excited during the fission process that would increase the TXE while reducing the TKE.

2. QRPA calculations for ^{240}Pu

The work presented here is strongly inspired by that of Péru and Martini[1] but also many others, especially in the context of the finite-amplitude method (FAM)[14–17]. Unlike the GCM+GOA approach described in the previous section, the QRPA is not limited to motion described by a few collective parameters, and can be used to describe both collective and non-collective quasiparticle (qp) excitations. Its main limitation is that it assumes harmonic motion. The QRPA is derived in many references[1,18].

For the present work, a code was written to solve the full QRPA matrix equation and applied to ^{240}Pu using the D1S finite-range interaction in a deformed harmonic-oscillator basis[19] to generate excitations of the system. Optimization was a crucial focus in the code development since QRPA calculations can rapidly become unwieldy as the size of the configuration space (i.e., the number of qp states) increases. The performance of the code was improved by exploiting both data parallelism and the systematic pre-calculation of often-used quantities identified through benchmarking.

The solutions to the QRPA equation were used to calculate a photoabsorption cross section on the ^{240}Pu ground state using the formalism in Oishi *et al.*[17] with a Lorentzian smearing width $\Gamma = 1$ MeV. Despite the code optimizations mentioned above however, additional restrictions had to be imposed to accommodate the available computational resources. Thus only shells $N_{\mathrm{sh}} = 0-10$ of the harmonic-oscillator basis were used, and only 2-qp excitations below 80 MeV were included. The resulting cross section is shown in Fig. 1, where it is compared to the ENDF/B-VII.1 evaluation, obtained with a Hauser-Feshbach calculation. While the QRPA results appear to be converged as function of the 2-qp cutoff energy, the cutoff at $N_{\mathrm{sh}} = 10$ is probably not adequate. Calculations that include up to the $N_{\mathrm{sh}} = 12$ shell would be more appropriate[1]. The FAM[14–17] should provide a more efficient alternative in this case.

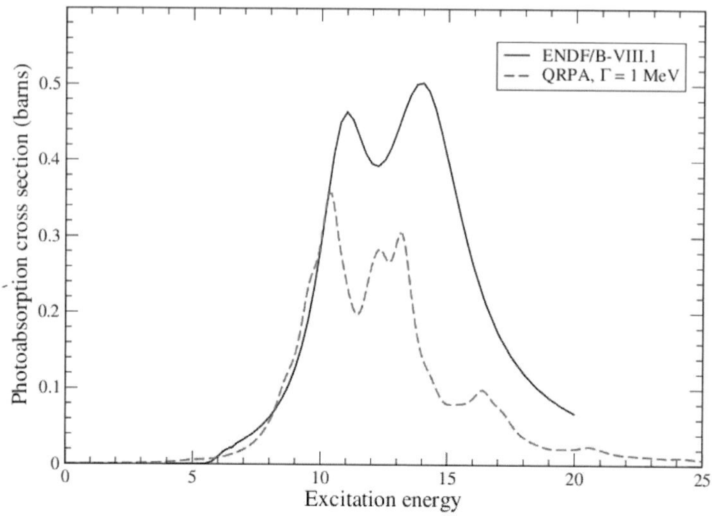

Fig. 1. Photoabsorption cross section on the ^{240}Pu ground state calculated using the QRPA and compared to the ENDF/B-VII.1 evaluation.

3. Exploring the SCIM with a schematic model

The SCIM was developed as part of the Ph.D. thesis of R. Bernard[2]. The model follows pioneering work from the 1970's onward seeking to include qp excitations within the GCM framework[20,21]. In the SCIM, the usual GCM ansatz in section 1 is extended to include 2-qp excitations, $|\Psi\rangle = \int dq\, f_0(q)\, |\Phi_0(q)\rangle + \sum_{i>0} \int dq\, f_i(q)\, |\Phi_i(q)\rangle$, where the index $i = 0$ refers to the HFB ground state, and $i > 0$ labels 2-qp excitations built on that ground state. A time-dependent version can also be introduced if needed. The variational principle with respect to the weight functions $f_i(q)$ (where $i = 0, 1, \ldots$) then leads to a Hill-Wheeler equation where the overlaps $H_{ij}(q, q') \equiv \langle \Phi_i(q) |H| \Phi_j(q') \rangle$ and $N_{ij}(q, q') \equiv \langle \Phi_i(q) | \Phi_j(q') \rangle$ are matrices. The same effective interaction that describes the HFB ground state also gives the coupling between ground state and 2-qp excitations and among 2-qp states. The full derivation of the SCIM can be found in Bernard *et al.*[2] and produces a Schrödinger-like equation with Hamiltonian

$$\mathscr{H} = \sum_j \left\{ \frac{1}{2} \left[\left[M^{-1}(q) \right]_{ij} \hat{P} \right]^{(2)} + \left[T_{ij}(q)\, \hat{P} \right]^{(1)} + V_{ij}(q) \right\} \qquad (1)$$

where $\hat{P} \equiv i\partial/\partial q$, and where we have introduced the symmetric ordered products of operators (SOPO's) defined for an arbitrary operator A as $\left[A^{(n)}(q)\,\hat{P}\right]^{(n)} \equiv \frac{1}{2^n}\sum_k \begin{pmatrix} n \\ k \end{pmatrix} \hat{P}^{n-k} A^{(n)}(q)\,\hat{P}^{(k)}$ and which depend on the symmetrized moment $A^{(n)}(q) \equiv i^n \int_{-\infty}^{+\infty} ds\, s^n \left\langle \Phi_i\left(q+\frac{s}{2}\right)|A|\,\Phi_j\left(q-\frac{s}{2}\right)\right\rangle$. In Eq. (1) the matrices $M(q)$, $T(q)$, and $V(q)$ can be expressed in terms of the symmetrized moments $H^{(n)}$ and $N^{(n)}$ and their derivatives with respect to q. The SCIM formalism thus eliminates the GOA, which implicitly assumes collective modes[22], and can therefore be used to describe non-adiabatic phenomena (e.g., the excitation of qp states during fission).

The eventual goal is to apply the SCIM to the fission problem with multiple constraints and all relevant 2-qp excitations. This program represents a natural extension to the work presented in section 1. As a more tractable step in that direction, we present calculations using a schematic model of the nucleus: the multi-$O(4)$ model[23,24]. This model can be thought of as a simplified version of the pairing+quadrupole model (with only $K=0$ terms), and for our purposes we will limit it to only one type of nucleon. We consider a "nucleus" consisting of N identical particles in n shells labeled by half-integer spin j, and where each shell consists of degenerate states of common energy e_j^0 and "deformation" d_j, labeled by the quantum number $m=-j,-j-1,\ldots,j-1,j$. Each m state can hold 2 nucleons so that a j shell contains at most $\Omega_j = j+1/2$ pairs of particles. The Hamiltonian of the system is

$$\hat{H} = \sum_j e_j^0 \hat{N}_j - \frac{1}{2}\chi\hat{D}^2 - \frac{1}{2}G\left(\hat{A}^\dagger\hat{A} + \hat{A}\hat{A}^\dagger\right) \tag{2}$$

where we have defined a number operator $\hat{N}_j \equiv \sum_m a_{jm}^\dagger a_{jm}$ for each shell j in terms of the particle creation and annihilation operators $a_{jm}^\dagger$ and a_{jm}, a "deformation" operator $\hat{D} \equiv \sum_j d_j \sum_m \sigma_{jm} a_{jm}^\dagger a_{jm}$ with $\sigma_{jm} = +1$ if $|m| < \Omega_j/2$ and -1 if $|m| > \Omega_j/2$, and an operator $\hat{A}^\dagger \equiv \sum_j \sum_{m>0} a_{jm}^\dagger a_{j\bar{m}}^\dagger$ with $\bar{m}$ designating the time-reversed state. Matrix elements of the Hamiltonian in Eq. (2) between many-body states can be calculated exactly using quasi-spin methods[23,24]. This makes the multi-$O(4)$ model particularly attractive as a test bed for the SCIM. A numerical diagonalization of Eq. (1) with the Hamiltonian in Eq. (2) was obtained for a system of 28 particles, 3 shells with $j = 27/2, 19/2, 7/2$, $e_j^0 = 0.0, 1.0, 3.5$, $\chi = 0.04$, and $G = 0.14$ in dimensionless "energy" units[23,24]. To do this, the deformation parameter $\left\langle\hat{D}\right\rangle$ was discretized and Eq. (1) was turned into a matrix diagonalization problem. The convergence of the eigenvalues with respect to the step size in $\left\langle \hat{D} \right\rangle$ was verified. In addition a study of the behavior of these eigenvalues as a function of the coupling between states was performed. This was accomplished by introducing a parameter κ multiplying the matrix elements connecting the ground state to 2-qp states, or connecting 2-qp states to one another. For $\kappa = 0$ many solutions were found with zero eigenvalue and zero eigenvector component in the ground state. These pure 2-qp excitations are not physical solutions in the $\kappa = 0$ case, which should consist entirely of superpositions of HFB ground states from the different $\left\langle \hat{D} \right\rangle$ values. Therefore, they were treated as spurious states. Furthermore, as κ increases from 0 to 1, many of the spurious solutions with 0 component in the ground states persist and "pollute" the entire energy spectrum. These spurious solutions have also been ignored.

In Fig. 2 we have plotted the cumulative number of levels as a function of excitation energy for the numerical example above. The figure shows the results of calculations obtained from 1) the exact solution of the multi-$O(4)$ model, 2) the GCM with particle-number projection, and 3) the SCIM solution using the approach described above. Below an excitation energy of

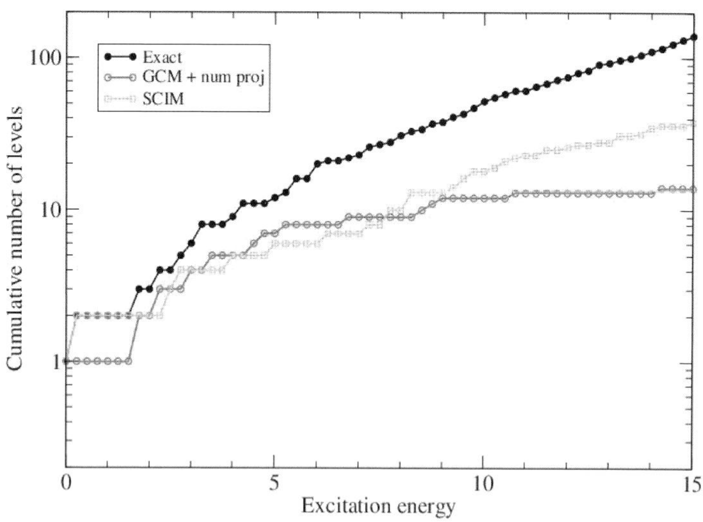

Fig. 2. Cumulative level density for the multi-$O(4)$ model example (see text) as a function of (unitless) excitation energy. Calculations from the exact solution, the GCM with particle-number projection, and the SCIM are compared.

9, both the GCM and SCIM solutions give comparable results that underestimate the cumulative level count for the exact solution by roughly similar amounts. For energies above 9, the GCM starts to run out of levels and approaches a plateau, whereas the SCIM curve continues to rise with a slope that is similar to the exact solution's.

4. Conclusion

The QRPA was used with a finite-range interaction to calculate the photoabsorption cross section for ^{240}Pu, and the results were compared to a standard Hauser-Feshbach calculation. In the future, limitations in the basis size and configuration space could be overcome through the use of the finite-amplitude method. Next, the SCIM was investigated using the schematic multi-$O\,(4)$ model. The cumulative level density obtained using the SCIM was compared to an exact solution of the multi-$O\,(4)$ model. Several improvements to the SCIM calculation will be investigated in future work: the effect of particle-number projection, the use of local QRPA excitations at each deformation instead of uncorrelated 2-qp states (the QRPA solutions would provide 2-qp excitations that are orthogonal to the local ground state to better separate vibrations in $\left\langle \hat{D} \right\rangle$ from other modes), and extensions to 4-qp excitations and terms beyond 2nd-order SOPO's.

Acknowledgments

This work was performed under the auspices of the U.S. Department of Energy by Lawrence Livermore National Security, LLC, Lawrence Livermore National Laboratory under Contract DE-AC52-07NA27344. This material is based upon work supported by the Department of Energy National Nuclear Security Administration under Award Numbers DE-NA0003180 and DE-NA0000979.

References

1. S. Péru and M. Martini, *Eur. Phys. J.* **A50**, 88 (2014).
2. R. Bernard, H. Goutte, D. Gogny and W. Younes, *Phys. Rev. C* **84**, 044308 (2011).
3. J. R. Nix, A. J. Sierk, H. Hofmann, F. Scheuter and D. Vautherin, *Nucl. Phys.* **A424**, 239 (1984).
4. N. Carjan, P. Talou and O. Serot, *Nucl. Phys.* **A792**, 102 (2007).

5. W. Younes, D. Gogny and N. Schunck, Proc. of the Fifth International Conference on ICFN5, Sanibel Island, Florida, USA, 4–10 Nov. 2012, p. 605, World Scientific, Eds. J. H. Hamilton and A. V. Ramayya (2013).
6. N. Schunck and L. M. Robledo, *Rep. Prog. Phys.* **79**, 116301 (2016).
7. J.-F. Berger, M. Girod and D. Gogny, *Comp. Phys. Comm.* **63**, 365 (1991).
8. D. L. Hill and J. A. Wheeler, *Phys. Rev.* **89**, 1102 (1953).
9. P. Bonche, J. Dobaczewski, H. Flocard, P.-H. Heenen and J. Meyer, *Nucl. Phys.* **A510**, 466 (1990).
10. L. M. Robledo and G. F. Bertsch, *Phys. Rev. C* **84**, 054302 (2011).
11. H. Goutte, J. F. Berger, P. Casoli and D. Gogny, *Phys. Rev.* **C 71**, 024316 (2005).
12. N. Dubray and D. Regnier, *Comp. Phys. Comm.* **183**, 2035 (2012).
13. W. Younes and D. Gogny, *Phys. Rev. Lett.* **107**, 132501 (2011).
14. P. Avogadro and T. Nakatsukasa, *Phys. Rev. C* **87**, 014331 (2013).
15. N. Hinohara, M. Kortelain and W. Nazarewicz, *Phys. Rev. C* **87**, 064309 (2013).
16. T. Nikšić, N. Kralj, T. Tutiš, D. Vretenar and P. Ring, *Phys. Rev. C* **88**, 044327 (2013).
17. T. Oishi, M. Kortelain and N. Hinohara, *Phys. Rev. C* **93**, 034329 (2016).
18. P. Ring and P. Schuck, *The Nuclear Many-Body Problem* (Springer, Heidelberg 1980).
19. W. Younes, *Comp. Phys. Comm.* **180**, 1013 (2009).
20. M. Didong, H. Müther, K. Goeke and A. Faessler, *Phys. Rev. C* **14**, 1189 (1973).
21. N. Tajima, H. Flocard, P. Bonche, J. Dobaczewski and P.-H. Heenen, *Nucl. Phys.* **A551**, 409 (1993).
22. P. G. Reinhard and K. Goeke, *Rep. Prog. Phys.* **50**, 1 (1987).
23. M. Kobayasi, T. Nakatsukasa, M. Matsuo and K. Matsuyanagi, *Prog. Theor. Phys.* **110**, 65 (2003).
24. N. Hinohara, T. Nakatsuakasa, M. Matsuo and K. Matsuyanagi, *Prog. Theor. Phys.* **115**, 567 (2006).

Novel Technique for Fission Yield Measurements Applying Ion Traps*

H. Penttilä, for the IGISOL group
University of Jyvaskyla, Department of Physics,
P.O. Box 35, FI-40014 University of Jyvaskyla, Finland

The ion guide technique [1] based on catching the primary ions from nuclear reactions in a noble gas is universal in the sense that ions of any element may be generated. In fission, elements from Ni to Dy are produced. An ion beam of any isotope can be formed in a time scale of $\sim$100 ms using the ion guide technique. Since all the isotopes in the beam are not only formed but also ionized in fission, identifying and counting them directly allows one to extract the beta decay independent fission yields.

The mass resolving techniques developed for ion traps [2] enable ion identification solely based on their mass. Combining the universality of the ion guide-based isotope separator IGISOL [3] and unambiguous identification of the ions with the JYFLTRAP Penning trap [4] the independent yields in 25 MeV proton-induced fission of natural U and Th have been measured [5] in the Accelerator Laboratory of the University of Jyvaskyla.

Since neutron-induced fission in some cases allows access to more neutron rich nuclei than proton-induced fission, an effort has been taken to develop a neutron source for induced fission studies at IGISOL [6]. In addition

*This work was supported by the Academy of Finland under the Finnish Centre of Excellence Programme 2012-2017 (Project No. 213503, Nuclear and Accelerator-Based Physics Research at JYFL) and the project No. 139382 (Precision Fission Studies for Practical Needs), and by the European Commission within the 7th Framework Programmes through Fission-2010-ERINDA (project No. 269499) and Fission-2013-CHANDA (project No. 605203).

to producing rare isotopes for spectroscopy purposes, the neutron source can be applied to fission yield studies. The status of this work will be summarized.

References

1. I.D. Moore, P. Dendooven and J. Ärje, *Hyperfine Int* **223**, 17–62 (2014).
2. H.-Jürgen Kluge, *International J. of Mass Spectrometry*, **349–350**, 26–37 (2013).
3. I.D. Moore *et al.*, *Nucl. Instr. and Meth. in Phys. Res. B* **317**, 208–213 (2013).
4. T. Eronen *et al.*, *European Physical Journal A* (2012) **48**:46.
5. H. Penttilä *et al.*, *European Physical Journal A* (2016) **52**:104.
6. D. Gorelov *et al.*, *Nucl. Instr. and Meth. in Phys. Res. B* **376**, 46–51 (2016).

Total Kinetic Energy Release in the Fast Neutron Induced Fission of ^{232}Th and ^{235}U

W. Loveland* and J. King
Chemistry Department, Oregon State University
Corvallis, OR 97331, USA
**E-mail: lovelanw@onid.orst.edu*

We have measured the total kinetic energy release (TKE), its variance and associated fission product distributions for the neutron induced fission of ^{232}Th and ^{235}U for E_n = 2-90 MeV. The neutron energies were determined on an event by event basis by time of flight measurements with the white spectrum neutron beam from LANSCE. The TKE decreases non-linearly with increasing neutron energy for both systems, while the TKE variances are sensitive indicators of nth chance fission. The associated fission product distributions show the decrease in TKE with increasing beam energy that is due to the increasing probability of symmetric fission — which has a lower associated TKE — and the decreasing TKE associated with asymmetric fission, presumably due to the decreasing importance of the A $\sim$ 132 shell structures.

Keywords: Fission; TKE; Variance of TKE; Fission Product Mass Distributions; ^{232}Th(n,f); ^{235}U(n,f).

1. Introduction

We report the total kinetic energy release (TKE), the variance (σ^2) of the TKE distribution, and the associated mass yield distributions in the fast neutron induced fission of ^{232}Th and ^{235}U(E_n = 2-90 MeV). (A preliminary version of the ^{235}U study has been reported before[1].) While these data have implications for nuclear weapons and reactors, our primary goal is to better understand the fission process. As we increase the energy input to these systems from $\sim$ 0 to $\sim$ 90 MeV, we find that the primary energy released in fission, the TKE, decreases by a few MeV (6.0 and 7.4 MeV, for ^{232}Th and ^{235}U, respectively). Clearly the dissipation of the incident neutron energy in the fissioning system takes place on a time scale ($\sim 10^{-21}$ s) that is short compared to the times for the large scale collective motion of fission ($\sim 10^{-18}$ s). The dissipated energy appears in the excitation energy of the fission fragments while the small decrease in the observed TKE reflects changes in the fission process in these reactions. Understanding and characterizing these processes will be the central theme of this work. By studying the ^{232}Th(n,f) and the ^{235}U(n,f) processes simultaneously, we hope to compare and contrast the physics of the different fissioning systems upon the fission product mass and energy distributions.

2. Experimental

This experiment was carried out at the Weapons Neutron Research Facility (WNR) at the Los Alamos Neutron Science Center (LANSCE) at the Los Alamos National Laboratory[2,3] over a 5 day period in December, 2014. "White spectrum" neutron beams were generated from an unmoderated tungsten spallation source using the 800 MeV proton beam from the LANSCE linear accelerator. The measured white spectrum near the target position is shown in Fig. 1. The proton beam is pulsed allowing one to measure the time of flight (energy) of the neutrons arriving at the experimental area. Note the beam intensities of $\sim 10^5$ n/s –a number that is higher than the beam used at radioactive beam facilities–but substantially less than that normally used in fission studies. Thus one has to run for long periods of time and/or to use large solid angle detection systems.

The detailed description of this experiment has been reported previously[1]. We refer the reader to[1] for details. In[1], assuming $\nu_{tot} = 2.43^4$ and $\nu_{pre} = 0.0^4$, we found that the measured thermal neutron induced post fission TKE was $E_{TKE}{}^{th.n} = 169.8 \pm 0.4$ MeV and that $\sigma_{TKE-th}{}^2 = 103.6 \pm 0.2$ MeV. The pre-fission TKE is $E_{TKE}{}^{*,th} = 170.7 \pm 0.4$ MeV. This

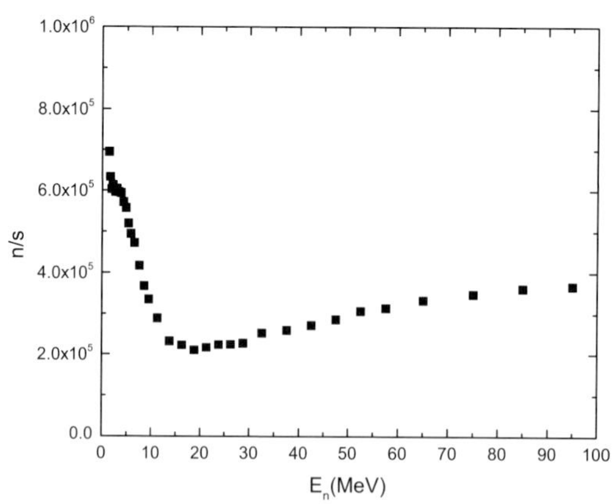

Fig. 1. Measured neutron beam intensity near the target location on the 15R beam line during this experiment

result is consistent with the previous measurements of 171.9 ± 1.4[5], 172.0 ± 2.0[6] and 170.1 ± 0.5[7]. For the spontaneous fission of ^{252}Cf, we found the average post fission light fragment TKE was 103.2 ± 1.1 MeV while the average heavy fragment TKE was 78.6 ± 1.5 MeV. This is to be compared to post neutron emission values[8] for the light fragment of 103.8 ± 0.5 MeV and heavy fragment energy of 79.4 ± 0.4 MeV. **Our measurements of the TKE should be regarded as absolute measurements.**

3. Results and Discussion

The TKE data for each system was binned into neutron energy bins. The resulting TKE distributions for each neutron energy bin were found to be Gaussian in shape with no observed deviations from a Gaussian shape. The mean values of the TKE for each bin for each system are shown in Fig. 2. The mean TKE values decrease non-linearly with increasing neutron energy in a manner that can be represented by a polynomial in neutron energy. These data are in good agreement with previous measurements at lower energies[9–13]. The Viola systematics[14], which applies only to pre-neutron emission data, would predict TKE values of 170.1 and 163.8 MeV, respectively, for the ^{235}U(n,f) and ^{232}Th(n,f) reactions. That "Viola ratio" (170.1/163.8) is remarkably similar to the observed TKE ratio for the two systems, perhaps indicating the same underlying physics. This observation of a non-linear decrease in TKE with increasing excitation energy is

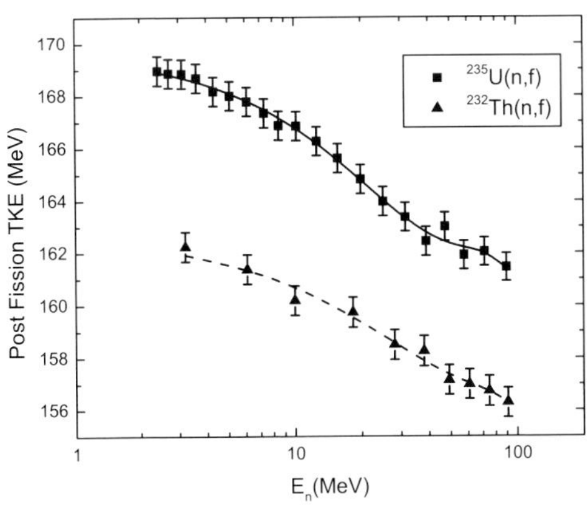

Fig. 2. Measured values of the mean TKE as a function of neutron energy for ^{232}Th (n,f) and ^{235}U(n,f).The lines are polynomial fits to the data, TKE(235) = 169.68-0.33881E_n-0.0052$E_n{}^2$-2.73475x10^{-5} $E_n{}^3$ and TKE(232)=162.56-0.20877E_n+0.00275$E_n{}^2$-1.3367x10^{-5} $E_n{}^3$.

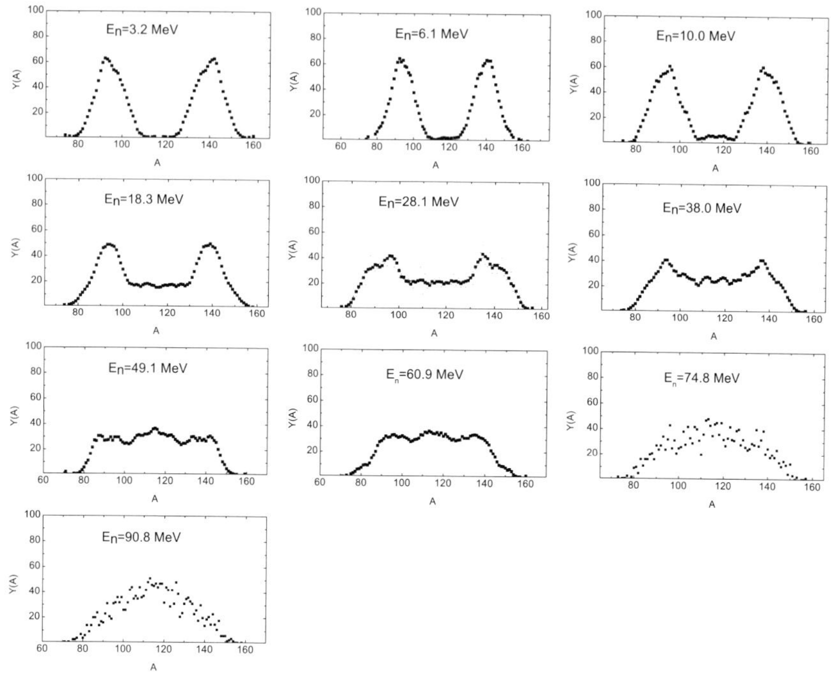

Fig. 5. (a) The measured pre-neutron emission fission product mass distributions as a function of neutron energy for the ^{232}Th(n,f) reaction.

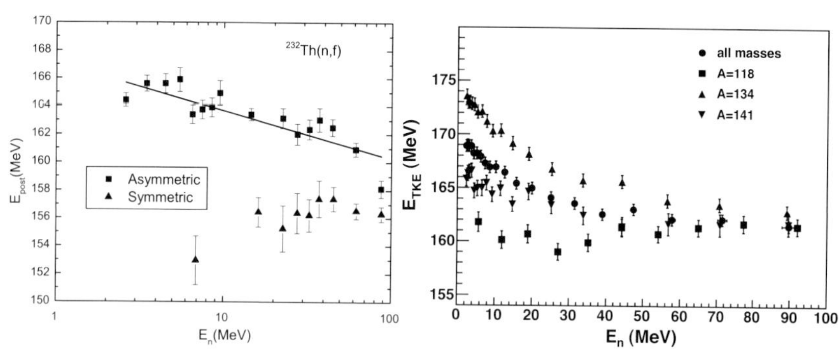

Fig. 6. (a) The mass sorted TKE values as a function of neutron energy for ^{232}Th(n,f). (b) The mass sorted TKE values as a function of neutron energy for ^{235}U(n,f).

than "post-dictive" estimates. Two semi-empirical models[18,19] for the TKE release as a function of neutron energy are compared to our data in Fig. 7.

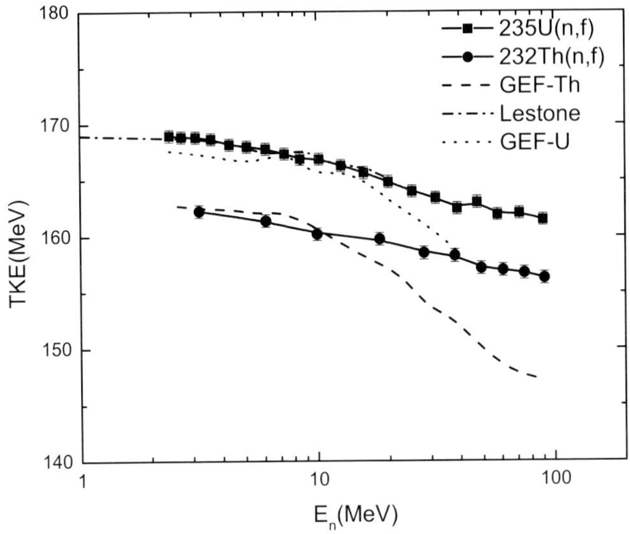

Fig. 7. Comparison of our measured TKE values with the predictions of[19] and[18].

Both models take into account multiple chance fission. The predictions of the GEF code differ from the data above $E_n = 20$ MeV while the predictions of[19] agree quite well with the data. One should note that more "fundamental" calculations of fission product yields[20] do exist for these systems. In these calculations, there is a washing out of the shell effect and changes in the pairing energies with increasing excitation energy. The calculation[20] is limited to predicting (correctly) the fission product mass distribution for ^{235}U(n,f) at E_n=14.5 MeV.

Acknowledgments

This work was supported in part by the U.S. Dept. of Energy, NNSA, under Grant DE-NA0002926 and the U.S. Dept. of Energy, Office of Science, Office of Nuclear Physics under award number DE-SC0014380. We gratefully acknowledge the contributions of Ricardo Yanez to this work.

References

1. R. Yanez *et al.*, arXiv:1605.09690.
2. P.W. Lisowski, C.D. Bowman, G.J. Russell and S.A. Wender, *Nucl. Sci. Eng.* **106**, 208 (1990).

3. P.W. Lisowski and K.F. Schoenberg, *Nucl. Instrum. Methods Phys. Res.*, Sect. A **562**, 910 (2006).
4. M. Chadwick *et al.*, *Nucl. Data Sheets* **112**, 2887 (2011).
5. H.W. Schmitt, J.H. Neiler and F.J. Walter, *Phys. Rev.* **141**, 1146 (1966).
6. M.J. Bennett and W.E. Stein, *Phys. Rev.* **156**, 1277 (1967).
7. G.F. Bertsch, W. Loveland, W. Nazarewicz and P. Talou, *J. Phys. G: Nucl. Part. Physics*, **42**, 077001 (2015).
8. H.W. Schmitt, W.E. Kiker and C.W. Williams, *Phys. Rev.* **137**, B837 (1965).
9. D.G. Madland, *Nucl. Phys. A* **772**, 113 (2006).
10. D. Duke, Ph.D. thesis, Colorado School of Mines, LA-UR-15-28829 (2015), unpublished.
11. J. Trochon, H.A. Yehia, F. Brisard and Y. Pranal, *Nucl. Phys. A* 318, 63 (1979).
12. A.I. Sergachev, V.G. Vorobyeva, D.B. Kuzminov, V.B. Mikhailov and M.Z. Tarasko, Yad. Fiz, **7**, 778 (1968).
13. J. Kiesewetter, K.-Th. Brinkmann, F.M. Baumann, H. Freiesleben and H. Sohlbach, *Nucl. Phys. A* **540**, 75 (1992).
14. V.E. Viola, K. Kwiatkowski and M. Walker, *Phys. Rev. C* **31**, 1550 (1985).
15. R.L. Ferguson, F. Plasil, F. Pleasanton, S.C. Burnett and H.W. Schmitt, *Phys. Rev. C* **7**, 2510 (1973).
16. U. Brosa, H.-H. Knitter, T. Fan, J. Hu and S. Bao, *Phys. Rev. C* **59**, 767 (1999).
17. X. Sun, C. Yu and N. Wang, *Phys. Rev. C* **85**, 014613 (2012).
18. GEF 2016/1.2, khs.erzhausen.de
19. J.P. Lestone and T.T. Strother, *Nucl. Data Sheets* **118**, 208 (2014).
20. M.D. Usang, F.A. Ivanyuk, C. Ishizuka and S. Chiba, *Phys. Rev. C* **94**, 044602 (2016).

Dynamical Approach to Low-Energy Nuclear Fission in Terms of Langevin Equation

Satoshi Chiba[1], Mark Dennis Anak Usang[1,2], Chikako Ishizuka[1], Hiroki Jojima[1] and Fedir Ivanyuk[1,3]

[1]*Tokyo Institute of Technology, Tokyo, Japan*
[2]*Malaysian Nuclear Agency, Kuala Lumpur, Malaysia*
[3]*Institute for Nuclear Research, Kiev, Ukraine*

Nuclear fission is the most basic phenomenon in many application of nuclear technology. The nuclear fission, on the other hand, continues to be an interesting subject for basic research point of views. It is interesting since it involves large-scale collective motion leading to violent rearrangement of nucleons, exhibiting quantum nature (shell effects) in low-energy region. Furthermore, many observables associated with fission are correlated to each other, and hence should be considered or treated in a consistent manner. It will require a deep understanding of nuclear fission.

We treat the nuclear fission as a fluctuation-dissipation process, and describe the fission in terms of multi-dimensional Langevin equation. So far, we have been using 3 collective coordinates, the elongation, fragment deformation and mass asymmetry. The potential energy surface is calculated by the liquid-drop model plus Strutinsky's prescription by using the two-center shell model parametrization of nuclear shape. The transport coefficients are calculated by macroscopic method, namely, Werner-Wheeler method for the inertial tensor, and wall-and-window formula for the friction tensor.

In this presentation, we will describe the current status of the method developed at Tokyo Tech. Firstly, we explain the general background of the theory, and how the current method can be compared with various fission observables such as mass and TKE distributions and neutron multiplicity. Nextly, we will describe how we are going to improve this method. We have introduced microscopic transport coefficients by a linear response theory with locally-harmonic approximation. Furthermore, we have introduced charge polarization and charge asymmetry degree of freedom for the Langevin method to have a capability to predict isotope distribution of fission fragments. We will also extend the 3-dimensional calculation to a 4-dimensional one. Here, we will set the deformation of the 2 fragments as independent variables. It will make the potential energy surface to more realistic, and excitation energy of the each fragment due to deformation can be calculated independently, which will hopefully lead to accurate prediction of the number of prompt neutrons.

Microscopic Description of the Low Energy Fission Process: A New Method

Verrière M.,[1] Dubray N.,[1] Regnier D.[1] and Schunck N.[2]
[1] *CEA, DAM, DIF, 91297 Arpajon, France*
[2] *Nuclear and Chemical Science Division, Lawrence Livermore National Laboratory, Livermore, CA 94551, USA*

Dynamical description of low energy fission is, in our full microscopic approach, decomposed in two steps. In the first step we generate the potential energy surface (PES) of the compound system we want to describe with the Hartree-Fock-Bogoliubov (HFB) method with a Gogny interaction. The second step uses the time dependent generator coordinate method (TDGCM) with the gaussian overlap approximation (GOA). The GOA holds in two assumptions: the overlap matrix between HFB states has a gaussian shape (with respect to the difference between coordinates of states in deformation space); and the expectation value of the collective hamiltonian between these states can be expanded up to order two, leading in this case to a Schrödinger-like equation.

In this work we replace TDGCM+GOA in the second step of our approach by an exact treatment of the TDGCM. The main equation of this method is the time-dependent Hill-Wheeler equation and involves two objects: the overlap matrix and the collective hamiltonian. We first calculate these matrices on a PES. Then, we build an "exact TDGCM" solver using a finite element method and a Crank-Nicolson scheme.

In this talk, we will present the time-dependent Hill-Wheeler equation and discretization schemes (in time and deformation space). The analytic calculation of overlap matrix and collective hamiltonian will be detailed. Finally, first results with an exact treatment of the TDGCM will be discussed.

TKE Measurements*

D.L. Duke[1], F. Tovesson[1], T. Brys[2], V. Geppert-Kleinrath[1], F.-J. Hambsch[2], A.B. Laptev[1], B. Manning[1], D. Mayorov[1], R. Meharchand[3], K. Meierbachtol[1], S. Mosby[1], B. Perdue[4], D. Richman[5], D. Shields[1,6] and M. Vidali[2]

[1]*Los Alamos National Laboratory, 87545 USA*
[2]*EC-JRC-Institute for Reference Materials and Measurements (IRMM)*
[3]*Institute for Defense Analyses*
[4]*Commonwealth Computer Research, Inc.*
[5]*Michigan State University*
[6]*Colorado School of Mines*

Advanced fission models and simulations require understanding how energy is partitioned during fission. The majority of energy in the fission process is released in the form of kinetic energy of the fission fragments. In this work, the average total kinetic energy (⟨TKE⟩) was measured for neutron induced fission on ^{235}U, ^{238}U, and ^{239}Pu targets over a broad range of incident neutron energies. These are important nuclear data which support both energy and defense applications. The majority of the previous work focused on neutron energies (E_n) below 9 MeV for ^{235}U and below 6 MeV for ^{239}Pu [1]. Few complete studies of ⟨TKE⟩ dependence on excitation energy have been made [2,3]. The Los Alamos Neutron Science Center (LANSCE) — Weapons Neutron Research (WNR) facility has capabilities to deliver neutron beams with energies from a few hundred keV up to hundreds of MeV. The ⟨TKE⟩ measurements were made at WNR using a Frisch-gridded ionization chamber and digital data acquisition system. The ⟨TKE⟩ data will be presented from E_n = 200 keV–30 MeV (for ^{235}U), E_n = 1.3–30 MeV (for ^{238}U), and E_n= 500 keV to 50 MeV (for ^{239}Pu). For

*This work is funded by the US Department of Energy and operated by Los Alamos National Security, LLC under contract DE-AC52-06NA25396.

all three measurements, the neutron-energy dependent $\langle \mathrm{TKE} \rangle$ has an overall decrease as a function of energy, yet exhibits structure at multichance fission thresholds which validates existing fission models [4]. Plans for further measurements will also be discussed.

References

1. D.G. Madland, *Nuclear Physics A*, **772**(34): 113–137, 2006.
2. R. Yanez *et al.*, *Phys. Rev. C* **89**, 051604 (2014).
3. C.M. Zöller, Ph.D. Thesis, Technische Hochschule Darmstadt. 1995.
4. J.P. Lestone and T.T. Strother, *Nuclear Data Sheets*, **118**(0): 208–210, 2014.

Time-Dependent Approach to Angular Distribution and Energy of Scission Neutrons

Takahiro Wada* and Tomomasa Asano

Department of Pure and Applied Physics, Kansai University,
3-3-35 Yamate-cho, Suita, 564-8680, Japan
**E-mail: wadataka@kansai-u.ac.jp*

We investigate the angular distribution of scission particles taking account of the effects of fission fragments. The time evolution of the wave function of the scission particle is obtained by integrating the time-dependent Schrödinger equation. The effects of the fission fragments are taken into account by means of the optical potentials. The angular distribution is strongly modified by the presence of the fragments. In the case of asymmetric fission, it is found that the heavy fragment has stronger effects. The angular distribution of the energy of scission neutrons is also discussed.

Keywords: Scission Neutron, Scission Proton, Time-Dependent Schrödinger Equation.

1. Introduction

Knowledge on how scission occurs is very important to understand the dynamics of nuclear fission. The kinetic energy of the fragments strongly depends on the scission configuration because it is essentially determined by the Coulomb repulsion between the fragments. The mass/charge distribution of the fragments also depends on the way how the neck breaks. At the moment of scission, the neck that has connected the two fission fragments ruptures. This neck rupture can be very quick because the number of nucleons involved in the neck region is relatively small. After the scission, the neck disappears being absorbed by the fragments leaving some nucleons behind in the neck region and these nucleons are observed as particle emission. The number of scission particles carries the information on the scission process. There have been attempts to estimate the number of scission neutrons theoretically [1, 2]. The results depend on the nuclear shape such as the neck radius before scission. If we extract the reliable number of scission neutrons from experiments, we can get information on the nuclear shape at the time of scission.

In measuring the scission particles (neutrons/protons) experimentally, the information on their angular distribution is essential. The emission of scission particles is normally supposed to be isotropic in the rest frame of the mother nucleus in the lowest order approximation. However, since they are emitted in the close vicinity of the fission fragments, the final angular distribution of the scission particles is modified by the fragments through scattering and re-absorption. In the previous study, we proposed a framework to calculate the angular distribution of scission neutrons by solving a time-dependent Schrödinger equation [3]. It was shown that the angular distribution of the scission neutron is strongly modified by the presence of the fragments, namely, the attractive nuclear potential enhances the yields around 0 and 180 degrees, while the absorptive potential diminishes them. However, the calculation was limited to the symmetric fission and only scission neutrons were considered. In this paper, we present results on the asymmetric fission and on the scission protons. We also present the results on the angular distribution of the mean energy of scission neutrons. In the next section, the formulation to calculate the angular distribution of the scission particles is revisited. In Section 3, results are presented for the case of the fission of ^{236}U. First, we discuss the angular distribution of mean energy of scission neutrons for the case of symmetric fission. Then we show the results for the case of asymmetric fission. Comparison of the results of the neutrons and protons is given. Finally, a summary is given.

2. Formalism

We start with a time-dependent Schrödinger equation (TDSE),

$$i\frac{\partial \psi}{\partial t} = H\psi \ , \quad H = -\frac{1}{2m}\nabla^2 + U \ , \quad \hbar = 1, \tag{1}$$

where ψ denotes the neutron wave function, H is the Hamiltonian, and U is the potential that represents the effect of the fission fragments. Assuming the axial symmetry, we solve the TDSE in two-dimensional grid space (ρ, z). The time development is obtained with the use of the mid-point integration,

$$\psi(t+\Delta t) = \psi(t) - i\Delta t\, H\psi(t+\Delta t/2)\,. \tag{2}$$

By decomposing ψ into the real and the imaginary part, $\psi = R + I$, the numerical solution is obtained using the leap-frog method [3]. The potential U has two centers and is parameterized in Woods-Saxon form centered at the position of each fragment,

$$U(\rho,z)=\frac{V_0+iW_0}{1+\exp\left(\dfrac{\sqrt{\rho^2+(z+B_2)^2}-R_{\mathrm{F2}}}{a}\right)}+\frac{V_0+iW_0}{1+\exp\left(\dfrac{\sqrt{\rho^2+(z-B_1)^2}-R_{\mathrm{F1}}}{a}\right)}, \quad (3)$$

where B_i is the distance between the center of scission neutron distribution and the i-th fragment, a is the diffuseness, and $R_{\mathrm{F}i}$ is the radius of the potential of i-th fragment. The distance B_i is given as $B_i = \eta R_{\mathrm{F}i}$, where η is the elongation parameter which is to be deduced from the systematics of the total kinetic energy. For the case of the emission of charged particles such as protons, we take account of a Coulomb field of the fragments assuming the uniform charge distribution inside the fragments [4].

The original emission of the scission neutrons is assumed to be isotropic, and we adopt a Gaussian wave packet for the initial wave function, $\psi(t=0) = C \exp(-\alpha(\rho^2+z^2))$, where C is a normalization constant and α denotes the width of the wave packet. We now set a sphere of radius R and calculate the neutron flux on this spherical surface,

$$\vec{j}(\vec{r},t)=\frac{1}{2im}\left(\psi\vec{\nabla}\psi^*-\psi^*\vec{\nabla}\psi\right). \quad (4)$$

We then calculate the number of outgoing scission particles per unit solid angle per unit time and integrate it with time to obtain the angular distribution: the number of particles per unit solid angle that passed the surface up to time t,

$$\frac{dN(\theta,t)}{d\Omega}=\int_0^t\frac{d^2N(\theta,t')}{d\Omega dt}dt'=\int_0^t\vec{j}(R,\theta,t')\cdot\vec{n}(R,\theta)R^2dt', \quad \vec{n}=\vec{e}_r. \quad (5)$$

The energy density and the energy flow are defined respectively as

$$\varepsilon(\vec{r},t)=\frac{h^2}{2m}\frac{\partial\psi^*}{\partial\vec{r}}\cdot\frac{\partial\psi}{\partial\vec{r}}+U(\vec{r},t)\psi^*\psi, \quad (6)$$

and

$$\vec{j}_\varepsilon(\vec{r},t)=\frac{\hbar}{2im}\left\{\frac{\hbar^2}{2m}\left(\frac{\partial\psi}{\partial\vec{r}}\cdot\vec{\nabla}\frac{\partial\psi^*}{\partial\vec{r}}-\frac{\partial\psi^*}{\partial\vec{r}}\cdot\vec{\nabla}\frac{\partial\psi}{\partial\vec{r}}\right)\right\}+U(\vec{r},t)\vec{j}(\vec{r},t). \quad (7)$$

The energy that is brought by scission neutrons to a certain angle θ is calculated in the same manner as in Equation (5),

$$\frac{dE(\theta,t)}{d\Omega}=\int_0^t\vec{j}_\varepsilon(R,\theta,t')\cdot\vec{n}(R,\theta)R^2dt', \quad \vec{n}=\vec{e}_r \quad (8)$$

and the average energy of scission neutron for each direction is given by

$$\langle\varepsilon(\theta)\rangle=\frac{dE(\theta,t)}{d\Omega}\bigg/\frac{dN(\theta,t)}{d\Omega}. \quad (9)$$

In order to avoid unrealistic reflections of the wave function at the border of the grid, we put a week absorbing potential outside of the sphere of radius R. We take a quadratic form for the absorbing potential. The strength is determined to minimize the effect of the reflection.

3. Results and discussions

As an example, we consider the fission of ^{236}U that corresponds to the neutron induced fission of ^{235}U. From a systematic study of the average total kinetic energy (TKE) of the fragments, Zhao *et al.* derived the elongation parameter η which is the ratio between the average distance between the fragments to the contact distance $r_0(A_1^{1/3} + A_2^{1/3})$, where A_1 and A_2 are the mass numbers of the fragments and r_0 = 1.17fm [5]. The average distance between the fragments is determined so that the corresponding point charge Coulomb energy is equal to the average TKE. They obtained η = 1.53 for the asymmetric fission in U region. In the calculation, we simply use the same elongation parameter for B_i as $B_i = \eta\, r_0 A_i^{1/3}$. The distance between the grid points is typically 0.1 fm in both z- and ρ-directions. The time step for the integration is typically Δt = 0.02 fm/c. The radius R is set to 50 fm and the integration is performed up to $t = 4\text{x}10^{-21}$s. The width of the initial wave packet is determined to give the average energy of neutron $<\varepsilon>$ = 1.5MeV. In calculating the time development, we take account of the motion of the fragments due to the Coulomb repulsion between the fragments. The pre-scission kinetic energy of 10 MeV is assumed.

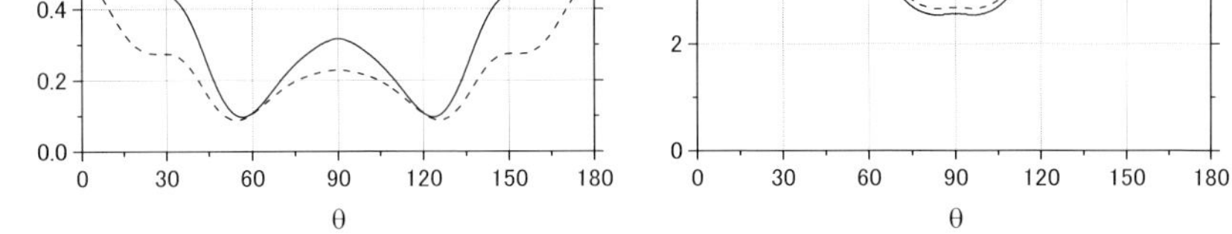

Figure 1. [left panel] Angular distribution of scission neutrons for symmetric fission: without re-absorption (solid line) and with re-absorption (dashed line). [right panel] Angular distribution of the energy of scission neutrons for the same case.

Figure 1 shows the calculated angular distributions of the number and the mean energy of scission neutrons for the case of symmetric fission. In the case of purely attractive potential, *i.e.*, $W_0 = 0$, the distribution is strongly peaked around 0 and 180 degrees. When we consider the re-absorption by the fragments with $W_0 = -2$ MeV, the yield around 0 and 180 degrees decreases significantly, resulting in a rather flat angular distribution. In the right panel, the angular distribution of the mean energy of scission neutrons is plotted. High energy component is observed around 0 and 180 degrees, which is emitted quickly after scission. It is also seen that the angular distribution of mean energy has little dependence on the strength of re-absorption. However, the calculated energy of neutrons is very high compared with the average energy of the initial wave packet which is 1.5 MeV. It is partly because we can only calculate up to a certain time after scission. Faster component with high kinetic energy is emitted first and the component with low kinetic energy comes later. It is also necessary to examine the effect of the initial distribution to the energy distribution.

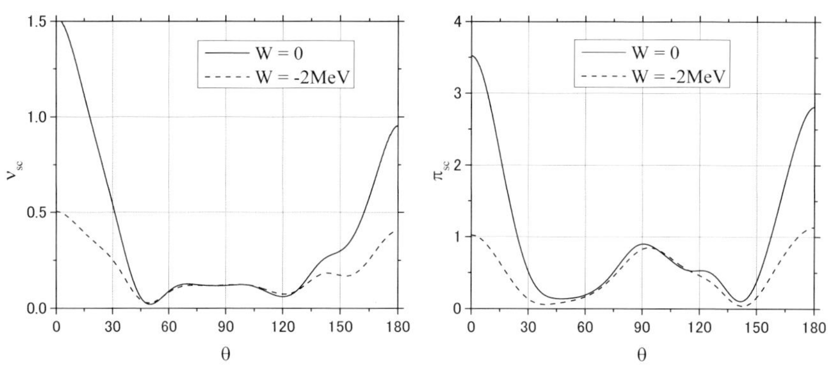

Figure 2. Angular distributions of the yield of scission particles for the case of asymmetric fission. [left panel] Angular distribution of scission neutrons without re-absorption (solid line) and with re-absorption (dashed line). [right panel] Angular distribution of scission protons.

Next, we investigate the case of asymmetric fission. The mass number of the heavy fragment is assumed to be $A_1 = 140$. The results are shown in the left panel of Figure 2; the solid curve denotes the result for $W_0 = 0$ (no absorption) case and the dashed curve denotes the one with $W_0 = -2$ MeV. As in the case of symmetric fission, the attractive nuclear interaction enhances the yields around 0 and 180 degrees and they decrease significantly when we take account of the absorption. We observe more neutrons in the direction of heavy fragment, which is contrary to the earlier work [2]. This is essentially due to the difference of the initial distributions of scission particles. In the right panel, we show the results for scission protons for the same system. It is seen that the yield around 90 degrees is significantly large in the case of protons due to the focusing effect of Coulomb field of the fragments. The difference between neutron and proton becomes clear when we compare the results more closely in the angular range between 60 and 120 degrees. In the case of scission neutrons, the angular distribution is shifted toward the heavy fragment side because of the larger radius of the attractive potential. On the other hand, in the case of scission protons, the angular distribution in this angular range shifted toward the light fragment side. This is because of the stronger Coulomb repulsion from the heavy fragment. It should be noted also that the angular distribution in this range does not change much even when we take account of the absorption.

4. Summary

The effects of the scattering and re-absorption by the fission fragments on the angular distribution of scission particles have been investigated in the framework of the time-dependent Schrödinger equation. The effects of the fragments are taken account in terms of the nuclear optical potentials and the Coulomb field. We solve the time-dependent Schrödinger equation of the emitted particle and the angular distribution is calculated by counting the outgoing flux. It has been demonstrated that the attractive nuclear potential enhances the yields around 0 and 180 degrees, while the absorptive potential diminishes them. The energy of scission neutron is discussed in connection with the angular distribution of the yield. In the case of asymmetric fission, the heavy fragment which is larger in size causes stronger effect to the angular distribution for both cases of scission neutrons and scission protons. For protons, the Coulomb field of the fragments modifies the angular distribution significantly, however, the effect of the nuclear interaction is found to be stronger. For more quantitative predictions, realistic initial distribution of the scission particle is necessary.

Acknowledgments

The authors would like to express their gratitude to Dr. N. Carjan for discussions.

References

1. N. Carjan, P. Talou, O. Serot, *Nucl. Phys.* **A 792**, 102 (2007).

2. M. Rizea, N. Carjan, T. Wada, *Physica Procedia* **47**, 27 (2013).
3. T. Wada, T. Asano, *Proc. 5th Int. Conf. on Fission and Properties of Neutron-Rich Nuclei* (World Scientific, 2014) 409.
4. T. Wada, T. Asano, N. Carjan, *Physica Procedia* **47**, 33 (2013).
5. Y. L. Zhao *et al.*, *Phys. Rev.* **C62**, 14612 (2000).

Energy Evolution of the Fission-Product Yields from Neutron-Induced Fission of ^{235}U, ^{238}U, and ^{239}Pu: An Unexpected Observation

A.P. Tonchev*, M.A. Stoyer, J.A. Becker, R. Macri, C. Ryan and S.A. Sheets

Lawrence Livermore National Laboratory, Livermore, California 94550, USA
**E-mail: tonchev2@llnl.gov*

M.E. Gooden, C.W. Arnold, E. Bond, T. Bredeweg, M.M. Fowler, G. Rusev, D.J. Vieira and J.B. Wilhelmy

Los Alamos National Laboratory, Los Alamos, New Mexico 87545, USA

W. Tornow[1,2], C.R. Howell[1,2], C. Bhatia[1,2], M. Bhike[1,2], F. Krishichayan[1,2], J.H. Kelley[1,3], B. Fallin[1,2] and S.W. Finch[1,2]

[1] *Triangle Universities Nuclear Laboratory, Durham, North Carolina 27708, USA*
[2] *Department of Physics, Duke University, Durham, North Carolina 27708, USA*
[3] *Department of Physics, North Carolina State University, Raleigh, North Carolina 27605, USA*

A consistent set of high-precision measurements have been performed to study the energy dependence of the fission product yields of ^{235}U, ^{238}U, and ^{239}Pu using monoenergetic neutrons between 0.5 and 14.8 MeV. The results confirm the progression towards symmetric fission at higher incident neutron energy, i.e., 14.8 MeV. However, at lower energies ($E_n < \sim 3$ MeV) the experimental data reveal a peculiar energy dependence of some of the fission-product yield from neutron-induced fission of ^{239}Pu: a positive slope up to about 4-5 MeV which then turns negative as the incident neutron energy increases. This latter finding at low-energy is in conflict with present theoretical predictions.

Keywords: Neutron-Induced Fission; Fission-Product Yields; Fission Ionization Chambers, Gamma-Ray Spectroscopy; Quasi-Monoenergetic Neutron Beams.

1. Introduction

Nuclear fission, the most pronounced collective nuclear-structure phenomenon, relates the fission product yields and their kinetic energies to the potential energy of the fissioning system[1]. The evolution of the fissioning system from the initial particle impact through intermediate saddle points to scission and finally to the configuration of separated fission fragments is governed by multi-dimensional potential-energy surfaces and the shell structure of the fragments. At low excitation energies shell and pairing effects in nuclei influence both the mass and energy distribution of the fission fragments. Hence, fine nuclear structure details are needed to describe experimental fission data. At higher excitation energies well above the barrier, shell and pairing effects become washed out and fission is well described by a liquid drop model. Therefore, only at low excitation energies, close to the fission barrier, do subtle effects show up when the excitation energy is varied. These variations are intimately connected with changes in the mass distributions which furthermore are affected by variations of the spin of the compound nuclear system. Models based on different phenomenological and microscopic approaches are used to describe the complexity of the fission process and are gauged on how well they describe the various experimental observables[2]. One of these observables is the dependence of the Fission Product Yields (FPYs) on incident neutron energy. The magnitude and the slope of the energy dependencies of the FPYs have significant impact on basic and applied physics. For example, r-process nucleosynthesis cannot be fully understood without a precise knowledge of the fission properties of the very neutron-rich isotopes of the heaviest elements, which are presently not accessible to direct measurements[3]. Precise knowledge of FPYs are important for nuclear reactors, nondestructive nuclear fuel investigation, accelerator-driven systems, and decay heat[4–7]. In addition, information on the FPYs has a fundamental impact on basic fission studies and the antineutrino spectrum[8]. The accuracy of purely theoretical predictions is insufficient for quantitative predictions of the energy dependence of fission yields with the precision that present nuclear applications require[9]. However, it is quite challenging to make quantitative predictions for the FPYs, especially within a microscopic approach based on time-dependent Hartree-Fock calculations, where the experimental magnitude of the energy dependence is comparable or even smaller than the uncertainties of the theoreticaly calculated FPYs[11,12].

This paper reveals the energy evolution of the fission-product yields from neutron-induced fission of ^{235}U, ^{238}U, and ^{239}Pu using quasi-monoenergetic neutrons. The main experimental results of this study have been published in Refs.[13–15]

2. Experimental Overview

The FPY measurements were performed employing the 10 MV Van de Graaff Tandem accelerator at the Triangle Universities Nuclear Laboratory (TUNL). Monoenergetic neutrons were produced utilizing the four nuclear reactions ^{7}Li$(p,n)^7$Be, ^{3}H$(p,n)^3$He, ^{2}H$(d,n)^3$He, and ^{3}H$(d,n)^4$He, providing high yields of fast neutrons[13,15]. These reactions cover different energy ranges in which the produced neutrons are truly monoenergetic and the contribution from off-energy neutrons are negligible[16]. A schematic of the experimental setup is shown in Fig. 1. The incident deuteron or proton beam passes through a 6.5 μm Havar foil, enters a helium buffer cell and after traversing a tritium loaded layer of titanium is finally stopped in a thin copper beam stop. At different irradiation configurations the tritiated target can be replaced by either ^{7}Li target or ^{2}H gas cell allowing production of quasi-monoenergetic neutrons at different energies.

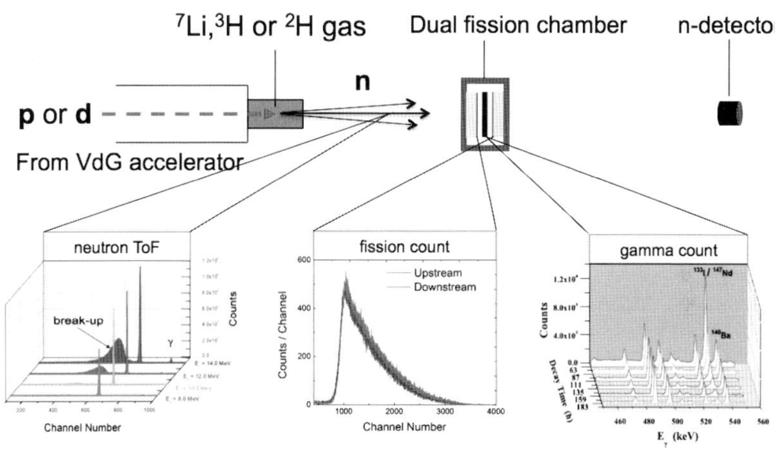

Fig. 1. Schematic of the FPY experimental setup at TUNL. A dual-fission chamber is positioned 4 cm downstream from the neutron production source while a liquid-scintillator neutron detector is positioned 3 m further downstream, serving as a secondary flux monitor.

A dual-fission chamber (DFC) chamber is used to provide the most accurate determination of the total number of fissions during the irradiation. The DFC is located in close proximity (4 cm) to the neutron production source. The chamber contains two thin reference foils and a thicker actinide activation target[16]. The activation target is contained in the center of the chamber while the thin reference foils are up- and down-stream from the activation target in identical ionization chambers. The thick activation target is of the same fissile material as the thin reference foils in the adjacent chambers. Three DFCs containing highly-enriched ^{239}Pu (99.954%), ^{235}U (99.998%), and depleted ^{238}U (99.998%), were dedicated to the three target isotopes. The thin reference foils in each chamber are 1.27 cm diameter deposits of $\sim$ 10 μg ^{239}Pu, 120 μg of ^{235}U and ^{238}U, while the thick activation targets are also 1.27 cm in diameter containing $\sim$ 220 mg ^{235}U and ^{239}Pu, and 440 mg of ^{238}U. A liquid scintillator based neutron detector was positioned approximately 3 m downstream of the neutron source as an additional neutron flux monitor to measure any neutron flux fluctuations. Auxiliary measurements were performed with pulsed proton or deuteron beams to obtain the neutron energy spectrum.

The absolute efficiency of the chambers were found to be 100 $\pm$ 2 % using a ^{252}Cf source which is in very good agreement with Ref.[17] After irradiation, the γ-ray activity produced in the thick activation target was measured continuously for 2-3 months with well shielded and calibrated high-purity germanium (HPGe) detectors with very accurately measured efficiencies. The fission products were identified from their characteristic γ-ray energies and half-lives, and their FPYs were evaluated from observed photopeak areas and dual-fission chamber counts using (Eq. (1)),

$$Y_i = \frac{N_i}{N_f}, \tag{1}$$

where N_i is the number of nuclei produced during activation as determined by γ-ray counting, and N_f is the total number of fissions in the activation target determined using a weighted average of the total fissions from each fission chamber and mass scaled to the thick activation target[13]. The advantage of using the DFC method is that the fission chamber determines the total number of fissions in the target without having to explicitly know either the fission cross section or the neutron flux, which greatly reduces the total uncertainty of the measurements.

FPY measurements have been made on the three isotopes of interest at eight incident neutron energies of 0.58, 1.37, 2.37, 3.6, 4.6, 5.5, 8.9 and 14.8 MeV. Typically, irradiation times ranged from 4-6 days depending on neutron flux and fission cross section. For each energy, approximately 16 cumulative FPYs were determined, nearly all being high-yield fission products, i.e., those occurring in the peaks of the mass distributions.

3. Results

The high-accuracy of the FPY measurements performed over an energy range from 0.58 to 14.8 MeV unveil a peculiar energy dependence. At low neutron energies the FPYs for asymmetric fission of ^{239}Pu and ^{238}U show a slight yield increase with increasing neutron energy from 0.5 to approximately 4-5 MeV. At $E_n > 5$ MeV the slope becomes negative for all three ^{235}U, ^{238}U, and ^{239}Pu fissile actinides. This is an important finding because it demonstrates that the high-yield fission products can have a dependence on incident neutron energy and hence have a significant impact on deduced fission yields. As an example, the experimental results from six high-yield fission fragments such as ^{95,97}Zr, ^{99}Mo, ^{140}Ba, ^{143}Ce, and ^{147}Nd from neutron induced fission of ^{239}Pu, ^{235}U, and ^{238}U are depicted in Fig. 2. As can been seen, the FPY of ^{147}Nd (bottom right panel) shows a positive energy trend in the $E_n < 5$ MeV region. This trend is almost linear up to about 4 MeV. The relative slope of $Y(^{147}Nd)/E_n = (5.6 \pm 1.6)\%/MeV$ measured

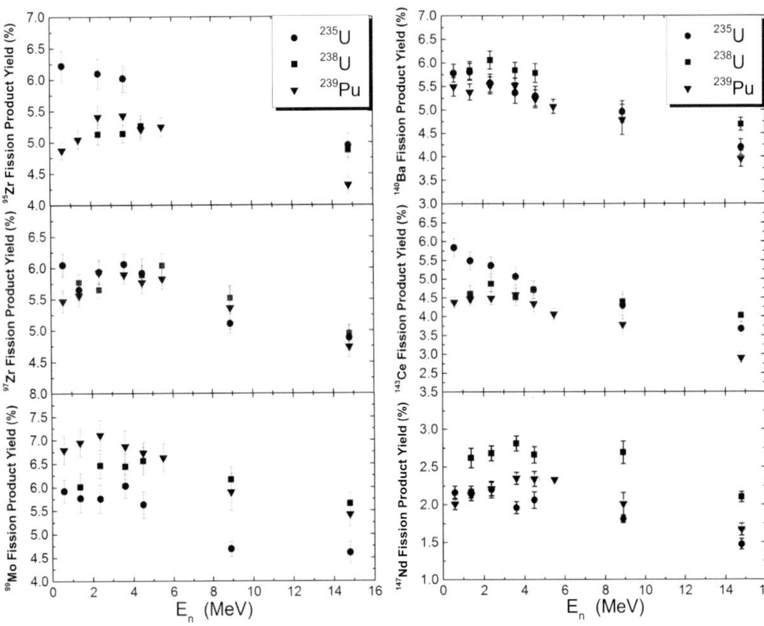

Fig. 2. Fission product yield of ^{95}Zr (left-top), ^{97}Zr (left-center), ^{99}Mo (left-bottom), ^{140}Ba (right-top), ^{143}Ce (right-center), and ^{147}Nd (right-bottom) from the fission of ^{235}U (circle), ^{238}U(squares), and ^{239}Pu (triangles) in the 0.5 MeV to 15 MeV neutron energy range.

from 0.58 to 3.6 MeV confirms the earlier assessment from critical assembly measurements for fission neutrons below $E_n = 2$ MeV[4,18]. At higher energies the slope of the FPY for ^{147}Nd turns over and becomes negative with relative slope of Y(^{147}Nd)/E_n = (-2.8 $\pm$ 0.7)%/MeV measured from 4.6 to 14.8 MeV. Compared to 4.6 MeV, the FPY of ^{147}Nd at E_n=14.8 MeV has decreased by 71%. The real surprise is that the positive slope is observed for all other high-yield fission products from 0.58 to 3.6 MeV in ^{239}Pu(n,f) and ^{238}U(n,f). The strongest overall energy dependence was seen in neutron induced fission of ^{239}Pu, however as been mentioned above, similar positive slopes were obtained from the study of ^{238}U as well.

In the case of ^{235}U(n,f) the energy dependency of the FPYs show constant or slightly negative slope from 0.5 to 5 MeV. The only common feature of all three fissile isotopes is the decrease of the FPY's with higher incident neutron energy ($E_n > 5$ MeV) , which is probably due to an increased symmetric fission component.

The mass-chain-yield data in the neutron-induced fission of ^{239}Pu at 1.37 and 14.8 MeV obtained from the present data and those at 1.5 and 14.0 MeV from the evaluated cumulative FPYs[21] are shown in Fig. 3. The associated figures for ^{235}U, ^{238}U, and ^{239}Pu are very similar, i.e., the fragments are formed in the well-known two-humped yield distribution, favoring an unequal division of masses. The mass-yield data show a slight increase of fission yields in the near symmetric mass region (valley) with increasing incident neutron energy in the 1.5 to 14 MeV range with only small changes in yield in other regions of the mass distribution. The increase of the symmetric FPY is accompanied by a decrease of the FPY for asymmetric fission, resulting in a substantial reduction in the peak-to-valley mass ratio with increasing neutron energy.

4. Conclusions

The unexpected trend of the FPY cannot be readily reproduced by existing, data-based phenomenological models. An even larger challenge for all fundamental theory-based models of nuclear fission is the positive slope found for high-yield fission products from neutron induced fission on ^{239}Pu and to a leser extend on ^{238}U. This is contrary to the low-energy slop of the same FPYs from ^{235}U, suggesting different interplay between the pairing and shell effects in these fissile nuclei at low-excitation energies[20]. Therefore, the new results presented here have important applied interest for yield extraction[22]. The FPY increases observed at low energies are also providing

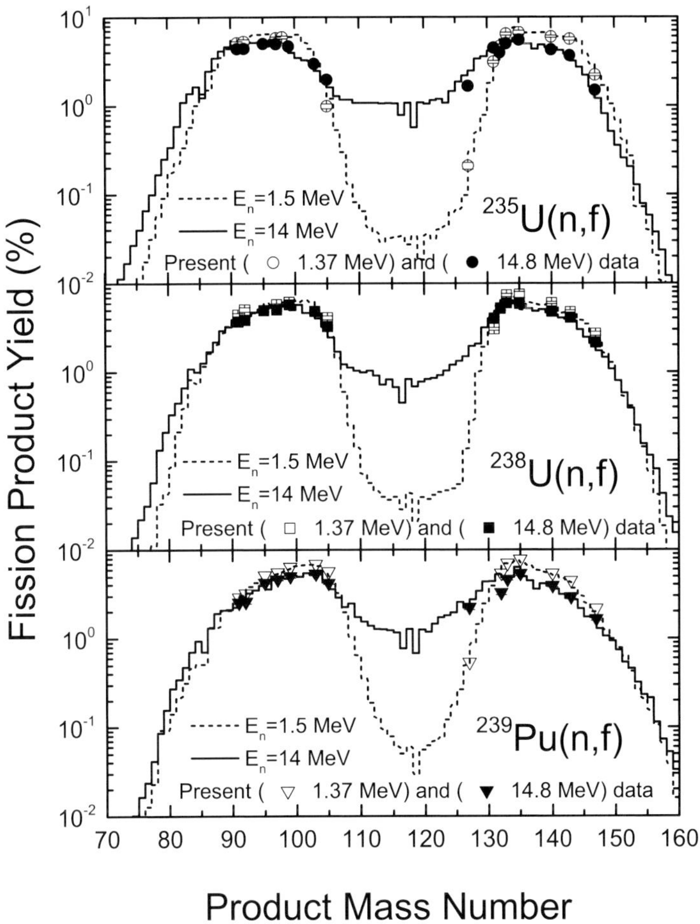

Fig. 3. Symbols: yields of fission products (%) as a function of mass number from neutron induced fission of ^{235}U (top), ^{238}U (central), and ^{239}Pu (bottom) at 1.37 and 14.8 MeV. Curves: evaluated fission product yield distribution from fast (1.5 MeV) and high-energy (14.0 MeV) neutrons is shown in continuous lines[21].

valuable new information to constrain fundamental science-based models of nuclei in general, and fission in particular. This study provides a comprehensive information of the FPY from the three major actinides over a wide range of fast neutron energies. Our intention is to further extend our measurements into the second chance fission where the energy dependency might be influenced by a complicated interaction between nuclear shape and temperature.

At LANSCE neutrons are generated via spallation off of a tungsten target at the end of an 800 MeV LINAC. At the Weapons Neutron Research Facility (WNR), the neutrons are used on various flight paths for a wide range of nuclear research. On the flight path 15° to the left of the target (15L), sits the Chi-Nu experiment. Chi-Nu is an experiment designed to be a high-precision, well-documented measurement of the fast-neutron induced prompt fission neutron spectra for the two major actinides: ^{239}Pu and ^{235}U. The focus of the results presented here is $^{235}\mathrm{U}(n, f)$. Chi-Nu is a joint effort by Los Alamos National Laboratory (LANL), and Lawerence Livermore National Laboratory (LLNL). Previous descriptions of the Chi-Nu experiment and its progress have been reported in various publications, see Refs. 6–10.

1.1. *Experimental Setup*

A Parallel-Plate Avalanche Counter (PPAC) housing $\approx$ 100 mg of ^{235}U was fabricated at LLNL and described in Ref. 11 to identify fission events and provide timing information. The PPAC is made of 10 plates that are perpendicular to the direction of the beam. Each plate actually consists of several layers, the first and last layer are platinum anodes, and in the middle of the stack that makes up each plate is the actinide sample. The PPAC is positioned in the center of the flight path cave, 21.5 m away from the WNR neutron production target and 106.7 cm above an 18 ft $\times$ 18 ft, thin aluminum floor. Incident neutron energies are determined on an event-by-event basis from the time of flight (T.O.F) for a neutron between the production target and the PPAC. Detection of a fission fragment in a PPAC cell also provides a start signal for the TOF measurement of the outgoing fission neutrons. Those neutrons are then detected in one of two arrays: an array of 22-Lithium Glass scintillation detectors (LiGl), 21 of which are doped with ^{6}Li, is used for measuring the low energy region of the PFNS (10 keV – 2.5 MeV), and an array of 54-liquid organic scintillators for the high energy region (500 keV – 15 MeV).

A significant source of background for measuring the prompt fission neutron spectrum in WNR is down-scattered neutrons, the shielding walls are placed a minimum of 3 m away from the arrays. In addition, beneath the thin aluminum floor there is a 2.1 m deep pit. To quantify the amount of room return and down-scattered neutrons seen by the Chi-Nu experiment, extensive modeling was done using the 'MCNPX-PoliMi' code which is a modified version of MCNPX the results of the modeling can be found in Ref. 12.

2. Data Analysis

2.1. ^{6}*Li Glass Array Data*

The data for ^{235}U were taken during the 2015 run cycle. A two-dimensional cut selecting fission neutrons is shown in Fig. 1, which is a histogram showing the outgoing T.O.F of fission neutrons and γ rays against the integral of the charge deposited in the LiGl detectors.

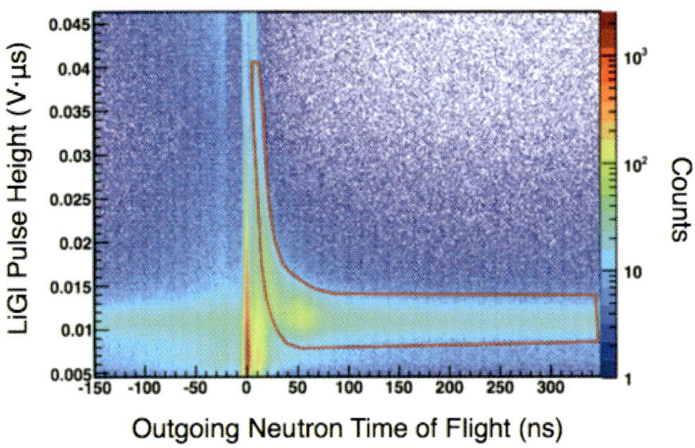

Fig. 1. A histogram showing TOF for the outgoing fission products versus the pulse height in the LiGl detector.The red polygon is a 2D cut attempting to select only the outgoing fission neutrons.

In this histogram, one can see the band in the x-direction at 0.01 Vμs, that is the Q-value band for the ^{6}Li$(n,\alpha)t$ reaction, with the resonance at 240 keV showing up as a more pronounced area on the band. One also notices that at $t=0$ we see a vertical band corresponding to fission γ rays. The data that remain after the cut are binned not just by their outgoing neutron energy, but also by the energy of the neutron that was incident on the ^{235}U. Such a histogram can be seen in the upper panel Fig. 2. Using a robust algorithm for measuring the random coincidence background, described in Ref. 13, the Chi-Nu background is well described and shown in the lower panel of Fig. 2.

2.1.1. *Low-Energy Results*

Using both panels of Fig. 2, one can project the data and the measured background into any arbitrary incident and outgoing bins a user desires.

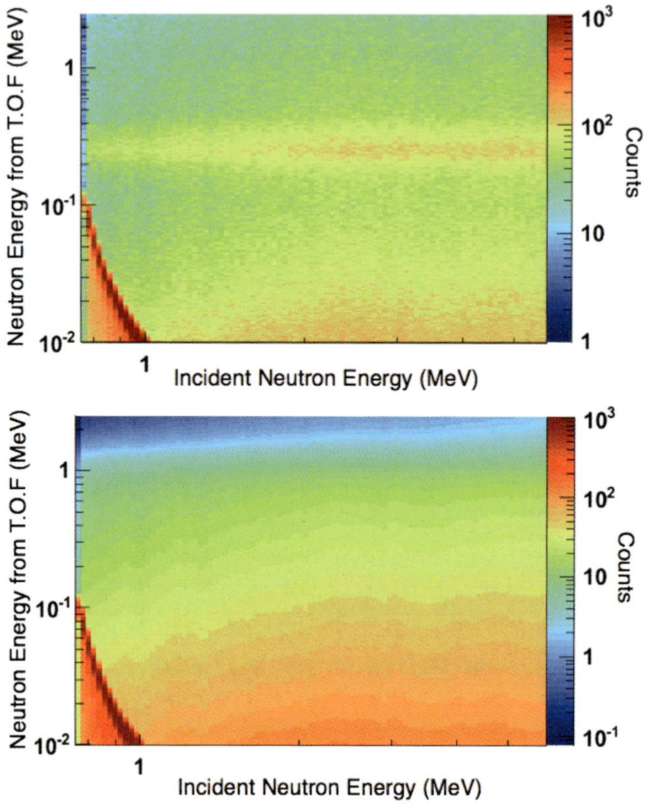

Fig. 2. Top Panel: A χ-matrix histogram for Chi-Nu data as was taken. The x-axis is incoming neutron energy, and the y-axis is the outgoing neutron energy. Bottom Panel: A χ-matrix histogram for the measured random background created using the algorithm in Ref. 13.

Combining the one dimensional projections of the foreground and the background for an incident neutron energy of (0.7 - 5.0 MeV), we see the result show in Fig. 3.

The PFNS presented in Fig. 4 is made by combining the data seen in Fig. 3 and the corresponding output from a detailed MCNP model.

2.2. *Liquid Scintillator Array Data*

Data were also taken on the PFNS of neutron-induced fission of ^{235}U with the liquid organic scintillator array, for outgoing neutron energies above 0.5 MeV. The analysis of these data is ongoing. These detectors use pulse-shape

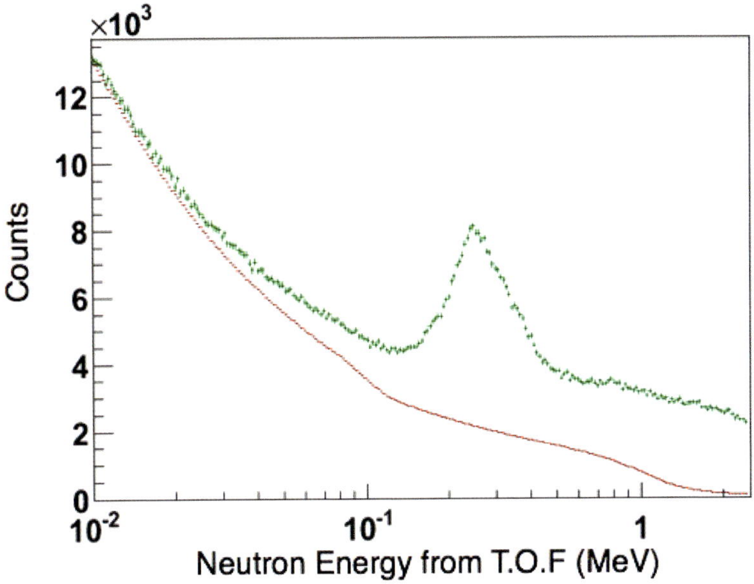

Fig. 3. One dimensional slice of χ-matrix histograms for the incident neutron energy range of (0.7 - 5.0 MeV). Measured total signal is shown by the black line, while measured random background is shown in red.

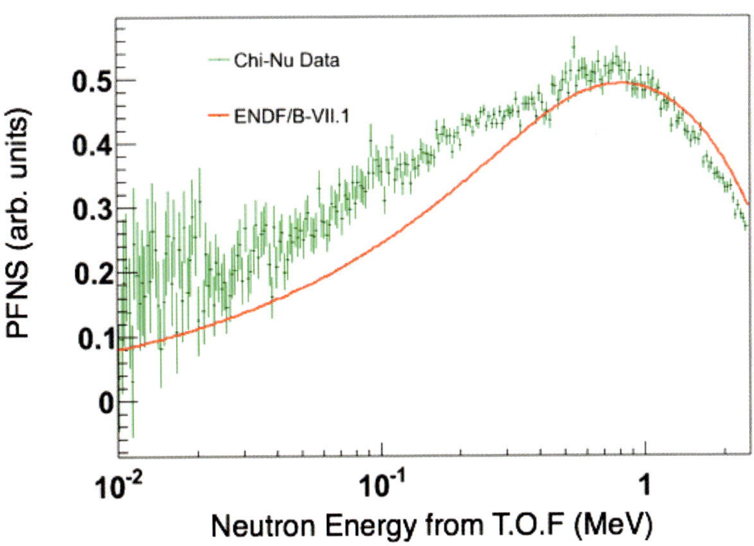

Fig. 4. Low energy portion of the PFNS for ^{235}U, the red curve is the PFNS from ENDF/B-VII.1. Both the data and ENDF are normalized over the plotted range.

discrimination to separate neutron detection from γ-ray detection. Like the LiGl array, the response of the array is being modeled extensively with a high-accuracy MCNP model, and the same background determination methods are being used. The results of both the low energy and high-energy arrays for the ^{235}U(n, f) are expected to be presented in the near future.

Acknowledgements

This work was performed under the auspices of the U.S. Department of Energy by Los Alamos National Laboratory under Contract DE-AC52-06NA25396 and the Lawrence Livermore National Laboratory under Contract DE-AC52-07NA27344. The experiment benefits from the use of neutrons at the Los Alamos Neutron Science Center.

References

1. R. Capote *et al.*, Prompt fission neutron spectra of actinides, *Nuclear Data Sheets* **131**, 1 (2016).
2. D. Neudecker *et al.*, The need for precise and well-documented experimental data on prompt fission neutron spectra from neutron-induced fission of ^{239}pu, *Nuclear Data Sheets* **131**, 289 (2016).
3. P. Staples *et al.*, Prompt fission neutron energy spectra induced by fast neutrons, *Nuclear Physics A* **591**, 41 (1995).
4. S. Noda *et al.*, Prompt fission neutron spectra from fission induced by 1 to 8 mev neutrons on ^{235}U and ^{239}Pu using the double time-of-flight technique, *Phys. Rev. C* **83**, p. 034604 (Mar 2011).
5. A. Chatillon *et al.*, Measurement of prompt neutron spectra from the ^{239}pu(n, f) fission reaction for incident neutron energies from 1 to 200 mev, *Phys. Rev. C* **89**, p. 014611 (Jan 2014).
6. R. C. Haight *et al.*, The lanl/llnl prompt fission neutron spectrum program at lansce and approach to uncertainties, *Nuclear Data Sheets* **123**, 130 (2015).
7. B. A. Perdue *et al.*, Development of an array of liquid scintillators to measure the prompt fission neutron spectrum at lansce, *Nuclear Data Sheets* **119**, 371 (2014).
8. R. C. Haight *et al.*, The prompt fission neutron spectrum (pfns) measurement program at lansce, *Nuclear Data Sheets* **119**, 205 (2014).

9. H. Y. Lee *et al.*, Prompt fission neutron spectrum study at lansce: Chi-nu project, in *Fission and Properties of Neutron-Rich Nuclei*, eds. J. H. Hamilton and A. V. Ramayya (World Scientific, 2013).
10. B. A. Perdue *et al.*, Development of neutron detector arrays for neutron-induced reaction measurements, *IEEE Transactions on Nuclear Science* **60**, 879 (2013).
11. C. Y. Wu *et al.*, A multiple parallel-plate avalanche counter for fission-fragment detection, *Nuclear Instruments and Methods in Physics Research Section A: Accelerators, Spectrometers, Detectors and Associated Equipment* **794**, 76 (2015).
12. T. N. Taddeucci *et al.*, Multiple-scattering corrections to measurements of the prompt fission neutron spectrum, *Nuclear Data Sheets* **123**, 135 (2015).
13. J. M. O'Donnell, A new method to reduce the statistical and systematic uncertainty of chance coincidence backgrounds measured with waveform digitizers, *Nuclear Instruments and Methods in Physics Research Section A* **805**, 87 (2016).

Measurement of Angular-Momentum-Dependent Fission Probabilities of ^{240}Pu

Johnathon Koglin[1,2], Jason Burke[1] and Igor Jovanovic[3]

[1] *Lawrence Livermore National Laboratory, 7000 East Ave, Livermore, CA 94550, United States*
[2] *Penn State University, 201 Old Main, University Park, PA 16802, United States*
[3] *University of Michigan, 500 S. State St, Ann Arbor, MI 49105, United States*

An experimental technique using the surrogate reaction method has been developed to measure fission probabilities of actinides as a function of the angular momentum state of the fissioning nucleus near the fission barrier. In this work, the ^{240}Pu$(\alpha, \alpha'\mathrm{f})$ reaction was used as a surrogate for ^{239}Pu(n,f). An array of 12 silicon telescopes positioned at 10° intervals from 40° to 140° detect the outgoing reaction particle for identification and excitation energy measurement. The angular momentum state of the transition nucleus is determined from the angular distribution of fission fragments. An array of 50 photovoltaic (solar) cells detects fission fragments with 10° granularity while being insensitive to light ions. Relative contributions from different angular momentum states are extracted from the measured distributions and compared across all α particle scattering angles to determine fission probability for a specific angular momentum state. The first experiment using this technique was recently completed using 36 MeV α particles incident on ^{240}Pu. First results of fission probabilities and angular distributions will be discussed. Recent efforts of ^{239}Pu(n,2n) cross section measurements using the surrogate technique with the NeutronSTARS array will also be discussed.

Toward Measuring Prompt Fission Products in Coincidence

W. A. Peters

Joint Institute for Nuclear Physics and Applications
at Oak Ridge National Laboratory
Oak Ridge, TN 37831, USA
E-mail: wapeters@nuclearemail.org

M. S. Smith and A. Galindo-Uribarri

Oak Ridge National Laboratory
Oak Ridge, TN 37831, USA

S. V. Paulauskas, K. Smith, C. Thornsberry, K. L. Jones and R. Grzywacz

University of Tennessee
Knoxville, TN 37996, USA

J. A. Cizewski and D. Walter

Rutgers University
New Brunswick, NJ 08903, USA

E. Temanson

University of Wisconsin-La Crosse
La Crosse, WI 54601, USA

Despite the discovery of fission nearly 80 years ago and its importance to nuclear energy, national security, and astrophysics, there are very few measurements of the correlations between multiple fission products. An experiment is underway at Oak Ridge National Laboratory to measure the energy and angle correlations between prompt fission neutrons, γ rays, and fragments. These measurements could reveal the existence of scission neutrons, the energy balance between these products, and other details needed for benchmarking advanced fission models. The setup includes an array of VANDLE neutron detectors and a γ-ray detector positioned opposite a suite of detectors to determine the mass of one fragment. Preliminary results with a spontaneous ^{252}Cf source show expected correlations between prompt fragments and neutrons along the fission axis. Further refinements will improve the resolution of the fragment mass and increase efficiency for triple-coincident events. Future experiments could be done with a neutron beam to study induced fission on thin targets.

Keywords: Fission, Prompt Neutron Correlations.

1. Introduction

Nuclear fission has been studied for about 80 years, yet remains poorly understood despite being important in the fields of nuclear energy, nuclear science, national security, and astrophysics. Fission occurs when certain heavy nuclei form an elongated shape and split at a scission point breaking into two, usually unequal, fragments. This can occur from the spontaneous decay of the nucleus, or can be induced by exciting a fissile nucleus through particle scattering, transfer reactions, or neutron capture.

To date, the fission-fragment mass distributions for many nuclei have been measured as a function of induced energy, along with separate measurements of neutron and γ-ray distributions[1]. The newly-developed ChiNu[2] detector system at Los Alamos National Laboratory (LANL) is designed to measure, with high accuracy, the neutron distribution, while the SPIDER[3] system at LANL is designed to measure fragment distributions. However, important correlations that expose details of the fission process and the balance of energy between the products cannot be determined when measuring only the average distribution of one fission product at a time. Future advances in modeling and comparisons with theory will depend on studies where multiple fission products are measured in time coincidence. While the neutron multiplicity has been measured as a function of the fragment mass[4], there is very limited data of coincidence measurements of multiple fission components[5,6], and no published work of photon-neutron-fragment correlations. The current experiment aims for a more complete characterization of fission events by measuring prompt fission neutrons, γ rays, and fragments in time-coincidence. This effort will lead to further studies that could completely characterize fission products as a function of fission-fragment mass and induced energy. Such a measurement could impact many areas of applied and fundamental nuclear physics.

2. Experimental Setup

To measure all the fission products in coincidence, our setup centers around a spherical vacuum chamber containing a spontaneous fission source. In the center of the spherical chamber, a ^{252}Cf source is mounted. A time-of-flight fragment identification arm (TOF arm) is attached to the chamber that contains the fragment detectors and defines the fission axis.

The TOF arm contains two transmission charged-particle detectors for timing and a final total-remaining-energy detector. This suite of detectors enables determination of a fragment's mass (M) via the "$2t1E$" method

Fig. 1. Photograph of fission correlation setup. The spherical chamber is connected to the TOF arm through a gate valve and has two linear drives holding the MCP assemblies. VANDLE modules surround the source chamber at a radius of 54.5 cm covering -5 to 100 degrees from the fission axis defined by the TOF arm.

($M = 2E(t^2/l^2)$), where E is the fragment energy from the charged particle detector in the TOF arm and t is the measured flight time over the path length l.

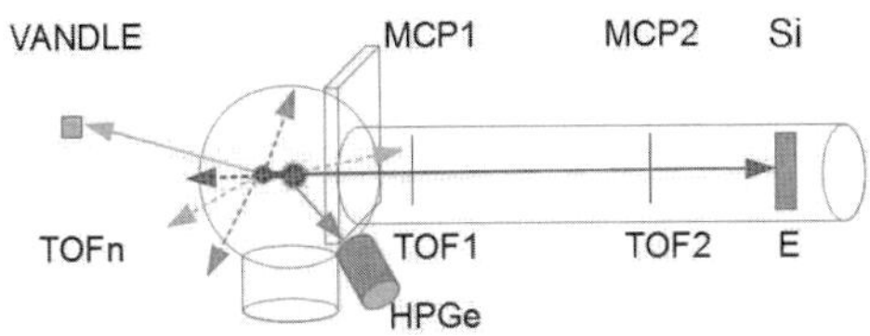

Fig. 2. Schematic of fission correlation with two MCP timing detectors and one Si total energy detector in the TOF arm. There is also a representation of a VANDLE module and a HPGe detector for measuring coincident neutrons and γ rays, respectively.

The VANDLE system (Versatile Array of Neutron Detectors at Low Energy)[7] was used for measuring the angle and energy of the neutrons via the time-of-flight method. An array of VANDLE modules covering more than 90° around the fission chamber measured coincident neutrons opposite the TOF arm. A single HPGe photon detector will be installed opposite the neutron detector array, to recored triple coincidence events of prompt gammas and neutrons and fragments. The current setup uses a 0.6 mCi ^{252}Cf source. Even though the expected triple coincidence rate is small, our statistics is limited only by our run time.

3. Preliminary Data

After 14 days of data we recorded over 100,000 neutron-fragment coincidence events. Figure 3 shows the calculated mass (in arbitrary units) versus the neutron TOF. The neutron peak is clearly shifted toward lower TOF for the heavy fragments. This is because when the heavier fragments are detected in the TOF arm, the lighter fragments are ejected in the opposite direction toward VANDLE. Since the fission energy is shared by the two fragments, the lighter fragment is moving faster in the lab frame. Thus, the prompt neutrons toward VANDLE, released by the lighter fragment, are boosted in the lab frame. The opposite is true when the lighter fragment is detected in the TOF arm. There is also a slight shift in the sharp peak associated with the prompt γ rays (near TOF channel 1000). This is due to small differences in the energy lost going through the MCP foils that is not taken into account when calculating the time of fission using the two MCP timing signals.

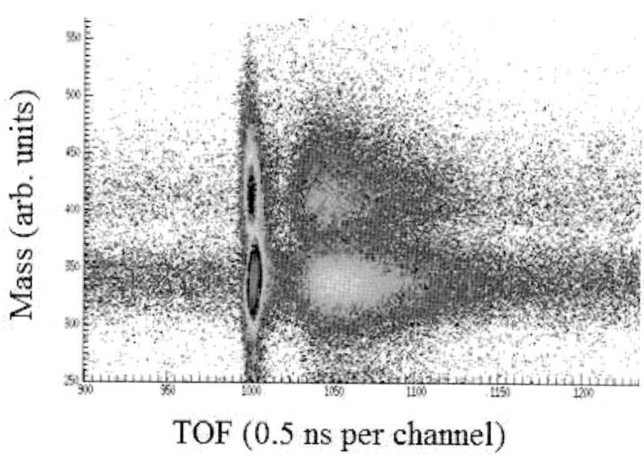

Fig. 3. Preliminary results of mass versus TOF of neutrons.

The differences of the neutron-energy spectra between the lighter and heavier fragment coincidences is amplified by gating on just the forward VANDLE modules. Figure 4 shows the TOF data for the VANDLE modules closest to the fission axis (defined by the TOF arm). The peak of the neutron TOF spectra changes by 3 ns, corresponding to 2.6 MeV and 3.4 MeV for coincidences with the lighter and heavier fragments, respectively. In the fragment reference frame, these neutron peaks correspond to 0.75 MeV for both fragments if boosted by 1.35 and 1.05 cm/ns for the lighter and heavier fragments, respectively. These two spectra are just two

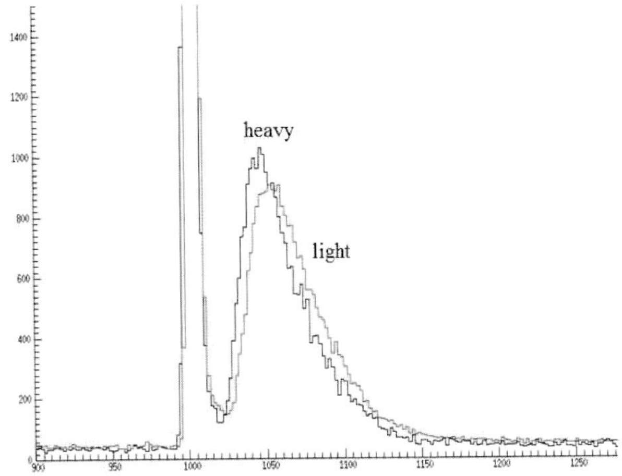

Fig. 4. Neutron TOF spectra on fission axis in coincidence with lighter or heavier fission fragments. The magnitude of the TOF spectrum in coincidence with the heavier fragments was normalized to the other for easier comparison.

subsets of data that are not discernible with a standard fission-neutron experiment where there is no correlated information about the fragment mass or angle.

4. Discussion and Future Plans

Analysis is ongoing to study differences in the neutron TOF at all angles and studying correlations of multiple neutrons as a function of fragment mass. After incorporating γ-ray detectors to confirm triple coincidence capabilities, there are a few upgrades that would be necessary for a complete characterization of all fission products with the desired resolution. An unmeasured amount of energy is lost through the "dead" layer of the final silicon detector and the thin nickel window on the source, in addition to the thin silicon-nitride MCP foils. This restricts our fragment mass resolution. Efforts to reduce the overall amount of material that the fragments must travel through can be done by reducing the number of windows or making them thinner.

In conclusion, this approach has the potential to provide needed data on triple-coincidence events to fully characterize relative changes to the energy, angle, and multiplicity of these fission products. With future upgrades, and installing the setup at a beam facility, these characterizations could be done for many fission targets and also investigate changes as a function of induced fission energy. The setup could also operate as a decay station for fissioning nuclei created by a radioactive ion beam facility like the future Facility for Rare Isotope Beams (FRIB) being built at Michigan State University[8].

Acknowledgments

This research is sponsored by the Laboratory Directed Research and Development Program of Oak Ridge National Laboratory, managed by UT-Battelle LLC for the US Department of Energy. This work was also supported in part by the U.S. Department of Energy under Contracts No. DE-NA0002132 (Rutgers,UTK), No. DE-AC05-00OR22725 (ORNL), No. DE-FG02-96ER40983 and No. DE-SC0001174 (UTK) and by the National Nuclear Security Administration. The authors wish to thank S.D. Mathews, D. Stracener, and the ORNL Radiation-Safety personnel for assisting in preparing the ^{252}Cf source.

References

1. E. Kwan *et al.*, *Nucl. Instr. and Meth. A* **688**, p. 55 (2012).
2. B. A. Perdue *et al.*, *IEEE Trans. on Nuc. Sci.* **60**, p. 879 (2013).
3. F. Tovesson *et al.*, *EPJ Web of Conf.* **62**, p. 055002 (2013).
4. A. Gook, F.-J. Hambsch and M. Vidali, *Phys. Rev. C* **90**, p. 064611 (2014).
5. H. Rossner *et al.*, *Phys. Rev. C* **45**, p. 719 (1992).
6. H. R. Bowman, J. C. D. Milton, S. G. Thompson and W. J. Swiatecki, *Phys. Rev.* **129**, p. 2133 (1963).
7. W. A. Peters, S. Ilyushkin, M. Madurga, C. Matei, S. V. Paulauskas, R. K. Grzywacz, D. W. Bardayan, C. R. Brune, J. Allen, J. M. Allen, Z. Bergstrom, J. Blackmon, N. T. Brewer, J. A. Cizewski, P. Copp, M. E. Howard, R. Ikeyama, R. L. Kozub, B. Manning, T. N. Massey, M. Matos, E. Merino, P. D. O'Malley, F. Raiola, C. S. Reingold, F. Sarazin, I. Spassova, S. Taylor and D. Walter, Performance of the versatile array of neutron detectors at low energy (vandle), *Nucl. Instr. and Meth. A* **836**, p. 122 (2016).
8. A. Gade, C. K. Gelbke and T. Glasmacher, *Nuclear Physics News* **24**, p. 28 (2014).

Search of Trineutron and Tetraneutron

Roman Ya. Kezerashvili
New York City College of Technology, The City University of New York
Brooklyn, NY 11201, USA
E-mail: rkezerashvili@citytech.cuny.edu

In light of a new experiment[2] which claims an identification of tetraneutron, we discuss the results of experimental search of trineutron and tetraneutron in different nuclear reactions. A summary of theoretical studies for trineutron and tetraneutron within variety of approaches are presented.

Keywords: Trineutron; Tetraneutron; Few-Body System.

1. Introduction

Exotic types of atomic nuclei consisting only of neutrons have been proposed and searched for with great interest for decades and whether the three- and four-neutron system exists or not is a long standing question. The recent enthusiasm in the search for the 4n system started with the experimental claim in Ref. 1 in 2002 that a bound 4n system could be formed in the breakup reaction: ^{14}Be $\rightarrow$ ^{14}Be + ^{4}n. A recently reported observation of the resonant tetraneutron state[2] and some theoretical results revive old questions: do trineutron and tetraneutron nuclei exist? Can a nucleus be made up of neutrons only? The answers to these questions, if such systems exist, would most certainly require a revision of modern realistic models of the nucleon-nucleon force and three-nucleon interaction[3,4] and more over to introduce a four-nucleon interaction[5]. Also it will impact our understanding of the structure and processes in neutron stars, and provide a bridge to describe very neutron-rich nuclei, in which multineutrons would play a major role through the many-neutron interactions that would be required to explain their existence. Below, I give a short summary and present the status of studies of trineutrons and tetraneutrons.

2. Trineutron

A three-neutron resonance or bound 3n system has not yet been firmly established. The weight of early experimental evidence up to 1987 reviewed in Ref. 6 is strongly against the existence of a bound state of the three-neutron system, and only controversial evidence of a three-neutron resonance was cited. Ajdačic[7] reported the possible existence of the trineutron through the reaction ^{3}H$(n,p)^3$n. Later, the same reaction was studied in Refs. 8 and 9 (see references herein) and no evidence for the existence of the trineutron was found. Searches for a bound state of the three neutron were conducted in reactions: ^{3}H$(\pi^-,\gamma)^3$n, ^{3}He$(\pi^-,\pi^+)^3$n, ^{4}He$(\pi^-,p)^3$n, and ^{7}He$(\pi^-,^4$He$)^3$n induced by negative pions, and multinucleon transfer reactions such as ^{7}Li(^{7}Li, ^{11}C)^{3}n, ^{7}Li(^{11}B, ^{15}O)^{3}n and ^{2}H(^{12}C, ^{13}N)^{3}n.

The most intensive search for a bound trineutron has been performed using a pionic double charge exchange (DCX) reaction ^{3}He$(\pi^-,\pi^+)^3$n. Early study by Sperrinde *et al.*[10] pointed to resonance in the three neutrons. The DCX process on ^{3}He was also investigated in Ref. 11, which while criticizing Ref. 10 pointed again to a three-neutron resonance around 12 MeV excitation. Without the need to invoke the 3n resonance state, the observed behavior of the cross section for this process can be explained by the final state interaction of three neutrons in continuum spectrum[12,13]. Interestingly, a calculation[14] predicts a resonance state with the width of 13 MeV in the three-neutron system. Although such a resonance would easily fit early interpretation of data on negative pion DCX on ^{3}He[10] and measurements[11], the experimental studies[15,16] of the process ^{3}He(π^-,π^+)3n found no experimental evidence of the existence of the ^{3}n or resonance state of three neutrons. To summarize, although the DCX reaction of negative pions on ^{3}He nucleus has been examined at various incident energies of pion, from the analysis of the invariant mass spectra for three neutrons, no evidence for the bound trineutron has been found.

All reactions mentioned above are plagued by the fact that although there is an expectation that a ^{3}n system is produced, at least one additional strongly interacting particle is in the final state. The latter fact increases interest to the reactions ^{3}H(π^-,γ)3n and ^{3}H(μ^-,ν_μ)3n that have only an extra photon or neutrino in the final state. In Refs. 17 and 18 reported trineutron mass spectrum that was observed in the stopped pion radiative capture reaction ^{3}H(π^-,γ)3n by measuring the photon energy with a high-resolution spectrometer. No indication of a bound trineutron is observed, as also in earlier studies. Theoretically the s-state radiative π-capture or μ-capture transition rates can be calculated quite accurately. The noticeable difference in μ^-–capture by a tritium nucleus obtained in calculations[19,20] by taking into account the final state interaction between three neutrons and a plane-wave approximation increases the interest in the experimental study of the channel of the complete break up of ^{3}H in μ^-−capture.

For a trineutron the bound state has been studied extensively in the last four decades resulting in a numerically precise solution of the Faddeev equations in momentum and coordinate space, and using the hyperspherical functions method. Early Faddeev calculations which predicted a trineutron contradicted the variational calculations which did not. This issue was addressed in Ref. 21 using the Faddeev method with the Reid potential. It was pointed out that the interaction potential must be enhanced by about a factor of 4 to bind ^{3}n near zero energy. The possibility of a trineutron bound state is investigated in Ref. 22 by solving the Faddeev equation in the coordinate space and have considered possible states of the trineutron including even the $^3P_2 -^3 F_2$ force. The author concluded that all states of ^{3}n are unbound for the Reid soft core potential. Independent of the theoretical framework, such as the Faddeev formalism[23,24], the method of the hyperspherical functions or variational calculations, most theoretical works do not predict a bound ^{3}n state in the three-neutron system.

Several more recent experiments have strengthened the evidence against the bound trineutron and have failed to discover a resonance structure that cannot be otherwise explained. As of yet, none of above mentioned reactions have provided evidence for a bound trineutron. Related to the three-neutron resonance authors[23] showed that three-neutron resonance pole trajectories connected to artificially enhanced nn forces could be found. However, the reliable results achieved, make it likely, that three-neutron resonance energies will have large imaginary parts and are physically not interesting. The study[25] shows that realistic nucleon-nucleon interaction models exclude any possible experimental signature of three-neutron resonances.

3. Tetraneutron

The search for the tetraneutron has been going on for more than fifty years. Several attempts have been made to find a bound tetraneutron system by using a uranium fission reaction[26–28] and the experimental observation of ^{4}n was claimed in the interaction of 100 MeV α−particle with uranium nucleus in Ref. 28. Early measurements of the ^{4}He$(\pi^-, \pi^+)^4$n reaction carried out in search of evidence for ^{4}n are summarized in Ref. 29. No bound ^{4}n was detected in these early works. Later the momentum spectrum from the pion DCX reaction was measured in Ref. 30 in a search for ^{4}n. The theoretical study[31] reported that the final-state interaction in the four-neutron system in continuum spectrum is so strong that the tetraneutron could not be observed in the kinematic region explored in Ref. 30. Pion spectra and total cross sections for pion DCX were also measured in Refs. 32–34 for different incident pion energies. The system of four bound neutrons was searched through using ion transfer reactions such as ^{7}Li(^{11}B, ^{14}O)^{4}n[35], ^{7}Li(^{7}Li, ^{10}C)^{4}n[36]. In the new millennium, an experimental search and theoretical study of tetraneutron transitioned to a new phase. In 2002 experimental study[1] reported on events exhibiting characteristics consistent with the detection of a bound tetraneutron. These results have been obtained by studying the breakup reaction of ^{14}Be through the fragmentation channel ^{10}Be + ^{4}n. The lifetime suggested by this measurement would indicate that the tetraneutron is a stable particle. The latter result has been reappraised in 2005[1] to encompass the possibility that the ^{4}n system might exist as a low-energy resonance as was suggested in Ref. 3. The pion DCX reaction as well as the ion transfer reactions, require considerable reconfiguration of the target nuclei and formation of ^{4}n should be strongly suppressed. In contrast, the nucleus ^{14}Be consists of a strongly bound core ^{10}Be and four weakly bound neutrons that could form a tetraneutron-like configuration, which might be shaken off in the ^{14}Be breakup. A candidate resonant tetraneutron state with the energy of 0.83$\pm$ 0.65(stat)$\pm$1.25(syst) MeV above the threshold of four-neutron decay has been found in the missing-mass spectrum obtained in the DCX ^{4}He(^{8}He,^{8}Be) at 186 MeV/u [2]. Motivated by the recent observation of tetraneutron candidates at RIKEN[2], there is a proposal[37] to investigate high energy forward scattering pion DCX reaction on ^{4}He at J-PARC. As opposed to the old experiments at LAMPF and TRIUMF, a rather high-energy pion beam, around 850 MeV, will be utilized with the benefit of reduction of the irrelevant background.

Several theoretical studies of pion DCX on ^{4}He have been reported. In Ref. 38 cross sections were calculated in a model in which two single charge exchange scatterings occur. This reaction was studied in the framework of a four-body hyperspherical functions method in Refs. 39 and 13 and existing experimental data were interpreted without bound or resonance state of four neutrons. No bound tetraneutron was found in Ref. 40 within the angular potential functions method, in Ref. 41 using the stochastic variational method and in Refs. 42 and 43 within the hyperspherical functions approach. In Ref. 43 using a local S-wave nn-potential possible bound and resonant states of three and four neutrons systems are sought as zeros of three- and four-body Jost functions in the complex momentum plane. It is found that zeros closest to the origin correspond to subthreshold $(3n)\frac{1}{2}^-$ and $(4n)0^+$ resonant states. The positions of these zeros are sensitive to the choice of the nn-potential. In contrast, calculations within the hyperspherical functions method led the authors[44] to the conclusion that the tetraneutron may exist as a resonance only for the NN potential that binds the dineutron.

The possible existence of the bound four neutron system was also discussed in the new millennium in theoretical studies by numerous authors, e.g., Refs. 5,45–49. In Ref. 46 it was proposed that, if tetraneutron existed, it could be formed by a bound state of a dineutron-dineutron molecule. The possibility for a tetraneutron to exist as a low-energy resonance state was studied in Ref. 49. In Ref. 45 the hyperspherical functions method have been used to argue against the existence of a tetraneutron. It was pointed out that due to the small probability for a pair of neutrons to be in the singlet even state, the two-body nuclear force cannot by itself bind four neutrons, even if it could bind a dineutron. In Ref. 5 Faddeev-Yakubovsky equations have been solved in configuration space using a realistic Reid 93 nn interaction to explore the possible existence of four-neutron resonances close to the physical energy region. The authors used two methods, namely, complex scaling and analytical continuation in the coupling constant to follow the resonance pole trajectories, which emerge out of artificially bound tetraneutron states. Four-neutron states evolving from the forced bound state end up in the unphysical third energy quadrant with negative real energy parts, thus having little effect on any physical process. It was demonstrated that the tetraneutron physics is entirely determined by nn S waves, namely $^1S_0$ one, which is controlled by the experimentally measurable $nn-$scattering length[5].

An unrealistic modification of the nucleon-nucleon force, three-nucleon force or introduction of unrealistic four-nucleon forces would be needed to bind a tetraneutron[3–5,45]. It is also important to mention that Ref. 3 shows that it does not seem possible to change any modern two- and three-nucleon interaction to bind a tetraneutron without destroying many other successful predictions of these interactions. This means that, should recent experimental claims[1,2] of a resonant tetraneutron state be confirmed, our understanding of nuclear forces will have to be significantly changed. Recently, in Ref. 4 a possibility of generating a four-neutron resonance with introducing of a phenomenological $T = 3/2$ three-neutron force in addition to a realistic NN interaction (Argonne AV8$'$ version of the NN

potential) is considered. Using these interactions the *ab initio* solution of the $4n$ Schrödinger equation is obtained using the complex scaling method with boundary conditions appropriate to the four-body resonances by inquiring what should be the strength of the three-neutron force in order to generate such a resonance. Authors[4] considered the most natural way to enhance a tetraneutron system near the threshold through an additional attractive isospin $T = 3/2$ term in the three-body force, but this modification was found to be inconsistent with other well-established nuclear properties and low energy scattering data. Therefore, it was demonstrated that a four-neutron resonance requires unrealistic changes in the three-neutron force to have observable resonance features. However, recent theoretical calculation[50] within various *ab initio* approaches with nucleon-nucleon interaction JISP16[51] predicts a low energy ^{4}n resonance near 0.8 MeV above threshold with a width of about 1.4 MeV. The calculation is performed without altering any of the properties of the realistic JISP16[51] NN interaction. The reported result is compatible with the recent experimental finding[2] for a resonant structure.

4. Conclusions

Today there is no unambiguous answer for the existence of the trineutron as a bound or resonance state. There are three claims of experimental observation of the ^{4}n or the resonance: one in a fission reaction[28], and the recent two observations of the resonant tetraneutron state in the breakup reaction of ^{14}Be into ^{10}Be and ^{4}n[1], and in the double-charge-exchange reaction ^{4}He(^{8}He,^{8}Be)[2]. However, in theoretical studies, except very recent one[50] with the JISP16[51] potential, no evidence for ^{4}n was obtained within the existing modern two- and three-nucleon interaction. If the experimental discoveries[1,2] will be confirmed, this will call into question current theoretical models of nuclear forces.

The negative result of numerous searches for trineutron and tetraneutron does not exclude the existence of heavier neutron clusters. There are experimental claims[52,53] of observations of multineutron clusters, most likely ^{6}n in the fission of ^{238}U nuclei. Theoretical investigations[45] of multineutrons have been carried out for a system of six, eight and ten bound neutrons using the hyperspherical functions method. Results of calculations show no bound states for these multineutron systems.

It should be interesting to study the DCX reaction of π^- on ^{6}He or ^{8}He. These halo nuclei have a strongly bound core ^{4}He and two or four weakly bound neutrons, respectively. The negative pion DCX reaction on these nuclei may exhibit different channels for final states:

$$\begin{aligned} \pi^- + {}^6\text{He} &\to \pi^+ + 6n \ (\text{or } {}^6\text{n}); \\ &\to \pi^+ + {}^3\text{n} + {}^3\text{n}; \\ &\to \pi^+ + {}^3\text{n} + 3\text{n}; \\ &\to \pi^+ + {}^4\text{n} + 2\text{n}. \end{aligned} \qquad \begin{aligned} \pi^- + {}^8\text{He} &\to \pi^+ + 8n \ (\text{or } {}^8\text{n}); \\ &\to \pi^+ + {}^4\text{n} + {}^4\text{n}; \\ &\to \pi^+ + {}^4\text{n} + 4\text{n}; \\ &\to \pi^+ + {}^6\text{n} + 2\text{n}. \end{aligned}$$

It seems that the most preferable kinematics should be a forward scattering DCX reaction and any observed channel can be interesting.

References

1. F. M. Marques *et al.*, *Phys. Rev. C* **65**, 0444006 (2002); arXiv:nucl-ex/0504009 (2005).
2. K. Kisamori *et al.*, *Phys Rev. Lett.* **116**, 052501 (2016).
3. S. C. Pieper, *Phys. Rev. Lett.* **90**, 252501 (2003).
4. E. Hiyama, R. Lazauskas, J. Carbonell and M. Kamimura, *Phys. Rev. C* **93**, 044004 (2016).
5. R. Lazauskas and J. Carbonell, *Phys. Rev. C* **72**, 034003 (2005).
6. D. R. Tilley, H. R. Weller and Hasan, *Nucl. Phys. A* **474**, 1 (1987).
7. V. Ajdačic *et al.*, *Phys. Rev. Lett.* **14**, 444 (1965).
8. S. T. Thornton *et al.*, *Phys. Rev. Lett.* **17**, 701 (1966).
9. C. Cernigoi *et al.*, *Nucl. Phys. A* **352**, 343 (1981).
10. J. Sperrinde *et al.*, *Phys. Lett. B* **32**, 185 (1970).
11. A. Stez *et al.*, *Nucl. Phys. A* **457**, 669 (1986).
12. R. I. Dzhibuti and R. Ya. Kezerashvili, *Sov. J. Nucl. Phys.* **39**, 264 (1984).
13. R. I. Jibuti and R. Ya. Kezerashvili, *Nucl. Phys. A* **437**, 687 (1985).
14. A. Csoto, H. Oberhummer and R. Pichler, *Phys. Rev. C* **53**, 1589 (1996).
15. M. Yuly *et al.*, *Phys. Rev. C* **55**, 1848 (1997).
16. J. Grater *et al.*, *Eur. Phys. J. A* **4**, 5 (1999); *Phys. Lett. B* **471**, 113 (1999).
17. J. A. Bistirlich *et al.*, *Phys. Pev. Lett.* **36**, 972 (1976).
18. J. P. Miller *et al.*, *Nucl. Phys. A* **343**, 347-355 (1980).
19. R. I. Dzhibuti and R. Ya. Kezerashvili, *Sov. J. Nucl. Phys.* **39**, 700 (1984); R. Ya. Kezerashvili, Preprint IFUP-TH 24/93 (1993).
20. J. Golak *et al.*, *Phys. Rev. C* **94**, 034002 (2016).

21. R. Offermann and W. Glöckle, *Nucl. Phys. A* **318**, 138 (1979).
22. Y. Sunami, *Prog. Theor. Phys.* **63**, 1997 (1980).
23. H. Witala and W. Glöckle, *Phys. Rev. C* **60**, 024002 (1999).
24. A. Hemmdan, W. Glöckle and H. Kamada, *Phys. Rev. C* **66**, 054001 (2002).
25. R. Lazauskas and J. Carbonell, *Phys. Rev. C* **71**, 044004 (2005).
26. J. P. Schiffer and R. Vanderbosch, *Phys. Lett.* **5**, 292 (1963).
27. S. Cierjacks *et al.*, *Phys. Rev.* **137**, B345 (1965).
28. V. A. Ageev *et al.*, *Ukr. J. Phys.* **31**, 1771 (1986).
29. S. Fiarman and W. E. Meyerhof, *Nucl. Phys. A* **206**, 1 (1973).
30. J. E. Urgan *et al.*, *Phys. Lett. B* **144**, 333 (1984).
31. R. Ya. Kezerashvili, *Sov. J. Nucl. Phys.* **44**, 542 (1986).
32. A. Stetz *et al.*, *Phys. Rev. Lett.* **47**, 782 (1981).
33. E. R. Kinney *et al.*, *Phys. Rev. Lett.* **57**, 3152 (1986).
34. T. P. Gorringe *et al.*, *Phys. Rev. C* **40**, 2390 (1989).
35. A. V. Belozyorov *et al.*, *Nucl. Phys. A* **477**, 131 (1988).
36. D. V. Aleksandrov *et al.*, *Yad. Fiz.* **47**, 3 (1988); *JETP Lett.* **81**, 43 (2005).
37. H. Fujioka *et al.*, arXiv:1609.00079 [nucl-ex] (2016).
38. W. R. Gibbs, B. F. Gibson, A. T. Hess and G. J. Stephenson, *Phys. Rev. C* **15**, 1384 (1977).
39. R. I. Jibuti, R. Ya. Kezerashvili and K. I. Sigua, *Sov. J. Nucl. Phys.* **32**, 795 (1980); *Phys. Lett. B* **102**, 381 (1981).
40. A. M. Gorbatov *et al.*, *Yad. Fiz.* **50**, 347 (1989).
41. K. Vagra and Y. Suzuki, *Phys. Rev. C* **52**, 2885 (1995).
42. A. M. Badalyan, T. I. Belova, N. B. Konyuhova and V. D. Efros, *Sov. J. Nucl. Phys.* **41**, 926 (1985).
43. S. A. Sofianos, S. A. Rakityansky and G. P. Vermaak, *J. Phys. G: Nucl. Part. Phys.* **23**, 1619 (1997).
44. I. F. Gutich, A. V. Nesterov and I. P. Okhrimenko, *Yad. Fiz.* **50**, 19 (1989).
45. N. K. Timofeyuk, *J. Phys. G: Nucl. Part. Phys.* **29**, L9 (2003).
46. C. A. Bertulani and V. Zelevinsky, *J. Phys. G* **29**, 2431 (2003).
47. L. V. Grigorenko, N. K. Timofeyuk and M. V. Zhukov, *Eur. Phys. J. A* **19**, 187 (2004).
48. I. V. Simenog, B. E. Grinyuk, Yu. M. Bidasyuk, *Ukr. J. Phys.* **51**, 954 (2006).
49. Y. A. Lashko and G. F. Filippov, *Phys. At. Nucl.* **71**, 209 (2008).
50. A. M. Shirokov *et al.*, *Phys. Rev. Lett.* **117**, 182502 (2016).

51. A. M. Shirokov, J. P. Vary, A. I. Mazur and T. A. Weber, *Phys. Lett. B* **644**, 33 (2007).
52. B. G. Novatskii *et al.*, *JETP Lett.* **96**, 280 (2012); **98**, 656 (2013).
53. V. M. Bystritskiy *et al.*, *Nucl. Inst. Methods Phys. Res. A* **834**, 164 (2016).

New Directions with Transfer Reactions

Kate L. Jones* for the VANDLE, GODDESS, HAGRiD, and CLARION-BareBall Collaborations

Department of Physics and Astronomy, University of Tennessee,
Knoxville, TN 37996, USA
**E-mail: kgrzywac@utk.edu*
http://www.phys.utk.edu/faculty/faculty-jones.html

Over the last two decades transfer reactions have seen a resurgence following developments in methods to use them with exotic beams. An important step in this evolution was the ability to perform the (d,p) reaction on fission fragment beams using the inverse kinematics technique, built on the experience with light beams. There has been renewed interest in using (^{9}Be, ^{8}Be) and (^{13}C, ^{12}C) reactions to selectively populate single-particle like states that can be studied via their subsequent γ decay. These reactions have been successfully utilized in the ^{132}Sn region. Additionally, our collaboration has recently performed experiments with GODDESS, a combination of the full ORRUBA detector and Gammasphere arrays. Another new direction is measuring neutrons from (d,n) reactions, performed in inverse kinematics, with the VANDLE array of plastic scintillators. Presented below is an overview of these new techniques and some of the early data from recent experiments.

Keywords: Transfer Reactions, Direct Reactions, Radioactive Ion Beams.

1. Introduction

As radioactive ion beams became available with energies around or above the Coulomb barrier, techniques were developed to use them with transfer reactions in inverse kinematics with light targets, typically of protons or deuterons. This opened up opportunities for detailed studies of exotic nuclei that were not previously available. This approach was very successful, with many groups studying exotic nuclei from fragmentation, reaccelerated Isotope Separator OnLine (ISOL), and in-flight facilities. Now, in the years leading up to the Facility for Rare Ion Beams (FRIB) online operations, it is a good time to develop new tools and techniques to use at FRIB. This proceeding briefly describes some of the developments made by our collaborations towards tools and techniques that can be used at future facilities.

2. Informing (n,γ) reaction calculations for astrophysics

Transfer reactions are a powerful tool for understanding the structure of the nucleus, through selectively populating certain states of interest, aiding in making spin-parity (J^{π}) assignments, and the extraction of spectroscopic factors, and thereby testing models of the nucleus. This information can also be used in nuclear astrophysics to inform nucleosynthesis simulations.

One application of the information derived from transfer reactions is in providing level information required for calculating direct neutron capture rates for element production. The rapid neutron capture process, commonly termed the r-process, occurs in a neutron-rich environment and creates approximately half of the stable elements heavier than iron[1]. The site of the r-process is still disputed, with the neutrino-driven wind of a core collapse supernova and neutron star mergers as two leading contenders. Having reliable nuclear data is critical to understanding the r-process and to defining the astrophysical site. The process proceeds through many neutron captures on seed nuclei in the nickel-iron region of the chart of nuclides, until the neutron-capture rate is in balance with that for photo-disintegration. The process can continue through β decay, which produces a higher-Z isobar. During the main r-process, a narrow band of isotopes running diagonally through medium-mass and heavy neutron-rich nuclei will be populated owing to an equilibrium between (n,γ) and (γ,n) reactions and the short half lives of these nuclei.

At late times in the r-process, which lasts of the order of 1 second, the (n,γ) (γ,n) equilibrium is broken as the number of free neutrons that are available to induce neutron capture decreases. During this freezeout period, individual neutron-capture rates become important to the final abundances of elements[2]. A sensitivity study[3] showed that the global r-process pattern is sensitive to the ^{130}Sn(n,γ) rate, which not only moves material from ^{130}Sn to ^{131}Sn, but also shifts the production of elements for rare earth and heavier elements. This second effect is due to the large absorption of neutrons in the case that the ^{130}Sn(n,γ) rate is high.

Our collaboration has made a series of studies at the Holifield Radioactive Ion Beam Facility (HRIBF)[4] using the (d,p) reaction on beams of tin isotopes from stable ^{124}Sn through to doubly-magic ^{132}Sn[5–8]. The unstable isotopes were produced as fragments from the proton-induced fission of uranium. Calculations of the direct semidirect (DSD) (n,γ) rate, that is the rate including both the direct capture and the effect of the giant dipole resonance, have been made for the neutron-rich isotopes ^{124,126,128,130,132}Sn[6].

These calculations relied on structure information, including spectroscopic factors, extracted from the (d,p) reactions.

In the particular case of ^{130}Sn, the DSD (n,γ) rate was very poorly known prior to these series of measurements, as the energies of the $3p_{1/2}$ and $3p_{3/2}$ states, which are the dominant contributors to the DSD rate, were unknown, as well as the spectroscopic factors. The HRIBF measurements reduced the uncertainty in this rate by orders of magnitude[7].

3. Measuring γ rays following transfer reactions

3.1. *Incorporating large γ arrays*

In addition to the work using the (d,p) reaction on neutron-rich tin isotopes, our collaboration has also measured γ rays emitted following the (^{9}Be,^{8}Be) and (^{13}C,^{12}C) reactions on short-lived beams of ^{130}Sn and ^{132}Sn[9]. This method has the advantage of providing more precise excitation energies, access to the lifetimes of states, and information from the γ decay cascades. The emitted particles, 2α particles following the breakup of ^{8}Be in the case of (^{9}Be,^{8}Be) and ^{12}C for the (^{13}C,^{12}C) reaction, were detected in the BareBall CsI(Tl) array[10]. Coincident γ rays were measured in the CLARION array of 11 Compton suppressed, segmented, high-purity germanium clover detectors[11]. The Particle IDentification (PID) plot for the ^{130}Sn beam experiment is shown in Figure 1. When both α particles are detected in the same CsI crystal, the combined signal is clearly separated in the PID plot, thus providing a clean tag for population of states in ^{131}Sn. Similarly, by requiring two single α signals, corresponding to events where the α particles are detected in neighboring crystals, a clean gate is available for analysis of the coincident γ rays.

These techniques are complimentary to using the (d,p) reaction, which also transfers a single neutron, but with different selectivity for populating single-particle like states. In particular, the (^{13}C,^{12}C) reaction, compared to (d,p) and (^{9}Be,^{8}Be), preferentially populates states with $j = \ell + \frac{1}{2}$, as spin-flip transitions are preferred in nucleon transfer reactions, and the last neutron in ^{13}C is in the $p_{1/2}$ orbital. Hence, looking at relative intensities of individual states from these different reactions can aid in defining the J^π of that state.

Placing silicon arrays inside large γ ray arrays provides an opportunity to simultaneously gather information from the charged ejectile following a transfer reaction and the subsequent γ rays. The Oak Ridge Rutgers University Barrel Array (ORRUBA)[12] of position sensitive silicon detects

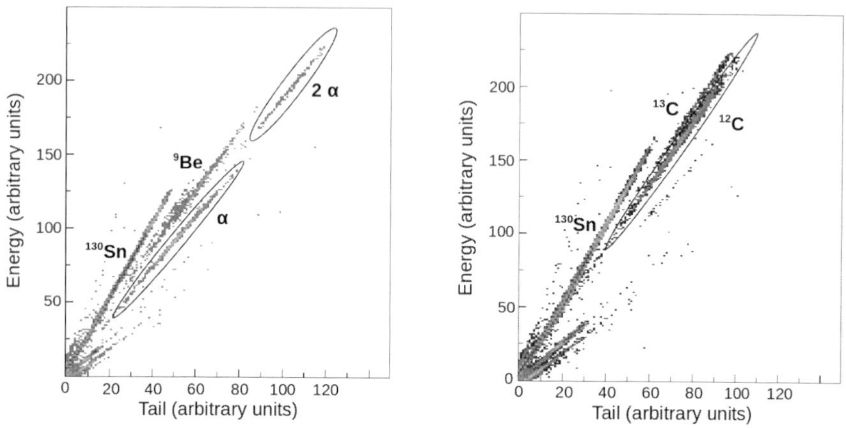

Fig. 1. Particle Identification spectra for reactions of ^{130}Sn on a ^{9}Be (left panel) and ^{13}C (right panel) target. The total energy deposited in a CsI crystal is plotted against the slow component (tail) of the signal to give separation between different light particles. Unreacted ^{130}Sn ions are sometimes scattered at large enough angles to be measured in BareBall. The α and 2α groups are more cleanly separated in the ^{9}Be target case than the ^{13}C and ^{12}C recoils in the ^{13}C target case.

can fit, in its entirety, inside the Gammasphere array of 110 high purity germanium detectors[13]. The combined instrumentation, named the Gammasphere ORRUBA Dual Detectors for Experimental Structure Studies (GODDESS)[14], has been used with beams provided by the ATLAS facility to measure reactions with ^{134}Xe, ^{95}Mo, and ^{3}He beams[15]. Work is in progress to combine ORRUBA with Gretina/GRETA[16] in a similar way to GODDESS, to make a powerful instrument for spectroscopic measurements using transfer reactions with reaccelerated beams at FRIB.

3.2. *Hybrid Array for Gamma Ray Detection (HAGRiD)*

The advantages of being able to measure γ rays that follow transfer reactions are such that our collaboration decided to build a scintillator array for this purpose, as well as for nuclear decay studies. The array needed to be flexible, as it will be used in multiple configurations, and transportable, so that it can be used at various laboratories where these experiments are performed. For these reasons scintillator detectors were preferred over germanium detectors, which require liquid nitrogen cooling and are prohibitively expensive. Lanthanum bromide was chosen for its relatively good energy resolution ($< 3\%$ at 662 keV), high intrinsic efficiency ($\sim 35\%$ at 1 MeV) and good timing resolution (~ 200 ps). The crystals selected are of the BrilLanCe 380™ type from Saint Gobain Crystals. As of the time of writing, the HAGRiD array comprises of 27 2″ × 2″ detectors, with a further 10 3″ × 3″ crystals on order. The crystals are currently being coupled to Hamamatsu R6231-100 photomultiplier tubes (PMTs) which include super bialkali photocathodes for increased quantum efficiency[17].

Fig. 2. The first 10 2″ × 2″ $LaBr_3(Ce)$ crystals of the HAGRiD array were delivered in October 2015.

The early implementation of HAGRiD, with 2″ × 2″ crystals coupled to a combination of Hamamatsu R7724, R7224-100, and R6231-100 PMTs, has been used in experiments with the Jet Experiments in Nuclear Structure and Astrophysics (JENSA) gas jet target[18] at the National Superconducting Cyclotron Laboratory (NSCL) and in β-delayed neutron emission studies at the HRIBF. A recent β-delayed neutron emission experiment at the NSCL with the Versatile Array of Neutron Detectors (VANDLE)[19,20] used HAGRiD with the optimized R6231-100 PMT's in a temporary coupling. Once the array is completed, its flexibility to be positioned around the JENSA gas target, other γ detectors, or neutron detectors will make it a valuable component of experiments studying the decay or reactions of exotic nuclei.

4. Proton transfer

The (d,n) reaction transfers a proton onto a nucleus in an analogous way to neutron transfer using the (d,p) reaction. The difficulty in using the (d,n) reaction is with measuring neutrons in place of protons, which can be detected with relative ease and good resolution using silicon, or other charged-particle detectors. Developing (d,n) reactions in inverse kinematics, such that they can be used with radioactive ion beams is important to placing single-proton spectroscopy on the same experimental footing as single-neutron spectroscopy.

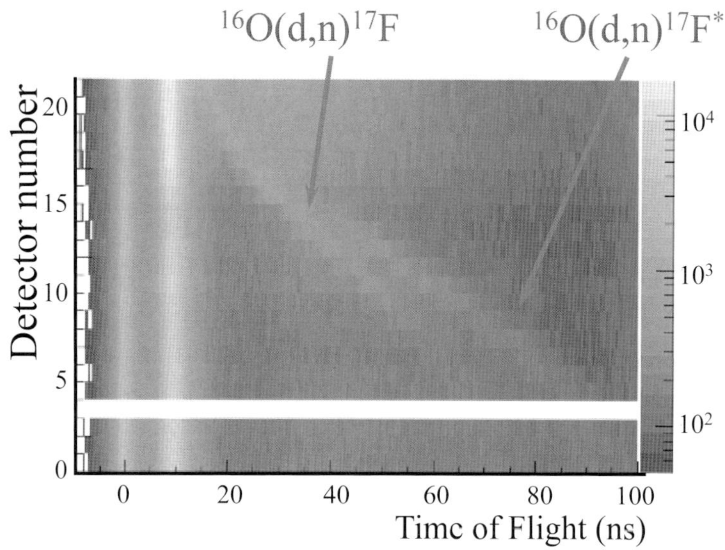

Fig. 3. Kinematics plot for the $^{16}O(d,n)^{17}F$ reaction. The detector number relates to the angle at which the neutron ejectile is measured, each detector covering roughly 3.4°.

We performed an experiment at the Nuclear Structure Laboratory (NSL), at the University of Notre Dame, using stable beams of ^{12}C and ^{16}O incident on a deuterated polyethylene target, with energies ranging from 23.5 to 41.7 MeV and at 64 MeV, respectively. Neutrons emerging from the (d,n) reaction were detected in 21 bars of VANDLE, placed at 0.5 m from the target, and 5 bars containing a xylene-based liquid scintillator. The time of flight of the neutrons in VANDLE was found from the difference between the timing signals from the bunched beam and VANDLE. A scintillator detector, used as a beamstop, provided an additional timing signal. The detectors were placed at angles from 60° (detector 21) to 170° (detector 1). The time of flight (ToF) is shown as a function of detector number, which varies smoothly with angle, in Fig. 3 for the ^{16}O(d,n) measurement. As VANDLE is made of plastic scintillator, a γ ray background is present, including the prompt flash of γ's when the beam hits the target, and a second, more intense flash from the beamstop. The target-induced γ flash gives a timing reference to calibrate the time of flight.

Two loci are apparent in Fig. 3, the lower from neutrons emerging from the ^{16}O(d,n) reaction populating the ground state of ^{17}F, and the upper is from population of the 3.1 MeV state in ^{17}F. The resolution in excitation energy in ^{17}F improves quadratically as the time of flight increases. These data are still under analysis. The next step in this program is to successfully measure the ^{7}Be(d,n)^{8}B reaction using a ^{7}Be beam from the TwinSol facility at the NSL. The secondary beam will have a lower intensity than what was used for the stable beam test, so extracting events corresponding to the (d,n) reaction from the γ ray background will be crucial. We are working on methods of recoil identification to make these measurements possible.

5. Summary

Building on the work performed using transfer reactions with fragmentation, in-flight, and ISOL beams, our collaboration has developed new tools and techniques to extend their reach for the next generation of radioactive ion beam facilities, such as FRIB. At this important moment, we have focused our efforts on developing techniques for measuring γ rays in coincidence with emergent particles following transfer, measuring proton transfer using the (d,n) reaction, and using the results of transfer reactions to inform nuclear astrophysics simulations. It is necessary that detector and target technology, experimental techniques, and theory continue to improve in parallel to radioactive beam developments in order to realize the full potential of new facilities.

Acknowledgments

This material is based upon work supported by the U.S. Department of Energy, Office of Science, Office of Nuclear Physics under contract number DE-FG02-96ER40983 and DE-SC000174, and research using resources of the Holifield Radioactive Ion Beam Facility, which was a DOE Office of Science User Facility operated by Oak Ridge National Laboratory, which is managed by UT-Batelle, LLS, for the U.S. Department of Energy (DOE). This work was funded in part by the National Nuclear Security Agency, Stockpile Stewardship Academic Alliance, and by the National Science Foundation. The author would like to thank the faculty, staff, and students at the NSL.

References

1. E. Burbidge, G. Burbidge, W. Fowler and F. Hoyle, *Rev. Mod. Phys.* **29**, p. 257 (1957).
2. R. Surman and J. Engel, *Phys. Rev. C* **64**, p. 035801 (2001).
3. J. Beun *et al.*, *J. Phys. G.* **36**, p. 025201 (2009).
4. J. R. Beene *et al.*, *J. Phys. G* **38**, p. 024002 (2011).
5. K. L. Jones *et al.*, *Phys. Rev. C* **70**, p. 067602 (2004).
6. B. Manning *et al.*, Priv. comm. and contribution in this volume (2017).
7. R. L. Kozub *et al.*, *Phys. Rev. Lett.* **109**, p. 172501 (2012).
8. K. L. Jones *et al.*, *Phys. Rev. C* **84**, p. 034601 (2011).
9. J. M. Allmond *et al.*, *Phys. Rev. Lett.* **112**, p. 172701 (2014).
10. A. Galindo-Uribarri, Exotic nuclei, *AIP Conference Proceedings* **1271**, 180 (2010).
11. C. J. Gross *et al.*, *Nucl. Instr. and Meth. in Phys. Res. A* **450**, 12 (2000).
12. S. D. Pain *et al.*, *Nucl. Instr. Meth. Phys. Res. A* **261**, 1122 (2007).
13. I.-Y. Lee, *Nucl. Phys. A* **520**, p. c641 (1990).
14. A. Ratkiewicz *et al.*, *EPJ Web of Conferences* **93** (2015).
15. S. D. Pain *et al.*, In this volume of proceedings (2017).
16. I. Y. Lee, M. A. Deleplanque and K. Vetter, *Reports on Progress in Physics* **66**, p. 1095 (2003).
17. K. Nakamura, Y. Hamana, Y. Ishigami and T. Matsui, *Nucl. Instr. and Meth. in Phys. Res. A* **623**, 276 (2010).
18. D. W. Bardayan *et al.*, *Physics Procedia* **66**, 451 (2015).
19. S. V. Paulauskas *et al.*, *Nucl. Instr. and Meth. in Phys. Res. A* **737**, 22 (2014).
20. W. A. Peters *et al.*, *Nucl. Instr. and Meth. in Phys. Res. A* **836**, 122 (2016).

Decays of the Three Top Contributors to the Reactor $\overline{\nu}_e$ High-Energy Spectrum, ^{92}Rb, ^{96gs}Y, and ^{142}Cs, Studied with Total Absorption Spectroscopy

B. C. Rasco,[1,2,3,4] M. Wolińska-Cichocka,[5,2,1] A. Fijałkowska,[6,3]
K. P. Rykaczewski,[2] M. Karny,[6,2,1] R. K. Grzywacz,[3,2,1] K. C. Goetz,[7,3]
C. J. Gross,[2] D. W. Stracener,[2] E. F. Zganjar,[4] J. C. Batchelder,[8,1]
J. C. Blackmon,[4] N. T. Brewer,[1,2,3] S. Go,[3] B. Heffron,[3,2] T. King,[3] J. T. Matta,[2]
K. Miernik,[6,1] C. D. Nesaraja,[2] S. V. Paulauskas,[3] M. M. Rajabali,[9]
E. H. Wang,[10] J. A. Winger,[11] Y. Xiao[3] and C. J. Zachary[10]

[1]*JINPA, Oak Ridge National Laboratory, Oak Ridge, TN 37831, USA*
[2]*Physics Division, Oak Ridge National Laboratory, Oak Ridge, TN 37831, USA*
[3]*Department of Physics and Astronomy, University of Tennessee, Knoxville, TN 37966, USA*
[4]*Department of Physics and Astronomy, Louisiana State University, Baton Rouge, LA 70803*
[5]*Heavy Ion Laboratory, University of Warsaw, PL-02-093 Warsaw, Poland*
[6]*Faculty of Physics, University of Warsaw, PL-02-093 Warsaw, Poland*
[7]*CIRE Bredesen Center, University of Tennessee, Knoxville, TN 37966, USA*
[8]*Department of Nuclear Engineering, University of California, Berkeley, Berkeley CA 94720, USA*
[9]*Department of Physics, Tennessee Technological University, Cookeville, TN 38505, USA*
[10]*Department of Physics and Astronomy, Vanderbilt University, Nashville, TN 37235, USA*
[11]*Department of Physics and Astronomy, Mississippi State University, Mississippi State, MS 39762*

We report total absorption spectroscopy measurements of ^{92}Rb, ^{96gs}Y, and ^{142}Cs β decays, which are the most important contributors to the high energy υ spectral shape in nuclear reactors. These three β decays contribute 43% of the $\overline{\nu}_e$ flux near 5.5 MeV emitted by nuclear reactors. This $\overline{\nu}_e$ energy is particularly interesting due to spectral features recently observed in several experiments including the Daya Bay, Double Chooz, and RENO Collaborations. Measurements were conducted at Oak Ridge National Laboratory by means of proton-induced fission of ^{238}U with on-line mass separation of fission fragments and the Modular Total Absorption Spectrometer. We observe a β-decay pattern that is similar to recent measurements of ^{92}Rb, with a ground-state to ground-state β feeding of 91(3)%. We verify the $^{96\text{gs}}$Y ground-state to ground-state β feeding of 95.5(20)%. Our measurements substantially modify the β-decay feedings of ^{142}Cs, reducing the β feeding to ^{142}Ba states below 2 MeV by 32% when compared with the latest evaluations. Our results increase the discrepancy between the observed and the expected reactor $\overline{\nu}_e$ flux between 5 and 7 MeV, the maximum excess increases from ~10% to ~12%.

Production of Heavy n-Rich Nuclei in Multinucleon Transfer Reactions

Cheng Li[1,2,*], Peiwei Wen[1,2], Long Zhu[3], Jingjing Li[1,2], Gen Zhang[1,2], Bing Li[1,2] and Feng-Shou Zhang[1,2,4,†]

[1] *Beijing Radiation Center, Beijing 100875, China*
[2] *The Key Laboratory of Beam Technology and Material Modification of Ministry of Education, College of Nuclear Science and Technology, Beijing Normal University, Beijing 100875, China*
[3] *Sino-French Institute of Nuclear Engineering and Technology, Sun Yat-sen University, Zhuhai 519082, China*
[4] *Center of Theoretical Nuclear Physics, National Laboratory of Heavy Ion Accelerator of Lanzhou, Lanzhou 730000, China*
[*] *E-mail: licheng@mail.bnu.edu.cn*
[†] *E-mail: fszhang@bnu.edu.cn*

The multinucleon transfer reactions are investigated by using the improved quantum molecular dynamics (ImQMD) model, the dinuclear system (DNS) model and the DNS+GRAZING model. In the framework of ImQMD, the total-kinetic-energy-mass distributions of primary binary fragments are studied. The isotope production cross sections of ^{136}Xe+^{208}Pb, ^{238}U+^{248}Cm and ^{64}Ni+^{238}U systems are analyzed via three models. The calculations show that although each model has its successes in description of multinucleon transfer reactions, it also has its limitations for the results.

Keywords: ImQMD Model; DNS Model; GRAZING Model; Multinucleon Transfer Reactions; Neutron-Rich Nuclei.

1. Introduction

Production of new heavy neutron-rich nuclei has attracted widespread interests in recent years[1–3]. The production of these exotic nuclei is very important for exploring the structure and decay properties of these nuclei and for understanding the r-process in nuclear astrophysics. Many new neutron-rich nuclei are produced by experiment with fragmentation method of projectile[4,5]. For example, there are 60 new neutron-rich isotopes in the atomic number range of $60 \leqslant Z \leqslant 78$ produced by using a 1 GeV/nucleon ^{238}U beam impinged on a ^{9}Be target by Kurcewicz *et al.* at GSI[4]. They found that the fragmentation plays a dominant role in the production of the isotopes of the elements above $Z = 72$. However, production cross sections of these nuclei are very small which are about a few ten pb orders of magnitude. Further producing of more neutron-rich nuclei with the fragmentation method face enormous challenges. Recently, a multinucleon transfer (MNT) reaction experiment of ^{136}Xe+^{198}Pt was performed by Watanabe *et al.* at GANIL[6]. They found that the MNT approach has huge advantages for the production of very neutron-rich nuclei with $Z \leqslant 77$ compared to above fragmentation approach. It demonstrates that the MNT route is a more promising path to produce new neutron-rich isotopes around $N = 126$ shell.

To describe the multinucleon transfer process, some phenomenological and microscopic models are proposed. The phenomenological models such as the DNS[7,8] and GRAZING[9,10] have shown reasonable success in predicting the production cross sections of heavy neutron-rich nuclei. To understand the dynamical mechanism in heavy-ion collisions, the microscopic dynamics models, such as the time-dependent Hartree-Fock (TDHF) model[11] and the ImQMD model[1,12] have been developed. In this work, the energy dissipation mechanism between the projectile and target is studied by using the ImQMD model. Then the isotope production cross sections of ^{136}Xe+^{208}Pb, ^{238}U+^{248}Cm and ^{64}Ni+^{238}U systems are investigated with ImQMD, DNS and DNS+GRAZING models.

The article is organized as follows. In section 2, we describe the theoretical models. The results and discussion are presented in section 3. We summarize the main results in section 4.

2. Models

2.1. *ImQMD model*

The ImQMD model is a semiclassical microscopic dynamics model which includes the mean-field and nucleon-nucleon collisions as well as the Pauli principle and is successfully applied to heavy-ion fusion reactions and intermediate-energy fragmentation reactions[13]. In the ImQMD model, the nuclear interaction potential energy $U_{\rm loc}$ is obtained from the integration of the Skyrme energy density functional $U = \int V_{\rm loc}(\mathbf{r})d\mathbf{r}$ without the spin-orbit term, which reads

$$\begin{aligned} V_{\mathbf{loc}} = {} & \frac{\alpha}{2}\frac{\rho^2}{\rho_0} + \frac{\beta}{\gamma+1}\frac{\rho^{\gamma+1}}{\rho_0^{\gamma}} + \frac{g_{sur}}{2\rho_0}(\nabla\rho)^2 \\ & + \frac{C_s}{2\rho_0}(\rho^2 - \kappa_s(\nabla\rho)^2)\delta^2 + g_\tau\frac{\rho^{\eta+1}}{\rho_0^{\eta}}. \end{aligned} \tag{1}$$

Here $\rho = \rho_n + \rho_p$ is the nucleons density. $\delta = (\rho_n - \rho_p)/(\rho_n + \rho_p)$ is the isospin asymmetry. The parameters named IQ2 are adopted in this work[1]. More descriptions of the ImQMD model see Refs.[1,13].

2.2. *DNS model*

In the DNS model, after capture happens, nucleon transfer is affected by the relative motion and the transfer process is considered as time-dependent. The evolution of the DNS is a diffusion process along the mass asymmetry degree and simultaneously in the relative distance between the centers of the interacting nuclei. The distribution probability can be obtained by solving a set of master equations numerically in the potential energy surface[7,8].

The total production cross section of a primary fragment with charge Z and mass number A can be calculated as follows:

$$\sigma_{tr}^{DNS}(Z,A,E) = \frac{\pi\hbar^2}{2\mu E}\sum_l (2l+1) \times P_c(E,l)P_{tr}^{DNS}(Z,A,E), \tag{2}$$

where P_c is capture probability and $P_{tr}^{DNS}(Z,A,E)$ is the production probability of a fragment. μ is the reduced mass of the colliding system.

2.3. *DNS+GRAZING model*

GRAZING model is designed in order to evaluate the observable in the MNT reactions that happen in the grazing region[9]. In the GRAZING model, exchange of particles is based on well-known form factors of the quantum coupled equations in a mean field approximation, while the colliding nuclei are considered to move on classical trajectories. Similar to the DNS model, the total production cross section can be expressed as

$$\sigma_{tr}^{GRAZING}(Z,A,E) = \frac{\pi\hbar^2}{2\mu E}\sum_l (2l+1) \times (1 - P_c(E,l))P_{tr}^{GRAZING}(Z,A,E). \tag{3}$$

It can be seen apparently from Eq. (2) and Eq. (3), the GRAZING model describes transfer reactions without capture, which happen at the scattering process. The DNS model describes the transfer reaction happens in the dinuclear evolution process after capture. These two models represents two kind of transfer mechanisms, which are mutually complementary. In this work, we try to simply combine DNS and GRAZING model (i.e., $\sigma_{tr}^{tot} = \sigma_{tr}^{DNS} + \sigma_{tr}^{GRAZING}$) for description of MNT reactions.

3. Results and Discussion

In Fig. 1, the density distributions for the reaction ^{136}Xe+^{208}Pb at $E_{c.m.} =$ 450 MeV are calculated by ImQMD model. The lifetime of the neck to be about 1000 and 500 fm/c for $b = 2$ and 7 fm, respectively. The lifetime of the neck decreases with increasing of impact parameters due to the contribution of the angular momentum of system. In addition, one can also find that there are two cores of density formed in the target-like fragments (TLFs) for $b = 2$ fm at $t = 1260$ fm/c. This is because a higher excited energy is obtained from the dissipation of the relative motion energy of the colliding nuclei in the central collisions. It implies that the TLFs may be fission in the later evolution.

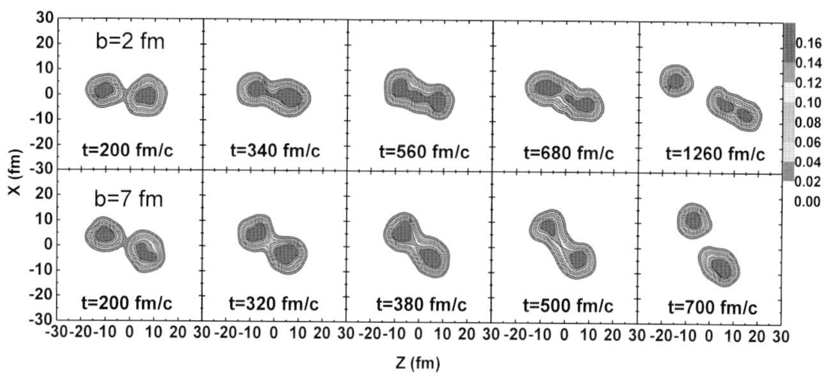

Fig. 1. Time evolution of density distribution for two typical events of the reaction ^{136}Xe+^{208}Pb at $b = 2$ and 7 fm.

The total-kinetic-energy-mass distributions of ^{136}Xe+^{208}Pb for primary binary fragments in different impact parameter regions are shown in Fig. 2. From the Fig. 2 one can see two behaviors: (1) The total kinetic energy (TKE) increased with increasing impact parameters. The central collisions ($0 < b < 3$ fm) most are deep-inelastic reactions. Both the deep-inelastic and quasielastic collision events appear in $3 \leq b < 7$ fm. This is because the dynamical fluctuation causes intense competition between the quasielastic and deep-inelastic reactions. The peripheral collisions correspond to quasielastic reactions. (2) The masses distribute in a rather broad range (80-250 u) in $0 < b < 7$ fm. There are a large number of nucleons transfers between the projectile and target. While in peripheral collisions, there are only a few nucleon transfers because the reaction mechanism is dominated by quasielastic collisions.

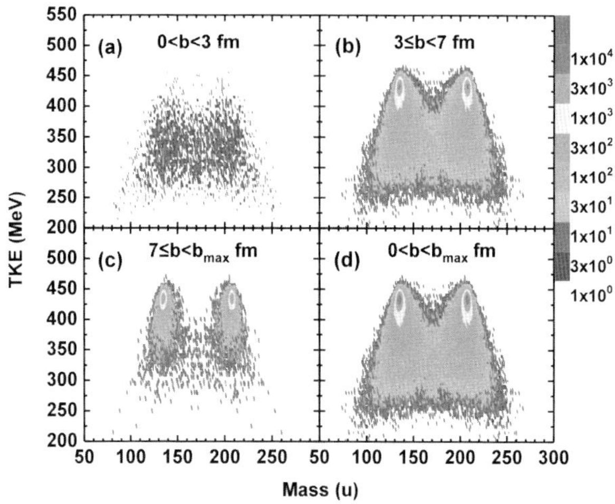

Fig. 2. TKE-mass distributions calculated by ImQMD model for ^{136}Xe+^{208}Pb at $E_{c.m.} = 450$ MeV in different impact parameter regions.

The isotopic production cross sections of ^{136}Xe+^{208}Pb at $E_{c.m.} = 450$ MeV are shown in Fig. 3. One can see that GRAZING is a suitable mode to estimate the production cross sections only for -1p to +2p. It grossly underestimates the production cross section by orders of magnitude in the case of more proton transfers. For the DNS model, the height of peaks keep consistent with experimental data at least on the orders of magnitude for -3p to +2p. The production cross sections decreases rapidly with increasing ΔZ in the region of charge numbers above 82 (Pb). For the ImQMD model, the height of peaks always keep consistent with experimental data for -3p to +3p, but the peak widths are usually greater than experimental data especially for $\Delta Z > 0$.

In Fig. 4, the production cross sections of Es (+3p), Fm (+4p) and Md (+5p) in ^{238}U+^{248}Cm at $E_{c.m.} = 800$ MeV are calculated by DNS model. One can see a good agreement between theoretical and experimental for larger proton pickup channels in approximately symmetrical system. The cross sections decrease drastically with increasing atomic numbers of the fragments. The survival probabilities of most of the primary fragments are quite low due to a dominant fission channel. The ^{261}Md and ^{262}Md could be synthesized in this reaction with cross sections of about 1.52 nb and 0.17 nb, respectively.

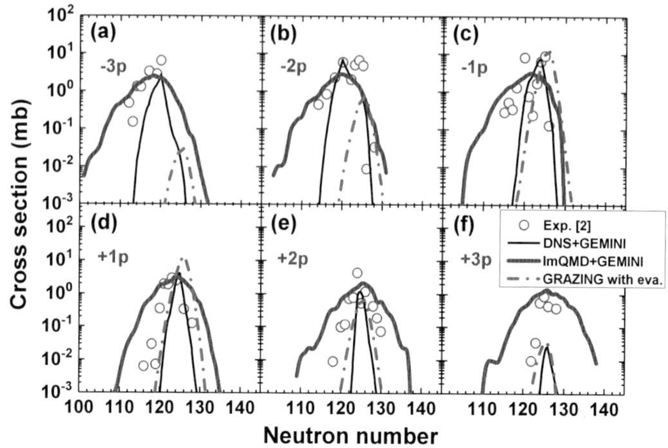

Fig. 3. Isotopic production cross sections (secondary fragments) of ^{136}Xe+^{208}Pb. The experimental data are from Ref. [2].

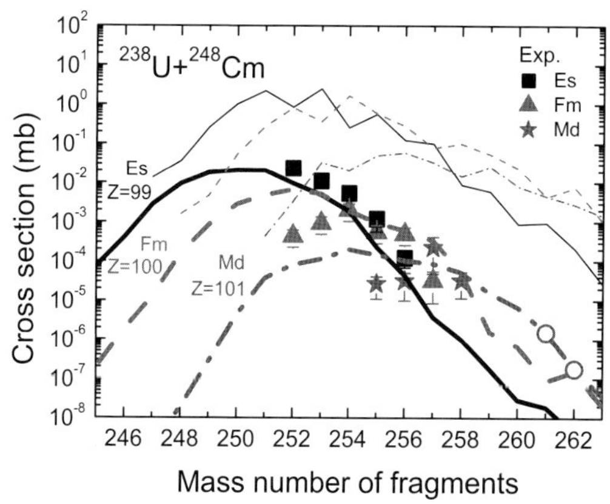

Fig. 4. Isotopic production cross sections of Es, Fm and Md in ^{238}U+^{248}Cm at $E_{c.m.}$ = 800 MeV. The thin and thick lines are distributions of primary and final fragments, respectively. The experimental data are taken from Ref. [14]. The circles denote the unknown nuclei.

In order to contain simultaneously the capture and scattering process, we combine the DNS and GRAZING model for description of MNT reactions. The isotopic production cross sections of ^{64}Ni+^{238}U at $E_{c.m.}$ =307.5 MeV are shown in Fig. 5. For the larger proton stripping channel (-4p),

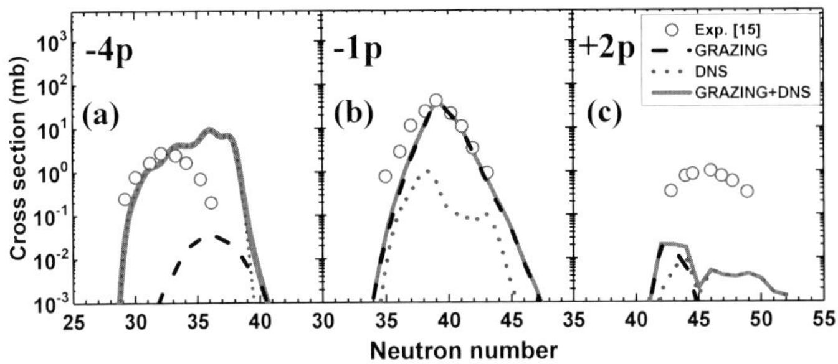

Fig. 5. Isotopic production cross sections (secondary fragments) of ^{64}Ni+^{238}U at $E_{c.m.}$ =307.5 MeV. The experimental data are from Ref.[15].

the DNS model agrees with experimental data better than GRAZING model. While for the -1p channel, the GRAZING model is better than DNS model. It hard to describe the transfer reaction with individual scattering or capture mechanism. The GRAZING+DNS model makes a large improvement on -4p and -1p channels. However for the +2p pickup channel, the GRAZING+DNS model underestimate substantially the experimental data. This combined model should be further develop for describing multinucleon transfer reactions.

4. Conclusions

The energy dissipation mechanism in MNT reaction of ^{136}Xe+^{208}Pb is studied by using the ImQMD model. The isotopic production cross sections of ^{136}Xe+^{208}Pb, ^{238}U+^{248}Cm and ^{64}Ni+^{238}U systems are calculated by the ImQMD, DNS and DNS+GRAZING models. We find that each model has shown reasonable success in predicting the isotopic production cross sections, but it also has its limitations for the results. For the ImQMD model, the height of peaks always keep consistent with experimental data, but the peak widths are usually greater than experimental data. The GRAZING model grossly underestimates the production cross section by orders of magnitude in the case of more proton transfers. The DNS model is successful in description of larger proton pickup channels for approximately symmetrical systems, but it underestimates the production cross section of larger proton pickup channels for asymmetry systems. The GRAZING+DNS model should be further develop for describing the proton pickup channels in multinucleon transfer reactions.

Acknowledgments

This work was supported by the National Natural Science Foundation of China under Grants No. 11635003, No. 11025524 and No. 11161130520; the National Basic Research Program of China under Grant No. 2010CB832903; the European Commission's 7th Framework Programme (Fp7-PEOPLE-2010-IRSES) under Grant Agreement Project No. 269131; the Project funded by China Postdoctoral Science Foundation (Grant No. 2016M600956).

References

1. C. Li, F. Zhang, J. J. Li, L. Zhu, J. L. Tian, N. Wang and F. S. Zhang, *Phys. Rev. C* **93**, 014618 (2016).
2. J. S. Barrett, W. Loveland, R. Yanez, S. Zhu, A. D. Ayangeakaa, M. P. Carpenter, J. P. Greene, R. V. F. Janssens, T. Lauritsen, E. A. McCutchan, A. A. Sonzogni, C. J. Chiara, J. L. Harker and W. B. Walters, *Phys. Rev. C* **91**, 064615 (2015).
3. V. I. Zagrebaev and W. Greiner, *Phys. Rev. Lett.* **101**, 122701 (2008).
4. J. Kurcewicz *et al.*, *Phys. Lett. B* **717**, 371 (2012).
5. O. B. Tarasov *et al.*, *Phys. Rev. Lett.* **102**, 142501 (2009).
6. Y. X. Watanabe *et al.*, *Phys. Rev. Lett.* **115**, 172503 (2015).
7. G. G. Adamian, N. V. Antonenko and W. Scheid, *Nucl. Phys. A* **618**, 176 (1997).
8. L. Zhu, Z. Q. Feng and F. S. Zhang, *J. Phys. G* **42**, 085102 (2015).
9. A. Winther, *Nucl. Phys. A* **572**, 191 (1994).
10. R. Yanez and W. Loveland, *Phys. Rev. C* **91**, 044608 (2015).
11. K. Sekizawa and K. Yabana, *Phys. Rev. C* **90**, 064614 (2014).
12. K. Zhao *et al.*, *Phys. Rev. C* **92**, 024613 (2015).
13. Z. Q. Feng, F. S. Zhang, G. M. Jin and X. Huang, *Nucl. Phys. A* **750**, 232 (2005).
14. M. Schädel *et al.*, *Phys. Rev. Lett.* **48**, 852 (1982).
15. L. Corradi *et al.*, *Phys. Rev. C* **59**, 261 (1999).

Beta-Delayed Neutron Measurements near ^{132}Sn with CARIBU*

S. Z. Taylor[1], K. Kolos[1], R. Grzywacz[1,2,3], S. V. Paulauskas[4], M. Madurga[5], T. King[1], M. Al-Shudifat[1], W. A. Peters[1,2], C. J. Gross[2], K. P. Rykaczewski[2], J. M. Allmond[2], M. Wolińska-Cichocka[2,3,6] and The VANDLE Collaboration

[1] *University of Tennessee, Knoxville, TN, 37996, USA*
[2] *Physics Division, Oak Ridge National Laboratory, Oak Ridge, TN, 37831, USA*
[3] *Joint Institute for Nuclear Physics and Applications, Oak Ridge, TN, 37831 USA*
[4] *National Superconducting Cyclotron Laboratory, Michigan State University, East Lansing, MI, 48824, USA*
[5] *ISOLDE, CERN, EP Division, CH-1211 Geneva 23, Switzerland*
[6] *Heavy Ion Laboratory, University of Warsaw, PL 08-093 Warsaw, Poland*

Measurement of spectroscopic information on beta-delayed neutrons of neutron rich fission fragments is of interest to the areas of nuclear structure, astrophysics, reactor design, and stockpile stewardship. The Versatile Array of Neutron Detectors at Low Energy (VANDLE) [1, 2, 3] used the time of flight (ToF) method to measure fission fragments of ^{252}Cf provided by CARIBU at Argonne National Lab. We measured ^{135}Sb and ^{136}Sb isotopes to explore the nuclear structure near the doubly magic nucleus ^{132}Sn. A new β detection system was developed for this experiment using silicon photo-multipliers (SiPMs) that allowed for a lower beta particle energy detection threshold and better timing resolution compared to previous VANDLE experiments. Additionally, a gamma-ray detection system paired with VANDLE enable measurements of gamma-rays emitted in coincidence with neutrons, and provides more complete information about the decay path. Measurement of β-n-γ coincidences are critical for accurate data interpretation.

*This work is funded by the U.S. Department of Energy NNSA under the Stewardship Science Academic Alliance program through DOE Cooperative Agreement No. DE-FG52-08NA28552.

References

1. C. Matei *et al.*, Proceedings of Science, NIC X, 138 (2008).
2. S. V. Paulauskas *et al.*, NIMA 797, 22 (2014).
3. W. A. Peters *et al.*, NIMA, (under review).

Beta-Delayed Neutron Studies Using Trapped Ions from CARIBU*

Kevin Siegl[1], N.D. Scielzo[2], A. Aprahamian[1], G. Savard[3], J.A. Clark[3], A.F. Levand[3], M. Burkey[4], S. Caldwell[4], A. Czeszumska[5], T.Y. Hirsch[3], K. Kolos[2], S.T. Marley[6], G.E. Morgan[6], E.B. Norman[7], A. Nystrom[1], R. Orford[8], S. Padgett[2], A. Pérez Galván[3], K.S. Sharma[9], S.Y. Strauss[1] and B.S. Wang[2]

[1] *University of Notre Dame, Notre Dame, IN 46556, USA*
[2] *Lawrence Livermore National Laboratory, Livermore, CA 94550, USA*
[3] *Argonne National Laboratory, Argonne, IL 60439, USA*
[4] *University of Chicago, Chicago, IL 60637, USA*
[5] *University of California, Berkeley, CA 94720, USA*
[6] *Louisiana State University, Baton Rouge, LA 70803, USA*
[7] *Lawrence Berkeley National Laboratory, Berkeley, CA 94720, USA*
[8] *McGill University, Montréal, QC H3A 0G4, Canada*
[9] *University of Manitoba, Winnipeg, MB R3T 2N2, Canada*

Confining radioactive ions in a radio-frequency quadrupole ion trap allows indirect measurements of beta-delayed neutron (BDN) emission. This is accomplished by determining the energy of the recoiling ion which can be significantly larger after neutron emission than from beta decay alone [1]. This method avoids most of the systematic errors associated with direct neutron detection but introduces dependencies on specifics of the decay and interactions of the ion with the RF fields. The decays of seven BDN precursors were studied using the Beta-decay Paul Trap (BPT) [2] to confine fission fragments from the Californium Rare Isotope Breeder Upgrade (CARIBU) facility [3] at Argonne National Laboratory. The analysis of these measurements and results for the branching ratios and neutron energy spectra will be presented.

*Supported by the NSF under grant PHY-1419765, and the U.S. DOE under the NEUP project 13-5485, contracts DE-AC02-06CH11357 (ANL) and DE-AC52-07NA27344 (LLNL), and award DE-NA0000979 (NNSA).

References

1. R.M. Yee *et al.*, *Phys. Rev. Lett.* **110**, 092501 (2013).
2. N. Scielzo *et al.*, *Nucl. Instrum. Methods Phys. Res. Sect. A* **681**, 94 (2012).
3. G. Savard *et al.*, *Nucl. Instrum. Methods Phys. Res. Sect. B* **266**, 4086 (2008).

Studying Beta-Delayed Neutron Emission via Recoil-Ion Spectroscopy*

Scott T. Marley

Louisiana State University, Baton Rouge, LA 70803

The study of β-delayed neutron emission is important to a variety of fields ranging from nuclear structure and astrophysics to nuclear energy. There is great interest in β-delayed neutron branching ratios of neutron-rich nuclei, which influence the final isotopic abundances of heavy elements formed during late stages of the rapid neutron-capture (r-) process. Accurate and precise measurements of β-delayed neutron precursors are required to help improve nuclear physics inputs to modern r-process simulations with the aim of constraining the astrophysical site of the r-process. Recoil-ion (time-of-flight) spectroscopy is a novel technique for studying β-delayed neutron emission without the challenges of neutron detection. Radioactive ions are confined to a small volume ($\sim$1 mm^3) using an ion trap and the time-of-flight of the recoiling daughter ion is measured in coincidence with other decay products. The difference in time-of-flight between recoils from β- and βn-decay allows for the delayed-neutron branching ratio to be determined. Additionally, the delayed-neutron energy spectrum can be reconstructed from the conservation of energy and momentum. Results from a recent campaign of measurements using this technique performed at the CAlifornium Rare Isotope Breeder Upgrade (CARIBU) facility at Argonne National Laboratory will be presented and future studies discussed.

---

*This work was supported by U.S. DOE under Contracts DE-AC52-07NA27344 (LLNL), DE-AC02-06CH11357 (ANL), DE-FG02-98ER41086 (Northwestern U.), DE-FG02-05ER41398 (RS), NSERC, Canada, under Application No. 216974, NSF grant number PHY0822648 (JINA), and the Louisiana Board of Regents Research Competitiveness Subprogram LEQSF(2016-17)-RD-A-09.

Nuclear Many-Body Perturbation Calculations Based on the Chiral N^3LO Potential*

F. R. Xu†, B. S. Hu and Z. H. Sun

School of Physics, Peking University, Beijing 100871, China
†E-mail: frxu@pku.edu.cn
www.phy.pku.edu.cn/~frxu/

Starting from the chiral N^3LO potential and using many-body perturbation theory (MBPT), we have calculated the structures of finite nuclei. In order to speed up the convergence of calculations, the similarity renormalization group (SRG) has been employed. The MBPT has been developed in the Hartree-Fock (HF) basis within the angular momentum coupling representation. The developed MBPT calculation has been successfully applied to closed-shell nuclei. It has been demonstrated that in the HF basis the corrections up to the third order in energy and up to the second order in radius can give converged results, while in the harmonic oscillator (HO) basis one would have to go to 30th-order corrections to obtain converged solutions. The calculated binding energies and radii of the ^{4}He and ^{16}O nuclei agree well with experimental data. The present MBPT calculations have been compared with other ab-initio calculations with the same potential.

Keywords: Realistic Nuclear Force; Many-Body Perturbation Theory; Hartree-Fock Basis, Nuclear Structures.

1. Introduction

Calculations from the first principles are currently hot topics for nuclear physics researches. Such calculations give more fundamental physics, but on the other hand, challenge existing theoretical methods. For such calculations, one should start from realistic nuclear forces, and rigorously treat the many-body problems. Usually, the computing task is huge for the first-principles (or called *ab-initio*) numerical calculations. In order to accelerate the convergences of the calculations, the realistic force which has hard repulsive core needs to be softened (or called

---

*This work is supported by the National Natural Science Foundation of China under Grants No. 11235001, No. 11320101004, and No. 11575007.

renormalized). The strong short-range repulsive core comes from the high-momentum components of the interaction, which causes convergence problems in the calculations of nuclear structures.

A traditional softening approach is the G-matrix renormalization in the Brueckner-Bethe-Goldstone theory [1–2] in which all particle ladder diagrams are summed to infinite orders. Recently, a new class of renormalization methods has been developed, including the Similarity Renormalization Group (SRG) [3], Okubo-Lee-Suzuki [4–9], and the Unitary Correlation Operator Method (UCOM) [10, 11]. The renormalizations soften realistic NN interactions and generate effective Hamiltonians, while all symmetries and observables are preserved in the low-energy domain. The renormalization process also generates effective multi-nucleon interactions, so-called "induced" forces (e.g., induced three-nucleon forces, induced-3NFs).

How to treat rigorously many-body correlations is another key issue for ab-initio calculations. There have been several *ab initio* many-body methods, such as No-Core Shell Model (NCSM) [12–16], Green's Function Monte Carlo (GFMC) [17–20], and Coupled Cluster (CC) [21–23]. However, due to the limits of computer capability, the NCSM and GFMC calculations are currently limited to light nuclei (e.g., lighter than 16O), while the CC calculations are limited to nuclei near double closed shells. In this proceedings paper, we show our recent *ab initio* calculations by developing the many-body perturbation theory (MBPT).

2. The Models and Calculations

The calculations are based on the realistic nuclear force. There exist several widely-used realistic forces, such as CD-Bonn [24], Nijmegen [25], Argonne V18 (AV18) [26], INOY[27] and chiral potential [28, 29]. In the present calculations, we use the chiral N^3LO potential [28, 29].

The intrinsic Hamiltonian of the A-nucleon system is written as

$$\hat{\mathrm{H}} = \sum_{i<j}^{A} \frac{(\vec{p}_i - \vec{p}_j)^2}{2mA} + \sum_{i<j}^{A} V_{NN,ij}, \tag{1}$$

where the notation is standard. The first term on the right is the intrinsic kinetic energy, and $V_{NN,ij}$ is the NN interaction including the Coulomb interaction for protons. We do not include 3NFs. The two-body interaction takes the chiral N^3LO potential [28, 29]. The bare N^3LO is renormalized by using the SRG technique to soften the short-range repulsion and short-range tensor components. The SRG method is based on a continuous unitary transformation that suppresses off-diagonal matrix elements and drives the Hamiltonian towards a band-diagonal form [3].

With the effective Hamiltonian established, we first perform the spherical HF (SHF) calculation and then calculate the MBPT corrections to the HF result. For simplicity of computational effort, we limit our investigations here to the spherical closed-shell nuclei ^{4}He and ^{16}O. The detailed formulation can be found in our previous publication [30].

The SHF is carried out within the HO basis. The HO basis is truncated by a cutoff according to the number $N_{\text{shll}} = \max(2n+l+1)$, where N_{shll} indicates how many major HO shells are included in the truncation. After the SHF calculation, the MBPT corrections are calculated in the SHF basis. In the present calculations, the basis spaces employed take $N_{\text{shll}} = 7$, 9, 11, and 13. We verify that such a truncation is sufficient for the converged calculations of the ground state energies for the magic nuclei ^{4}He and ^{16}O.

Fig. 1 shows the MBPT calculated ground-state energy of ^{4}He. The calculations were done with the chiral N^{3}LO interaction which was renormalized by SRG. We see that good convergences are obtained. The softening parameter $\lambda = 3.0$ fm^{-1} seems to be insufficient to produce an interaction soft enough for good convergence in MBPT. $\lambda \approx 2.0 - 2.5$ fm^{-1} can reasonably reproduce the experimental ^{4}He ground-state energy with the NN-only interaction (without requiring 3NFs).

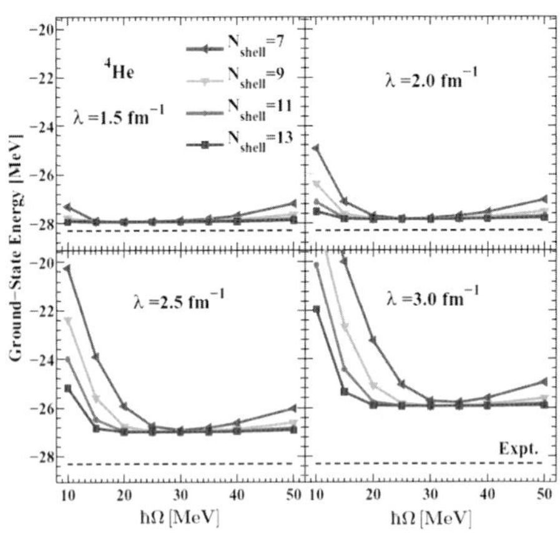

Fig. 1. The HF-MBPT calculations of ^{4}He ground-state energy through third order as a function of oscillator parameter Ω with the chiral N^{3}LO potential [28, 29] renormalized by SRG at different softening parameters $\lambda = 1.5, 2.0, 2.5, 3.0$ fm^{-1}. The dashed line represents the experimental ground-state energy.

Fig. 2 shows the radius calculations with $\lambda = 2.0\ fm^{-1}$. Tables 1 and 2 give the details of the HF-MBPT calculations with different λ values. We see that both second- and third-order corrections to energy decrease with decreasing λ. This is easily understood because MBPT mainly treats intermediate-range correlations and these correlations are weakened with decreasing λ. With sufficiently small λ, higher-order corrections to the energy can be neglected. The second-order correction to the radius is already small, which decreases with decreasing λ in ^4He.

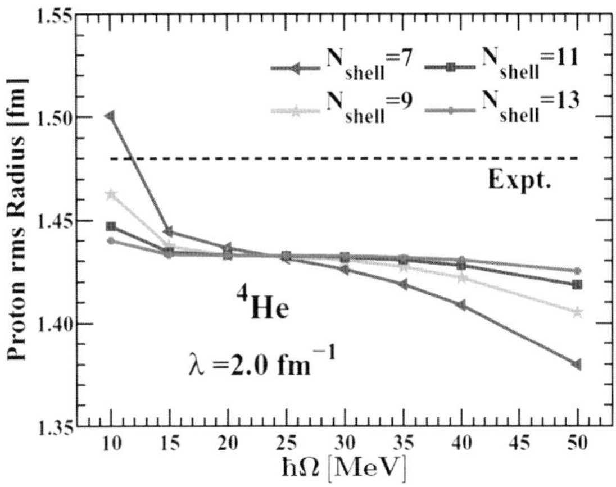

Fig. 2. Point-proton *rms* radius of ^4He. The chiral N^3LO potential [28, 29] is softened by the SRG method.

Table 1. ^4He ground-state energy (in MeV), analyzed in order-by-order HF-MBPT calculations with the chiral N^3LO potential softened at different SRG-softening parameter values (λ in fm^{-1}). PT2 and PT3 represent the second- and third-order corrections to energy, respectively. We take $N_{shell} = 13$ and $\Omega = 35$ MeV.

	$\lambda = 1.5$	2.0	2.5	3.0
Expt. [31]	-28.30	-28.30	-28.30	-28.30
NCSM [32]	-28.20	-28.41	-27.43	-26.80
SHF	-25.75	-21.86	-15.85	-10.28
PT2	-1.79	-5.09	-9.65	-13.78
PT3	-0.39	-0.90	-1.52	-1.95
SHF+PT2+PT3	-27.93	-27.85	-27.03	-26.01

Table 2. Similar to Table 1, but for the calculations of the point-proton *rms* radius (in fm) of ^{4}He.

	$\lambda = 1.5$	2.0	2.5	3.0
Expt.	1.477	1.477	1.477	1.477
SHF	1.677	1.652	1.714	1.816
PT2	0.007	0.001	-0.021	-0.065
$\Delta r_{\text{c.m.}}$	-0.226	-0.222	-0.227	-0.235
SHF+PT2+ $\Delta r_{\text{c.m.}}$	1.458	1.431	1.466	1.516

Fig. 3 shows the energy calculations for ^{16}O. The convergence behavior is similar to that in ^{4}He. However, calculations with small λ values (<2.0 fm^{-1}) give over binding, compared with data. This phenomenon should be more obvious for heavier nuclei. The main reason is that the three-body and higher-order forces are omitted in the calculations. The emergence of induced 3NFs and beyond is related to the SRG softening parameter. A larger λ value evolves a harder effective *NN* potential and weaker induced correlations (e.g., weaker induced 3NFs). But a large λ value may not sufficiently soften the short-range correlations of the realistic force, leading to demands for an excessively large model space and increased dependence on higher-order corrections. While a small λ value may sufficiently soften the potential, the contribution from induced three-body force may be not ignorable. Within SRG, $\lambda \sim 2.0 - 2.5$ fm^{-1} seems to be an optimal range in which the *NN* interaction can be softened reasonably and the combined three-body (initial plus induced) effects are greatly reduced [3, 32, 33].

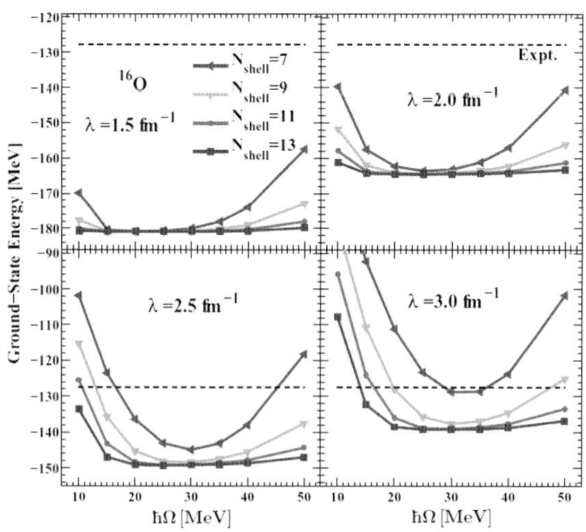

Fig. 3. Similar to Fig. 1, but for ^{16}O.

The calculation of the radius for ^{16}O is displayed in Fig. 4. Reasonable convergence is obtained for $N_{\text{shell}} = 11$ and 13. But the calculated radius is smaller than the experimental value. It seems that other ab initio results yield radii that are systematically smaller than experiment [34, 35]. The situation is similar to that in ^4He. We may conclude that the good convergence can be obtained for the MBPT calculations within the HF basis at small λ values. It has pointed out that in the HF basis the fourth- and higher-order MBPT corrections are known to be negligible in some cases [33].

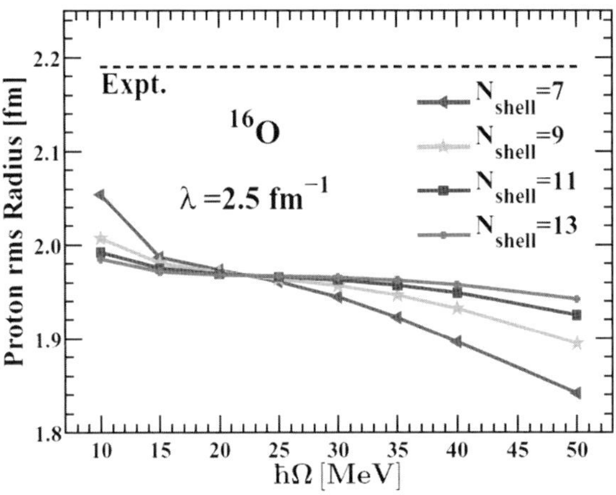

Fig. 4. Similar to Fig. 2, but for ^{16}O.

3. Conclusions

We have performed the HF-MBPT calculations with the chiral N^3LO potential which is softened by using the SRG method. First, we made a spherical HF calculation to establish the HF basis for the MBPT calculation. The spherical symmetry preserves the angular momentum. The angular momentum coupled scheme reduces significantly the model dimension and save the computational resources. The calculations have been applied to the closed-shell nuclei, ^4He and ^{16}O. The results are consistent with other *ab initio* calculations with the same *NN* potential. The MBPT convergences are discussed in order by order, showing that corrections up to the third order in energy and up to the second order in radius appear to be reasonable when one performs the MBPT calculations within the HF single-particle basis.

Acknowledgments

This work was in collaboration with Prof. James Vary at Department of Physics and Astronomy, Iowa State University.

References

1. K. A. Brueckner, *Phys. Rev.* **97**, 1353 (1955).
2. J. Goldstone, *Proc. R. Soc. London A* **239**, 267 (1957).
3. S. K. Bogner, R. J. Furnstahl and R. J. Perry, *Phys. Rev. C* **75**, 061001(R) (2007).
4. S. Okubo, *Prog. Theor. Phys.* **12**, 603 (1954).
5. K. Suzuki and S. Y. Lee, *Prog. Theor. Phys.* **64**, 2091 (1980).
6. K. Suzuki, *Prog. Theor. Phys.* **68**, 246 (1982).
7. K. Suzuki and R. Okamoto, *Prog. Theor. Phys.* **70**, 439 (1983).
8. K. Suzuki, *Prog. Theor. Phys.* **68**, 1999 (1982).
9. K. Suzuki and R. Okamoto, *Prog. Theor. Phys.* **92**, 1045 (1994).
10. R. Roth, H. Hergert, P. Papakonstantinou, T. Neff and H. Feldmeier, *Phys. Rev. C* **72**, 034002 (2005).
11. R. Roth, T. Neff and H. Feldmeier, *Prog. Part. Nucl. Phys.* **65**, 50 (2010).
12. P. Navratil, J. P. Vary and B. R. Barrett, *Phys. Rev. C* **62**, 054311 (2000).
13. P. Navratil, J. P. Vary and B. R. Barrett, *Phys. Rev. Lett.* **84**, 5728 (2000).
14. P. Navratil, S. Quaglioni, I. Stetcu and B. R. Barrett, *J. Phys. G: Nucl. Part. Phys.* **36**, 083101 (2009).
15. M. A. Caprio, P. Maris and J. P. Vary, *Phys. Lett. B* **719**, 179 (2013).
16. B. R. Barrett, P. Navratil and J. P. Vary, *Prog. Part. Nucl. Phys.* **69**, 131 (2013).
17. S. C. Pieper, V. R. Pandharipande, R. B.Wiringa and J. Carlson, *Phys. Rev. C* **64**, 014001 (2001).
18. S. C. Pieper, R. B. Wiringa and J. Carlson, *Phys. Rev. C* **70**, 054325 (2004).
19. M. Pervin, S. C. Pieper and R. B. Wiringa, *Phys. Rev. C* **76**, 064319 (2007).
20. L. E. Marcucci, M. Pervin, S. C. Pieper, R. Schiavilla and R. B. Wiringa, *Phys. Rev. C* **78**, 065501 (2008).
21. G. Hagen, T. Papenbrock, D. J. Dean and M. Hjorth-Jensen, *Phys. Rev. Lett.* **101**, 092502 (2008).
22. G. Hagen, T. Papenbrock and D. J. Dean, *Phys. Rev. Lett.* **103**, 062503 (2009).
23. G. Hagen, T. Papenbrock, D. J. Dean and M. Hjorth-Jensen, *Phys. Rev. C* **82**, 034330 (2010).

24. R. Machleidt, *Phys. Rev. C* **63**, 024001 (2001).
25. V. G. J. Stoks, R. A. M. Klomp, C. P. F. Terheggen and J. J. de Swart, *Phys. Rev. C* **49**, 2950 (1994).
26. R. B. Wiringa, V. G. J. Stoks and R. Schiavilla, *Phys. Rev. C* **51**, 38 (1995).
27. P. Doleschall, *Phys. Rev. C* **69**, 054001 (2004).
28. D. R. Entem and R. Machleidt, *Phys. Rev. C* **68**, 041001(R) (2003).
29. R. Machleidt and D. R. Entem, *Phys. Rep.* **503**, 1 (2011).
30. B. S. Hu, F. R. Xu, Z. H. Sun, J. P. Vary, and T. Li, *Phys. Rev. C* **94**, 014303 (2016).
31. G. Audi, F. Kondev, M. Wang, B. Pfeiffer, X. Sun, J. Blachot, and M. MacCormick, *Chin. Phys. C* **36**, 1157 (2012).
32. E. D. Jurgenson, P. Maris, R. J. Furnstahl, P. Navrátil, W. E. Ormand, and J. P. Vary, *Phys. Rev. C* **87**, 054312 (2013).
33. A. Tichai, J. Langhammer, S. Binder, and R. Roth, *Phys. Lett. B* **756**, 283 (2016).
34. R. Roth, P. Papakonstantinou, N. Paar, H. Hergert, T. Neff, and H. Feldmeier, *Phys. Rev. C* **73**, 044312 (2006).
35. A. Ekstrom, G. R. Jansen, K. A. Wendt, G. Hagen, T. Papenbrock, B. D. Carlsson, C. Forssen, M. Hjorth-Jensen, P. Navrátil, and W. Nazarewicz, *Phys. Rev. C* **91**, 051301(R) (2015).

Single Neutron Structure of Neutron-Rich N = 50 Nuclei

D. Walter*, A. Ratkiewicz, T. Baugher, J.A. Cizewski and B. Manning
Department of Physics and Astronomy, Rutgers University, New Brunswick, NJ 08901, USA
**E-mail: dwalter@physics.rutgers.edu*

F.M. Nunes and S. Ahn
Department of Physics and Astronomy, Michigan State University, East Lansing, MI 48824, USA

S.D. Pain
Physics Division, Oak Ridge National Laboratory, Oak Ridge, TN 37830, USA

G. Cerizza, C. Thornsberry and K.L. Jones
Department of Physics and Astronomy, University of Tennessee, Knoxville, TN 37996, USA

Single-particle transfer reactions are typically measured at energies where only a peripheral reaction can occur, without probing the interior of the nuclear wave function. At low energies (≈5 MeV/u), spectroscopic factors cannot be reliably extracted without a detailed description of the bound-state potential. Mukhamedzhanov and Nunes have proposed a method to constrain the shape of the bound state potential by combining transfer reaction measurements at two different energies. The external contribution of the wave function is extracted using a peripheral reaction, and is combined with a higher energy measurement which probes the nuclear interior more deeply. These two measurements should constrain the single-particle asymptotic normalization coefficient, ANC, and enable spectroscopic factors to be deduced with uncertainties dominated by the cross-section measurements rather than the bound-state potential. Published measurements of ^{86}Kr(d,p) at 5.5 MeV/u were used to determine the external contribution of this reaction. At less-peripheral energies, ^{86}Kr(d,p) at 35 MeV/u has been measured in inverse kinematics at the NSCL using the ORRUBA and SIDAR arrays of silicon strip detectors. Preliminary analysis shows that the single-particle ANC can be constrained. The details of the analysis and prospects for measurements with neutron-rich beams will be presented.

Keywords: Nuclear Structure; Transfer Reactions; ^{87}Kr; (d,p); Spectroscopic Factors.

1. Introduction

Recent r-process network calculations have highlighted the need to obtain better measurements of nuclear properties such as masses, half lives and neutron capture rates to understand the observed r-process abundances and to constrain the site of these events[1]. Therefore, it is important to measure the properties of neutron-rich nuclei near closed shells and near the r-process waiting points. Single-nucleon transfer reactions are a traditional and reliable tool used to probe the single-particle structure of nuclei[2,3]. The extracted spectroscopic properties can be used to constrain nuclear structure models used to predict properties of even more unstable nuclei.

Reliably extracting spectroscopic information is difficult in exotic, neutron-rich nuclei partly because of the unknown shape of the bound state potential. Mukhamedzhanov and Nunes[4] have proposed measuring the same reaction at two very different energies; at low energy (≈5 MeV/u) to probe the asymptotic region of the nucleus, and at higher energy (≈40 MeV/u) to probe deeper into the interior of nucleus. This approach predicts that the single-particle ANC, $b_{\ell j}$, and the associated bound state potential parameters can be constrained when combinig the two results. The asymptotic normalization coefficient for the nucleus ($C_{\ell j}$) can be extracted with the peripheral reaction. However, the spectroscopic factor will have a strong dependence on the choice of bound-state parameters. At higher energy, there is a larger contribution from the interior of the nucleus. The combination of the two measurements should produce uncertainties in the extracted spectroscopic factor, S, dominated by experimental uncertainties in the cross sections, rather than the unknown radius and diffuseness of the bound state potential.

2. Experiment Details

The neutron transfer reaction ^{86}Kr(d,p)^{87}Kr is used to verify the combined method of extracting spectroscopic information. The low-lying excitations in ^{87}Kr are expected to be single-neutron excitations above the N=50 shell closure. The low-energy analysis is based on published studies of the ^{86}Kr(d,p) reaction in normal kinematics at 5.5 MeV/u by Haravu *et al.*[5] This analysis was combined with the resuls from the same reaction measured at 35 MeV/u.

The first measurement of the (d,p) reaction in inverse kinematics with 35 MeV/u ^{86}Kr beams was conducted at the National Superconducting Cyclotron Laboratory (NSCL). An array of position sensitive silicon strip detectors were coupled to the S800 magnetic spectrograph[6] to measure charged particles and heavy recoils, respectively. SIDAR (SIlicon Detector Array)[7] and ORRUBA (Oak Ridge Rutgers University Barrel Array)[8], covered an angular range between 60° and 165° in the laboratory frame. Elastically scattered products were detected at angles forward of 90° and were used for beam normalization. The reaction was done in inverse kinematics, impinging on a thin ($\approx$0.8 mg/cm^2) deuterated polyethylene (C_2D_4) foil. The ejectile protons were detected at backward angles (>90°) in the laboratory. Protons were detected in coincidence with heavy recoils in the S800, and population of low-lying states in ^{87}Kr was observed.

2.1. *Results*

The measured ground state angular distribution from the NSCL data is displayed in Fig. 1a. Due to complications during the experiment, most of the excited states in ^{87}Kr could not be resolved except for a state at an excitation energy of E = 2.5 MeV, assumed to be the $7/2^+$ state based on previous work[5]. The angular distribution for this state is displayed in Fig. 1b.

Global optical parameters are used for the theoretical calculations within a finite-range adiabatic wave approximation (FR-ADWA) framework[9]. Due to the unknown shape of the bound state potential, each

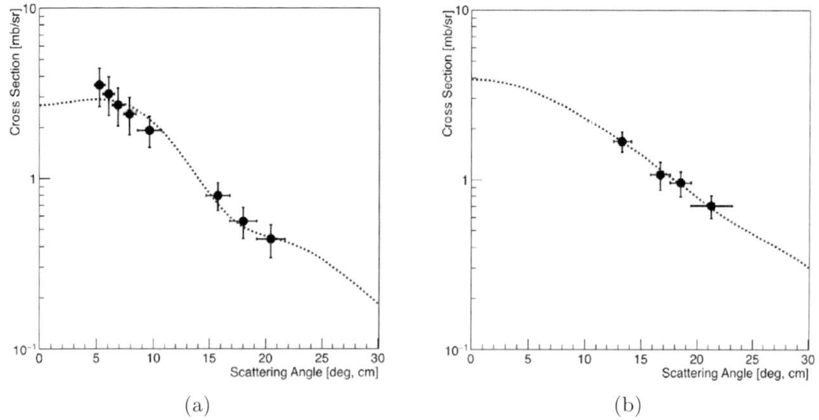

Fig. 1. ^{86}Kr(d,p) reaction at 35 MeV/u differential cross sections (points) as a function of center of mass angle and compared to FR-ADWA calculations (lines) using r_0=1.3 fm and a=0.676 fm parameters for the neutron bound state. The theoretical cross sections were scaled using a least squares fit to the data points. (a) $5/2^+$ ground state of ^{87}Kr with S = 0.41. (b) $7/2^+$ 2.5 MeV state of ^{87}Kr with S = 0.34.

angular distribution was fit with a variety of FR-ADWA calculated cross sections to produce a range of single-particle ANC, $b_{\ell j}$, by changing only the radius (r_0) and diffuseness (a) of the bound state potential. The same method was applied to the low energy results to produce Fig. 2 and Fig. 3.

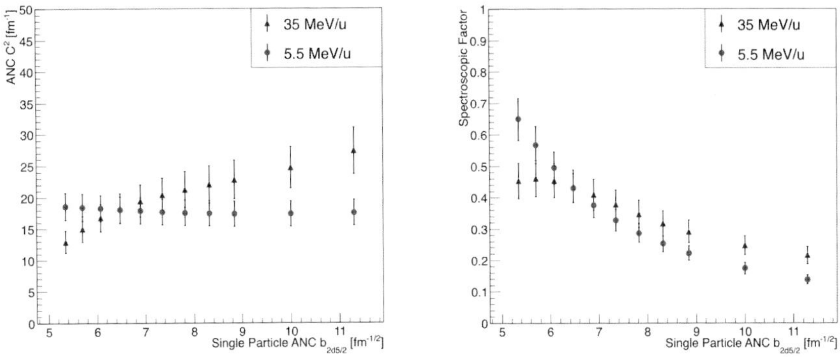

(a) (b)

Fig. 2. Results from the FR-ADWA analysis of the ^{87}Kr ground-state measurements at 5.5 MeV/u (circles) and 35 MeV/u (triangles). (a) Nuclear ANC C^2 values as a function of single-particle ANC $b_{\ell j}$. (b) Spectroscopic factors as a function of single-particle ANC b_{ℓ}, j.

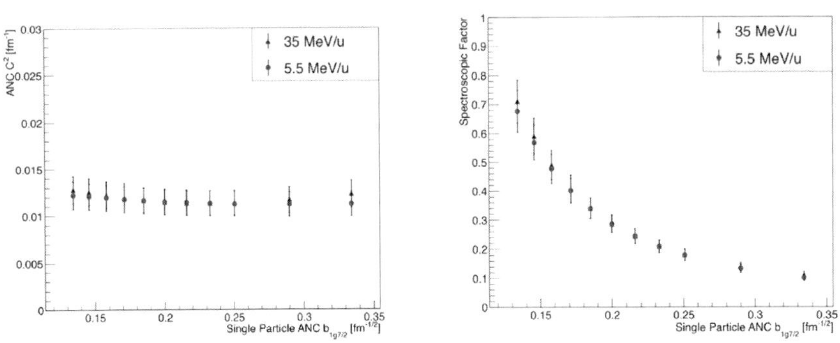

(a) (b)

Fig. 3. Results from the FR-ADWA analysis of the ^{87}Kr 2.5 MeV $1g_{7/2}$ state measurements at 5.5 MeV/u (circles) and 35 MeV/u (triangles). (a) Nuclear ANC C^2 values as a function of single-particle ANC $b_{\ell j}$. (b) Spectroscopic factors as a function of single-particle ANC b_{ℓ}, j.

The combination of the two measurements constrains the value of the single-particle ANC $b_{\ell j} \approx 6.5$ fm$^{-1/2}$ for the ground state populated in ^{86}Kr(d,p)^{87}Kr, consistent with radius and diffuseness of $r_0 \approx 1.3$ fm and $a \approx 0.68$ fm. In contrast, the single-particle ANC for the E = 2.5 MeV excited state cannot be constrained with the analysis in Fig. 3. The transfer to this state involves a larger angular momentum transfer of $\ell = 4$ opposed to $\ell = 2$ for the ground state reaction. It appears that the excited state reaction, even at 35 MeV/u, remains peripheral because of the higher centrifugal barrier to transfer.

3. Summary and Outlook

The ^{86}Kr(d,p) reaction was measured for the first time in inverse kinematics at 35 MeV/u at the National Superconducting Cyclotron Laboratory with the ORRUBA and SIDAR detector systems coupled to the S800 spectrometer. The results from this measurement were combined with those from an earlier study at 5.5 MeV/u to test the validity of the method proposed by Mukhamedzhanov and Nunes[4] to constrain the parameters of the neutron bound-state potential. The present results demonstrate that for relatively low angular momentum transfer ($\ell \leq 2$) spectroscopic factors can be deduced with uncertainties dominated by statistics rather than the unknown bound-state potential.

Future prospects of the combined method with radioactive ion beams in inverse kinematics include the ^{84}Se(d,p)^{85}Se neutron transfer reaction at 40 MeV/u at the NSCL. This measurement would be combined with a previous measurement of ^{84}Se(d,p) at 4.5 MeV/u completed at the Holifield Radioactive Ion Beam Facility at Oak Ridge National Laboratory[10].

Acknowledgments

This work was supported in part by the National Science Foundation NSF-PHY-1404218 (Rutgers) and NSF-PHY-0555893 (MSU) and the U.S. Department of Energy under Contracts No. DE-NA0002132 (Rutgers), No. DE-AC05-00OR22725 (ORNL), No. DE-FG02-96ER40983 (UTK), No. DE-SC0001174 (UTK), and No. DE-FG02-96ER40955 (TTU).

References

1. M. R. Mumpower, R. Surman, G. C. McLaughlin and A. Aprahamian, *Prog. Part. Nucl. Phys.* **86**, 86 (2016), [Erratum: *Prog. Part. Nucl. Phys.* **87**, 116 (2016)].

2. G. Satchler, *Direct Nuclear Reactions* (Clerendon Press, 1983).
3. N. Austern, *Direct Nuclear Reaction Theories* (John Wiley and Sons Inc., 1970).
4. A. M. Mukhamedzhanov and F. M. Nunes, *Phys. Rev. C* **72**, p. 017602 (Jul 2005).
5. K. Haravu, C. L. Hollas, P. J. Riley and W. R. Coker, ^{86}Kr$(d,p)^{87}$Kr, *Phys. Rev. C* **1**, 938 (Mar 1970).
6. D. Bazin, J. Caggiano, B. Sherrill, J. Yurkon and A. Zeller, The {S800} spectrograph, *Nuclear Instruments and Methods in Physics Research Section B: Beam Interactions with Materials and Atoms* **204**, 629 (2003).
7. D. W. Bardayan *et al.*, *Phys. Rev. C* **63**, p. 065802 (May 2001).
8. D. W. Bardayan *et al.*, *Nuclear Instruments and Methods in Physics Research Section A: Accelerators, Spectrometers, Detectors and Associated Equipment* **711**, 160 (2013).
9. N. B. Nguyen, F. M. Nunes and R. C. Johnson, Finite-range effects in (d,p) reactions, *Phys. Rev. C* **82**, p. 014611 (Jul 2010).
10. J. S. Thomas *et al.*, *Phys. Rev. C* **76**, p. 044302 (Oct 2007).

Nuclear Astrophysics

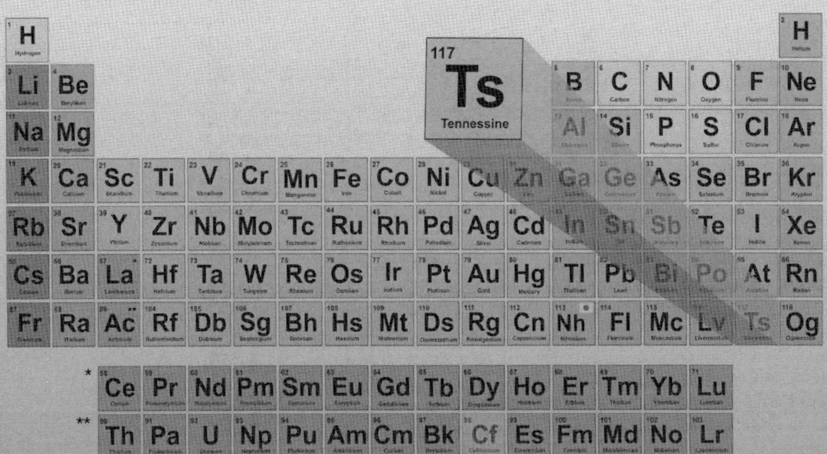

Nuclear Clusters in Astrophysics

Carl Wheldon* and Tzany Kokalova Wheldon
School of Physics and Astronomy, University of Birmingham,
Edgbaston, Birmingham, B15 2TT, United Kingdom
**E-mail: c.wheldon@bham.ac.uk*
www.ph.bham.ac.uk

Nuclear clusters are intricately linked to astrophysical processes involving light nuclei. Here, several examples of the role played in the evolution of the universe by these exotic structures in light nuclei are given, concluding with an outlook for future direct measurements.

Keywords: Nuclear Clusters; Astrophysical Processes, Radioactive Beams; Direct Measurements; Particle Spectroscopy, Ejectile-Tagged Recoil Break-Up; Resonant Scattering; Thick Target in Inverse Kinematics (TTIK).

Introduction

Nuclear energy levels in which the protons and neutrons clump into sub units are called cluster states, and their special structure plays a key role in reactions important for many astrophysical processes. The most ubiquitous cluster is the α particle, due to its high binding energy ($B.E.$=28.296 MeV) and high break-up threshold (S_p=19.814 MeV [1]). The presence of a 'pre-formed' light nucleus such as an α particle increases both the decay probability and also the formation probability of channels and reactions involving such species. In addition to the purely astrophysical interest, nuclear cluster states also enable precise tests of nuclear models.

The simplest cluster state is the ^{8}Be ground-state resonance that lies 91.8 keV above the two-α separation threshold [1]. Despite its short-lived nature, Γ =5.57(25) eV [2] ($\equiv 5.88\times10^{-17}$ s), it has played a key role in both testing early *ab initio* models due to the relatively small number of nucleons and is a pivotal stepping stone in helium burning that takes place in red giant stars. A good example of the former is the Green's function Monte-Carlo calculations of Wiringa *et al.* [3] as shown in Fig. 1. Here, the double-peaked nature of the intrinsic density is clearly evident, interpreted as a two-α structure.

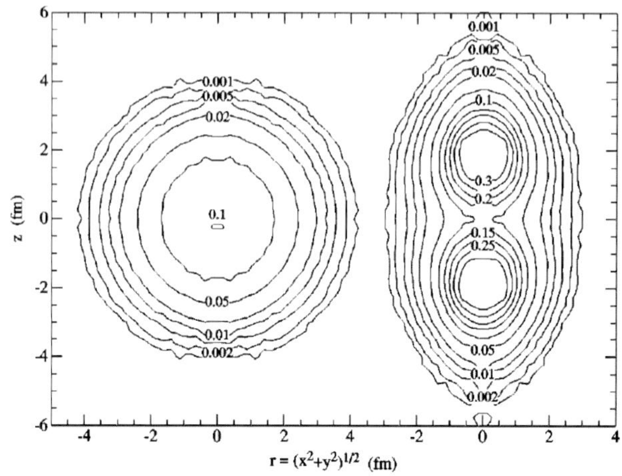

Fig. 1. Contour plot of the 0^+ ^{8}Be ground state density plotted in cylindrical coordinates with z as the quantisation axis. Left: standard density; right: intrinsic density — the two-peaked distribution of the two-α structure is clear. Reproduced from Ref. [3].

The most famous cluster state in nuclei is the Hoyle state in ^{12}C [4], named after Sir Fred Hoyle, the astronomer who predicted its existence [5] based on astrophysical arguments. Helium burning is the process that leads to the creation of ^{12}C from α particles in stars at temperatures of $\geq 10^7$ K. The first step in this triple α process, as alluded to above, is the $\alpha + \alpha \rightleftharpoons {}^8$Be reaction, resulting in an equilibrium quantity of ^{8}Be. The Hoyle state lies 287 keV above the $\alpha+{}^8$Be threshold; energy which can be obtained from the thermal environment within a star. Thus, the Hoyle state provides a resonance route to ^{12}C. Furthermore, its 0^+ spin and parity negate any centrifugal barrier and the 0.04% of its total Γ=9.3 eV width that corresponds to γ decay, ultimately provides a route to the stable ground-state of ^{12}C. The production of ^{12}C in stars via this multi-step process is enhanced by eight orders of magnitude due to the existence and properties of the Hoyle resonance.

Oxygen-16 and stellar burning

Though significant work is underway to fully explain all the properties of ^{12}C, and in particular the Hoyle state, much is already known, and the aims now are to increase precision and explore the higher excitation energy regimes to observe the rotational spectrum of the Hoyle state in full [6–9]. As a result, attention has turned to ^{16}O and heavier α-conjugate nuclei such as ^{20}Ne *etc.* Cluster states in these nuclei are important, again in the multi-α threshold region, as these can boost the stellar burning reaction rates in which, as the stellar temperature increases, successive α-capture reactions on heavier products of earlier reactions, can take place such as ^{12}C(α,γ)^{16}O and ^{16}O(α,γ)^{20}Ne...[14]. In the search for a Holye-state analogue, most recent attention has been focused on a candidate that lies just above the four-α threshold ($S_{4\alpha}$=14.437 MeV [1]) at $\approx$15.1 MeV [10, 11]. However, the properties of this state are proving difficult to study due the increased density of states at this relatively large excitation energy — an issue that becomes ever more acute with the increasing multi-α break-up thresholds as heavier $N\alpha$ nuclei are probed. However, as for ^{12}C, for a structure to play a significant role in an astrophysical environment, its spin should be 0 to minimize the barrier and yield favorable reaction rates. Therefore, the powerful technique of measuring reactions at 0° can be used where the contributions of 0^+ states peak and all others are minimized. This is already shedding light on the pertinent excitation energy regime in ^{16}O as can be seen in the spectrum of Fig. 2 [12], obtained using the ^{16}O(α,α') reaction at 200 MeV at iThemba Laboratory, South Africa. The K600 spectrometer for recording the scattered α ejectile at 0° was coupled with the CAKE large-acceptance position-sensitive silicon detector array to register the break-up

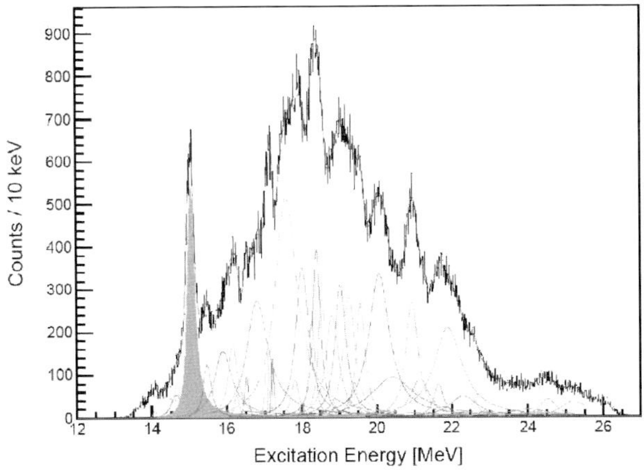

Fig. 2. Spectrum from ^{16}O(α,α') scattering at 200 MeV using the K600 spectrometer. The spectrum shown is for coincidences, detected in the CAKE silicon detector array, with the α_1 break-up particles to the first excited (4.4 MeV) state in ^{12}C. The prominent shaded peak is the 15.097 MeV four-α-cluster candidate. The contributions from other states in the region are shown by unfilled distributions. Adapted from Ref. [12].

particles — ejectile-tagged recoil break-up. The candidate Hoyle analogue is clearly visible in Fig. 2 as the shaded peak. This analysis is ongoing to elucidate the properties of this and nearby levels in this energy region.

In the next heaviest α-conjugate nucleus, ^{20}Ne, the first steps have been made to explore the already high excitation energies that are needed to reach above the five-α threshold in ^{20}Ne at 19.167 MeV [1]. In Ref. [13] the authors report resonances produced in the ^{16}O(α,^{20}Ne*) resonant reaction during which beam energies from 23.0 to 29.0 MeV were used to build up an excitation function. Significant resonance strength in both the ^{8}Be+^{8}Be+α and ^{8}Be+^{12}C(0_2^+)[Hoyle] channels was observed above 24.5 MeV providing a basis for future work.

Nuclear molecules and exotic clusters

While much current astrophysics-focused cluster investigations have concentrated on even-even N=Z nuclei and the energy regions around the multi-α thresholds, the so called 'exotic-cluster' nuclei and nuclear molecules are also receiving increase attention due to their roles in key astrophysical processes. The term nuclear molecules is generally applied to clustered nuclear states bound by one or more valence neutrons [15, 16]. Here, clusters that are otherwise unbound become stable through the exchange of a neutron, adding extra binding. Again, the simplest example is found in the ^{9}Be nucleus [17], a nucleus formed in primordial nuclear synthesis [18]. However, due to the large widths of states, even structures at low excitation energy are still not fully understood [19]. As discussed earlier, such nuclei provide a test bed for nuclear models. Beryllium-9 is of interest for many reasons. It provides a test of cosmological theories predicting big-bang abundances [18]. Also, not only is the structure of ^{9}Be interesting in its own right from the manifestation of diverse features [17, 20], but it provides insight into the structure of its unstable mirror partner, ^{9}B [19, 21]. The neutron-rich beryllium isotopes exhibit some of the clearest signatures of nuclear molecular behavior [22, 23] and now these are being extended to heavier molecular configurations such as ^{14}C [24, 25] and ^{18}O [26, 27], with stability predictions being a key test of theory [28].

Exotic clustering is the regime away from the valley of stability, in which, for example, neutron-deficient clusters contribute. One such case is ^{19}Ne. Over the last decade interest in this nucleus above the α-decay threshold (S_α=3.528 MeV [1]) has surged due to its role in the break out from the hot-CNO cycles which takes place through the ^{15}O(α,γ)^{19}Ne reaction [29]. This break-out triggers the rapid proton capture (rp) process which is the cause of X-ray bursts during which a colossal amount of energy is released within a few seconds from the rapid conversion of mass-50 to mass-110 isotopes [30]. Additionally, ^{19}Ne resonances above the proton threshold (S_p=6.410 MeV [1]) are important in the destruction of ^{18}F via ^{18}F(p,α)^{15}O. Fluorine-18 ($t_{1/2}$=110 m [31]) is a key radionuclide, which, from its γ-emission can be observed to provide evidence of nucleosynthesis [32].

Clearly, the nature of exotic clustering implies that one of the clusters can be radioactive. In the case of ^{19}Ne investigating the possible ^{15}O+α structure directly is difficult as ^{15}O is short-lived; $t_{1/2}$=122 s [33]. Recently, this and other nuclei have become accessible through the use of the Thick Target in Inverse Kinematics (TTIK) technique coupled with resonant scattering [34]. Here, a radioactive beam is elastically scattered from an extended gas target, *e.g.* ^{4}He, which also acts as an energy-loss medium for the beam. The lighter scattered particles are able to reach an array of position-sensitive silicon detectors due to their lower energy loss. The nature of this technique makes it ideally suited to low-intensity radioactive beams (10^4–10^5 particles per second) with a large energy profile. The slowing down of the beam through the gas means with one beam energy, a large excitation energy regime can be studied, making this a very efficient technique. The strength of any resonances using this method is directly related to their α structure in the case where ^{4}He gas is used as the target.

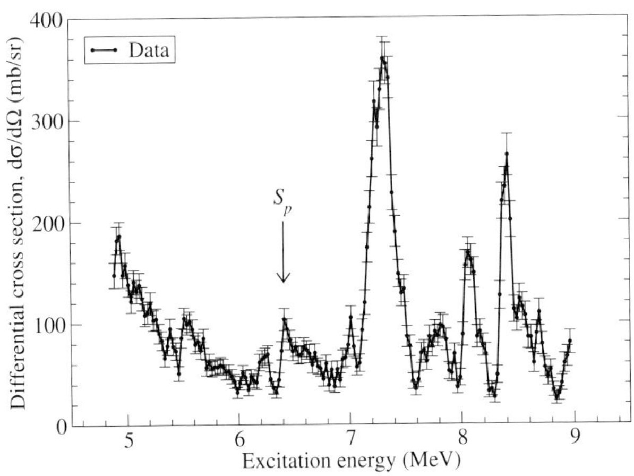

Fig. 3. Resonant scattering with the TTIK method using a radioactive beam ^{15}O produced by Exotic at LNL incident on a thick ^{4}He gas target. The spectrum was produced from measuring elastically scattered α particles at 0° in a position-sensitive silicon detector.

Figure 3 shows a provisional spectrum for ^{19}Ne for energies spanning the proton threshold. This was obtained by studying the *direct* ^{15}O(α,α) reaction at the EXOTIC facility [35, 36] at Legnaro National Laboratory, Italy. The excitation function, obtained by detecting the scattered α particles at 0°, shows prominent features indicative of states with significant α clustering.

Many other systems are accessible via this method and with the advent of sophisticated R-matrix fitting software [37, 38], the information that can be extracted from resonant scattering spectra is increasing.

Outlook

Nuclear clusters lie at the heart of many astrophysical processes and the increasing precision of both measurements and theory coupled with the availability of ever more intense radioactive beams is making investigations of direct processes viable. New analysis tools [39, 40] are improving the ability to extract information on the structure of levels, thus enabling the influence on stellar processes to be quantitatively assessed.

Acknowledgments

Some of the work presented has received funding from the European Union's Horizon 2020 research and innovation program under the Marie Skłodowska-Curie grant agreement No 659744 and the UK STFC under grant number ST/L005751/1.

References

[1] M. Wang, G. Audi, A.H. Wapstra, F.G. Kondev, M. MacCormick, X. Xu and B. Pfeiffer, The AME2012 atomic mass evaluation (II). *Chin. Phys. C* **36**, 1603 (2012).

[2] R.E. Warner, A. Okihana, M. Fujiwara, N. Matsuoka, S. Kakigi, S. Hayashi, K. Fukunaga, J. Kasagi and M. Tosaki, ^{7}Li(α,2α)^{3}H reaction from 77 to 119 MeV, *Phys. Rev. C* **45**, 2328 (1992).

[3] R.B. Wiringa, Steven C. Pieper, J. Carlson and V.R. Pandharipande, Quantum Monte Carlo calculations of $A = 8$ nuclei, *Phys. Rev. C* **62**, 014001 (2000).

[4] M. Freer and H.O.U. Fynbo, The Hoyle State in ^{12}C, *Prog. in Part. and Nucl. Phys.* **78**, 1 (2014).

[5] F. Hoyle, D.N.F. Dunbar, W.A. Wenzel and W. Whaling, A state in ^{12}C predicted from astrophysical evidence, *Phys. Rev.* **92**, 1095 (1953).
[6] M. Freer *et al.*, Consistent analysis of the 2^+ excitation of the ^{12}C Hoyle state populated in proton and α-particle inelastic scattering, *Phys. Rev. C* **86**, (2012) 034320.
[7] W.R. Zimmerman *et al.*, Unambiguous identification of the second 2^+ state in ^{12}C and the structure of the Hoyle state, *Phys. Rev. Lett.* **110**, 152502 (2013).
[8] M. Freer *et al.*, Evidence for a new ^{12}C state at 13.3 MeV, *Phys. Rev. C* **83**, 034314 (2011).
[9] D.J. Marín-Lámbarri, R. Bijker, M. Freer, M. Gai, Tz. Kokalova, D.J. Parker and C. Wheldon, Evidence for triangular $\mathcal{D}_{3h}$ symmetry in ^{12}C, *Phys. Rev. Lett.* **113**, (2014) 012502.
[10] Y. Funaki, T. Yamada, H. Horiuchi, G. Röpke, P. Schuck and A. Tohsaki, Alpha-particle condensation in ^{16}O studied with a full four-body orthogonality condition model calculation, *Phys. Rev. Lett.* **101**, 082502 (2008).
[11] C. Wheldon *et al.*, High-resolution measurement of absolute α-decay widths in ^{16}O, *Phys. Rev. C* **83**, 064324 (2011) and references therein.
[12] R. Neveling *et al.*, Characterization of a potential 4-α cluster state, in *Proc. Nuclear Reaction Mechanisms 2015*, (CERN conference proceedings, Elsevier, 2016) p. 79.
[13] Tz. Kokalova *et al.*, Yield measurements for resonances above the multi-α threshold in ^{20}Ne, *Phys. Rev. C* 87, 057309 (2013).
[14] P. Dyer and C.A. Barnes, The ^{12}C(α,γ)^{16}O reaction and stellar helium burning.
[15] W. von Oertzen, Martin Freer and Yoshiko Kanada-En'yo, Nuclear clusters and nuclear molecules, *Phys. Rep.* **432**, 43 (2006).
[16] Martin Freer, The clustered nucleus — cluster structures in stable and unstable nuclei, *Rep. Prog. Phy.* **70**, 2149 (2007).
[17] W. von Oertzen, Two-center molecular states in ^{9}B, ^{9}Be, ^{10}Be and ^{10}Be, *Z. Phys. A* **354**, 37 (1996).
[18] R.N. Boyd and T. Kajino, Can ^{9}Be provide a test of cosmological theories?, *Astro. J.* **336**, L55 (1989).
[19] R. Smith *et al.*, Evidence for a 3.8 MeV state in ^{9}Be, *Phys. Rev. C* **94**, 014320 (2016).
[20] Wolfram von Oertzen, Nuclear clusters and covalently bound nuclear molecules, *Prof. of Theoretical, Phys. Suppl.* No. **146**, 169 (2002).

[21] C. Wheldon *et al.*, Spectroscopy of ^{9}B via high-resolution ejectile-tagged recoil break-up, *Phy. Rev. C* **91**, 0243308 (2015).
[22] M. Freer *et al.*, α:$2n$:α molecular band in ^{10}Be, *Phys. Rev. Lett.* **96**, 042501 (2006).
[23] M. Freer *et al.*, Exotic molecular states in ^{12}Be, *Phys. Rev. Lett.* **82**, 1383 (1999).
[24] M. Freer *et al.*, Resonances in ^{14}C observed in the ^{4}He(^{10}Be,α)^{10}Be reaction, *Phys. Rev. C* **90**, 054324 (2014).
[25] H. Yamaguchi *et al.*, Experimental investigation of a linear-chain structure in the nucleus ^{14}C, *Phys. Lett. B* **766**, 11 (2017).
[26] E.D. Johnson *et al.*, Extreme α-clustering in the ^{18}O nucleus, *Eur. Phys. J. A* **42**, 135 (2009).
[27] W. von Oertzen *et al.*, Molecular and cluster structures in ^{18}O, *Eur. Phys. J. A.* **43**, 17 (2010).
[28] N. Itagaki, S. Okabe, K. Ikeda and I. Tanihata, Molecular-orbital structure in neutron-rich C isotopes, *Phys. Rev. C* **64**, 014301 (2001).
[29] M. Wiescher *et al.*, Astrophysical nuclear reactions and the break-out from the hot CNO cycles, *Prog. Part. and Nucl. Phys.* **59**, 51 (2007).
[30] W.P. Tan *et al.*, ^{15}O(α,γ^{19}Ne) breakout reaction and impact on X-ray bursts, *Phys. Rev. Lett.* **96**, 242503 (2007) and references therein.
[31] H. Schrader, Measurement of the half-lives of ^{18}F, ^{56}Co, ^{125}I, ^{195}Au and ^{201}Tl, *Appl. Radiat. Isot.* **40**, 381 (1989).
[32] A.M. Laird *et al.*, Is γ-ray emission from novae affected by interference effects in the ^{18}F(p,α)^{15}O reaction?, *Phys. Rev. Lett.* **110**, 032502 (2013).
[33] G. Azuelos, J.E. Kitching and K. Ramavataram, Half-lives and branching ratios of some T=1/2 nuclei, *Phys. Rev. C* **15**, 1847 (1977).
[34] K.P. Artemov *et al.*, [*translation*] An effective method for investigating α-clustered states, *Sov. J. Nucl. Phys.* **52**, 408 (1990).
[35] V.Z. Maidikov *et al.*, The optics of the exotic beam line at LNL, *Nucl. Phys. A* **746**, 389c (2004).
[36] M. Mazzocco *et al.*, Upgrade of the facility EXOTIC for the in-flight production of light Radioactive Ion Beams, *Nucl. Instr. and Meth. in Phys. Res. B* **317**, 223 (2013).
[37] N. Larson, SAMMY8: A code system for multilevel R-matrix fits to neutron data using Bayes' equations, ORNL/TM-9179/R5 (2000) unpublished.
[38] R.E. Azuma *et al.*, AZURE: An R-matrix code for nuclear astrophysics, *Phys. Rev. C* **81**, 045805 (2010).

[39] Sam Bailey, New analytical techniques for the investigation of alpha clustering in nuclei, PhD Thesis, University of Birmingham (2017), `http://etheses.bham.ac.uk/7150/`

[40] N. Curtis and J. Walshe, Rex: A Monte Carlo simulation of thick gas target resonant scattering reactions, *Nucl. Instr. and Meth. in Phys. Res. A* **797**, 44 (2015).

Experiences and Prospects of Nuclear Astrophysics in Underground Laboratories

M. Junker*

Istituto Nazionale di Fisica Nucleare
Laboratori Nazionali del Gran Sasso
67100 L'Aquila, Italy
**E-mail: matthias.junker@junker@lngs.infn.it*
www.lngs.infn.it

Impressive progress has been made in the course the last decades in understanding astrophysical objects. Increasing precision of nuclear physics data has contributed significantly to this success, but now a better understanding of several important findings is frequently limited by uncertainties related to the available nuclear physics data. Consequently it is desirable to improve significantly the quality of these data. An important step towards higher precision is an excellent signal to background ratio of the data. Placing an accelerator facility inside an underground laboratory to reduce the cosmic ray induced background by six orders of magnitude is a powerful method to reach this goal. Nevertheless careful reduction of environmental and beam induced background must still be considered. Experience in the field of underground nuclear astrophysics has been gained since 20 years due to the pioneering work of the LUNA Collaboration (Laboratory for Underground Nuclear Astrophysics) operating inside the underground laboratories of the Laboratori Nazionali del Gran Sasso (LNGS) in Italy. Based on the success of this work presently also several other projects for underground laboratories dedicated to nuclear astrophysics are being pursued worldwide. This contribution gives a survey of the past experience in underground nuclear astrophysics as well as an outlook on future developments

Keywords: Experimental Nuclear Astrophysics, Underground Laboratory
PACS: 26.20.-f, 26.30.Ca, 26.65.+t, 29.20.Ba.

1. Introduction

Accurate knowledge of thermonuclear reaction rates is important in understanding the energy generation, the neutrino luminosity and the element synthesis in stars. Due to the Coulomb barrier (height E_c) of the entrance channel, the reaction cross section $\sigma(E)$ drops nearly exponentially with decreasing energy. Thus it becomes increasingly difficult to measure $\sigma(E)$ and to deduce the astrophysical S factor defined by the equation

$\sigma(E) = S(E)/E \cdot \exp(-2\pi\eta)$, where $2\pi\eta = 31.29 \cdot Z_1 \cdot Z_2\sqrt{(\mu/E)}$ denotes the Sommerfeld, Z_1 and Z_2 are the nuclear charges of the interacting particles in the entrance channel, μ is the reduced mass in amu, and E is the center-of-mass energy in keV.[1]

Experimental techniques have improved significantly to push the limit of measurements close to or even inside the thermal energy region in stars, which is determined by the Gamow window $E_0 \pm \delta E_0$. It frequently lies far below the height of the Coulomb barrier, approximately at $E_0/E_c = 0.01$. Even in those cases in which direct measurements prove to be feasible, uncertainties in low energy stopping power data or the insufficient understanding of the electron screening effect introduce significant systematic uncertainties.[2]

As a result extrapolations of data from higher energies using e.g. R-matrix fits[3] are often required. In order to minimize the risks related to the extrapolation, the data sets used should cover a wide range of beam energies. Furthermore is important to aim for direct measurements at energies as close as possible to the Gamow Peak maintaining small statistical and systematical errors. Scattering experiments or transfer reactions provide important additional information. Indirect methods like the Troian Horse technique can be exploited as an alternative tool to determine $\sigma(E)$.[2]

In the following we will focus on direct experiments at underground accelerators as a powerful tool to obtain nuclear cross sections inside the Gamow peak.

2. Background in detectors and its reduction

Due to their small reaction rate nuclear astrophysics experiments have to compete with background originating from cosmic rays (CR) and environmental radioactivity. At earth surface CR are mainly composed of muons. As an effect of their high energy, CR can penetrate also massive shielding even though the energy deposition in a specific detector is related to its material and shape. Also secondary particles caused by CR interaction in surrounding materials interact with the detector. While the direct interactions of CR can be efficiently rejected by active shielding (e.g. with plastic scintillators), events related to secondary particles (e.g. neutrons) are difficult to suppress and limit the efficiency of active shielding. Only the very massive shielding available in underground laboratories can provide a significant reduction of the CR flux of several orders of magnitude.[4] This approach is commonly exploited by low event rate experiments in the field of Dark Matter research and Neutrino physics while presently only the LUNA[a] experiment performs accelerator-based nuclear astrophysics experiments in an environment deep underground.[5]

Environmental radioactivity is another source of background which reaches gamma energies of up to 3.5 MeV. It is caused by radionuclides belonging to natural radioactive series (^{214}Bi, ^{226}Ra, ^{214}Pb, ^{208}Tl...), long-lived natural radionuclides (^{40}K, ^{87}Rb ..), radionuclides of cosmogenic origin (^{3}H, ^{14}C, ^{22}Na, ^{7}Be..) and radionuclides of artificial origin (^{95}Nd, ^{137}Cs, ^{90}Sr, ^{144}Ce..). These nuclides can be found outside but also inside the detectors themselves. The related background can be reduced by careful material selection, by cleaning and by massive passive shielding with high purity lead and copper. As CR induce secondary particles when interacting with passive shielding, the reduction of the CR flux inside an underground laboratory makes massive Pb and Cu shielding more efficient underground than above ground. Furthermore, the higher the rock coverage of a site the higher the reduction of the underground CR flux and the higher the suppression of background achieved by lead and copper shielding in the gamma energy range up to 3.5 MeV.[6]

The interaction of the ion beam with traces of material on the target can give rise to beam induced background which is specific for each reaction. In particular materials with atomic numbers lower than or similar to the target material are candidates for beam induced backgrounds. Consequently each reaction must be treated individually considering the involved beam energy and particle spectrum as well as the Q-value and the resonance structure of the background reactions. The type of the target (gaseous or solid) and the target production process are of fundamental importance. In some particular cases background can be induced also by target nuclei scattered by the impinging projectiles.[7]

The excellent signal to background ratio in the high resolution spectra of HPGe detectors favors the identification of the background, but they provide a limited efficiency with respect to high density materials like BGO or suffer consistent intrinsic background (e.g. $LaBr_3$). On the other hand detectors with low resolution disfavor a reliable background recognition. In any case, the spectra obtained when operating underground significantly improve the identification and subsequent suppression of beam induced background.

[a]Laboratory for Underground Nuclear Astrophysics; `http://luna.lngs.infn.it`

3. The LUNA experiment at LNGS

Presently, the LUNA Collaboration operates the worldwide only underground accelerator inside the Laboratori Nazionali del Gran Sasso (LNGS) which is part of the Italian Istituto Nazionale di Fisica Nuclear (INFN)[b]. The underground laboratories consist in three halls (100 m long, 20 m wide, 18 m high). They are covered by 1400 m of rock corresponding to 3.800 meter of water equivalent (m.w.e.) which reduce the cosmic muon flux be six orders of magnitude.

Located long side the motorway Teramo – Rome the underground site is accessible even for trucks of extra width and length. Inside the laboratories lifting equipment up to 40 t, ventilation and 3 MW electrical power provided by redundant sources are available. Plants relevant for safety (e.g. ventilation and fire detection) and selected experiment specific information are monitored by a laboratory wide supervision system. Firefighters and guards are available permanently in the underground control room.

LNGS hosts clean rooms usable for material cleaning, assembling and decontamination, a chemistry laboratory equipped, among others, with an ICPMS facility, electronic and mechanical workshops and an experienced engineering division. The LNGS facility STELLA (SubTErranean Low Level Assay) is available for measurements of low radioactivity samples. LNGS is a reference point for a vivid international scientific community involving more than 950 scientist[8].

The LUNA collaboration started its activities at LNGS in 1992 installing a home made 50 kV accelerator and is operating a 400 kV Singletron machine since year 2000.[9] Currently the measurements at LUNA are focused on nuclear synthesis in intermediate-mass asymptotic giant branch (AGB) or super AGB stars, fast rotating massive stars, massive and supermassive stars in close binary systems, stellar collisions and classical novae.[10] Reviews of the scientific activities of LUNA and their impact on astrophysics, solar and neutrino physics can be found in a topical issue on Underground Nuclear Astrophysics and Solar Neutrinos of the European Physical Journal A.[11–17]

[b]http://www.lngs.infn.it

4. Upcoming possibilities in underground nuclear astrophysics

The results obtained by the LUNA-Collaboration prove underground nuclear astrophysics as a powerful tool not only for pushing measurements to lower energies but also, and even more, for providing high quality data which can be used to reduce the uncertainties of unavoidable extrapolations.

Stimulated by the success of the LUNA experiment several next generation facilities for underground nuclear astrophysics have been proposed in the last years. A series of workshops has been organized in Europe (Barcelona 2009, Dresden 2010, Gran Sasso 2011[18], Canfranc 2012, Gran Sasso 2013, 2016[20])[c]. In its Long Range Plan 2010[19] NUPECC recommends to "fully exploit the currently existing large-scale research infrastructures (...) and (to) perform limited-size upgrades to ensure the best use of the large investments made in the past: (...)" and in particular "the nuclear astrophysics underground accelerator LUNA at INFN Gran Sasso, and the exploration of advanced new facilities."[19]

In 2012 and 2013 the Italian Ministry of Research has granted INFN with financial support dedicated to the LUNA-MV project. In March 2016 INFN assigned the contract for constructing and installing the LUNA-MV accelerator to High Voltage Engineering Europe (HVE). The single ended machine with a maximum Terminal Voltage (TV) of 3.5 MV will deliver intense Proton, Helium and Carbon beams (1, 0.5 and 0.15 mA respectively) with well defined energy definition (0.01 % of TV) and stability (0.001 $\%h^{-1}$ of TV). The installation at LNGS is scheduled to be concluded by December 2018, after it has been fully tested in the factory and, subsequently, at the LUNA-MV site.

As the impact of the neutrons produced by the LUNA-MV accelerator was of concern, this issue has been evaluated by an external committee of experts. It has been shown that the neutron background of the proposed experiments can be reduced to a level which is lower than the natural neutron flux inside LNGS. For this the machine and all experimental setups are installed inside a room made out of 80 cm concrete walls and shielding

---

[c]Web pages of the workshops on next generation underground nuclear astrophysics installations:
http://www.hzdr.de/db/Cms?pNid=2088
http://nuclear.fis.ucm.es/nucastro//na_lsc_2012/workshop-2012.html
https://agenda.infn.it/conferenceDisplay.py?confId=5450
https://agenda.infn.it/conferenceDisplay.py?ovw=True\&confId=11579

doors. Presently the building and the related technical plants have been fully engineered and are currently being tendered.

A scientific proposal "LUNA–MV 2018–2022" has been approved by the Scientific Committee of LNGS in spring 2016. It is mainly devoted to the study of Helium and Carbon burning and in particular the reactions $^{14}N(p,\gamma)^{15}O$, the bottleneck reaction of the CNO cycle; $^{12}C+^{12}C$, the trigger of carbon burning as well as $^{13}C(\alpha,n)^{16}O$ and $^{22}Ne(\alpha,n)^{25}Mg$ for the production of isotopes A$\geq$90 in the frame work of the s–process.[20]

In 2009 CUNA (Canfranc Underground Nuclear Astrophysics) has been proposed at the Laboratorio Subterráneo de Canfranc (LSC) located under the Pyrenees (2.500 m.w.e.).[d] The accelerator-based nuclear astrophysics project aims for the installation of a multi MV accelerator coupled with a RF source delivering proton and alpha beams to investigate the sources of neutrons in the stars and other reactions of astrophysical interest, e.g. $^{13}C(\alpha,n)^{16}O$ and $^{22}Ne(\alpha,n)^{25}Mg$

An engineering design study for an independent experimental hall, distant from the rest of the caves at the LSC has been completed. Detailed simulations and measurements have been performed to study the natural and the accelerator induced neutron background at the CUNA site and in the other experimental areas of LSC. Design studies for high efficiency 4π neutron and gamma detectors have been performed, too.

CUNA is part of the strategic plans for the LSC facility up to 2020/2021 and has a strong support by LSC. Furthermore expertise on neutron and gamma detection is available within collaboration and part of the equipment already exists. Only the accelerator itself is not yet financed.[20]

In the USA the DIANA project (Dakota Ion Accelerators for Nuclear Astrophysics), a collaboration of University of Notre Dame, University of North Carolina, Western Michigan University, Colorado School of Mines and Lawrence Berkeley National Laboratory aimed to build a nuclear astrophysics underground accelerator facility inside the Sandford Underground Research Facility SURF[e] covered by 4.136 m.w.e. DIANA was designed to achieve large laboratory reaction rates by delivering ion beam currents up to 100 mA to a high density super-sonic jet-gas target as well as to a solid target. It was proposed to consist of two accelerators, 50–400 kV and 0.4–3 MV, which would have covered a wide range of ion beam energies with sufficient energy overlap to connect the results to measurements

[d] `http://www.lsc-canfranc.es`
[e] `http://www.sanfordlab.org`

above-ground. DIANA planned to address scientific issues in the field of solar neutrino sources and the metallicity of the sun, carbon-based nucleosynthesis and neutron sources for the production of trans-Fe elements.

In 2013 DIANA was positively reviewed by NSF but postponed due to US budget sequestration in 2013. Since then the smaller scale facility CASPAR (Compact Accelerator System for Performing Astrophysical Research) has been developed to focus on reactions in stellar Helium burning ^{13}C,^{22}Ne,^{25}Mg,^{26}Mg(α,n) and ^{22}Ne,^{26}Mg (α,γ)) as well as to the potential beam induced background reactions ^{10,11}B,^{17,18}O(α,n)^{13}N. CASPAR consists in a 1 MV electrostatic accelerator (HVEC JN Model) which previously was installed at Nuclear Structure Laboratory at University of Notre Dame. The belt driven single stage accelerator can deliver positive ion beams with a particle intensity of approximately 150μA on an exchangeable solid target or a windowless extended gas target.

In 2015 the cavity inside SURF has been rehabilitated, beneficial occupancy of accelerator vault was in August 2015. Accelerator system and target tests will be concluded by the start of 2017 and first beams will be delivered on target by early of 2017.[20]

In Cina the JUNA[f] (Jinping underground lab for Underground Nuclear Astrophysics) project aims at direct measurements of (α,γ), (γ,n), (p,γ) and (p,α) reactions in hydrostatic Hydrogen and Helium burning. In particular it is planned to study the reactions ^{25}Mg(p,γ)^{26}Al, ^{12}C(α,γ)^{16}O, ^{19}F(p,γ)^{16}O and ^{13}C(α,n)^{16}O close to or even inside the relevant Gamow window

The key techniques, which include the laboratory inside the Jinping deep underground facilities at a depth of 6720 m.w.e., a 400 kV accelerator system capable to produce up to 10 mA of Proton and Helium beam and up to 5 mA of He^{++}, as well as target systems and detector stations are currently being developed. Neutron and Gamma background measurements at the underground location are being performed. First experiments are expected to be performed in 2018.[21]

References

1. Iliadis C., *Nuclear Physics in Stars* (Wiley-VCH, 2007).
2. Tumino A. *et al.*, *Few-Body Sys.* **54**, 745 (2013).
3. Lane A. M. and Thomas R. G., *Rev. Mod. Phys.* **30**, 257 (1958).

---

[f]http://www.juna.ac.cn

4. A. Bettini, *Eur. Phys. J. Plus* **127**, 114 (2012).
5. A. Best *et al.*, *Nucl. Instr. Meth.* A **812**, 1 (2016).
6. Heusser G., *Ann. Rev. Nucl. Part. Sci.* **45**, 543 (1995).
7. Anders M., Trezzi D., Bellini A. *et al.*, *EPJ* A **49**, 28 (2013).
8. R. Antolini (Ed.), *Nazionali del Gran Sasso, Annual Report*, ISBN-978-88-907304-8-1 https://www.lngs.infn.it/images/REIS/Annual_Report/annual_report_2011.pdf (2011).
9. A. Best and A. Caciolli, *Eur. Phys. J.* A **52**, 72 (2016).
10. R. Depalo, F. Cavanna *et al.*, *Phys. Rev.* C **94**, 055804 (2016).
11. Costantini H., Formicola A., Imbriani G., Junker M., Rolfs C. and Strieder F., *Rep. Prog. Phys.* **72**, 086301 (2009).
12. Broggini C., Bemmerer D., Guglielmetti A. and Menegazzo R., *Ann. Rev. Nucl. Part. Sci.* **60**, 53 (2010).
13. A. Formicola, P. Corvisiero and G. Gervino, *Eur. Phys. J.* A **52**, 73 (2016).
14. C. Gustavino, M. Anders, D. Bemmerer, Z. Elekes and D. Trezzi, *Eur. Phys. J.* A **52**, 74 (2016).
15. A. Boeltzig, C.G. Bruno, F. Cavanna *et al.*, *Eur. Phys. J.* A **52**, 75 (2016).
16. M. Aliotta, M. Junker, P. Prati and F. Strieder, *Eur. Phys. J.* **52**, 76 (2016).
17. G. Bellini, C. Broggini and A. Guglielmetti, *Eur. Phys. J.* A **52**, 88 (2016).
18. Summary of the Round Table "LUNA-MV at LNGS", https://luna.lngs.infn.it/images/presentations/LUNA-MV-2011/LUNA-MV_2010_Proceedings.pdf, LNGS, Feb. 10–12, 2011.
19. Perspectives of Nuclear Physics in Europe, *NuPECC Long Range Plan 2010*, European Science Foundation http://www.nupecc.org/lrp2010/Documents/lrp2010_final_hires.pdf (2010).
20. Scientific workshop in occasion of the celebration of the first 25 years of LUNA at Gran Sasso, https://agenda.infn.it/internalPage.py?pageId=1&confId=11579, LNGS Dec. 1–2, 2016.
21. W. Liu, "Progress of Nuclear Astrophysics Project in China and Underground Project", *Oral presentation at the 14th International Symposium on Nuclei in the Cosmos*, http://nic2016.jp/ 19–24 June 2016, Toki Messe, Niigata, Japan.
22. Proceedings of the round table workshop "LUNA-MV at Gran Sasso" Feb. 10–11, 2011, http://agenda.infn.it/getFile.py/access?resId=0&materialId=3&confId=3083

A Free-Neutron Target for Nuclear Reaction Studies

R. Reifarth*, K. Göbel, T. Heftrich and M. Weigand
Goethe-Universität Frankfurt
Frankfurt am Main, Germany
** E-mail: reifarth@physik.uni-frankfurt.de*
www.exp-astro.de

Beatriz Jurado
CENBG
Gradignan, France

Franz Käppeler
Karlsruhe Institute of Technology
Karlsruhe, Germany

Yuri A. Litvinov
GSI Helmholtzzentrum für Schwerionenforschung
Darmstadt, Germany

We discuss the possibility to build a neutron target for nuclear reaction studies in inverse kinematics utilizing a storage ring and radioactive ion beams. The proposed neutron target is a specially designed spallation target surrounded by a large moderator of heavy water (D_2O).

Keywords: Neutron Absorption, Direct Reactions, Reactor Design and Components, Storage Rings.

1. Introduction

Neutron capture cross sections of unstable isotopes are important for neutron induced nucleosynthesis as well as for technological applications[1–3]. Traditionally, time-of-flight setups are used to measure neutron-induced cross sections. However, only isotopes above a certain facility-dependent half-life cut can be investigated[4]. Ion storage rings equipped with hydrogen or helium gas targets are well-suited tools to investigate charged-particle induced reactions on unstable isotopes[5]. It has been shown that a combination of a radioactive beam facility, an ion storage ring and a high flux reactor would allow a direct measurement of neutron induced reactions over a wide energy range on isotopes with half lives down to minutes[6]. Here, the possibility to replace the rather demanding reactor with a specially designed spallation neutron source is discussed. If a spallation target is surrounded by a moderator of heavy water (D_2O), similar neutron densities in a close-by ion beam pipe can be obtained as in a research reactor. It might therefore be feasible to build such at modern nuclear physics facilities like LANSCE at the Los Alamos National Laboratoy[7,8].

2. Geometry

The center of the simulated setup is a tungsten spallation target. Compared to other elements, tungsten has the advantage of a high density combined with a very high melting point. The target is a cylinder aligned in the direction of the proton beam (z-axis). The radius of the cylinder is 1.5 cm and the length is 10 cm. This cylinder is inside a proton beam pipe of 2.5 cm radius, which is pure vacuum. The material of the beam pipe should have only minor effects on the neutrons. A second (ion) beam pipe -again pure vaccum- is oriented under 90° with respect to the proton beam pipe, see Fig. 1. The two pipes do not intersect each other, since the ion beam pipe is parallel to the y-axis at z=0 and x=7.5 cm, see Figs. 2-3. The whole setup is surrounded by a heavy-water moderator of different thicknesses. The center of the spherical moderator is also the center of the spallation target.

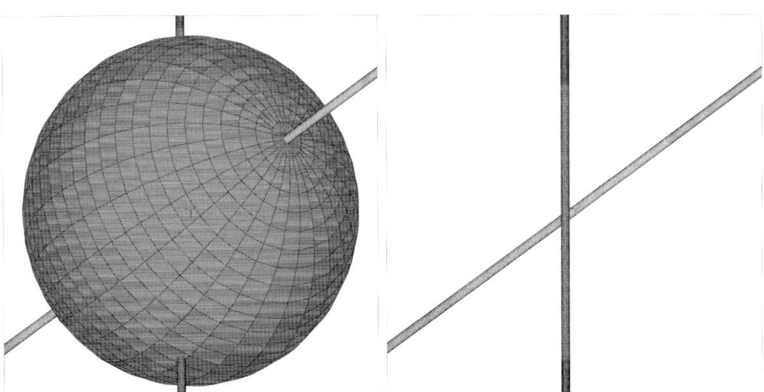

Fig. 1. Everything seen under 45° towards flight direction of the protons and ions. Left: Circular water moderator, vertical ion beam pipe and 30° proton beam pipe. Right: Vertical ion beam pipe and 30° proton beam pipe.

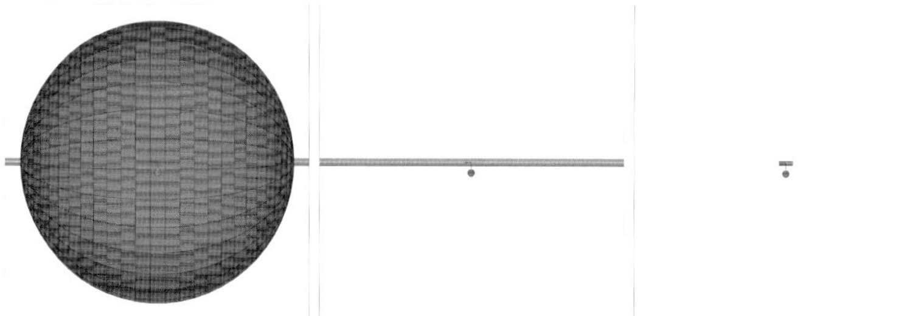

Fig. 2. Everything seen in flight direction of the ions. Left: Circular water moderator, and ion beam pipe which is perpendicular to the paper through the center of the moderator and proton beam pipe horizontal. Middle: Ion beam pipe (dot) is perpendicular to the proton beam pipe. Right: Ion beam pipe and tungsten target.

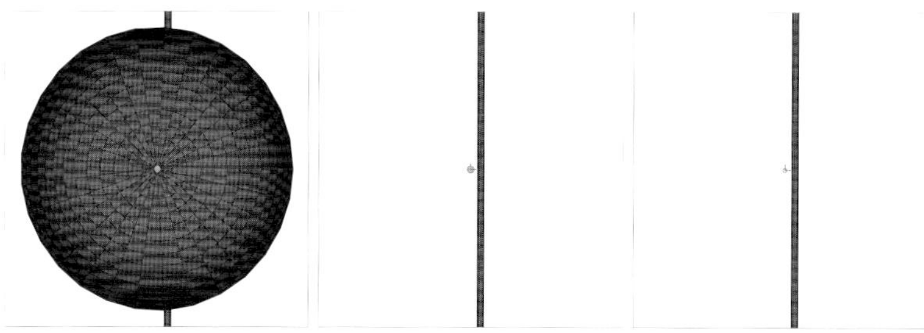

Fig. 3. Everything seen in flight direction of the protons. Left: Circular water moderator, vertical ion beam pipe and perpendicular to the center proton beam pipe. Middle: Vertical ion beam pipe and perpendicular proton beam pipe. Right: Vertical ion beam pipe and tungsten target.

3. Simulations and Results

The simulations were performed using GEANT-3.21[9] using the GCALOR package[10]. For each moderator thickness (Radius 0.5 m, 1.0 m, 2.0 m) neutrons with different energies were radomly started in the tungsten spallation target. The direction of the neutrons was chosen isotropically. The effect of the angle correlation with the proton beam can be neglected. Primary neutron energies from $10^{-2} - 10^{9}$ eV were simulated. A separate simulation for each energy decade was performed and the energy distribution within a decade was assumed to be $1/E$. Each simulation run included 10^6 neutrons. The most important quantity to be observed is the time a neutron spends on average in the ion beam pipe, where it acts as a neutron target for the ions. This quantity depends on the velocity of the neutron, the passing angle and a number of times the neutron actually crosses the ion beam pipe. For each neutron emitted from the spallation target, the time spent inside the ion tube is summed up and stored into a histogram.

Illustrative examples are given in Fig. 4. It is interesting to observe, that with higher primary neutron energy, the number of tresspassings increases as well as the number of events with at least one pass through the beam pipe. In addition, very short times of only a few 10^{-9} s are only observed for higher primary neutron energies. At lower energies, the neutrons are basically so slow that even a single pass requires a few 10^{-6} s.

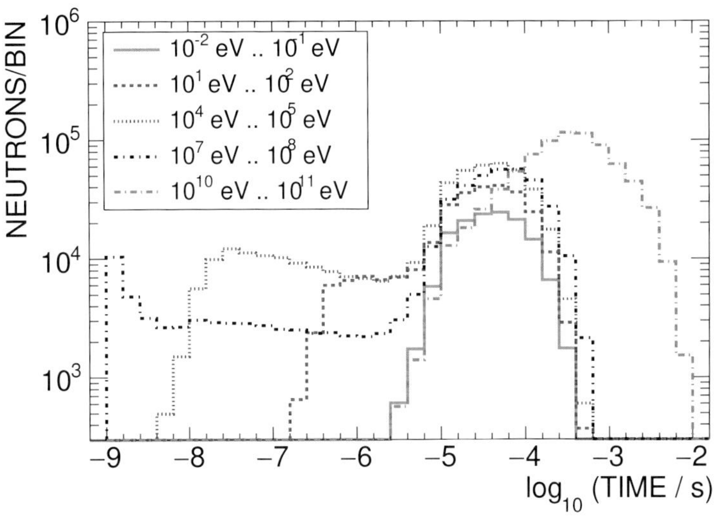

Fig. 4. The times, spent by neutrons inside the ion pipe surrounded by a moderator of 2 m radius are plotted. The different curves correspond to different primary neutron energies. If no neutron at all went through the pipe, the event is not shown in this histogram (zero-suppressed). 30%-50% of all neutrons pass the ion beam pipe at least once.

Fig. 5 shows the average time a neutron spends in the ion beam pipe as a function of its original energy. The most important result is that over a wide range of energies, these times don't depend much on the original energy. This means, the neutrons are quickly moderated and spend significant time trapped inside the volume of the moderator until they finally escape. This explanation is supported by the fact that bigger moderator volumes increase the average time spent inside the ion beam pipe. The significant increase at neutron-energies above 10 MeV is caused by the fact that the corresponding neutrons can produce more neutrons via (n,2n) reactions or even spallation reaction in the moderator.

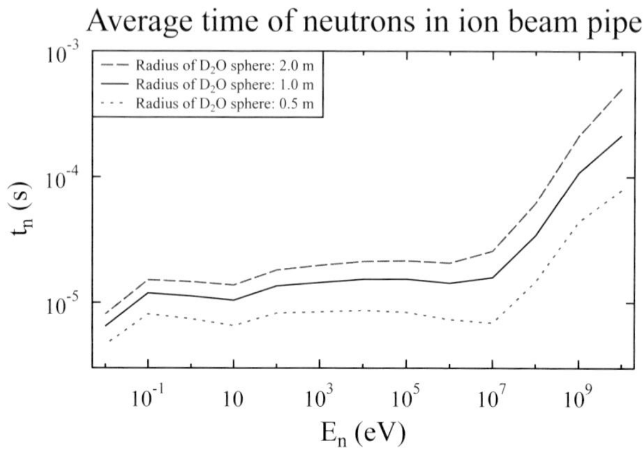

Fig. 5. Average time, which a neutron spents inside the ion pipe. For this average, all neutrons are taken into account, even if they did not pass the neutron beam pipe at all. The energy on the x-axis corresponds to the lower edge of the energy interval. Each interval corresponds to one energy decade, hence the point at 10^{10} eV corresponds to neutron energies between 10 GeV and 100 GeV.

It is very important to mention that even for a relatively small moderator of only 50 cm in radius, the neutrons spend on average several microseconds in the ion beam pipe. An ambitious moderator of 2 m radius would even lead to about 20 μs. The longer the average neutron time inside the beam pipe, the higher the density of the neutron target for a given proton current.

4. Conclusions and Possible Realization

This set of simulations was driven by the specific opportunities at the LANSCE-accelerator at LANL. However, it is certainly not restricted to this facility. Other institutions like CERN, GSI, NSCL/MSU, MYRRHA are very intersting options too.

A proton current of 1 μA corresponds to about 6 10^{12} protons/s. The neutron yield for a tungsten spallation target as described here with a beam energy of 800 MeV is about 25. This means, about 1.5 10^{14} neutrons/s are produced at 1 μA. An average time of 10 μs results then in average number of 1.5 10^9 neutrons inside the tube. The tube described here has a cross section of 20 cm^2, which means, a neutrons density of:

$$n_{area} \approx 10^8 \frac{\mathrm{n}}{\mu\mathrm{A\ cm}^2} \tag{1}$$

can be reached with a moderator of 1 m radius.

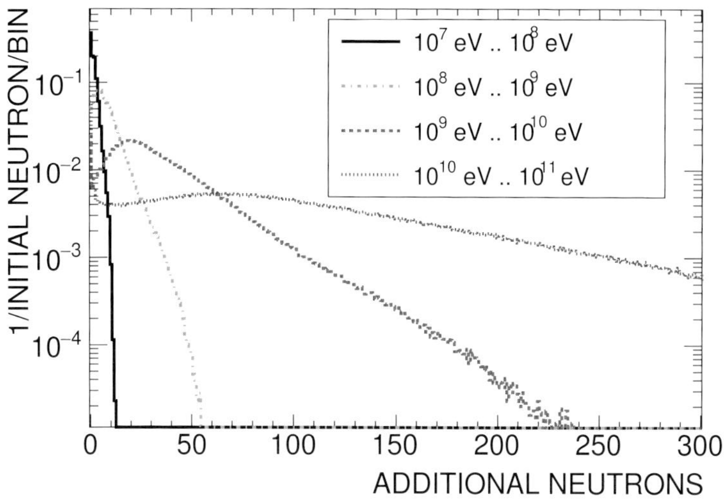

Fig. 6. Number of additional neutrons per initial neutron produced inside the moderator as a function of the initial neutron energy. Above a few MeV, (n,2n) reactions are possible and at higher energies even spallation reactions.

The Lujan target operates with an average current of about 100 μA. The average times of neutrons inside the ion beam pipe for a D_2O sphere of 2 m radius is about 20 μs. Such a proton current in combination with this huge moderator results in a neutron target of 1.5 10^{10} n/cm^2, which is almost the same as described in[6] resulting from the assumption of neutron fluence in the reactor of 10^{14} n/cm^2/s and 0.5 m interaction length.

The LANSCE accelerator serves two targets, the Lujan and the WNR target, with H^- ions,[7]. At the exit of the accelerator, the ions are bend south. The WNR target is directly bombarded with the negative ions[8], while the protons for the Lujan target are first stored in the proton storage ring PSR. On entry into this ring, the H^- ions get completely stripped to bare protons.

However, the LANSCE accelerator is capable of accelrating H^- and H^+ ions simultanously without any losses for either of the ion species. This has been sucessfully shown in the past. The magnet at the end of the accelerator bends H^- ions south and H^+ ions north, were additional experimental facilities could be run. This could be the ideal place for a combination of a ion storage ring and the discussed neutron target. In particular the close proximity of the isotope production facility (IPF), just at the beginning of the LANSCE accelerator, makes this idea very attrative. One has to keep in mind that the amount of radioactive material necessary for an experiment using the inverse-kinematics approach is orders of magnitude **smaller** than the amount of material needed for a traditional time-of-flight experiment as DANCE[4,11,12]. Typically the storage ring need to be filled every few minutes with about 10^7 ions[5] resulting in a consumption of about 1.5 10^{10} atoms the course of a day.

As already discussed in[6], several neutron-induced reactions can be directly measured with such a facility. Contrary to measurements in direct kinematics, which are based on the detection of the light products of the reactions, this technique is based on the detection of the heavy, projectile-like residues of the reactions. This represents a significant advantage for the measurement of (n,γ) cross sections of γ-emitting samples or of fissile nuclei, because there is no need to discriminate the γ-rays from the investigated reaction from the decay γ-rays. Also the measurement of (n,p) and (n,α) cross sections at low energies is made possible, as there is no need to detect the emitted protons and α-particles. Therefore, the proposed technique is highly advantageous also for the measurement of cross sections on relatively long-lived or even stable nuclei.

In addition to the neutron capture reactions, which are also very interesting for nuclear astrophysics, in particular two reaction types are very important for technological and defense applications:

- Neutron-induced **fission or (n,f)** can be investigated above $\approx$10 MeV energy in the center of mass, which corresponds to about 10 MeV neutron energy in regular kinematics. An interesting case is for instance the isomeric state ^{235}U^m with an half life of only 26 min.
- Neutron-induced **neutron removal or (n,2n)** have a threshold of typically several MeV (neutron binding energy). This is ideally matched by the proposed setup, since it is much easier to run with beams at higher energies. An example would be the chain of Tm isotopes. ^{169}Tm is stable. Provided that a beam is available, the proposed setup make measurements of (n,2n) reactions for all isotopes between ^{165}Tm (30 h) on the neutron-deficient side and ^{173}Tm (8 h) on the neutron-rich side.

References

1. R. Reifarth, C. Lederer and F. Käppeler, Neutron reactions in astrophysics, *Journal of Physics G Nuclear Physics* **41**, p. 053101 (2014).

2. K. Göbel, J. Glorius, A. Koloczek *et al.*, Nucleosynthesis simulations for the production of the p-nuclei ^{92}Mo and ^{94}Mo in a Supernova type II model, *EPJ Web of Conferences* **93**, p. 03006 (2015).
3. A. Koloczek, B. Thomas, J. Glorius *et al.*, Sensitivity study for s process nucleosynthesis in {AGB} stars, *ADNDT* **108**, 1 (2016).
4. A. Couture and R. Reifarth, Direct measurements of neutron capture on radioactive isotopes, *ADNDT* **93**, p. 807 (2007).
5. B. Mei, T. Aumann, S. Bishop *et al.*, First measurement of the $^{96}\mathrm{Ru}(p,\gamma)^{97}\mathrm{Rh}$ cross section for the p process with a storage ring, *Phys. Rev. C* **92**, p. 035803 (2015).
6. R. Reifarth and Y. A. Litvinov, Measurements of neutron-induced reactions in inverse kinematics, *Phys. Rev. ST Accel. Beams* **17**, p. 014701 (2014).
7. P. W. Lisowski, C. D. Bowman, G. J. Russell and S. A. Wender, The los alamos national laboratory spallation neutron sources, *Nucl. Sci. Engineering* **106**, p. 208 (1990).
8. A. Michaudon and S. Wender, *Performance of the LANSCE-WNR Facility as an Intense Pulsed Neutron Source for Neutron Nuclear Physics*, tech. rep., Los Alamos National Laboratory, New Mexico, USA (1990), Report LA-UR-90-4355.
9. J. Apostolakis, *CERN program library long writeup, W5013*, tech. rep., CERN, GEANT library (1993), http://wwwinfo.cern.ch/asd/geant/.
10. C. Zeitnitz and T. Gabriel, The geant/calor interface and benchmark calculations for zeus calorimeters, *Nucl. Instr. Meth. A* **349**, p. 106 (1994).
11. E.-I. Esch, R. Reifarth, E. M. Bond *et al.*, Measurement of the $^{237}\mathrm{Np}(\mathrm{n},\gamma)$ cross section from 20 meV to 500 keV with a high efficiency, highly segmented 4π BaF_2 detector, *Phys. Rev. C* **77**, p. 034309 (2008).
12. M. Weigand, T. A. Bredeweg, A. Couture *et al.*, $^{63}\mathrm{Ni}$ (n,γ) cross sections measured with DANCE, *Phys. Rev. C* **92**, p. 045810 (2015).

Nuclear Astrophysics with Radioactive Beams*

Sean Liddick[1], A. Spyrou[1,3,4], B.P. Crider[1], F. Naqvi[1], A.C. Larsen[5], M. Guttormsen[5], M. Mumpower[6,7], R. Surman[6], G. Perdikakis[8,1,4], D.L. Bleuel[9], L. Crespo Campo[5], A. Couture[7], A.C. Dombos[1,3,4], R. Lewis[1,2], S. Mosby[7], S. Nikas[8,4], C.J. Prokop[1,2], S. Siem[5], T. Renstrøm[5] and S.J. Quinn[1,3,4]

[1]*National Superconducting Cyclotron Laboratory, Michigan State University, East Lansing, MI 48824, USA*
[2]*Department of Chemistry, Michigan State University, East Lansing, MI 48824, USA*
[3]*Department of Physics and Astronomy, Michigan State University, East Lansing, MI 48824, USA*
[4]*Joint Institute for Nuclear Astrophysics, Michigan State University, East Lansing, MI 48824, USA*
[5]*Department of Physics, University of Oslo, N-0316 Oslo, Norway*
[6]*Department of Physics, University of Notre Dame, Notre Dame, IN 46556, USA*
[7]*Los Alamos National Laboratory, Los Alamos, NM, 87545, USA*
[8]*Department of Physics, Central Michigan University, Mount Pleasant, MI, 48859, USA*
[9]*Lawrence Livermore National Laboratory, 7000 East Avenue, Livermore, CA 94550-9234, USA*

The astrophysical r process is responsible for the synthesis of about half of the isotopes of the heavy elements. Despite its well-known role in nucleosynthesis, the astrophysical site where it takes place has not been unambiguously determined. One of the important inputs is neutron capture cross sections which are well constrained near stable isotopes where experimental data are available; however, moving far from the valley of stability, uncertainties grow by orders of magnitude. This is due to the complete lack of experimental constraints, as the direct measurement of a neutron-capture reaction on a short-lived nucleus is extremely challenging. Indirect

*This research was supported by Michigan State University and the Facility for Rare Isotope Beams and was funded in part by the NSF under Contracts No. PHY-1102511, No. 1350234, No. 1430152, No. 0822648, and No. 1419765, and the U.S. Department of Energy under Contracts No. DE-SC0013039, No. DE-AC52-07NA273449, and the National Nuclear Security Administration under Grants No. DE-NA0000979 and No. DE-FC03-03NA00143. G. P. would like to acknowledge support from the College of Science and Technology of Central Michigan University.

techniques have been developed to address this experimental shortcoming. The most recent indirect technique, the so called beta-Oslo method, takes advantage of the beta decay of a short-lived radioactive nucleus to populate high energy states in the daughter. The resulting gamma-ray cascade is monitored and the nuclear level density and gamma-ray strength is extracted. Recent experimental results from the National Superconducting Cyclotron Laboratory in the A = 70 mass region will be presented.

Nuclear Astrophysics Studies with an Isomeric $^{26}Al^m$ Beam

S. Almaraz-Calderon*, B. Asher, N. Gerken and K. Hanselman

Department of Physics, Florida State University,
Tallahassee, FL 32308, USA
**E-mail: salmarazcalderon@fsu.edu*

K.E. Rehm, M. Avila, B.P. Kay, R. Talwar, A.D. Ayangeakaa, C. Dickerson, C.R. Hoffman, C.L. Jiang, S. Kuvin, O. Nusair, R.C. Pardo and C. Ugalde

Physics Division, Argonne National Laboratory,
Argonne, IL 60439, USA

A.A. Chen

Department of Physics and Astronomy, McMaster University,
Hamilton ON L8S 4M1, Canada

C. Deibel and D. Santiago-Gonzalez

Department of Physics and Astronomy, Louisiana State University,
Baton Rouge, LA 70803, USA

Nuclear isomers usually suggest unexpected nuclear structure properties and may play an important role in nuclear astrophysics. One of the most interesting cases is the 0^+($T_{1/2}$=6.3 s) isomer in ^{26}Al, located just 228-keV above the ground state (5^+, $T_{1/2}$=740 ky). Proton captures on both, the ground state and the isomeric state, have a direct impact on the abundance of ^{26}Al in the Galaxy. We have developed a high-quality isomeric $^{26}Al^m$ beam via In-flight technique. By tunning the production energy we can favor the population of the isomeric state or the ground state in ^{26}Al. In this work, our efforts to develop and characterize an isomeric $^{26}Al^m$ beam are discussed as well as plans for experimental studies using this isomeric beam to constrain the destruction rate of Galactic ^{26}Al in relevant astrophysical environments.

Keywords: Nuclear Astrophysics; Nuclear Structure; Isomeric Beams.

1. Introduction

The radioisotope ^{26}Al is of great importance in nuclear structure and nuclear astrophysics. The possibility of using isomeric beams to perform in-beam reactions has long been considered as a potentially powerful method (e.g. Ref. [1] and references therein). Several experiments have been carried out taking advantage of high-spin isomeric states to selectively perform high-spin spectroscopy studies (e.g. [2–4]). In ^{26}Al, the large spin difference between the ground state, ^{26}Al^{g} (J$^{\pi}$=5^{+}) and the isomeric state, ^{26}Al^{m} (J$^{\pi}$=0^{+}), makes a unique case for nuclear structure and nuclear astrophysics studies. The detection of ^{26}Al in the Galaxy [5, 6] is of crucial importance for nuclear astrophysics. Since the half-life of ^{26}Al ($T_{1/2}$ ~7.2×10^{5} yr) is small compared to the timescale of Galactic chemical evolution (~10^{10} yr), its observation provides direct evidence of ongoing nucleosynthesis in the stars. In order to correctly interpret the observations, the different processes leading to the production and destruction of ^{26}Al in the Galaxy need to be understood. The presence of a low-lying isomer in ^{26}Al ($T_{1/2}$=6.3 s) greatly complicates the calibration of its nucleosynthesis. The ground state of ^{26}Al, ^{26}Al^{g} (J$^{\pi}$=5^{+}) decays via positron emission to the first excited 2^{+} state in ^{26}Mg followed by the decay to the ground state of ^{26}Mg emitting a 1.809-MeV γ-ray. Its isomeric state, ^{26}Al^{m} (J$^{\pi}$=0^{+}), located 228-keV above the ground state undergoes a super-allowed β^{+} transition directly to the ground state of ^{26}Mg bypassing the emission of the 1.809-MeV γ-ray.

The observational evidence points to massive stars as the main productions sites of ^{26}Al in the Galaxy [6–8]. In astrophysical scenarios like Asymptotic Giant Branch (AGB) stars, Classical Novae (CN) and Core Collapse Supernovae (CCSN), the destruction of ^{26}Al is expected to be dominated by proton captures on the ground and the isomeric state in ^{26}Al. The $^{26}Al^{g}(p,\gamma)^{27}Si$ has been the object of several experimental studies [9–18]. In contrast, experimental information is very scarce for the $^{26}Al^{m}(p,\gamma)^{27}Si$ reaction [19, 20].

Here we present the development of a high-quality ^{26}Al^{m} isomeric beam for nuclear structure and nuclear astrophysics studies.

2. Development of an isomeric ^{26}Al^{m} beam

The experiment was performed at the ATLAS accelerator In-Flight facility at Argonne National Laboratory (ANL). The ^{26}Al beam was produced In-Flight [21] via the $^{26}Mg(p,n)^{26}Al$ reaction. A 6.1 MeV/u primary ^{26}Mg beam was used to bombard a gas cell filled with hydrogen at 1 atm. In-Flight radioactive beam facilities have the advantage that the energy at which the production reaction occurs can be tuned to favor the population of a specific reaction channel.

The energy of the primary beam and the target thickness were chosen to maximize the production of the isomeric state over the ground state as measured by Ref. [22, 23], where the cross section of neutrons from the ground and excited states in the $^{26}Mg(p,n)$ reaction was measured. By averaging the cross section of the ground state and the excited states reported in Refs. [22, 23] in different target thicknesses, we determined the optimum production energy and target thickness to perform the experiment.

After the production gas cell, a re-buncher and a 22° bending magnet were used to separate a ^{26}Al^{13+} beam due to its magnetic rigidity. The main contaminants of the beam were lower energy charge states of the primary ^{26}Mg. A 12-MHz radio-frequency (RF)-sweeper [24] was further used to improve the purity of the beam. The ^{26}Al beam was then delivered to the experimental area for its characterization.

The yield of ^{26}Al$^{m}(0^{+})$ was determined using a rotating wheel setup which allows to measure the β^{+} annihilation radiation from the 0^{+} state which will dominate the decay due to its short 6.35 sec half-life. The radioactive ^{26}Al beam was stopped in a 100 mg/cm^2 thick Au-catcher foil located at the bottom of the wheel. After the implantation time (6 s or 15 s) the foil was rotated by 180° and placed in between two face-to-face NaI detectors where the two coincident 511-keV photons from the β^{+} decay of ^{26}Al^{m} were detected. A schematics of the rotating wheel setup is shown in Fig. 1. The counting time was the same as the implantation time. The detection efficiency of the 511-keV gamma-rays was determined after the experiment using GEANT-4 simulations and a calibrated ^{22}Na source mounted on the rotating wheel [25]. During the counting time, the total yield of radioactive ^{26}Al beam was measured in the focal plane detector of the Split-Pole Spectrograph (SPS) or in a silicon detector placed at 0° where the beam was identified by its mass, charge and total energy.

A two-dimensional spectrum for the two NaI detectors is shown in Fig. 2(a), where the coincidences between the two 511-keV gamma-rays can be seen. Fig. 2(b) shows the decay curve of the coincident radiation as a function of time, where the measured half-life of $T_{1/2}$=(6±1) sec is in agreement with the accepted half-life of ^{26}Al^{m} of 6.3 sec, confirming the presence of the isomer.

A spectrum of the silicon detector placed at 0° is shown in Fig. 3. It shows the purity of the ^{26}Al beam (≥98%) achieved in this experiment. The main contaminant of the beam was a lower energy ^{26}Mg^{11+} from the primary beam with the same magnetic rigidity than the ^{26}Al^{13+} beam. It is important to point out that the ^{26}Mg^{12+} was completely eliminated by the RF-sweeper.

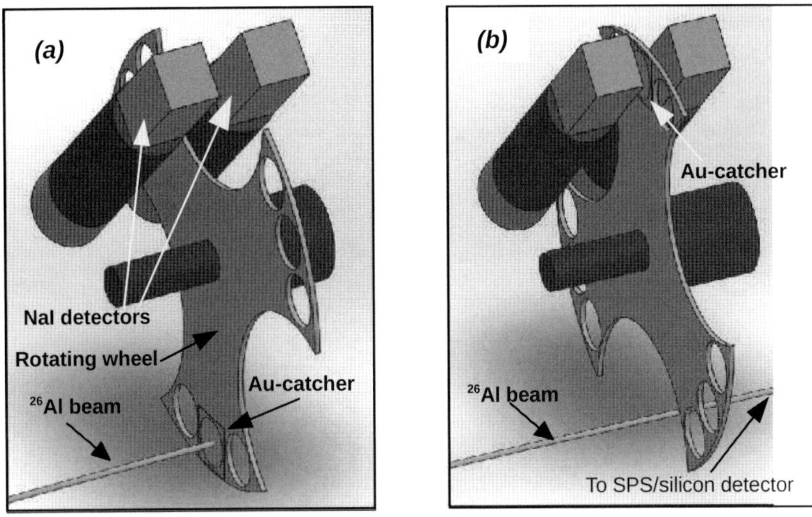

Fig. 1. Schematics of the rotating wheel setup. (a) The ^{26}Al beam is stopped in a Au-catcher foil. (b) After 15 sec of irradiation time, the wheel is rotated by 180° and it is placed between two face-to-face NaI detectors where the decay radiation is measured. During the counting time (15 sec), the beam is sent to the Split-Pole Spectrograph (SPS) or to a silicon detector placed at 0° for its identification.

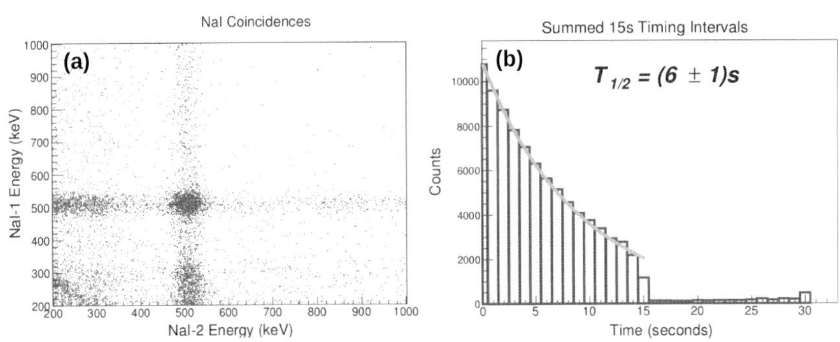

Fig. 2. Decay radiation of the ^{26}Al. (a) 511-keV coincidences between the two NaI detectors. (b) A measurement of the half-life of the coincident radiation confirms the presence of the isomeric beam.

The cross section for production of ^{26}Al measured in Ref. [23] averaged over the energy range of the production gas cell and plotted as the ratio of the isomeric 0^+ to the total ^{26}Al, as a function of the incident energy of ^{26}Mg in MeV/u is shown in Fig. 4. Our experimental value measured by

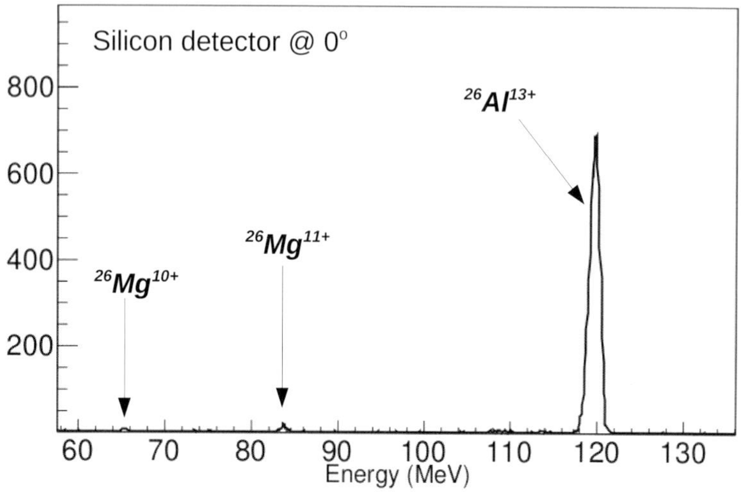

Fig. 3. Purity and energy resolution of the ^{26}Al beam as measured in the silicon detector placed at 0°. The main contaminants of the beam are lower charged states of the primary ^{26}Mg beam.

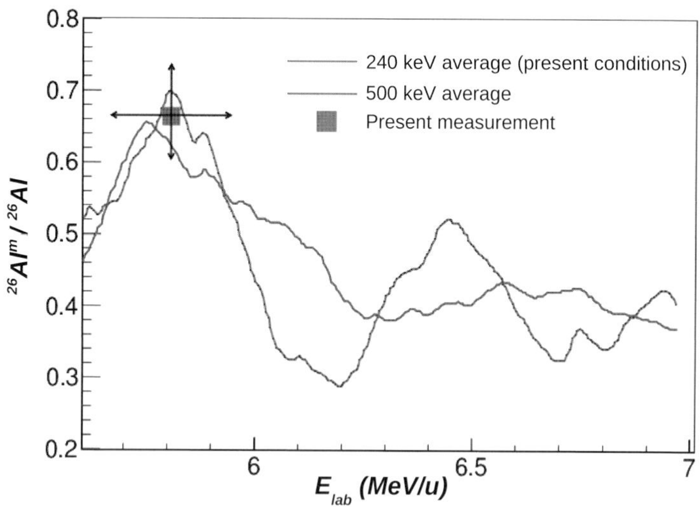

Fig. 4. Ratio of $^{26}Al^m$ to total ^{26}Al as a function of incident energy of the primary ^{26}Mg beam. The solid lines show averaged cross section of the $^{26}Mg(p,n)^{26}Al$ reactions as measured by Skelton [23]. The line above the data point shows the present experimental condition. The line below the data point shows a greater target thickness if the gas cell is cool down with liquid nitrogen. The data point shows the present experimental measurement.

the rotating wheel-SPS-silicon detector technique described above is also shown in the plot. Our result is in very good agreement with the predicted data. The effect of changing the production energy or the target thickness is also shown in Fig. 4 where the different lines represent different target thicknesses. The target thickness is modified by changing the pressure and cooling down the gas cell. This has the advantage that there is no need to change the primary beam energy in order to obtain a different isomer-to-ground-state ratio which usually is a time consuming process, while changing the gas pressure only take a few minutes.

3. Summary

We have developed and characterized a high-quality ^{26}Al isomeric beam with high intensity ($\sim 1\times 10^5$ pps), purity ($\geq 98\%$), energy resolution ($\Delta E/E \leq 1\%$), and isomeric content (70%), to study nuclear reactions induced by the isomeric component of the ^{26}Al radioactive beam. Such development have allowed us to perform a measurement of the $^{26}Al^m(d,p)^{27}Al$ reaction in inverse kinematics at ANL. The analysis of the experiment is on-going. We populated isobaric analog states to the $^{26}Al^m(p,\gamma)^{27}Si$ reaction, allowing for the extraction of spectroscopic information of key low-lying resonances in ^{27}Si via its mirror symmetry. The $^{26}Al^m(p,\gamma)^{27}Si$ reaction is relevant for understanding the destruction of Galactic ^{26}Al in AGB stars, CN and CCSN. Future experiments with different transfer reactions and at higher energies to change to isomer-to-ground-sate ratio are planned.

Acknowledgments

This work was partially supported by the State of Florida, and the US Department of Energy, Office of Nu- clear Physics, under Contract No. DEAC02-06CH11357. This research used resources of ANL's ATLAS facility, which is a DOE Office of Science User Facility.

References

[1] G.D. Dracoulis *et al.*, *Rep. Prog. Phys.* **79**, 076301 (2016).
[2] H. Watanabe *et al.*, *Nucl. Phys. A* **746**, 540c (2004).
[3] T. Morikawa *et al.*, *Phys. Lett. B* **350**, 169 (1995).
[4] S.A. Milne *et al.*, *Phys. Rev. Lett.* **117**, 082502 (2016).
[5] W.A. Mahoney, *Astrophys. J.* **262**, 724 (1982).
[6] R. Diehl, *Astron. Astrophys.* **298**, 445 (1995).

[7] N. Prantzos, *Astrophys. J.* **405**, L55 (1993).
[8] N. Prantzos and R. Diehl, *Phys. Rep.* **267**, 1 (1996).
[9] L. Buchmann *et al.*, *Nucl. Phys. A* **415**, 93 (1984).
[10] R. B. Vogelaar, Ph.D. Thesis, California Institute of Technology, 1989.
[11] T.F. Wang *et al.*, *Nucl. Phys. A* **499**, 546 (1989).
[12] R.B. Vogelaar *et al.*, *Phys. Rev. C* **53**, 1945 (1996).
[13] C. Ruiz *et al.*, *Phys. Rev. Lett.* **96**, 252501 (2006).
[14] G. Lotay *et al.*, *Phys. Rev. Lett.* **102**, 162502 (2009).
[15] G. Lotay *et al.*, *Phys. Rev. C* **84**, 035802 (2011).
[16] A. Parikh *et al.*, *Phys. Rev. C* **84**, 065808 (2011).
[17] V. Margerin *et al.*, *Phys. Rev. Lett.* **115**, 062701 (2015).
[18] S. D. Pain *et al.*, *Phys. Rev. Lett.* **114**, 212501 (2015).
[19] C. M. Deibel *et al.*, *Phys. Rev. C* **80**, 035806 (2009).
[20] G. Lotay *et al.*, *Phys. Rev. C* **80**, 055802 (2009).
[21] K.E. Rehm *et al.*, *Nuc. Inst. Meth. Phys. A* **647**, 3 (2011).
[22] G. Doukellis and J. Rapaport, *Nucl. Phys. A* **467**, 511 (1987).
[23] R.T. Skelton, R.W. Kavanagh and D.G. Sargood, *Phys. Rev. C* **35**, 45 (1987).
[24] R.C. Pardo *et al.*, *Nuc. Inst. Meth. Phys. A* **790**, 1 (2015).
[25] O. Nusair, Ph.D. Thesis, Johann Wolfgang Goethe-Universitat in Frankfurt am Main (2015).

Isomeric Character of the Lowest 4^+ State in ^{44}S*

I. Wiedenhöver[1], J.J. Parker IV[1], P. Cottle[1], J. Baker[1], D. McPherson[1], M.A. Riley[1], D. Santiago-Gonzalez[1,a], A. Volya[1], V.M. Bader[2,3], T. Baugher[2,3], D. Bazin[3], A. Gade[2,3], T. Ginter[2], H. Iwasaki[2,3], C. Loelius[2,3], C. Morse[2,3], F. Recchia[2], D. Smalley[2], S.R. Stroberg[2,3], K. Whitmore[2,3], D. Weisshaar[2], A. Lemasson[4], H. Crawford[5], A.O. Macchiavelli[5] and K. Wimmer[2,6,b]

[1]*Department of Physics, Florida State University, Tallahassee, FL, 32306, USA*
[2]*National Superconducting Cyclotron Laboratory, Michigan State University, East Lansing, MI, 48824, USA*
[3]*Department of Physics and Astronomy, Michigan State University, East Lansing, MI, 48824, USA*
[4]*Grand Accélérateur National d'Ions Lourds (GANIL), CEA/DSM-CNRS/IN2P3, Caen, France*
[5]*Lawrence Berkeley National Laboratory, Nuclear Science Division, Berkeley, CA 94720-8153, USA*
[6]*Department of Physics, Central Michigan University, Mt. Pleasant, MI 48859, USA*

Previous experimental results established the lowest 4^+ state in the N = 28 nucleus ^{44}S and suggested that this state may exhibit a hindered E2-decay rate inconsistent with being a member of the collective ground state band. Recent theoretical investigations by Utsuno *et al.* [2] and Egido *et al.* [3] supported this suggestion and propose a characterization with a dominant K = 4 component, despite substantial triaxiality. We populate this 4^+ state via two-proton knockout from a beam of ^{46}Ar projectiles and measure its lifetime using the recoil distance method with the GRETINA γ-ray spectrometer. Our experiment establishes an isomeric level lifetime and allows to deduce a hindered B(E2) value for the $4^+ \rightarrow 2_1^+$ transition. This value confirms that the lowest 4^+ state in ^{44}S is not a collective

---

*Supported in part by the National Science Foundation, The Department of Energy (DOE) National Nuclear Security Administration, award number DE-NA0000979 and the DOE Office of Science.

[a]Current address: Department of Physics and Astronomy, Louisiana State University, Baton Rouge, LA 70803, USA.
[b]Current address: The University of Tokyo, Hongo, Bunkyo-ku, Tokyo 113-0033, Japan.

rotational excitation and suggests a K = 4 isomer. The unusual properties of this state are discussed in the context of recent theoretical calculations.

References

1. D. Santiago-Gonzalez *et al.*, *Phys. Rev. C* **83**, R061305 (2011).
2. Y. Utsuno, N. Shimizu, T. Otsuka, T. Yoshida and Y. Tsunoda, *Phys. Rev. Lett.* **114**, 032501 (2015).
3. J. Luis Egido, M. Borrajo, T. R. Rodríguez, *Phys. Rev. Lett.* **116**, 052502 (2016).

Fission II

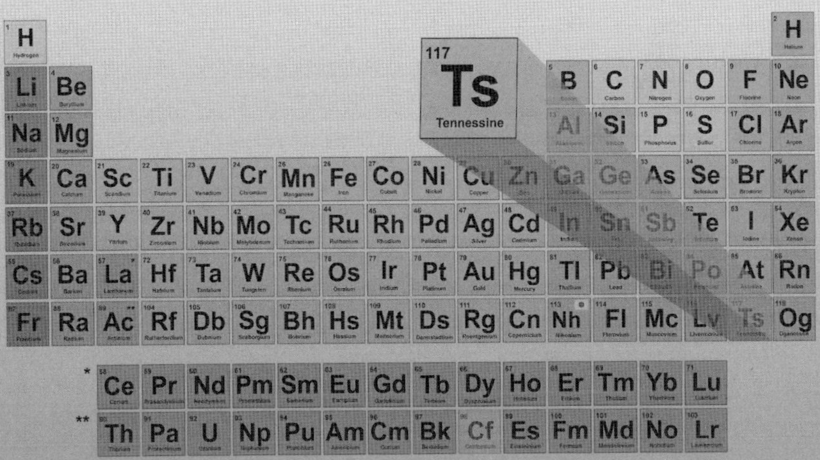
117
Ts
Tennessine
H H
Li Be B C N O F Ne
Na Mg Al Si P S Cl Ar
K Ca Sc Ti V Cr Mn Fe Co Ni Cu Zn Ga Ge As Se Br Kr
Rb Sr Y Zr Nb Mo Tc Ru Rh Pd Ag Cd In Sn Sb Te I Xe
Cs Ba La Hf Ta W Re Os Ir Pt Au Hg Tl Pb Bi Po At Rn
Fr Ra Ac Rf Db Sg Bh Hs Mt Ds Rg Cn Nh Fl Mc Lv Ts Og
* Ce Pr Nd Pm Sm Eu Gd Tb Dy Ho Er Tm Yb Lu
** Th Pa U Np Pu Am Cm Bk Cf Es Fm Md No Lr

Prompt Fission Neutron and γ Correlations

Patrick Talou

Nuclear Physics Group, Theoretical Division
Los Alamos National Laboratory, Los Alamos, NM 87545, USA
talou@lanl.gov

Prompt neutrons and γ rays emitted shortly after scission provide valuable information to constrain fission physics models and their parameters. Numerical results obtained with event generators that simulate the de-excitation of the primary fission fragments can indeed be directly compared to various post-scission observables. The CGMF code implements the Hauser-Feshbach equations in a Monte Carlo approach to simulate the decay of the primary fission fragments that are produced in various configurations in energy, spin and parity. The neutron and γ decay channel widths are computed at each stage of the decay chain, thereby following successive emissions of particles on an event-by-event basis. Experimental as well as theoretical studies of correlations among the emitted particles shed some light on the mechanisms of excitation energy sorting at scission, the production of angular momentum in the fission fragments, and on the statistical nature of the fission fragment decay. Nuclear spectroscopy of the fission fragments is crucial for the accurate prediction of isomeric ratios and late-time prompt γ-ray emissions. The CGMF code was recently integrated into the MCNP6.2 transport code, providing new and powerful tools for the interpretation of complex experimental data.

Keywords: Prompt Fission Neutrons and γ Rays; Monte Carlo Hauser-Feshbach; MCNP6.2; Transport Code.

1. A Very Rich and Complex Process

Nuclear fission is a very rich and complex physical process, as can be attested by the vast literature that has accumulated on the subject since its discovery more than 75 years ago. This body of fission-related literature has been growing significantly in the last decade or so, driven by advances in experimental facilities and detectors, and in theoretical models put to work in ever more powerful computer simulations. Although various stages of the fission process can be cleanly differentiated in time and phase-space, many correlations persist among fission observables. Those correlations are most often ignored by models and/or evaluations that narrowly focus on one particular aspect of the fission process. Examples abound: fission cross sections, fission fragment angular distributions, prompt fission neutron and γ-ray multiplicity and spectrum, independent and cumulative fission yields, etc., are all modeled and evaluated independently from each other. Understanding these correlations provides a powerful venue to predict fission observables beyond the range of available experimental data.

In transport simulations, such as the ones performed with the well-known `MCNP` Monte Carlo transport code, data tables from the U.S. ENDF/B library of evaluated nuclear data are often used and sampled to simulate a particular reaction channel and outgoing products. Evaluated data tables are a very convenient way to perform very fast sampling of reaction probabilities, which is crucial for efficient transport simulations of complicated scenarios. On the other hand, such evaluated tables cannot contain very detailed information about multi-dimensional distributions, hence ignoring many correlations. The development of the `CGMF` code was intended to provide advanced simulation capabilities to address these shortcomings and study correlated fission data in particular.

2. Prompt $n-\gamma$ Correlation Studies

Measuring all fission observables at once has been the San Graal of nuclear fission experimentalists for a very long time. This idea is somewhat ill-advised as measuring various quantities at once is often pursued at the expense of an accurate measurement on only one of those observables. As an example, an accurate measurement of the total prompt neutron multiplicity requires capturing almost all neutrons with nearly one-hundred percent efficiency, in a Gd-loaded tank for instance, while measuring the neutron energy spectrum requires high-resolution neutron energy detection, in a time-of-flight experiment, with much reduced efficiency. Those two requirements are difficult to combine, to say the least. A promising approach, as described in this contribution, is to study high-quality single data through a physics model that describes more than one quantity consistently, thereby introducing correlations through model calculations and experimental constraints. However, experimental studies of specific correlations are crucial to probe finer details of the fission process and validate some of the assumptions made in the fission event generators. We are describing some of those ongoing studies below, followed by a description of the nuclear physics models implemented in the `CGMF` code.

2.1. *Experimental Studies*

At Los Alamos National Laboratory, several experimental setups have been devised to study various aspects of the nuclear fission process. The DANCE setup[1] ("Detector for Advanced Neutron Capture Experiments") consists of 160 BaF_2 crystals arranged in a compact 4π configuration, which act as a calorimeter. Originally developed to measure neutron capture cross sections on very small targets and/or very radioactive samples, it has been used more recently to study multiplicity-dependent prompt fission γ-ray spectra and to distinguish fission from capture reactions for various measurements. In these past two years, an array of 22 stilbene scintillators named NEUANCE (for "NEUtrons inside DANCE") was built and arranged in a compact geometry to fit inside the DANCE target cavity. Thanks to this arrangement, more intricate correlations between neutrons and γ rays can be studied.

The Chi-Nu detector setups[2] have been built to measure the prompt fission neutron spectra of important actinides very accurately: the first setup is made up of 54 NE213 liquid scintillators aimed at measuring the high-energy tail of the neutron spectrum up to 15 MeV, while the second setup consists of 22 ^{6}Li glass detectors to measure the low-energy tail down to about 10 keV outgoing neutron energy. Although the focus of the Chi-Nu project is not on $n-n$ correlations, these two arrays can indeed be used to study those correlations in angle and energy. They are located at the WNR facility at the Los Alamos Neutron Science Center, a white neutron source providing incident neutrons from a few hundred keV up to several hundred MeV, thereby offering an opportunity to study fission neutrons as a function of incident neutron energy.

In the last decade or so, the University of Michigan and the `MCNPX – PoliMi` development team have also invested time and effort into measuring neutron and γ correlations for nonproliferation applications. One of their setups, as described in[3], consists of 16 EJ-309 liquid scintillators arranged in a compact geometry around a spontaneous fissioning source of ^{240}Pu. Neutron-neutron correlations in energy and angle were measured and compared with `MCNPX – PoliMi` simulations.

Another experiment of interest is being conducted by Dale *et al.*[4] at Idaho National Laboratory to study energy and angle of neutrons in photofission reactions. Finally, the Versatile Array of Neutron Detectors at Low Energy (VANDLE) could also be used to investigate some of the fission neutron correlations, as discussed in this Conference[5].

Note that various other experimental facilities and setups, not being discussed here, are being devised worldwide to study such correlated signatures of fission.

2.2. *Theoretical Developments*

The physics models implemented in the CGMF code have been described at length in other publications[6–8], so we will only briefly recall the salient ideas here. The CGMF code implements a Monte Carlo version of the Hauser-Feshbach statistical theory of nuclear reaction that follows the decay of excited primary fission fragments on an event-by-event basis. It first samples a fission fragment distribution in mass, charge, and total kinetic energy, Y(A,Z,TKE), corresponding to a particular fission reaction, e.g., spontaneous fission of ^{252}Cf or 10-MeV neutron-induced fission of ^{239}Pu. From a given state in excitation energy, spin and parity of a fission fragment, CGMF calculates the probabilities of emission of neutrons and γ rays from the energy, spin and parity-dependent neutron and γ-ray decay widths obtained for this specific nucleus. Neutron widths are calculated using a global optical model potential, while the γ-ray widths are obtained in the strength function formalism, considering E1, E2 and M1 electromagnetic transitions. Those decay probabilities are sampled using the Monte Carlo technique, and a particular decay path is chosen. This process is repeated until a ground-state or metastable state is reached. From this sequence of binary decays, CGMF records the initial configuration of the fission fragment, including the momentum vector of the fragment in the laboratory frame of reference, the total number of prompt neutrons and γ rays emitted, as well as the energy and angle of each emitted particle in the sequence. This process is then reiterated for another sampled initial fission fragment configuration until enough statistics are obtained. The required statistics strongly depend on the particular observable that one is interested in. For instance, the average prompt neutron multiplicity can be accurately obtained within a few tens of thousands of fission events, while determining precisely the high-energy tail of the γ-ray spectrum when only 15 γ rays are emitted requires significantly longer.

CGMF requires various input quantities:

- Fission fragment yields Y(A,Z,KE): this important ingredient is often reconstructed from available experimental data, although recent calculations of fission yields using the macroscopic-microscopic approach of Randrup and Möller[9] is being considered as one intends to perform systematic calculations across the nuclear chart.

- Initial excitation energy distributions in the fragments; the energy sorting mechanisms at work near scission have a large influence on the observed saw-tooth curve of $\overline{\nu}$(A). Several options can be considered in CGMF, such as the ones discussed in Morariu *et al.*[10]
- Initial (J, π) distributions in the fragments; this distribution is important to properly describe prompt fission γ-ray spectra[7] and the population of isomeric states in fission fragments[8,11].

Both initial excitation energy and spin distributions can be tuned in CGMF to better represent some experimental data, and in particular, $\overline{\nu}$(A) and the average prompt fission γ-ray spectrum. In addition to these input data and parameters, the Hauser-Feshbach equations implemented in CGMF require the calculation of neutron and γ decay widths, through optical potential and γ-ray strength function calculations. Those are performed in standard fashion, and will not be repeated here.

As mentioned above, the output of a CGMF run is a potentially long suite of fission event records, which provide prompt neutron and γ data in energy, angle and multiplicity with respect to a particular initial fission fragment pair configuration. Such data can be easily analyzed and many correlations inferred through dedicated scripts. Only selected results will be presented here, mostly obtained in the case of ^{252}Cf spontaneous fission, to illustrate some important points relate to the physics of the fission process.

The prompt fission neutron multiplicity distribution P(ν) is shown in Fig. 1. This evaluated result of Santi and Miller[12] has been obtained by fitting several accurate and consistent experimental data sets for this quantity. CGMF calculations reproduce the average and width of the distribution relatively well, with default input fragment yields. The P(ν) distribution is particularly sensitive to the kinetic energy distribution of the fragments, and a fine-tuning of this distribution would lead to even better results. A very accurate representation of this quantity is crucial for nuclear assay simulations and measurement interpretations.

Of interest to detector response simulations is the dependence of the prompt fission neutron spectrum to the neutron multiplicity. Albeit difficult to measure, CGMF predictions, shown in Fig. 2, indicate a small hardening of the neutron spectrum with increasing multiplicity. A higher neutron multiplicity corresponds to higher excitation energies, hence higher temperatures in the fission fragments.

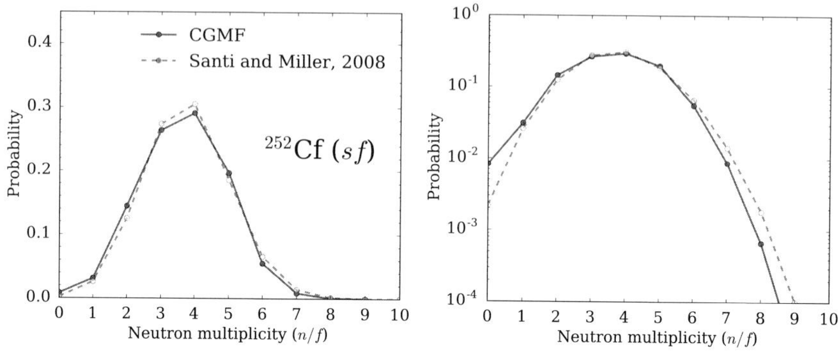

Fig. 1. Prompt fission neutron multiplicity distribution P(ν) calculated with `CGMF` and compared with evaluated data[12] for the spontaneous fission of ^{252}Cf.

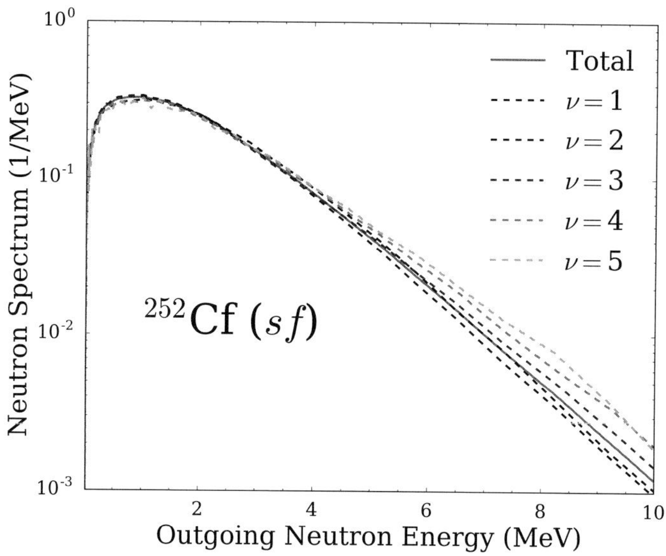

Fig. 2. Neutron-multiplicity dependent prompt fission neutron spectra (PFNS), in the case of ^{252}Cf spontaneous fission, show a hardening of the spectrum with increasing neutron multiplicity.

Recent experiments by Göök *et al.*[13] have measured the prompt fission neutron spectrum and neutron angular distribution as a function of the fission fragment mass. This is an extremely interesting quantity as it places stringent constraints on the physics models describing prompt fission neutron spectra. The experimental data for thermal neutron-induced fission

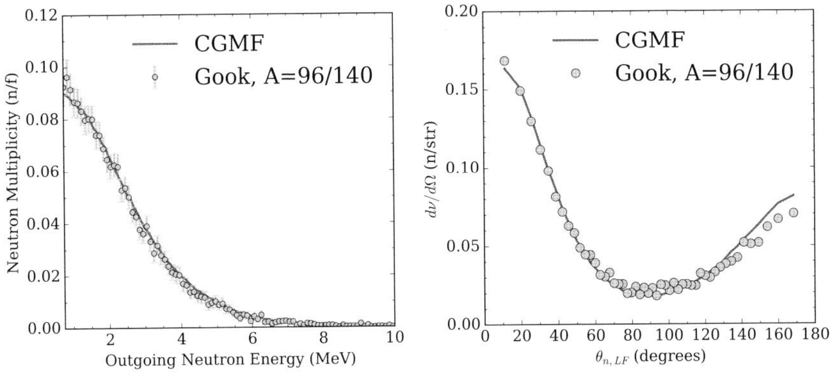

Fig. 3. Prompt fission neutron energy distribution (left) and neutron-light fragment angular distribution (right) in the thermal neutron-induced fission reaction on ^{235}U, for a particular fragment mass split (96/140). CGMF-calculated results agree remarkably well with the recent measurements by Göök *et al.*[13].

on ^{235}U are compared with CGMF results in Fig. 3. The agreement is remarkable for both PFNS and angular distributions. Note that CGMF does not assume any additional neutron source beyond the ones coming from the fully accelerated fragments. The agreement between data and model calculations indicate that such an extra neutron source would most likely be very small, or indistinguishable from the fission fragment evaporation.

Prompt fission γ rays provide a complementary picture of the fission fragment de-excitation process. The observation that the average total γ-ray energy is significantly greater than the average neutron separation energy showed early on that γ-ray emissions could compete effectively with neutron decay in some configurations. From this observation, average initial fission fragment spins have been inferred[14] at about $7 \pm 2\hbar$. CGMF calculations, with model parameters tuned to reproduce prompt fission γ-ray spectra[7] and isomeric production ratios[11], indicate similar values.

Contrary to multiplicity-dependent PFNS that point to only a rather small dependence of the average outgoing neutron energy to the neutron multiplicity, CGMF calculated results, in excellent agreement with DANCE measured data, show that the PFGS is significantly softer with increasing γ-ray multiplicity. This result could lead to significant biases in the interpretation of experimental prompt fission γ-ray multiplicity measurements, which are very sensitive to the detection threshold. By cutting out γ rays below a certain energy threshold, the inferred $P(N_\gamma)$ can be significantly biased.

As a final example of CGMF results to be discussed in this short contribution, we can point to the time dependence of emission of prompt fission γ rays due to the presence and population of isomers in the fission fragment yields[8]. Up to 7% of prompt γ rays evaporated in the thermal neutron-induced fission reactions of ^{235}U and ^{239}Pu are predicted to be emitted between 10 ns and 5 μs since the time of fission. Experimental data on the prompt fission γ-ray multiplicity and total energy are often obtained with different γ-ray energy thresholds and time coincidence windows, which therefore requires a careful analysis of the model-calculated results.

3. Integration into the MCNP6.2 Transport Code

In recent years, significant efforts have been made to incorporate fission physics models as implemented in the CGMF code as fission event generators in the transport code MCNP6.2[15]. In this running mode, CGMF is called every time a fission reaction occurs, and returns the characteristics in energy and momentum of the neutrons and γ rays emitted. It therefore provides all the information needed to simulate a most realistic representation of the kinematics of the fission reaction. As part of this effort, the FREYA code[16] was also integrated into MCNP6.2 for fast simulations of neutron and γ emissions. We are now developing benchmark simulation tests and experiments to perform appropriate verification and validation of these new features. In particular, several experimental setups at Los Alamos and University of Michigan (UM) are being simulated with the new MCNP6.2 features and tools, as well as with the MCNPX – PoliMi code[17] developed at UM.

4. Conclusion

Recently, important developments in the study of prompt fission neutrons and γ rays and their correlations have been made both experimentally and theoretically. New experimental setups are being devised to measure more than one fission observable at once, hence providing new and unique correlated data. On the other hand, physics model codes, e.g., CGMF, that can generate prompt data on a fission event-by-event basis are being developed. Such codes produce a vast amount of data that can be compared with the rich literature on fission observables. By predicting more than one fission observable at a time, such models and their input parameters can be better constrained and their predictive power enhanced.

Acknowledgements

This work was performed at Los Alamos National Laboratory, under the auspices of the National Nuclear Security Administration of the US Department of Energy at Los Alamos National Laboratory under Contract No. DE-AC52-06NA25396. This work was partly supported by the Office of Defense Nuclear Nonproliferation Research & Development (DNN R&D), National Nuclear Security Administration, US Department of Energy.

References

1. M. Jandel *et al.*, *Capture and fission with DANCE and NEUANCE*, Eur. Phys. J. A **51**, 179 (2015). DOI: 10.1140/epja/i2015-15179-7.
2. H. Y. Lee *et al.*, *Prompt Fission Neutron Spectrum Study at LANSCE: Chi-Nu Project*, in proceedings of the 5th International Conference on Fission and Properties of Neutron-Rich Nuclei, Eds. J. H. Hamilton and A. V. Ramayya, World Scientific, p. 430 (2014).
3. M. J. Marcath, T. H. Shin, S. D. Clarke, P. Peerani and S. A. Pozzi, *Neutron angular distribution in plutonium-240 spontaneous fission*, Nucl. Inst. Meth. Phys. Res. A **830**, 163 (2016).
4. D. S. Dale, O. Kosinov, T. Forest, J. Burggraf, S. Stave, G. Warren and V. Starovoitova, *Two neutron correlations in photofission*, Int. J. Mod. Phys. Conf. Ser. **44**, 1660240 (2016), DOI: 10.1142/S2010194516602404.
5. W. A. Peters *et al.*, *Performance of the Versatile Array of Neutron Detectors at Low Energy (VANDLE)*, Nucl. Inst. Meth. Phys. Res. A **836**, 122 (2016), and *this Conference.*
6. T. Kawano, P. Talou, M. B. Chadwick and T. Watanabe, *Monte Carlo Simulation for Particle and γ-Ray Emissions in Statistical Hauser-Feshbach Model*, J. Nucl. Sci. Tech. **47**, 462 (2010).
7. I. Stetcu, P. Talou, T. Kawano and M. Jandel, *Properties of prompt fission γ rays*, Phys. Rev. C **90**, 024617 (2014).
8. P. Talou, T. Kawano, I. Stetcu, J. P. Lestone, E. McKigney and M. B. Chadwick, *Late-time emission of prompt fission γ rays*, Phys. Rev. C **94**, 064613 (2016).
9. J. Randrup and P. Möller, *Brownian Shape Motion on Five-Dimensional Potential-Energy Surfaces: Nuclear Fission-Fragment Mass Distributions*, Phys. Rev. Lett. **106**, 132503 (2011).
10. C. Morariu, A. Tudora, F.-J. Hambsch, S. Oberstedt and C. Manailescu, *Modelling of the total excitation energy partition*

including fragment deformation and excitation energies at scission, J. Phys. G: Nucl. Part. Phys. **39**, 055103 (2012).

11. I. Stetcu, P. Talou, T. Kawano and M. Jandel, *Isomer production ratios and the angular momentum distribution of fission fragments*, Phys. Rev. C**88**, 044603 (2013).
12. P. Santi and M. Miller, *Reevaluation of Prompt Neutron Emission Multiplicity Distributions for Spontaneous Fission*, Nucl. Sci. Eng. **160**, 190 (2008).
13. A. Göök, F.-J. Hambsch and M. Vidali, *Prompt neutron multiplicity in correlation with fragments from spontaneous fission of* ^{252}Cf, Phys. Rev. C **90**, 064611 (2014); and private communication for ^{235}U data.
14. J. B. Wilhelmy, E. Cheifetz, R. C. Jared, S. G. Thompson and H. R. Bowman, *Angular Momentum of Primary Products Formed in the Spontaneous Fission of* ^{252}Cf, Phys. Rev. C **5**, 2041 (1972).
15. MCNP6 Development Team, *Initial MCNP6 Release Overview — MCNP6 version 1.0*, Los Alamos National Laboratory, LA-UR-13-22934 (2013).
16. J. M. Verbeke, J. Randrup and R. Vogt, *Fission Reaction Event Yield Algorithm,* FREYA *— For event-by-event simulation of fission*, Comp. Phys. Comm. **191**, 178 (2015).
17. S. Pozzi *et al.*, *MCNPX-PoliMi for nuclear nonproliferation applications*, Nucl. Inst. Meth. in Phys. Res. A **694**, 119 (2012).

Prompt Neutron Emission Correlations with Fission Fragment Properties

F.-J. Hambsch, T. Gamboni, W. Geerts, A. Göök, S. Oberstedt and M. Vidali

EC-JRC Directorate for Nuclear Safety and Security (DN2S),
B-2440 Geel, Belgium
E-mail: franz-josef.hambsch@ec.europa.eu

A. Al-Adili, K. Jansson and S. Pomp

Division of Applied Nuclear Physics - Department of Physics and Astronomy
Uppsala University - Uppsala, Sweden
E-mail: ali.al-adili@physics.uu.se

The investigation of the dynamics of the nuclear fission process has been a standing research topic at the JRC-Geel during the past decades. Recently the focus was put on the de-excitation of fission fragments through the emission of prompt neutrons and gamma-rays.

To this end new detector systems were developed at JRC-Geel, e.g. a position sensitive ionization chamber used in conjunction with the neutron scintillator array SCINTIA. The array has been tested using the spontaneous fission of ^{252}Cf. The goal is to study correlations of fission fragments with prompt neutron emission in the resolved resonance region. No strong fluctuations of the average prompt neutron multiplicity for the strongest resonances in ^{235}U were observed. From the present data the mass-dependent neutron multiplicity, ν(A), was generated. The ν(A) distribution shows a more pronounced dip around the doubly magic mass A = 132 and at very low masses around A ~ 80 compared to the literature. In addition, a steeper slope for ν(TKE) is observed. Cross checking with fragment data clearly shows a narrower mass and total kinetic energy (TKE) distribution.

The 2E-2v spectrometer VERDI (VElocity foR Direct mass Identification) became operational. For ^{252}Cf(sf) superior mass resolution is observed compared to a twin Frisch-grid ionization chamber. For post-neutron mass distributions still some issues need to be solved and ν(A), being the difference of pre- and post-neutron mass distributions, is still deviating from literature data. Eventually, VERDI will provide a complementary measurement technique to assess ν(A) and ν(TKE). In addition, an experimental campaign to measure ν(A) as a function of incident neutron energy for different actinides has been started. First tests show promising results.

Keywords: Neutron Induced Fission; Prompt Neutron Emission; 2E-2ν Spectrometer VERDI.

1. Introduction

Studies of correlations of prompt neutron and γ-ray emission with fission fragment properties have recently seen a renaissance in theoretical modeling as well as experimental investigations.

It started out with the manifestation of the GEF code showing that an increased excitation energy in neutron induced fission leads to an increase of the prompt neutron number as a function of mass mainly for the heavy fragment [1]. Very sparse experimental information is available, but the experiments seem to support the results of GEF.

Hence, an improved knowledge of the properties of prompt fission neutrons, their multiplicities, as well as their energy and angular distributions is at stake to answer all kinds of questions related e.g. to the formation of the fission fragments, the sharing of excitation energy among them and the time scale of the process. The same holds for prompt γ-ray emission, which is not discussed in this paper. For nuclear modeling and improved evaluation of nuclear data the knowledge about fluctuations in the prompt neutron multiplicity as a function of incident neutron energy is of importance not only for the major actinides ^{235}U and ^{239}Pu [2, 3].

2. Prompt neutron emission in resonance induced fission of ^{235}U

Fluctuations in fission fragment mass and total kinetic energy (TKE) in both isotopes have been observed in resonance neutron induced fission [4, 5]. Independently, fluctuations in the number of emitted neutrons have also been identified [6].

New correlation measurements have been started of fission fragments and prompt neutron emission. The experimental setup is schematically illustrated in Fig. 1. It consists of two parts, a position sensitive twin ionization chamber for fission fragment detection and an array of scintillation neutron detectors (SCINTIA). A detailed description of the position sensitive ionization chamber can be found in Ref. [7].

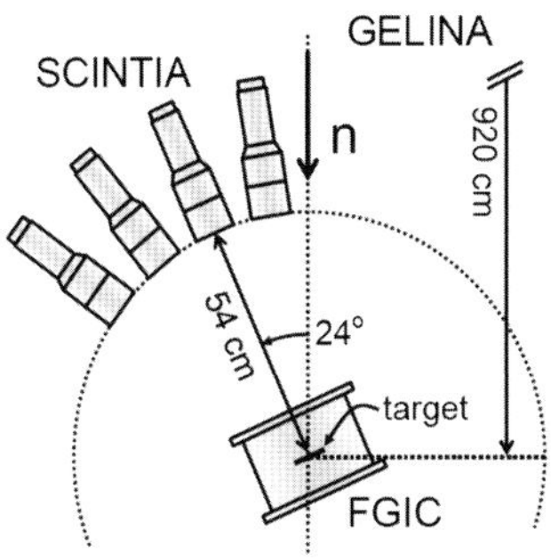

Fig. 1. Schematical setup of SCINTIA.

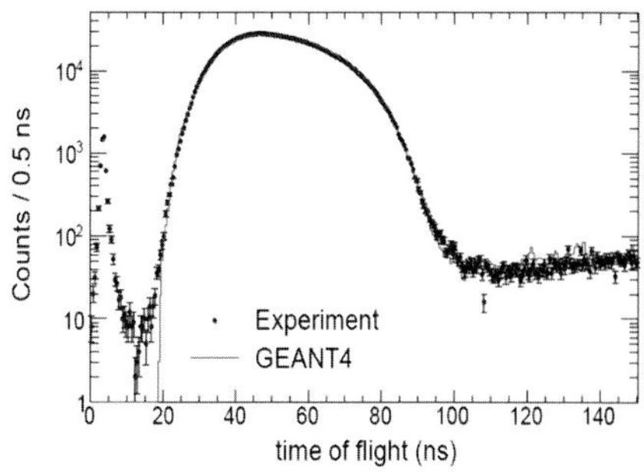

Fig. 2. Comparison between measured and simulated neutron spectrum from $^{252}Cf(sf)$.

2.1. *Preliminary results*

With the ionization chamber pulse heights and emission angles of the fission fragments are determined. The SCINTIA array measures the prompt fission neutron spectrum. Fig. 2 shows a spectrum measured using $^{252}Cf(SF)$ during detector calibration and is compared to a simulation with GEANT4. From the ionization chamber signals the mass and kinetic energies of the fission fragments are deduced as a function of resonance energy. Fig. 3 shows as example the difference in the mass distribution at the strong resonance at 19.3 eV compared to the one at thermal.

A sizable increase in yield in the region $A_H \in (125, 135)$ u is observed. In this mass region TKE(A) is at its maximum, which explains the observed increase in TKE for this resonance when integrating over the mass distribution. Contrary to the strongly mass dependent TKE(A), the total average neutron

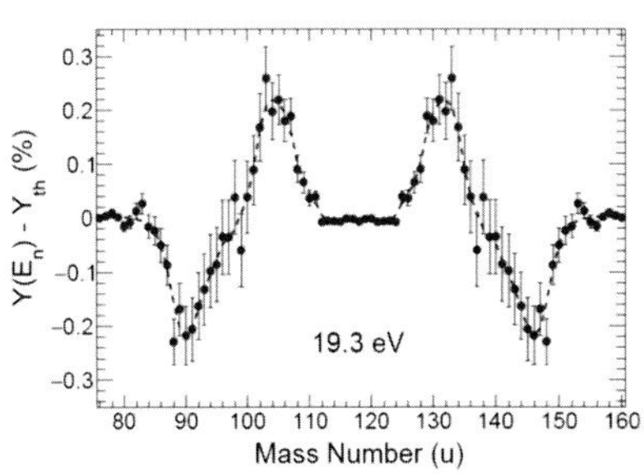

Fig. 3. Difference in mass yield to the thermal neutron induced fission of the 19.3 eV resonance. The dashed line shows the data smoothed over two mass units, drawn to guide the eye.

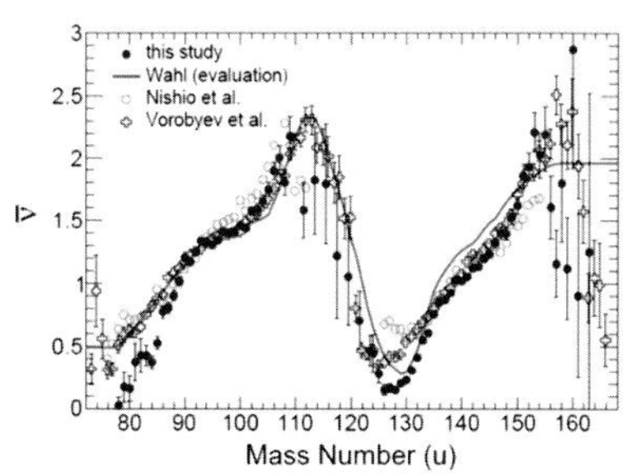

Fig. 4. The average number of neutrons emitted per fragment as a function of the fragment mass compared to data from Refs. [8, 9] and the evaluation by Wahl [10].

multiplicity is only weakly dependent on mass. However, in the mass region $A_H \in (125, 135)$ u the average number of neutrons per fission shows a pronounced minimum. Hence, the increased yield in this mass region should have an impact also on the total neutron multiplicity per fission for certain resonances.

The average neutron multiplicity as a function of the fragment mass is shown in Fig. 4. For comparison, experimental data from Refs. [8, 9] and the evaluation from Ref. [10] are also show. The general shape is reproduced in this work, however the minima around mass number ~80 u for the light fragments and ~130 u for the heavy fragments are more pronounced in the present data. In Fig. 5 the average neutron multiplicity per fission is plotted as a function of the fragment TKE. As expected, a close to linear dependence is observed, except for at low TKE. A least square fit, indicated by the full black line in Fig. 5, results in $-\partial TKE / \partial\nu = 11.86$ MeV/n. This value is substantially lower than the values 16.7 – 18.5 MeV/n determined in earlier studies, performed at thermal incident neutron energies [8, 9, 11, 12]. This inverse slope is linked to the energy cost to emit a neutron. It is not expected that the small increase in the excitation energy of the compound nucleus in this experiment would have an influence on the energy cost to emit a neutron. To understand the strong difference compared to the earlier studies, the fission fragment TKE distribution from the present measurement is compared to the data of Refs. [8, 9, 11] in Fig. 6. The

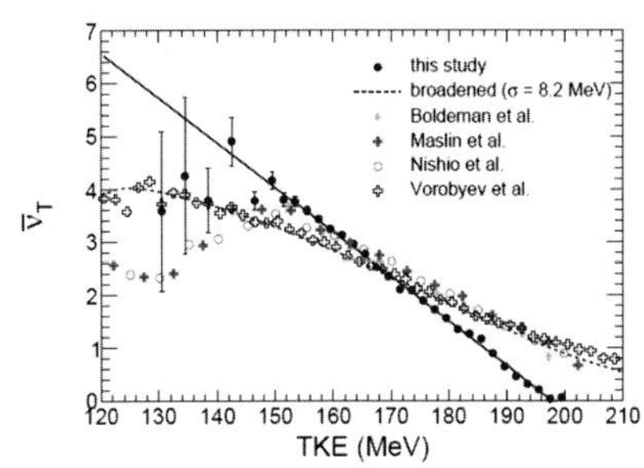

Fig. 5. Dependence of prompt neutron multiplicity per fission on the fragment TKE obtained in this study, compared to data from Refs. [8, 9, 11, 12]. The data of Ref. [9] has been re-binned for better clarity of the figure. The full black line is a linear least squared fit to the present data, which corresponds to an inverse slope of $-\partial TKE/\partial\nu = 11.86$ MeV/n. The dotted black line is the result obtained by folding the present data set with a Gaussian TKE resolution function with $\sigma = 8.2$ MeV.

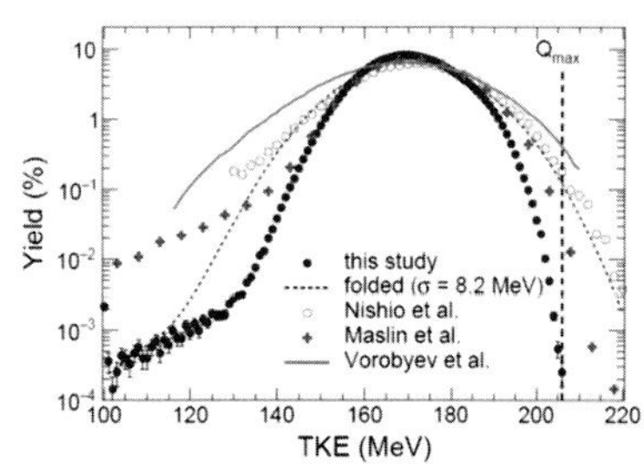

Fig. 6. The fragment TKE Distribution compared to data of Maslin *et al.* [11] and Nishio *et al.* [8]. The dotted black line is the result obtained by folding the present data set with a Gaussian resolution function with $\sigma = 8.2$ MeV, necessary to reproduce the data of Nishio *et al.* The vertical dashed line labeled Q_{max} indicates the maximum available energy for the fragmentation where the heavy fragment is the doubly magic ^{132}Sn nucleus.

distribution from Refs. [8, 9, 11] are substantially broadened. A high yield is found above the maximum available energy, as determined by the Q-value for the fragmentation where the heavy fragment is the doubly magic ^{132}Sn nucleus.

Since the TKE cannot physically exceed the maximum available energy the TKE distributions from Refs. [8, 9, 11] must suffer from substantial resolution broadening. The dashed line in Fig. 6 represents the result of folding the present data with a Gaussian resolution function in order to reproduce the distribution of Nishio *et al.* [8]. If the same broadening is applied when determining ν(TKE) the dotted line in Fig. 5 is obtained. The convolution reproduces qualitatively features in the literature data, i.e. a less steep slope and ν different from zero at the maximum available energy. There are discrepancies at low TKE where the data of Refs. [8, 11] show a strong decrease in ν, that cannot be explained by the resolution broadening. However, examining Fig. 6, tails in the TKE distributions can be observed at low TKE values. Events belonging to this tail are likely due to scattering of the fission fragments in the target foil and/or surrounding materials, as already noted by Maslin *et al.* [11]. The neutron emission from such energy degraded fission events is expected to be close to the average value. This is consistent with the observed decrease in ν at lower TKE as the tail becomes more and more dominant in the yield. The tail is present also in the data from this study, although at much lower intensities compared to the data of Refs. [8, 11]. Due to the presence of the tail we cannot conclude that the decrease in slope of ν below TKE ~ 140 MeV is of a physical origin. Instead it seems more likely that this decrease is an experimental artefact.

3. VERDI the 2E-2v detector system

VERDI (VElocity foR Direct particle Identification) is a double energy and double velocity (2E-2v) spectrometer [13]. The velocity provides information about the pre-neutron mass of each fission fragment when isotropic prompt-neutron emission from the fragments is assumed. The kinetic energy, in combination with the velocity, provides the post-neutron mass. From the difference between pre- and post-neutron masses, the number of neutrons emitted by each fragment can be determined.

The energy measurement is achieved by 32 silicon detectors located at a distance of about 50 cm from the target. They are arranged in two arrays, each housing 16 detectors on their respective side of the target. These detectors measure the kinetic energy and the arrival (stop) time of the incoming fission products. Emission (start) time of the fragments is determined by micro channel plate (MCP) detectors. The electrons sputtered from the surface of the actinide sample are deflected onto the MCPs by electrostatic mirrors. A complete description of the setup using one electron detector and the mentioned silicon arrays is found in Ref. [13].

Comparison of the present results using a $^{252}Cf(SF)$ source is done to the data from Ref. [14]. The latter data are taken using an ionization chamber. There are still small issues with the energy calibration, so the results shown are preliminary. The energy and mass distributions have been symmetrized as VERDI at the moment has an instrumental asymmetry between the different sides of about 10% when it comes to detection efficiency of light and heavy fragments, respectively. The preliminary pre-neutron energy distribution (Fig. 7) is in excellent agreement with the data from Ref. [14], and so is the heavy fragment peak (low energy) of the post-neutron energies (not shown). Both pre-neutron (Fig. 8) and post-neutron (not shown) masses are in good agreement with Ref. [14], showing a slightly higher peak-to-valley ratio. This indicates a higher resolution of the VERDI setup compared to a standard ionization chamber.

Since both pre- and post-neutron masses are available we can easily derive the average neutron multiplicity $\nu(A)$. In Fig. 9, ν as a function of M_{pre} is shown. Discrepancies, mainly around the symmetry region where the present data indicate a negative neutron multiplicity are visible. A negative $\nu(A)$ would mean absorption of a prompt neutron from the complementary fragment. Since this process seems very unlikely the most probable cause for the negative $\nu(A)$ is to be searched in the calibration. No tweaking, e.g., by using a known average neutron multiplicity for ^{252}Cf has been made, and still a rather good agreement with the data of Ref. [14] is achieved except for the fission fragment mass range

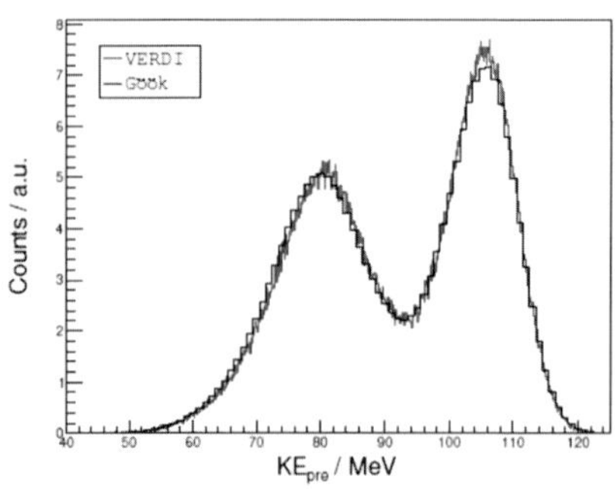

Fig. 7. Pre-neutron emission fission fragment kinetic energy measured by VERDI compared to data from Ref. [14].

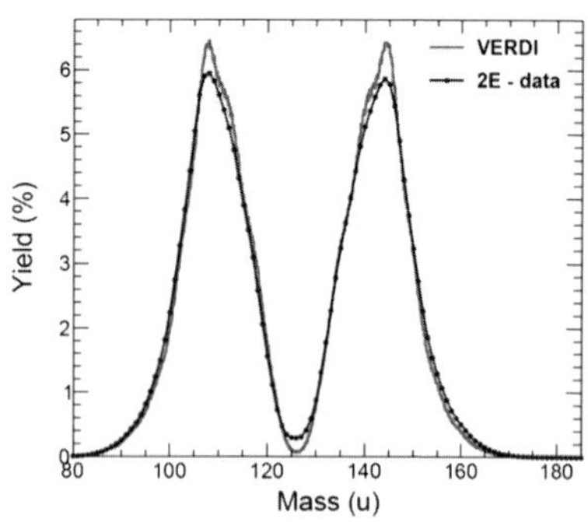

Fig. 8. Pre-neutron emission fission fragment mass measured by VERDI compared to data from Ref. [14].

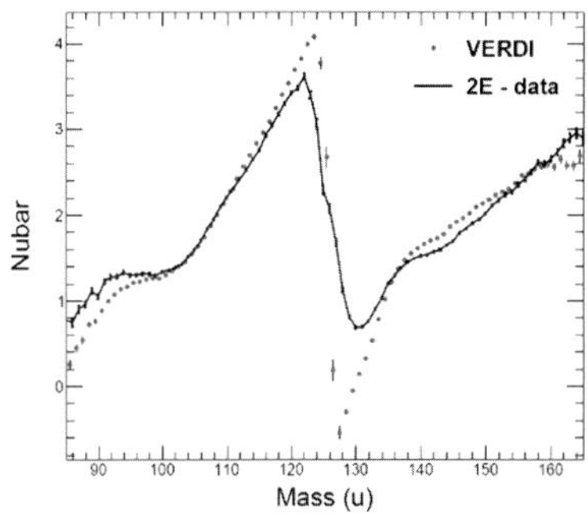

Fig. 9. Neutron multiplicity measured by VERDI compared to data from Ref. [14].

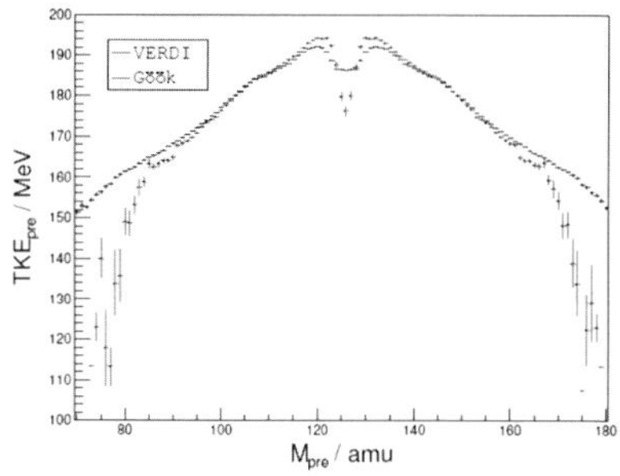

Fig. 10. Pre-neutron TKE versus mass measured by VERDI and compared to data from Ref. [14].

A < 100 and A > 155. The obtained total average neutron multiplicity was 3.62 which is within 4% of the recommended value of 3.759 for ^{252}Cf.

The TKE as a function of M_{pre} also agrees well with Ref. [14] (Fig. 10) except for masses far from symmetry, and the average TKE $\langle TKE_{pre} \rangle = 184.8$ MeV is not too far off from the recommended value 184.1 MeV [20]. The excellent timing properties of the MCP detectors allow for a better mass resolution as clearly seen in the symmetry region. Similarly VERDI measures a slightly higher TKE_{post} with, $\langle TKE_{post} \rangle = 182.1$ MeV compared to 181.4 MeV from Ref. [14].

4. Prompt neutron emission in neutron induced fission of ^{235}U at MeV incident neutron energy

In this work we focus on the prompt fission neutron multiplicity, ν(A), as a function of E_{exc} and especially how the fragment mass-dependence manifests itself in (A, E_n). A few earlier experiments supports a heavy-fragment extra-neutron emission but are, surprisingly, ignored in some contemporary fission calculation codes. The first experimental efforts in the planned long-term project are described along with the first preliminary results. The experimental setup, at JRC-Geel, consists of a Frisch-Grid Ionization Chamber (FGIC) together with two liquid scintillators (NE-213 equivalent) which were characterized by Kornilov *et al.* [15]).

Fig. 11 shows one of the few existing measurements on the so-called "saw-tooth" shape at different incident neutron energies (E_n) [16]. The data implies that the extra neutron emission observed at higher E_{exc} emerges mainly from the heavier fragments. However, the limited number of available data has led to that

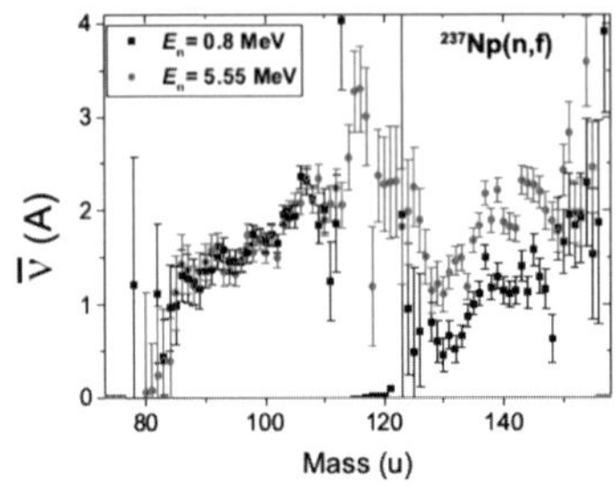

Fig. 11. The ν(A) for ^{237}Np(n,f) at different En [16]. The data shows a higher ν but seemingly only from the heavy fragments.

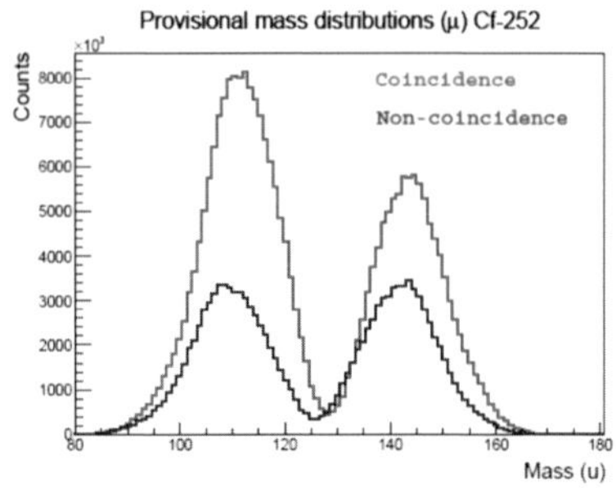

Fig. 12. The provisional mass distributions, from the ratio of the detected fragment energies. The non-coincidence (lower curve), the coincidence data between the ND and the chamber (upper curve).

many fission codes as well as experimental data analysis, are based on an average increase instead. This was discussed in Refs. [17, 18]. Particularly in the 2-E data this assumption has been shown to have significant effect on the measured mass yields [19]. To answer the data-needs the Uppsala group and JRC-Geel have initiated a measurement campaign. The first experimental study cases were ^{252}Cf(SF) and ^{235}U(n_{th},f). The Cf source had a strength of about 15000 fissions/sec. During the beam-measurements, two different ^{235}U samples where used, with 0.1 mg/cm^2 and 0.2 mg/cm^2, respectively. All samples are thin enough to allow for both fragments to escape. Only fragments emitted at grazing angles are sometimes lost. For the ^{235}U case, a 0.5 MeV neutron beam was produced via the ^{7}Li(p,n) reaction and then thermalized by a block of paraffin (12 cm) put in front of the FGIC.

The preliminary results presented here are from ^{252}Cf(SF) for which the analysis has progressed most. The ^{252}Cf(SF) data are analyzed using the so-called 2-E technique where pre-neutron masses are iteratively calculated based on known prompt fission neutron multiplicities. The neutron multiplicity is calculated by the ratio of the two mass distributions, namely the coincidence and non-coincidence events [14].

The pre-neutron mass distribution from ^{252}Cf(SF) not shown here is compared to the data of Ref. [14]. The agreement is rather satisfactory although the analysis is not finalized yet.

The provisional mass (μ) distributions, are calculated from the ratio of detected energies (corrected for energy losses and pulse-height defect) and are shown in Fig. 12. They differ from the final pre-neutron masses since no

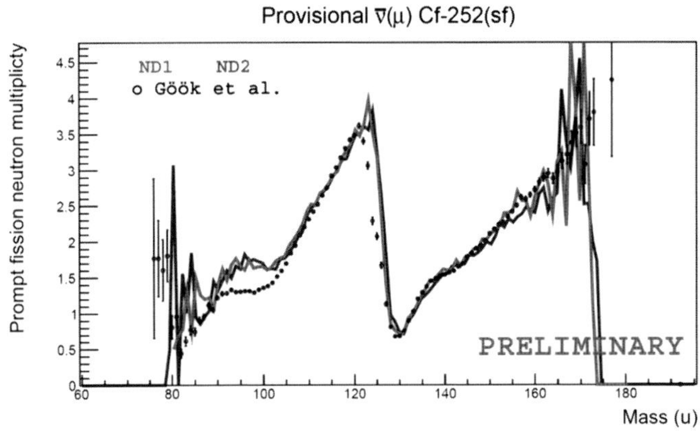

Fig. 13. The ratio of the two distributions of Fig. 12 gives the provisional saw-tooth shape of ^{252}Cf(sf) [14].

iterative procedure has been applied yet (including to account for the prompt neutron emission). At this moment the provisional $\nu(\mu)$ distribution is determined from the ratio of those two provisional (coincidence and non-coincidence) mass distributions. The data are compared to $\nu(A)$ data from Ref. [14] in Fig. 13. The analysis is currently being finalized to obtain pre-neutron emission masses after the iterative procedure, and to obtain $\nu(A)$ after applying the corrections mentioned in Ref. [14]. In doing so it is necessary to make an initial assumption of $\nu(A)$ in order to obtain the pre-neutron masses. Fig. 13 shows that the impact of the initial assumption made on the $\nu(A)$ on the final result will be small.

5. Conclusions

Several new detector systems have been developed in recent years at JRC-Geel, which now deliver improved results as has been exemplified in the present paper. Correlation measurements of prompt neutron emission with fission fragment mass determination have shown that with improved resolution also shortcomings published in literature can be identified. The 2E-2v detector systems presently under development in different institutes have the potential to achieve even greater resolution but the way to get there is rather bumpy. Presented results by VERDI show superior resolution but the analysis is still in progress. Also correlations with prompt neutron emission and fission fragment properties at higher incident neutron energies are difficult to measure, but we have confidence that we are able to do so in the near future.

Acknowledgments

Part of the presented work has been financed by the Swedish Radiation Safety authority (SSM). Part of the presented work has also been supported by the CHANDA trans-national access program and the JRC-EUFRAT program. The authors would like to thank the operators of the GELINA and Van de Graaff facilities of JRC-Geel for providing the necessary neutron beams.

References

1. K.-H. Schmidt, B. Jurado, C. Amouroux and C. Schmitt, *Nucl. Data Sheets* **131**, 107 (2016).
2. C. Lubitz and R. Roussin, *NEA/WPEC* **18** (1999).
3. C. De Saint Jean and R. McKnight, *NEA/WPEC* **34** (2014).
4. F.-J. Hambsch, H.-H. Knitter, C. Budtz-Jørgensen and J.-P. Theobald, *Nuclear Physics* **A491**, 56 (1989).
5. F.-J. Hambsch, I. Ruskov and L. Dematté, in *Proc. Theory-1*, *EUR* 24802 (Sinaia, Romania, 2011), p. 41.
6. R.E. Howe, T.W. Phillips and C.D. Bowman, *Phys. Rev. C* **13**, 195 (1976).
7. A. Göök, W. Geerts, F.-J. Hambsch, S. Oberstedt, M. Vidali and S. Zeynalov, *Nucl. Instr. and Meth.* **A830**, 366 (2016).
8. K. Nishio, Y. Nakagome, H. Yamamoto and I. Kimura, *Nucl. Phys. A* **632**, 540 (1998).
9. A. Vorobyev, O. Shcherbakov, A. Gagarski, G. Val'ski and G. Petrov, *EPJ Web of Conf.* **8**, 03004 (2010).
10. C. Wahl, *Atomic Data and Nuclear Data Tables* **39**, 56 (1988).
11. E.E. Maslin, A.L. Rodgers and W.G.F. Core, *Phys. Rev.* **164**, 1520 (1967).
12. J.W. Boldeman, A.R. de L. Musgrove and W.R. L., *Aust. J. Phys.* **24**, 821 (1971).
13. M.O. Frégeau *et al.*, *Nucl. Instr. Meth. A* **817**, 35 (2016).
14. A. Göök *et al.*, *Phys. Rev. C* **90**, 064611 (2014).
15. N.V. Kornilov *et al.*, *Nucl. Meth. Instr. A* **599**, 226 (2009).
16. A. Naqvi *et al.*, *Phys. Rev. C* **34**, 218 (1986).
17. A. Al-Adili *et al.*, *Nucl. Data Sheets* **119**, 342 (2014).
18. A. Al-Adili *et al.*, *EPJ Web Conf.* **122**, 01007 (2016).
19. A. Al-Adili *et al.*, *Phys. Rev. C* **86**, 054601 (2012).
20. F. Gönnenwein, in: The nuclear fission process (Ed. C. Wagemans, CRC Press, Boca Raton, Florida, 1991), Chap. 8, pp. 287–473.

Recent Results with FREYA

R. Vogt
Nuclear and Chemical Sciences Division, Lawrence Livermore National Laboratory, Livermore, CA 94551, USA
E-mail: vogt2@llnl.gov

J. Randrup
Nuclear Science Division, Lawrence Berkeley National Laboratory, Berkeley, CA 94720, USA
E-mail: jrandrup@lbl.gov

P. Talou
Theoretical Division, Los Alamos National Laboratory, Los Alamos, NM 87545 USA
E-mail: an_author@laboratory.com

For many years, the state of the art for handling fission in radiation transport codes has involved sampling from average distributions. Such "average" fission models have limited interaction-by-interaction capabilities. Energy is not explicitly conserved and no correlations are available because all particles are emitted isotropically and independently. However, in a true fission event, the energies, momenta and multiplicities of emitted particles are correlated. The FREYA (Fission Reaction Event Yield Algorithm) code generates complete fission events. Event-by-event techniques such as those of FREYA are particularly useful because it is possible to obtain complete kinematic information on the prompt neutrons and photons emitted during the fission process. It is therefore possible to extract any desired correlation observables. We describe FREYA and compare our results with neutron-neutron and neutron-light fragment correlation data. We also present preliminary results studying the sensitivities of various neutron observables to the yield distribution $Y(A, Z, \mathrm{TKE})$ used as input to the code. All results are shown for the spontaneous fission of ^{252}Cf.

Keywords: Fission; Correlations; Monte Carlo.

1. Introduction to FREYA

The computational model FREYA[1–7] generates complete fission events, *i.e.* it provides the the full kinematic information on the two product nuclei as well as all the emitted neutrons and photons. In its development, an emphasis had been put on speed, so large event samples can be generated fast, and FREYA therefore relies on experimental data supplemented by simple physics-based modeling. In its standard version, to treat a given fission case, FREYA needs the fission fragment mass distribution $Y(A)$ and the total kinetic energy TKE(A) for the particular excitation energy considered. $Y(A)$ is taken either directly as the measured yields or as a five-Gaussian fit to the data which makes it possible to parametrize its energy dependence[3].

In order to generate an event, FREYA first selects the mass split based on $Y(A)$. The fragment charges are then sampled from the normal distributions suggested by experiment[3]. The linear and angular momenta of the two fragments and their internal excitations are subsequently sampled. After their formation, the fully accelerated fragments de-excite first by neutron evaporation and then by photon emission. In addition to spontaneous fission, FREYA treats neutron-induced fission up to $E_n = 20\,\mathrm{MeV}$; the possibility of pre-fission evaporation is considered as well as pre-equilibrium neutron emission (which plays a role at the highest energies). For full details, see Refs. 1–6.

FREYA contains a number of adjustable parameters that control various physics aspects:

- dTKE — an overall shift of $\overline{\mathrm{TKE}}$ relative to the input TKE(A), used to adjust to the average neutron multiplicity $\overline{\nu}$;
- e_0 — the overall scale of the Fermi-gas level density parameters;
- x — the advantage in excitation energy given to the light fragment;
- c_S — the ratio of the 'spin temperature' to the 'scission temperature';
- c_T — the relative statistical fluctuation in the fragment excitations;
- $Q_{\min}$ — the energy above the neutron separation threshold where photon emission takes over from neutron emission;
- $t_{\max}$ — the maximum half-life of a level during the photon decay process which stops when the level half-life exceeds $t_{\max}$.

The correlations described in Sec. 2 are rather robust relative to the variation of the above parameters with the strongest dependence being on the excitation energy sharing, embodied by the parameter x. See Ref. 5 for details, including sensitivity of the correlations to these physics-based parameters. In Sec. 3 preliminary results on a study of the sensitivity of the FREYA results to the input yields, keeping the parameters fixed, is described.

2. Correlations

Two examples of correlated neutron results with FREYA are now discussed. On the left-hand side of Fig. 1 two-neutron correlations are compared with

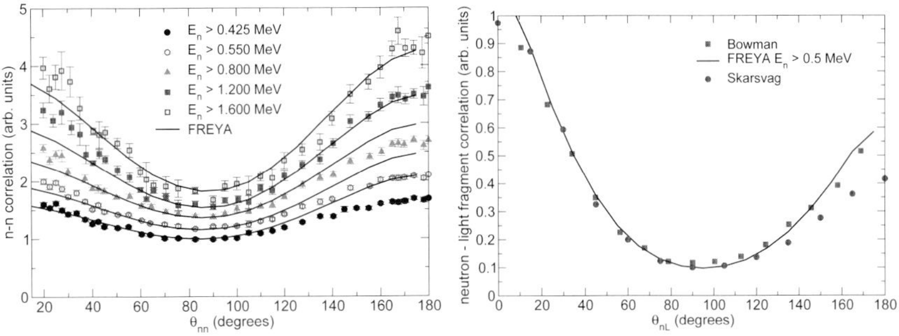

Fig. 1. FREYA calculations of angular correlations are compared to ^{252}Cf(sf) data. (Left) Two-neutron correlations are compared to data from Ref. 8. (Right) Neutron-light fragment correlations are compared to data from 10 (squares) and 11 (circles). The minimum kinetic energy of the neutrons is 0.5 MeV. See Ref.[5] for full details.

data from Ref. 8. Two-neutron angular correlations result in a characteristic shape with peaks at $\theta_{nn} = 0$ and 180°. The peak at $\theta_{nn} = 0$ occurs when both neutrons are emitted by the same fragment. If both neutrons are emitted by the light fragment, the 0° peak is higher. The peak at 180° arises when one neutron is emitted by each fragment. Even though the neutrons are emitted isotropically in the fragment rest frame, the boost to the lab frame assures that the neutrons follow the parent fragments. The neutron multiplicity affects the correlations: the higher ν, the weaker the correlation[2]. See Refs. 5,9 for other comparisons to neutron-neutron correlation data.

Neutrons can also be correlated with a particular fragment. Bowman *et al.* presented the angular correlation between all measured neutrons and the identified light fragment. While the correlation is made with the light fragment, it was not possible to determine which fragment emitted the neutron[10]. If all neutrons come from the light fragment, there is a strong peak at $\theta_{nL} = 0$ with essentially no signal in the opposite direction since the neutrons will follow the parent fragment due to the boost. Likewise, if all detected neutrons arise from the heavy fragment, the correlation is effectively reflected around $\theta_{nL} = 90°$. The shape of the correlation including all emitted neutrons retains the largest peak at zero degrees while, in the backward direction, the signal is reduced. This is because more neutrons are emitted by the light fragment because it gets more intrinsic excitation energy. The model results are compared to the data on the right-hand side of Fig. 1. The agreement of our correlation with the shape of the data is very good. See Ref. 5 for more details.

3. Sensitivity Study

Relative to the FREYA code described in the literature[1–6], a modified version that samples A, Z, and TKE directly from the input yield function $Y(A, Z, \mathrm{TKE})$ is employed here[12]. In the present preliminary study, the adjustable FREYA parameters $x, e_0, d\mathrm{TKE}, c_S$ and c_T are kept constant at nominal values in order to bring out the effect of the variation in the input yield functions $Y(A, Z, \mathrm{TKE})$. Consequently no attempt has yet been made to match the relatively well determined mean neutron multiplicity for each of the yield functions.

To generate a sample of yield functions, $\{Y^{(n)}(A, Z, \mathrm{TKE})\}$, a generalized least-square analysis of the various components that enter into the definition of the yields is performed. A particular yield function is assumed to have the form

$$Y(A, Z, \mathrm{TKE}) \;=\; Y(A) \times P(Z\,|\,A) \times P(\mathrm{TKE}\,|\,A)\,. \tag{1}$$

For each mass split, the TKE distribution is assumed to be Gaussian

$$Y(\mathrm{TKE}\,|\,A) \sim \exp\left(-[\mathrm{TKE} - \overline{\mathrm{TKE}}(A)]^2/2\sigma^2_{\mathrm{TKE}}(A)\right)\,, \tag{2}$$

with $\int d\mathrm{TKE}\, Y(\mathrm{TKE}\,|\,A) = 1$. A least-square analysis of the available experimental data on $Y(A)$, $\overline{\mathrm{TKE}}(A)$, and $\sigma^2_{\mathrm{TKE}}(A)$ entering in Eqs. (1) and (2) was performed to obtained the mean distributions and their covariance matrices. Six experimental data sets were considered to obtain $Y(A)$[13–18], six for $\overline{\mathrm{TKE}}(A)$[13–15,17,19,20], and three for $\sigma^2_{\mathrm{TKE}}(A)$[13,14,17]. No correlations between the different quantities were assumed here.

For a particular fragment mass, the charge distribution is obtained from Wahl systematics[21],

$$Y(Z\,|\,A) \;=\; (1/2)F(A,Z)\,N(A)\,\left[\mathrm{erf}(V^+) - \mathrm{erf}(V^-)\right]\,, \tag{3}$$

where $V^{\pm} = (Z - Z_p \pm 0.5)/(\sqrt{2}\sigma_z)$, $N(A)$ is a normalization factor ensuring $\int dZ\, P(Z\,|\,A) = 1$, and $F(A,Z)$ gives the odd-even staggering.

A Principal Component Analysis (PCA)[22] was then performed on each of the evaluated covariance matrices for each partial distribution in Eq. (1). The PCA decomposition leads to sampled realizations $Y(A)^{(i)}$, $\overline{\mathrm{TKE}}(A)^{(j)}$, and $\sigma^2_{\mathrm{TKE}}(A)^{(k)}$, and the sampled 3D yield functions $Y^{(n)}(A, Z, \mathrm{TKE})$ with $n = (ijk)$ then follow from Eq. (1).

Following the sampling of the yields, the excitation energy is distributed among the fragments and neutron and photon emission proceeds as in the standard FREYA.

The sensitivity of the neutron multiplicity distribution and two-neutron correlations resulting from employing a sample of 1000 yield functions constructed as described above is now described. A total of one million events were generated for each yield function and the results were compiled and analyzed.

Starting from 10^6 complete fission events for each of the $N = 1000$ yield functions, $Y^{(n)}(A, Z, \mathrm{TKE})$, $n = 1, \ldots, N$, the associated neutron multiplicity distribution, $P^{(n)}(\nu)$, is defined as

$$1 = \sum_{\nu \geq 0} P^{(n)}(\nu) , \quad \bar{\nu}^{(n)} = \sum_{\nu \geq 0} \nu \, P^{(n)}(\nu) , \tag{4}$$

where $\bar{\nu}^{(n)}$ is the mean neutron multiplicity resulting from the particular yield function $Y^{(n)}(A, Z, \mathrm{TKE})$.

The resulting mean multiplicity distribution, shown on the left-hand side of Fig. 2, is then

$$\langle P(\nu) \rangle = \frac{1}{N} \sum_{n=1}^{N} P^{(n)}(\nu) . \tag{5}$$

The multiplicity distribution $P(\nu)$ can be characterized by its factorial moments,

$$\mathcal{M}_m \equiv \sum_{\nu \geq 0} \nu(\nu - 1) \cdots (\nu - m + 1) \, P(\nu) . \tag{6}$$

These quantities were extracted for all the N multiplicity distributions, $\{ P^{(n)}(\nu) \}$. The right hand side of Fig. 2 shows the resulting average values and the associated dispersions (indicated as error bars) divided by $m!$ to tame the rapid increase in the moments as a function of order m.

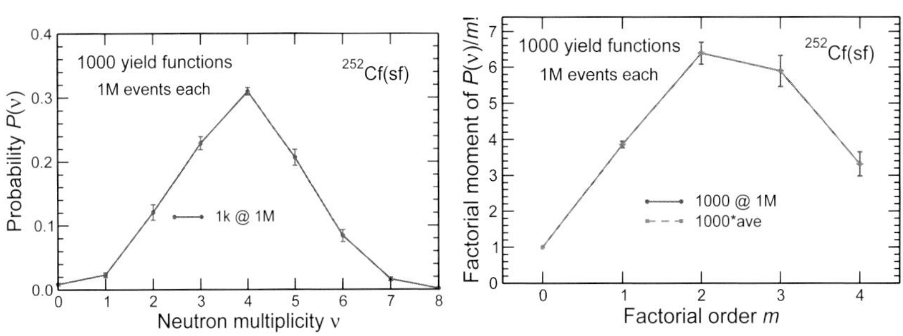

Fig. 2. The neutron multiplicity distribution $P(\nu)$ (left) and the factorial moments of the neutron multiplicity distribution, divided by $m!$.

Part of the dispersion obtained is due to the finite event sample (10^6) generated for each yield function. It is therefore important that the samples be sufficiently large. To verify that this is indeed the case, the entire procedure has been repeated using the same yield function all 1000 times employing the average of the 1000 yield functions as input. In this case there is no effect on the extracted observables due to the variation of the FREYA input, the entire dispersion is solely caused by the finite number of events. The corresponding results are shown in the curve on the right-hand side of Fig. 2. It is evident that the variance arising from the finite event sample is significantly smaller than the total variance, which can therefore be regarded as being well determined.

The neutron-neutron angular correlations are rather insensitive to the input yields $Y(A, Z, \mathrm{TKE})$, as shown in Fig. 3. These angular correlations are dominated by the kinematic effects of the boost given to the evaporated neutrons by the moving fragments. As a consequence, $C_{\mathrm{nn}}(\theta_{12})$. does not depend much on the specific emitter fragments. Of greater importance is the relative excitation energy given to the light and heavy fragments, controlled in part by the FREYA parameter x, as shown in Ref. 5. (Note that in Fig. 3 the result is shown as a function of $\cos(\theta_{12})$ while the correlation function on the left-hand side of Fig. 1 is shown as a function of the separation angle, θ_{nn} where $\theta_{nn} = \theta_{12}$.)

This preliminary study of ^{252}Cf(sf) illustrates how the uncertainty in the input yield function $Y(A, Z, \mathrm{TKE})$, resulting from the inherent uncertainty in the experimental data, manifests itself in various neutron observables.

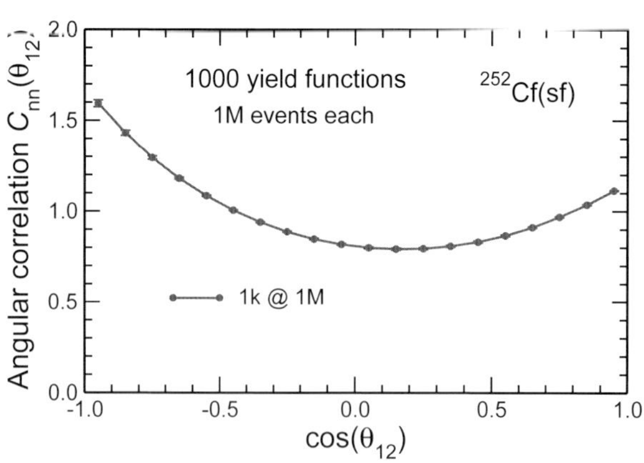

Fig. 3. The two-neutron angular correlation function as a function of $\cos(\theta_{12})$.

In particular, the procedure employed to determine the distribution of yield functions did not attempt to match the overall mean neutron multiplicity $\overline{\nu}$. As a consequence, the values of $\overline{\nu}$ obtained for the various sampled yield functions exhibit a variability that significantly exceeds the rather narrow evaluated uncertainty on this key quantity. It would therefore be preferable to constrain the yield functions to consider only those that produce $\overline{\nu}$ values within the evaluated range. Such an analysis is underway.

Note that ^{252}Cf(sf) is an especially good test case because the data used to generate the input yield functions $Y(A, Z, \mathrm{TKE})$ are of good quality with high statistics. For all other actinides, with the possible exception of ^{235}U(n$_{\mathrm{th}}$,f) and ^{239}Pu(n$_{\mathrm{th}}$,f), the available input data are either not of as high quality or, in some cases, arise from models when no data are available at a given energy. Thus they do not provide as strong constraints on the yields and, therefore, also not on the neutron observables, as do those of ^{252}Cf(sf) which, as shown here, are already insufficient to agree with the constraints on $\overline{\nu}$.

4. Summary

Event-by-event models of fission, such as `FREYA`, provide a powerful tool for studying correlations in fission. The calculations are robust, being relatively insensitive to the input parameters which can be constrained by other data. The agreement of our neutron correlation calculations with the available data is good and does not lend strong support for the requirement of scission neutrons to explain the correlations. However, further data on these correlations based on fission of other isotopes and, for neutron-induced fission, at higher incident neutron energies would be welcome to help verify the results in Sec. 2. The sensitivity study in Sec. 3 for the input yields will be extended to the input model parameters in the future.

Acknowledgments

This work was performed under the auspices of the U.S. Department of Energy under Contracts No. DE-AC02-05CH11231 (JR), DE-AC52-06NA25396 (PT), and DE-AC52-07NA27344 (RV). The support of the Office of Defense Nuclear Nonproliferation Research and Development in DOE/NNSA is also gratefully acknowledged.

References

1. J. Randrup and R. Vogt, *Phys. Rev. C* **80**, 024601 (2009).

2. R. Vogt and J. Randrup, *Phys. Rev. C* **84**, 044621 (2011).
3. R. Vogt, J. Randrup, D. A. Brown, M. A. Descalle and W. E. Ormand, *Phys. Rev. C* **85**, 024608 (2012).
4. R. Vogt and J. Randrup, *Phys. Rev. C* **87**, 044602 (2013).
5. R. Vogt and J. Randrup, *Phys. Rev. C* **90**, 064623 (2014).
6. J. Randrup and R. Vogt, *Phys. Rev. C* **89**, 044601 (2014).
7. J. Randrup and R. Vogt, in progress.
8. A. M. Gagarski *et al.*, *Bull. Russ. Acad. Sciences: Physics*, **72**, 773 (2008).
9. S. A. Pozzi *et al*, *Nucl. Sci. Eng.* **178**, 1 (2014).
10. H. R. Bowman, J. C. D. Milton, S. G. Thompson and W. J. Swiatecki, *Phys. Rev.* **126**, 2120 (1962); *Phys. Rev.* **129**, 2133 (1963).
11. K. Skarsvag and K. Bergheim, *Nucl. Phys.* **45**, 72 (1963).
12. J. Randrup, P. Talou and R. Vogt, in preparation (2016).
13. C. Budtz-Jørgensen and H.-H. Knitter, *Nucl. Phys. A* **490**, 307 (1988).
14. J. Hambsch and S. Oberstedt, *Nucl. Phys. A* **617**, 347 (1997).
15. E. M. Kozulin, A. A. Bogachev, M. G. Itkis, I. M. Itkis, G. N. Knyazheva, N. A. Kondratiev, L. Krupa, I. V. Pokrovsky and E. V. Prokhorova, *Inst. and Exp. Techniques* **51**, 44 (2008).
16. C. Romano, Y. Danon, R. Block, J. Thompson, E. Blain and E. Bond, *Phys. Rev. C* **81**, 014607 (2010).
17. A. Göök, F.-J. Hambsch and M. Vidali, *Phys. Rev. C* **90**, 064611 (2014).
18. Sh. Zeynalov, F.-J. Hambsch and S. Oberstedt, *J. Korean Phys. Soc.* **59**, 1396 (2011).
19. S. L. Whetstone, Jr., *Phys. Rev.* **131**, 1232 (1963).
20. G. K. Mehta, J. Poitou, M. Ribrag and C. Signarbieux, *Phys. Rev. C* **7**, 373 (1973).
21. A. C. Wahl, Los Alamos Technical Report, LA-13298 (2002).
22. M. E. Rising, A. K. Prinja and P. Talou, *Nucl. Sci. Eng.* **175**, 188 (2013).

Measurements of Correlated Fission Data with DANCE and NEUANCE

G. Rusev*, B. Baramsai, E. M. Bond, T. A. Bredeweg, M. Jandel, D. J. Vieira, C. L. Walker and J. B. Wilhelmy

Chemistry Division, Los Alamos National Laboratory, Los Alamos, New Mexico 87545, USA
**E-mail: rusev@lanl.gov*

A. Couture, S. Mosby and J. L. Ullmann

Physics Division, Los Alamos National Laboratory, Los Alamos, New Mexico 87545, USA

A. C. Hayes, T. Kawano, I. Stetcu and P. Talou

Theoretical Division, Los Alamos National Laboratory, Los Alamos, New Mexico 87545, USA

A. Favalli, K. D. Ianakiev and M. L. Iliev

Nuclear Engineering and Nonproliferation Division, Los Alamos National Laboratory, Los Alamos, New Mexico 87545, USA

The 4π multi-detector γ-ray calorimeter DANCE at the Los Alamos Neutron Science Center and its recent upgrade with an array of neutron detectors NEUANCE are described. DANCE has proven its ability to measure prompt fission γ rays with a large number of γ-ray multiplicities. The prompt fission γ rays may exceed a multiplicity of 25 and total energy of 20 MeV. NEUANCE adds the option of neutron measurement in addition to γ-ray measurement with DANCE and thus to study the fission γ-n correlations. In addition, a pair of silicon detectors measuring the kinetic energy of the fission fragments enables research on the γ-n-TKE correlations. Preliminary results of correlated data from ^{252}Cf(sf) and ^{235}U(n, f) measurements are presented.

Keywords: DANCE; NEUANCE; Prompt Fission γ Rays; n-γ Correlations; n-γ-TKE Correlations.

1. Motivation

Nuclear fission has been known for decades. Large varieties of integral measurements have been performed driven by applications like nuclear reactors and defense. The fission process seen as collective reorganization of nuclear matter is described in a number of macroscopic and microscopic models. Detailed measurements of the emitted radiation from the fission process — fission fragments, neutrons and γ rays — are required to test the theory predictions. Measurements of correlated data can show the competition between the variety of de-excitation modes and verify fission models, but also provide data for nuclear forensics, non-proliferation and stockpile stewardship. Multi-detector systems have to be employed in these measurements to disentangle the multiplicity and energy distributions of the prompt fission neutrons and γ rays.

The Detector for Advanced Neutron Capture Experiments (DANCE) has been in operation for more than a decade at the Los Alamos Neutron Science CEnter (LANSCE) with its primary purpose to measure the γ rays from neutron-capture reactions. Studies of the prompt fission γ rays (PFG) can not only help obtaining more accurate (n,γ) cross sections from fissile actinides, but provide valuable information for the correlation of the PFG energy and multiplicity distribution. Therefore, a program has been initiated[1,2] to measure correlated data at DANCE leading to upgrade with a segmented neutron detector NEUANCE (NEUtron detector at dANCE).

In this report, we will present a PFG measurement of $^{235}U(n,f)$ at DANCE in Section 2. NEUANCE will be described in Section 3, and a recent measurement of the n-γ-TKE correlations in $^{252}Cf(sf)$ involving DANCE, NEUANCE and a pair of silicon detectors will be discussed in Section 4.

2. Prompt fission gamma rays measured with DANCE

DANCE is a 4π γ-ray calorimeter consisting of 160 BaF_2 detectors. A photograph of DANCE is shown in Fig. 1. The BaF_2 crystals are in four different shapes so as to build a sphere without gaps[3]. The only material between the crystals is the lightproof optical foil, ensuring that a Compton-scattered γ ray between two neighboring crystals will not lose energy. A γ ray may cause a few BaF_2 crystals to fire (mostly two), but an add-back procedure can restore the initial γ-ray energy in the off-line analysis. The efficiency of DANCE has been simulated to be 86% for a single γ ray with energy of 1 MeV[4]. A ^6LiH shell with a thickness of 6 cm, installed in the inner cavity of DANCE, surrounds the target to absorb the low-energy neutrons scattered from the sample. The ^6LiH shell and the aluminum beam pipe can cause Compton scattering or pair creation of the target γ rays. We

Fig. 1. A Photograph of DANCE.

apply direct[5] and inverse[6] deconvolution methods based on Geant4 simulations for DANCE[7] to correct for the detector response. The observables DANCE provides are the γ-ray energy E_γ, the total γ-ray energy from the reaction, and the γ-ray multiplicity.

DANCE is located at the Manuel J. Lujan, Jr. Neutron Scattering Center at LANSCE[8]. Spallation neutrons produced by irradiation of a tungsten target with 800-MeV protons with a repetition rate of 20 Hz are used in the measurements. The low-energy part of the neutron-flux distribution is additionally enhanced by a water-moderator block with a thickness of 2.54 cm. The neutrons are collimated to a narrow beam by a double-truncated copper collimator forming a spot at the target position with a full width at half maximum of 0.7 cm. A detailed description of the neutron facility at the Lujan Center and the properties of the beam can be found in Refs.[9,10]. Measuring the time-of-flight from the neutron-production target to the first fired DANCE detector provides the energy of the neutron that caused fission. The γ-ray spectra can be collected for a given resonance or a range of neutron energies.

The capability of DANCE to measure PFG has been demonstrated in a number of experiments[11–14]. Here we present recent results[14] from a measurement on ^{235}U(n, f). A parallel-plate avalanche counter (PPAC) containing about 100 μg/cm^2 of ^{235}U identified the fission events. Gamma rays measured with DANCE in coincidence with the PPAC are shown in

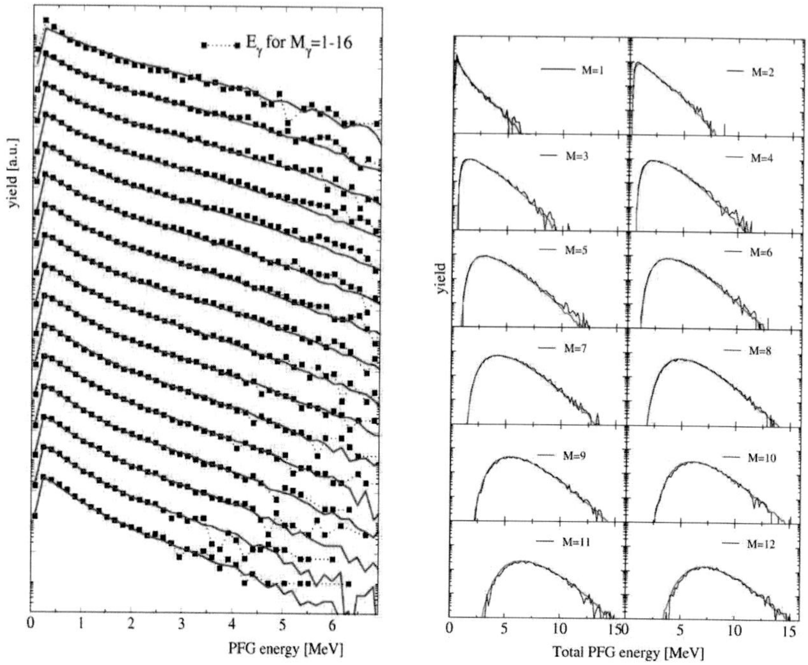

Fig. 2. Spectra of E_γ (left) and E_γ^{total} (right) of PFG from $^{235}U(n,f)$ measured with DANCE. The lines on the left and smooth lines on the right depict a fit with a parameterized model. The figures are taken from Ref. [14].

Fig. 2 for various number of γ-ray multiplicities. A parameterized model, depicted in smooth line, has been developed[14] to fit the data.

3. Neutron-gamma correlation measurements

A neutron detector array NEUANCE was developed to enhance the capabilities of DANCE by simultaneous measurement of the prompt fission neutrons and γ rays[1]. NEUANCE consists of 21 stilbene crystals with dimensions of 23 mm $\times$ 23 mm $\times$ 100 mm coupled to Hamammatsu photomultiplier tubes H-11200. The stilbene crystals are arranged in two rings around the beam pipe with 7 and 14 detectors, respectively. Photographs of the NEUANCE array are shown in Fig. 3. NEUANCE is installed in the inner cavity of DANCE with the ^{6}LiH shell removed. Neutrons are identified in off-line analysis using pulse-shape discrimination. Measured efficiency of NEUANCE for detecting a single neutron is about 14%.

Fig. 3. Photographs of NEUANCE (left) and NEUANCE installed in DANCE (right).

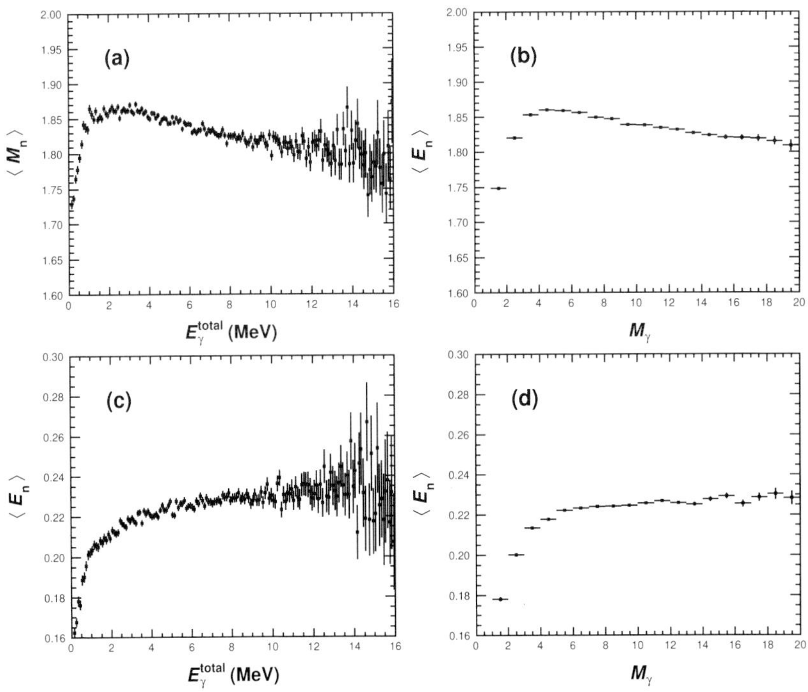

Fig. 4. Correlation graphs between the γ-ray total energy E_γ^{total} and multiplicity M_γ and the neutron average energy E_n and multiplicity M_n from ^{252}Cf(sf). The figure is taken from Ref.[15].

A new data acquisition system was built based on the 14-bit, 500 MS/s CAEN VX1730B digitizers[16] to handle the signals from different type of detectors: BaF_2, stilbene, and slow auxiliary detectors as HPGe or Si detectors. The DANCE and NEUANCE detectors are aligned in time using the Compton-scattered γ rays between two neighboring detectors. Building coincidence events between DANCE and NEUANCE detectors allows us to study the correlations between γ rays and neutrons.

Results from a measurement with a sealed 1 μCi ^{252}Cf with DANCE and NEUANCE demonstrating correlations between various neutron and γ-ray observables are shown in Fig. 4. Further work on correcting these graphs for the detector response is required.

4. Neutron-gamma-TKE correlation measurement

The kinetic energy of the fission fragments can be measured with a pair of Si detectors. We performed a measurement of ^{252}Cf(sf) at DANCE with NEUANCE to study the correlation of γ rays and neutrons with the total kinetic energy of the two fission fragments. A 500 μCi ^{252}Cf source deposited on a 0.1 μm thick Parylene foil was installed in a canister with two F-series Ortec Si detectors at the two ends. The canister was placed in the beam line at the DANCE target position. Coincidence events when the two Si detectors registered fission fragments and DANCE a γ-ray cascade are shown in Fig. 5. CGMF calculations predict the E_γ^{total} vs. TKE dependence

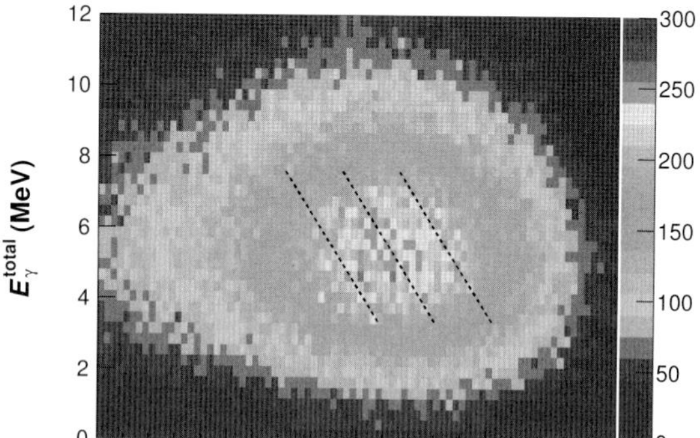

Fig. 5. Measurement of the total γ-ray energy released from ^{252}Cf(sf) versus the sum kinetic energy of the fission fragments deposited in the two Si detectors. The dotted lines depict three possible bands predicted by CGFM calculations.

to form three bands with the middle one being most intense. Figure 5 shows a similar tendency, but the ^{252}Cf deposition appears to be thick and smears the mass distribution. We plan to repeat the measurement with an improved ^{252}Cf sample.

5. Summary

The 4π γ-ray array DANCE and the neutron detector array NEUANCE enable measurements of neutron-gamma correlated data. DANCE alone provides measurement of the single γ-ray energy, the total γ-ray energy released from fission and the γ-ray multiplicity. NEUANCE adds up measurement of the neutron energies and multiplicity in coincidence with the DANCE events. Correlation graphs can be built from any of the observed quantities. In addition, installing two or more Si detectors inside the beam line provides a measurement of the kinetic energy of the fission fragments.

Acknowledgments

This work has been supported by the U.S. Department of Energy through the LANL/LDRD Program, the U.S. Department of Energy, Office of Science, Nuclear Physics under Early Career Award No. LANL20135009, and partly supported by the Office of Defense Nuclear Nonproliferation Research and Development, National Nuclear Security Administration, U.S. Department of Energy.

This work benefited from the use of the LANSCE facility at the Los Alamos National Laboratory. This work was performed under the auspices of the U.S. Department of Energy by Los Alamos National Security, LLC, under Contract No. DE-AC52-06NA25396.

References

1. M. Jandel, B. Baramsai, E. M. Bond *et al.*, Capture and fission with DANCE and NEUANCE, *Eur. Phys. J. A* **51**, 179 (2015).
2. M. Jandel, B. Baramsai, T. A. Bredeweg *et al.*, Current and Future Research at DANCE, *EPJ Web of Conferences* **93**, 02019 (2015).
3. M. Heil, R. Reifarth, M. M. Fowler *et al.*, A 4π BaF_2 detector for (n,γ) cross-section measurements at a spallation neutron source, *Nucl. Instrum. Methods Phys. Res. A* **459**, 229 (2001).
4. M. Jandel, T. A. Bredeweg, E. M. Bond *et al.*, Neutron capture cross section of ^{241}Am, *Phys. Rev. C* **78**, 034609 (2008).

5. G. Rusev, M. Jandel, M. Krtička *et al.*, Cascade γ rays following capture of thermal neutrons on ^{113}Cd, *Phys. Rev. C* **88**, 057602 (2013).
6. G. Rusev, M. Jandel, C. W. Arnold *et al.*, Detector-Response Correction of Two-Dimensional γ-Ray Spectra from Neutron Capture, *EPJ Web of Conferences* **93**, 07002 (2015).
7. M. Jandel, T. A. Bredeweg, A. Couture *et al.*, GEANT4 simulations of the DANCE array, *Nucl. Instrum. Methods Phys. Res. B* **261**, 1117 (2007).
8. P.W. Lisowski and K. F. Schoenberg, The Los Alamos Neutron Science Center, *Nucl. Instrum. Methods Phys. Res. A* **562**, 910 (2006).
9. M. Mocko, G. Muhrer, F. Tovesson and J. Ullmann, Nuclear Cross-Section Measurements at the Manuel Lujan Jr. Neutron Scattering Center, *Proceedings of the International Topical Meeting on Nuclear Research Applications and Utilization of Accelerators, Vienna, May 48*, **AP/IE-03** (2009).
10. M. Mocko and G. Muhrer, Fourth-generation spallation neutron target-moderator-reflector-shield assembly at the Manuel Lujan Jr. neutron scattering center, *Nucl. Instrum. Methods Phys. Res. A* **704**, 27 (2013).
11. A. Chyzh, C. Y. Wu, E. Kwan *et al.*, Evidence for the stochastic aspect of prompt γ emission in spontaneous fission, *Phys. Rev. C* **85**, 021601(R) (2012).
12. J. L. Ullmann, E. M. Bond, T. A. Bredeweg *et al.*, Prompt γ-ray production in neutron-induced fission of ^{239}Pu, *Phys. Rev. C* **87**, 044607 (2013).
13. A. Chyzh, C. Y. Wu, E. Kwan *et al.*, Systematics of prompt γ-ray emission in fission, *Phys. Rev. C* **87**, 034620 (2013).
14. M. Jandel, G. Rusev, E. M. Bond *et al.*, Prompt Fission Gamma-Ray Studies at DANCE, *Phys. Procedia* **59**, 101 (2015).
15. M. Jandel, G. Rusev, B. Baramsai *et al.*, Correlated fission data measurements with NEUANCE, *Nucl. Instrum. Methods Phys. Res. A* **in preparation** (2017).
16. S. Mosby, F. Tovesson, A. Couture *et al.*, A fission fragment detector for correlated fission output studies, *Nucl. Instrum. Methods Phys. Res. A* **757**, 75 (2014).

Fission Activities at the CERN n_TOF Facility

M. Diakaki

Dept. of Physics, National Technical University of Athens, 15780 Athens, Greece
European Organisation for Nuclear Research (CERN), Geneva, Switzerland
E-mail: mdiakaki@cern.ch

L. Audouin, Y. Chen, L.-S. Leong and L. Tassan-Got

CNRS/IN2P3-IPN, Orsay, France

E. Berthoumieux, E. Dupont and F. Gunsing

IRFU, CEA, Université Paris-Saclay, Gif-sur-Yvette, France

M. Bacak, M. Calviani and A. Tsinganis

European Organisation for Nuclear Research (CERN), Geneva, Switzerland

N. Colonna and M. Mastromarco

Istituto Nazionale di Fisica Nucleare (INFN) - Sezione di Bari, Bari, Italy

I. Duran and E. Leal-Cidoncha

Universidade de Santiago de Compostela, Spain

C. Guerrero

Universidad de Sevilla, Spain

C. Paradela

EC JRC, IRMM, Geel, Belgium

A. Stamatopoulos and R. Vlastou

Dept. of Physics, National Technical University of Athens, 15780 Athens, Greece

D. Tarrio

Department of Physics and Astronomy, Uppsala University, Sweden

A.G. Smith and T. Wright
University of Manchester, Manchester, M13 9PL, U.K.

and the n_TOF Collaboration
www.cern.ch/ntof

A major part of the scientific program of the CERN n_TOF facility involves the study of neutron-induced fission reactions, which is essential for nuclear technology, as well as for the development of the theoretical models of fission. An overview of the fission studies performed and the corresponding detection systems is given, with a short description of the latest highlights.

Keywords: n_TOF Facility; Fission Cross Section; Fission Fragments.

1. Introduction

High-quality data of neutron induced fission reactions of actinide nuclei are needed in the field of nuclear reactor technology, in order to improve the safety and efficiency of conventional reactors, as well as to develop the new generation of nuclear reactor systems, with the aim of safer and cleaner energy production. On the other hand, highly accurate fission data are needed in order to develop the theoretical models, and thus better understand one of the most complicated phenomena of nuclear physics, involving large-scale collective motion and shell and pairing effects. The n_TOF (Neutron Time-Of-Flight) facility at CERN was constructed in order to address the pressing needs of nuclear data for advanced nuclear technologies, for nuclear astrophysics and for other fields of basic and applied nuclear physics. The n_TOF collaboration, consisting of about 150 researchers from 40 institutes, is responsible for the extensive program of measurements at this facility, in order to produce a significant collection of high-quality neutron-induced reaction data for many isotopes, a major part of which is fission data on actinides. A comprehensive overview of all the nuclear data activities at the facility was recently published[1].

2. The n_TOF facility

The CERN n_TOF facility is an innovative neutron source[2,3] which combines unique features, such as high instantaneous flux, a neutron beam covering a wide energy range (meV-GeV), high energy resolution, low background and no overlap between two sequential neutron pulses. The white neutron beam at n_TOF is produced via the spallation of protons with a momentum of 20 GeV/c on a thick lead target. The neutrons produced are exploited in two measurement areas with different characteristics. The first experimental area (EAR1) was available since the very beginning of the facility and is situated at 185 m from the neutron production target at a nearly forward direction with respect to the proton beam. It covers a wide neutron energy range from thermal to GeV, with high instantaneous fluxes and a remarkable neutron energy resolution of $\Delta E/E = 10^{-4}$ for thermal neutrons and 0.06 for neutrons with energies close to GeV. A detailed description of the neutron source and EAR1 can be found in Ref. 4. The second experimental area (EAR2) became operational in 2014 and is situated at ~19 m above the spallation target, thus using the neutron beam perpendicular with respect to the proton beam. Due to the position of the experimental area and the short flight path, it is characterised by fluxes of about 25 times higher than those in EAR1, at the cost of lower neutron energy resolution. However, EAR2 extends the measurement capabilities of the facility to targets of low mass and/or to reactions with low cross sections within a reasonable time. More details on EAR2 can be found in Refs. 5,6 and a comparison of the neutron fluxes available in the two areas can be found in Fig. 3 of Ref. 1.

3. Detection setups for fission studies

Different detection systems have been used over the years for the fission studies at n_TOF, a short description of which is given below. The measurements where the detectors have been used can be found in Table 1.

- **FIC:** The Fast Ionisation Chamber (FIC)[7] consists of a stack of parallel plate cells operated with a fast argon mixture. For each actinide target, one of the two fission fragments (FF) is detected with a solid angle acceptance of $\sim 2\pi$, and the signal has no amplification. The stack of actinide target cells is contained in a sealed housing at a pressure of 720 mbar, following the requirements for the measurement of very short-lived actinides. It presents high detection efficiency and fast response, and has been used at n_TOF since the beginning of its operation.
- **Micromegas:** The Micromegas detector (MGAS) is an ionisation chamber with an extra amplification stage separated from the region where the ionisation takes place with the so-called *micromesh*[8]. The *microbulk* version[9] allows for the minimisation of the detector material, very important for in-beam neutron measurements and has been widely used at the n_TOF facility for fission cross section measurements,

neutron beam monitoring and neutron induced charged particle reactions. Other advantages of Micromegas detectors are their versatility in detecting different kinds of radiation, robustness and radiation hardness, as well as a high signal-to-noise ratio.

- **PPAC:** The Parallel Plate Avalanche Counter (PPAC) is a stack of parallel plate avalanche counters which allows for the detection of both FFs of each target in coincidence, providing a very efficient background rejection of alpha particles and high energy reaction products. Furthermore, the setup is characterised by very low mass which makes the detector practically insensitive to the γ-flash. These two characteristics make the detector capable of performing measurements up to 1 GeV. Furthermore, the angle of the FFs is extracted, thus providing the FF angular distribution for studying the fission anisotropy with respect to the neutron energy[10].
- **Silicon Detectors:** Silicon detectors have been very recently used at n_TOF for the ^{235}U fission cross section measurement. The detectors used were 5x5 cm^2 and 200 μm thickness and two detectors were put in front of and behind the targets in order to cover the forward and backward emission of particles. The main advantage of silicon detectors is the energy resolution but their use for in-beam neutron measurements is often avoided due to neutron damage.
- **Fission Tagging setup:** Systems combining the 4π Total Absorption Calorimeter (TAC), consisting of 40 BaF_2 crystals, and especially constructed compact fission chambers are used at n_TOF for the simultaneous measurement of the fission and capture cross-sections of fissile isotopes. Fission events are "tagged" with the fission chamber detector and the γ-rays detected in coincidence are assigned to a fission event, rather than capture. The first attempt was to use Micromegas as fission tagging detector, for the measurement of the ^{235}U(n,γ)/(n,f) reaction[11]. The measurement was successful, but a severe background issue occurred above 100 eV due to neutron induced reactions on copper (structural material of the microbulk Micromegas) and the data analysis was rather complex due to the use of active and passive targets. A new fission tagging detector was developed in order to address these issues; a novel multiplate fission chamber, able to host up to 14 actinide targets in a compact design, coupled to very fast electronics was developed and tested[12]. Furthermore, a major TAC upgrade took place in 2016, in order to improve the detector signals and response to the γ-flash.

- **STEFF:** The Spectrometer for Exotic Fission Fragments (STEFF)[13] is a system of γ-ray and fission detectors used to study the fission process, i.e. detect and characterise the FFs in coincidence with γ-rays using the 2E-2v method (mass resolution $\sim$ 4 a.m.u. - reported from a previous measurement at ILL-Grenoble). The setup used at n_TOF consists of a central chamber which contains the actinide target under study and a MCP start detector, and is surrounded by 12 NaI detectors. Four arms exist for the detection and characterisation of the FFs, equipped with SED detectors for the timing and thus the determination of the FF velocity and Bragg axial ionization chambers for the determination of the FF energy.

4. Fission cross section measurements overview

The experimental campaign of the facility is divided in three phases, namely Phase I (2001-2004), Phase II (2009-2012) and Phase III (2014-2018). The fission measurements include mainly cross section measurements on most of the long-lived actinides (432 y and above), as well as FF anisotropies[14]. In most cases, the n_TOF data are characterised by improved accuracy, higher resolution and a wider energy range than previous data. In some cases, they have also helped to solve existing discrepancies between previous experimental and/or evaluated cross-sections. Combinatorial setups (STEFF, fission tagging setup) are used for the study of the fission reaction by detecting fission gamma rays with respect to the neutron energy. A list of fission measurements and related references are given in Table 1.

Among the many interesting results of the n_TOF fission program, the latest highlights are briefly presented below.

The fission tagging setup with the novel compact fast fission chamber (Sec. 3) was used for the first time for the ^{233}U (n,γ)/(n,f) measurement at the end of 2016. A photo of the fission detector mounted in the centre of the TAC for this measurement is shown in Fig. 1. The first results are very promising for both detection systems, the upper neutron energy limit is expected to be significantly increased and the analysis procedure is simplified compared to the previous experiments of this type, thanks to the use of the same set of measurements for the capture and fission γ-rays.

Another recent highlight is the installation of STEFF in EAR2 during the 2015 and 2016 campaign for commissioning and a measurement on ^{235}U. Thanks to the improved timing resolution of the MWPC SED detectors used at n_TOF compared to the previous measurement at ILL,

Table 1. The fission measurements at n_TOF from 2001 up-to-date. For phases I and II only EAR1 was available.

Reaction	Experimental setup	reference
Phase I (2001-2004)		
^{nat}Pb (n,f)	PPAC	[64] [1]
^{209}Bi (n,f)	PPAC	[64] [1]
^{233}U (n,f)	FIC	[68,69] [1]
^{234}U (n,f)	FIC	[70] [1]
^{234}U (n,f)	PPAC	[71,72] [1]
^{235}U (n,f)	PPAC	[104] [1]
^{236}U (n,f)	FIC	[73] [1]
^{238}U (n,f)	FIC/PPAC	[74,75] [1]
^{237}Np (n,f)	FIC	[77] [1]
^{237}Np (n,f)	PPAC	[71] [1]
^{241}Am (n,f)	FIC	[78] [1]
^{243}Am (n,f)	FIC	[80] [1]
^{245}Cm (n,f)	FIC	[81] [1]
Phase II (2009-2012)		
^{232}Th (n,f) FFAD	PPAC	[89,102] [1]
^{234}U (n,f) FFAD	PPAC	[72] [1]
^{235}U (n,f) FFAD	PPAC	[103] [1]
^{235}U (n,γ)/(n,f)	TAC/MGAS	[87] [1]
^{238}U (n,f) FFAD	PPAC	[103] [1]
^{242}Pu (n,f)	MGAS	[108] [1]
Phase III (2014-2016)		
^{233}U (n,γ)/(n,f)	TAC/fission chamber	[12]
^{235}U (n,f)	Silicon (EAR2)	[15]
^{235}U (n,f) FF	STEFF (EAR2)	[16]
^{237}Np (n,f)	MGAS (EAR2)	[17]
^{237}Np (n,f)	PPAC (EAR1)	[17]
^{240}Pu (n,f)	MGAS (EAR2)	[116] [1]

the final mass resolution at n_TOF is expected to be improved. Despite the demanding γ-ray environment of EAR-2, the rejection capabilities of the system make it possible to discriminate one fission γ-ray event over a thousand of background events, thanks to the coincidence with a single or both FFs (Fig. 2). The installation of STEFF at n_TOF is described in [18].

Other recently performed measurements aim to the extraction of high accuracy data for important reactions in order to resolve previous discrepancies. The first one is the measurement of the ^{237}Np(n,f) cross section in both experimental areas with two different experimental setups (PPAC and Micromegas) in order to minimize the systematic uncertainties and give a

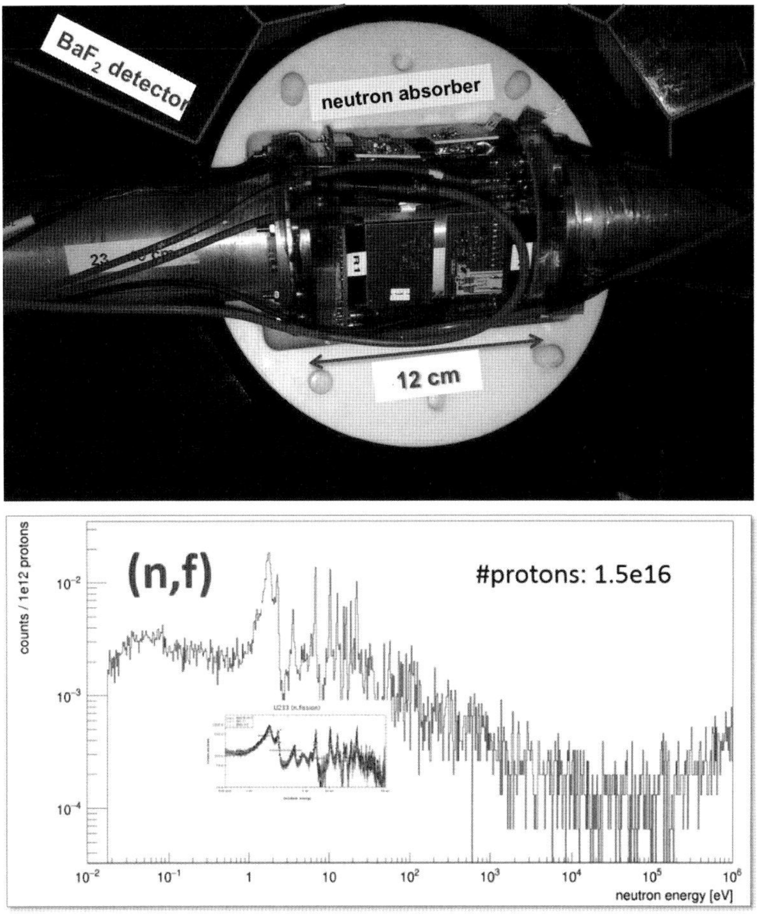

Fig. 1. Top: The novel compact fission tagging detector, containing 14 ^{233}U targets, placed in the center of the TAC, for the ^{233}U (n,γ)/(n,f) measurement. The detector and electronics are put inside a neutron absorber in order to protect the TAC from scattered neutrons. Bottom: Preliminary normalised counting rate of the fission tagging detector, from a very small fraction of the total statistics. The resolved resonance region is visible and is compatible with the evaluated resonances (inset).

definitive answer to the discrepancies found for this cross section mainly above threshold[17]. The second one is the measurement of the ^{235}U(n,f) cross section with silicon detectors relative to the reactions ^{6}Li(n,t) and ^{10}B(n,α), aiming mainly to give an answer to the various hints that there is an overestimation of the evaluations in the energy region 10-30 keV[15]. The

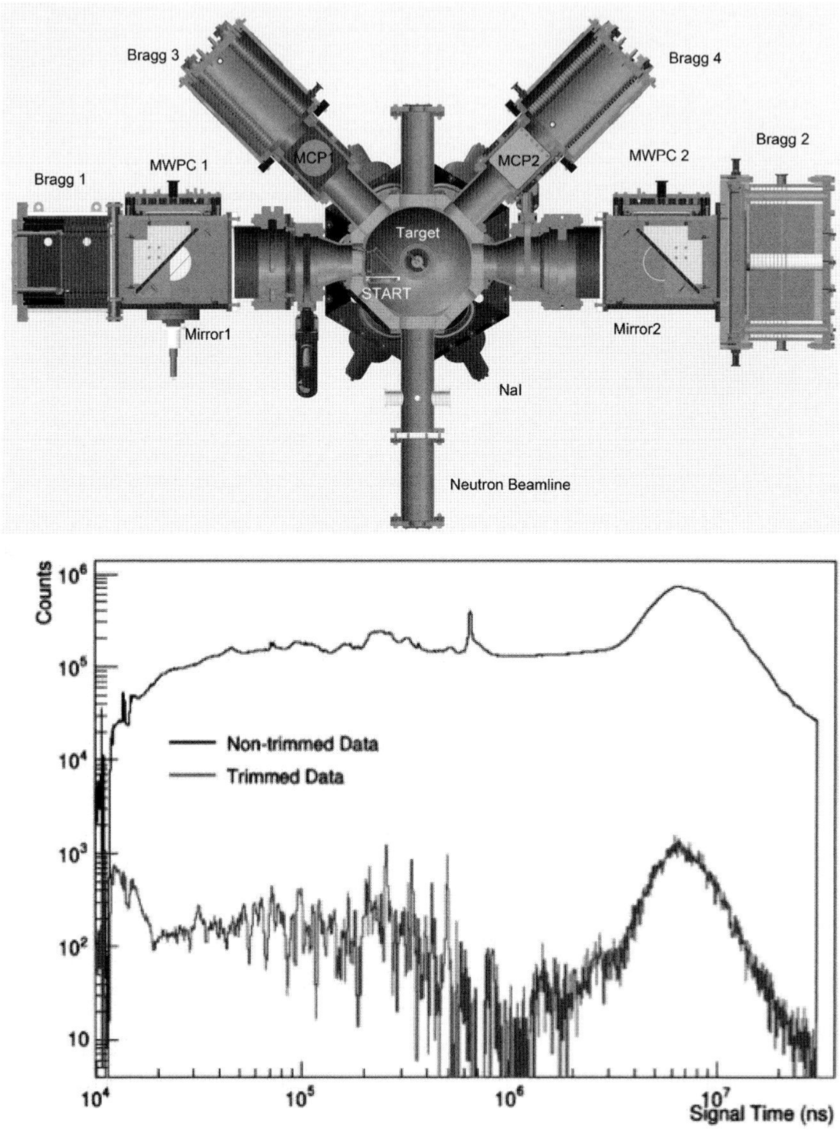

Fig. 2. Top: Schematics of the STEFF detector. Bottom: Comparison of the γ-ray counts vs neutron time-of-flight without (upper curve) and with (lower curve) FF coincidence.

first results show that the detectors behaved satisfactorily and presented very good α-FF discrimination up to hundreds of keV neutron energy.

5. Conclusion

Since 2001, n_TOF has provided an important contribution to the field of neutron induced fission studies on actinides, with several high-quality measurements. Results obtained so far are available to the international community via the EXFOR database and the n_TOF Twiki page[19] and can help improve the current evaluated databases and theoretical models of the neutron induced fission reaction on actinides.

References

1. F. Gunsing *et al.*, *Eur. Phys. J. Plus* **131:371** (2016).
2. C. Rubbia *et al.*, CERN/LHC/98-02 (EET).
3. C. Borcea *et al.*, *Nucl. Instrum. Methods Phys. Res. Sect. A* **513**, p. 524 (2003).
4. C. Guerrero *et al.*, *Eur. Phys. J. A* **49**, p. 27 (2013).
5. C. Weiss *et al.*, *Nucl. Instrum. Methods Phys. Res. Sect. A* **799**, p. 90 (2015).
6. S. Barros *et al.*, *J. Instrum.* **10**, p. P09003 (2015).
7. M. Calviani *et al.*, *Nucl. Instrum. Methods Phys. Res. Sect. A* **594**, 220 (2008).
8. I. Giomataris *et al.*, *Nucl. Instrum. Methods Phys. Res. Sect. A* **376**, p. 29 (1996).
9. S. Andriamonje *et al.*, *Journal of the Korean Physical Society* **59**, p. 1597 (2011).
10. D. Tarrio *et al.*, *Nucl. Data Sheets* **119**, p. 35 (2014).
11. C. Guerrero *et al.*, *Eur. Phys. J. A* **48:29** (2012).
12. M. Bacak *et al.*, in *Proceedings of the International Conference on Nuclear Data for Science and Technology*, (Bruges, Belgium, 2016).
13. E. Murray *et al.*, *Nucl. Data Sheets* **119**, p. 217 (2014).
14. A. Tsinganis *et al.*, *Physics Procedia* **64**, p. 130 (2015).
15. M. Barbagallo *et al.*, (2014), CERN-INTC-2014-048 / INTC-P-416.
16. T. Wright *et al.*, in *Proceedings of the International Conference on Nuclear Data for Science and Technology*, (Bruges, Belgium, 2016).
17. L. Audouin *et al.*, (2015), CERN-INTC-2015-007 / INTC-P-431.
18. S. G. Warren, Development of the STEFF detector for the neutron Time Of Flight facility (n TOF), CERN, PhD thesis, University of Manchester, (Manchester,England, 2016).
19. http://twiki.cern.ch/NTOFPublic/DataDissemination.

Calculations of Fission-Fragment Isotopic Yields*

P. Möller

Theoretical Division, Los Alamos National Laboratory,
Los Alamos, New Mexico 87545, USA
E-mail: moller@lanl.gov
http://t2.lanl.gov/molleretal/

C. Schmitt

Grand Accélérateur National d'Ions Lourds,
CEA/DSM–CNRS/IN2P3, 14076 Caen, France
E-mail: christelle.schmitt@ganil.fr

Information of isotopic fission-fragment yields is important for both fundamental and applied nuclear physics. In the first applications of the Brownian shape motion (BSM) method fission fragment yields were obtained versus nucleon number A. Charge yields were obtained by obvious transformations assuming constant proton Z to neutron N ratios in the emerging fragments, identical to that of the compound nucleus. In these applications the method was implemented as a Metropolis random walk on five-dimensional (5D) potential-energy surfaces calculated on a discrete grid. The surfaces were functions of five nuclear shape degrees of freedom, namely elongation, spheroidal deformation of each of the nascent fragments, neck diameter, and left-right shape asymmetry. We envisioned that if we could devise a method to calculate these surfaces versus Z and N in the emerging fragments, rather than overall shape asymmetry, then a straightforward implementation of the standard BSM method as a random walk, now on 6D surfaces, would yield fragment distributions $Y(Z, N)$ versus both neutron and proton numbers. Once the fragment identities can be taken into account, pairing effects arising from odd-odd or even-even nucleon divisions could also be introduced. We review the steps recently taken to obtain such surfaces and discuss some results.

Keywords: Fission Charge Yields, Brownian Motion, Odd-Even Staggering.

1. Introduction

The Brownian shape-motion method for calculating fission-fragment charge and mass yields was introduced by Randrup *et al.*[1] It is based on Metropolis

---

*Proc. The 6th International Conference on *Fission and properties of neutron-rich nuclei* (ICFN5), November 6–12, 2016 Sanibel Island, Florida, United States. World Scientific, to appear, (LANL Preprint LA-UR-17-xxxxx).

random walks on nuclear potential-energy surfaces that have, in our work, been previously calculated versus five shape coordinates for more than five million different shapes. Once these surfaces are available this method allows a fission yield distribution for a specific energy to be calculated on a single CPU in minutes to hours for a heavy nucleus. For a full specification of the model we refer to Refs.[2,3] for the calculation of the 5D potential-energy surfaces and for the dynamics and sensitivity studies to Refs.[1,4] It was subsequently found that to accurately describe the evolution of yields with excitation energy, it is necessary to account for the damping of shell corrections with energy, see Refs.[5,6] for details. The model has been extensively benchmarked with respect to experimental data, in particular to the 70 charge yields measured at GSI[7], with encouraging results[6,8]. Furthermore, we have in Ref.[9] presented calculations of fission yields for 987 nuclei and in graphical form summarized the corresponding symmetry or asymmetry of the calculated fission yields, so that the model predictions can be thoroughly tested in future experiments.

Isotopic $Y(Z,N)$ fission yield data are important in fundamental and applied science applications. To extend the original BSM model to provide such isotopic yields with odd-even staggering included we have proceeded in two steps. First we calculate charge yields with odd-even staggering included, but ignore the neutrons. In the second step we generalize further the calculation of the potential-energy so that rather than just depending on the asymmetry of the overall shape of the nucleus it depends on $Z_{\rm f1}$ and $N_{\rm f1}$ in fragment "1". Because of particle conservation the proton and neutron numbers $Z_{\rm f2}$ and $N_{\rm f2}$ in fragment "2" are not independent variables but obviously given by $Z_{\rm f2} = Z - Z_{\rm f1}$ and $N_{\rm f2} = N - N_{\rm f1}$ where Z and N are the proton and neutron numbers of the fissioning compound system. The enhancements and a full specification of the new model are given in Ref.[12] Light-particle or γ-ray emission, before and after fission, are not included.

2. Odd-even staggering in fission-fragment charge yields

It is possible to calculate charge yields using a shortcut rather than performing the much more elaborate calculation of the isotopic $Y(Z,N)$ yields and subsequently sum this function over N for constant Z values. In the short-cut approach we calculate 5D potential-energy surfaces with a selection of shape asymmetries that correspond to integer values of the proton numbers $Z_{\rm f1}$ and $Z_{\rm f2}$ in the nascent fragments. To account for pairing effects in the emerging fragments we add $2 \times \Delta$, where Δ is the pairing-gap parameter, when the asymmetry split corresponds to odd-odd charge division and when the shape has zero neck radius (a scission shape). We choose $\Delta = 1.0$ MeV. Earlier in the shape evolution towards scission we add a smaller fraction of $2 \times \Delta$, which depends on the neck radius. When the shape is sufficiently compact that a neck has still to develop no odd-even effect is added. The calculated yields are not sensitive to the exact value of Δ nor to exactly how the pairing effect depends on neck radius as shown in Ref.[12] We also discuss in this reference that the BSM model has to be slightly generalized to allow for the correlated transfer of paired two-proton configurations which has been observed in many nuclear reaction studies.

In Fig. 1 we have applied the method to the calculation of charge yields for the four even-even $^{234(2)240}$U isotopes of uranium. In both the experimental data and the calculated results we observe the interesting evolution of the location of the maximum yield from $Z = 54$ towards $Z = 52$ with increasing neutron number. Moreover, in concert, the wide region of very

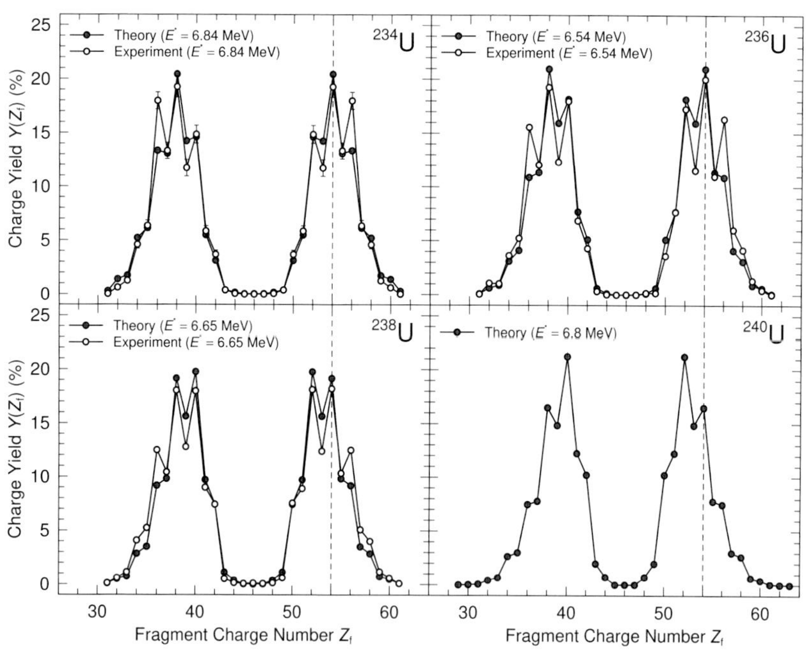

Fig. 1. Calculated fission-fragment charge distributions for four uranium isotopes at thermal or near thermal energies, compared to experimental evaluations, where available. Adapted from Ref.[10]

low yield in the region of symmetric fission gradually disappears. The figure clearly suggests that future experiments on ^{240}U would be highly desirable, to confirm that the observed trend versus N continues.

In Fig. 2 we show an experimental fission-yield evaluation for 14 MeV neutrons incident on a ^{235}U target leading to the compound nucleus ^{236}U at 20.55 MeV excitation energy, compared to our calculation. There appears to be a considerable difference. However, at the considered excitation, well above the neutron threshold, so called multi-chance fission is very likely. That is, fission from ^{236}U, as well as from its daughter and grand-daughter (^{235}U and ^{234}U) at lower excitation energy, do contribute to the measured set of fission events. In Fig. 3 we show the calculated probabilities of multi-chance fission versus incident neutron energy. We have used this information to calculate the fragment yields for ^{236}U, ^{235}U, and ^{234}U in the proportions given in Fig. 3 and at the appropriate lower excitation energies associated with the neutron emission. The different contributions, properly weighted are shown separately in Fig. 4, and finally summed up for comparison with the evaluated data. We now find consistency between

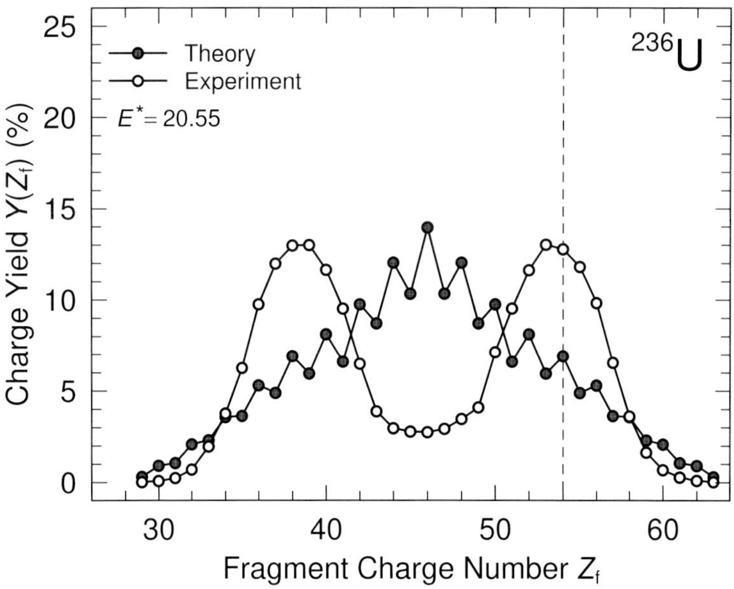

Fig. 2. Evaluated and calculated fission-fragment charge yields for ^{236}U for incident neutron energy E_n = 14 MeV, corresponding to a fissioning system excitation energy of E^* = 20.55 MeV. In the calculation we have assumed that all fission events originate from a ^{236}U compound nucleus, with no pre-fission neutrons emitted. From Ref.[10]

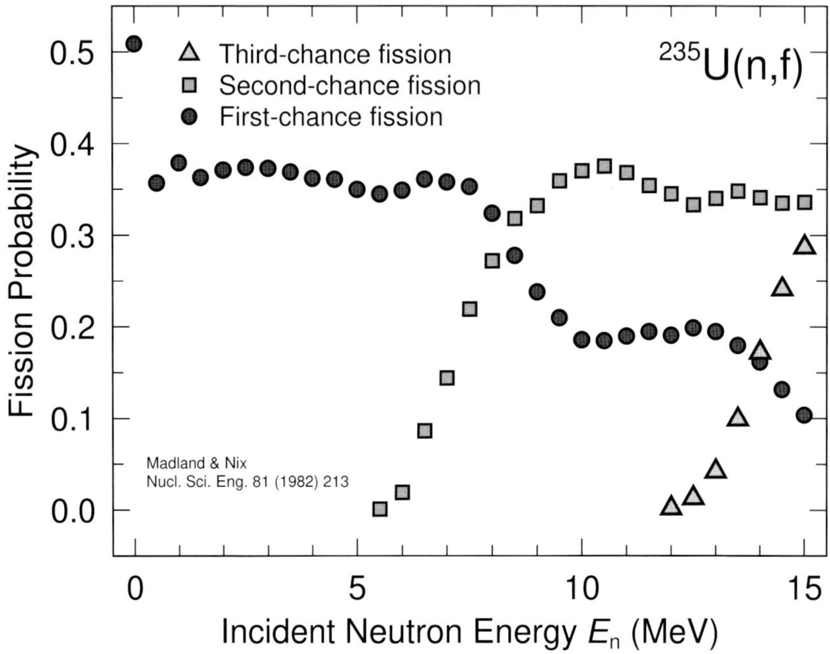

Fig. 3. Calculated fission probabilities for first-, second-, and third-chance fission versus incident neutron energy for neutron-induced fission of ^{235}U. Adapted from Ref. [11]

the calculated and experimental yield, except that there is no discernible staggering in the experimental evaluation. This lack of staggering is due to the limited mass and charge resolutions in the experiments.

3. Calculations of "2D" yields $Y(Z, N)$

To obtain a potential energy surface $E(Q_2, d, \epsilon_{f1}, \epsilon_{f2}, Z_{f1}, N_{f1})$ versus two asymmetry variables, namely Z_{f1} and N_{f1} in the fragments rather than a single "shape asymmetry" we observe that since the proton and neutron single-particle potentials are almost independent of each other we can calculate the neutron and proton shell-plus-pairing corrections independently. We therefore calculate the neutron shell corrections for integer spacing in N_{f1} and store them, and similarly for the proton shell corrections. When the proton shell correction is calculated we also store the macroscopic energy. For a specific charge split Z_{f1} it corresponds to a neutron number $N(Z_{f1}) = Z_{f1} \times N/Z$ where Z and N are the compound-nucleus proton and

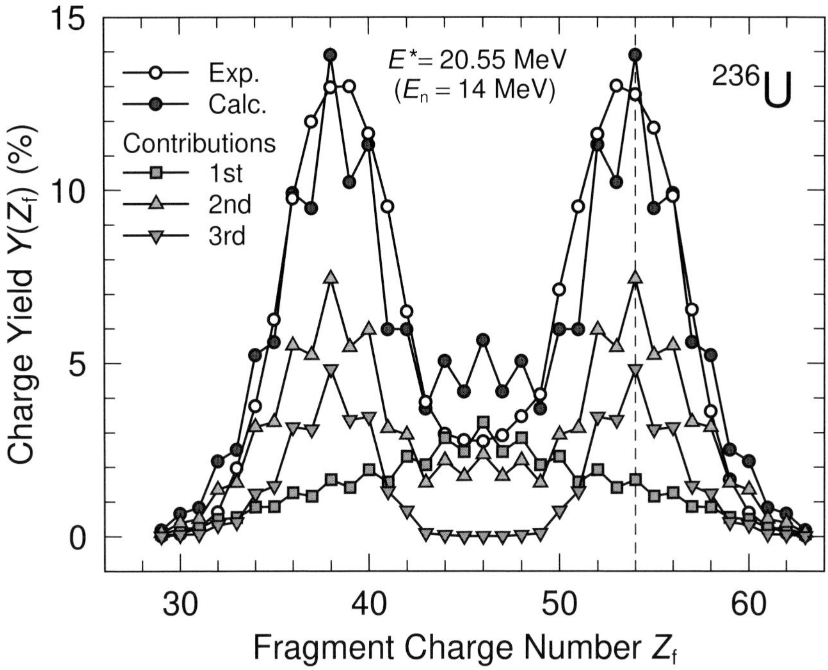

Fig. 4. Calculated fission-fragment charge distributions at 14 MeV incident neutron energy, corresponding to 20.55 MeV excitation after neutron absorption. The yields from first-, second-, and third-chance fission are shown at their relative contributions and added up to a total yield. An experimental evaluation is also shown. From Ref.[10]

neutron numbers. Then

$$\begin{aligned} & E(Q_2, d, \epsilon_{\mathrm{f}1}, \epsilon_{\mathrm{f}2}, Z_{\mathrm{f}1}, N_{\mathrm{f}1}) = \\ & \quad ESHELLZ(Q_2, d, \epsilon_{\mathrm{f}1}, \epsilon_{\mathrm{f}2}, Z_{\mathrm{f}1}) + ESHELLN(Q_2, d, \epsilon_{\mathrm{f}1}, \epsilon_{\mathrm{f}2}, N_{\mathrm{f}1}) + \\ & \quad EMACR(Q_2, d, \epsilon_{\mathrm{f}1}, \epsilon_{\mathrm{f}2}, Z_{\mathrm{f}1}, N(Z_{\mathrm{f}1}) + \Delta E \end{aligned}$$

The quantity ΔE accounts for how the actual macroscopic energy at $Z_{\mathrm{f}1}$ and $N_{\mathrm{f}1}$ differs from the stored one. In our approach it is obtained as the macroscopic energy for separated spherical fragments for the actual proton-neutron division we consider minus the macroscopic energy for separated fragments with the proton-neutron split of the stored macroscopic energy; full details are in Ref.[12]

In Figs. 5 and 6 we show calculated isotopic yields $Y(Z, N)$ for ^{234}U at thermal neutron energies and at 11 MeV excitation energy, respectively. To our knowledge, this is the first time that isotopic fragment yield

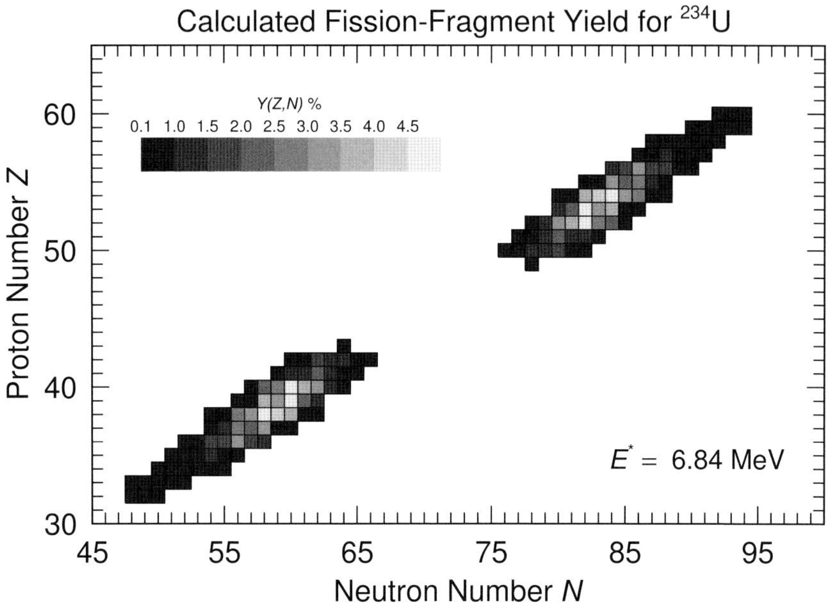

Fig. 5. Calculated fission-fragment yields $Y(Z, N)$ versus *both* neutron and proton number at thermal neutron energy for ^{236}U.

distributions are obtained over the full range of the fragment production from a dynamical model for fission at low excitation energy. Although we just show two examples here, one should note that the method can be applied to any heavy fissioning system. It is hoped that the analysis of such figures for more systems will permit to understand what drives the fragment split in fission. In particular, due to the still incomplete experimental information, the respective role of neutron and proton shells remains controversial[7]. Calculations for other fissioning systems are in progress. They are planed to be compared to recent, sophisticated experiments at SOFIA/GSI[13] and VAMOS++/GANIL[14].

Acknowledgment

This work was carried out under the auspices of the National Nuclear Security Administration of the U.S. Department of Energy at Los Alamos National Laboratory under Contract No. DE-AC52-06NA25396 (PM). We also acknowledge support through FUSTIPEN (French-U.S. Theory Institute for Physics with Exotic Nuclei) under DOE grant number DE-FG02-

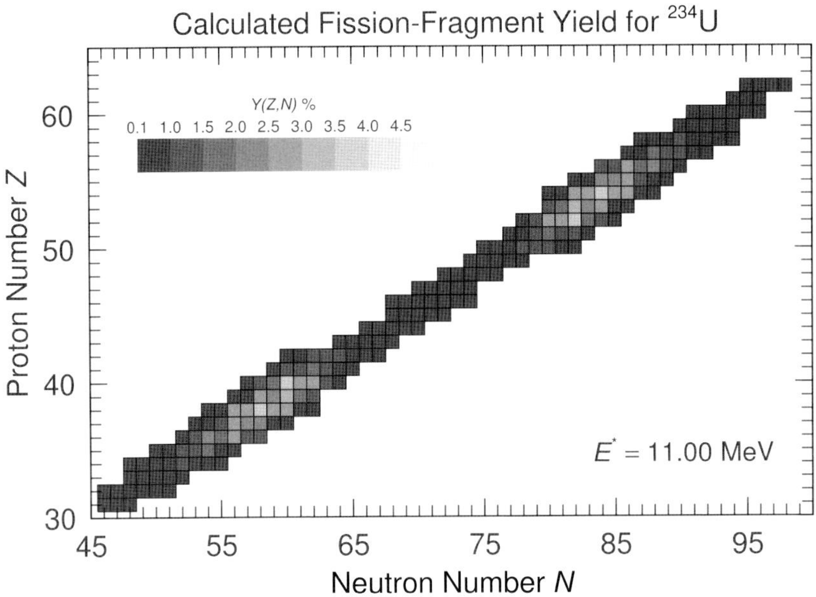

Fig. 6. Calculated fission-fragment yields $Y(Z, N)$ versus *both* neutron and proton number for fission of ^{236}U at $E^* = 11$ MeV excitation energy.

10ER41700. This work also benefited from discussions with T. Ichikawa, A. Iwamoto, D. Madland, J. Randrup, K.-H. Schmidt and A. J. Sierk.

References

1. J. Randrup and P. Möller, *Phys. Rev. Lett.* **106** (2011) 132503.
2. P. Möller, D. G. Madland, A. J. Sierk and A. Iwamoto, *Nature* **409** (2001) 785.
3. P. Möller, A. J. Sierk, T. Ichikawa, A. Iwamoto, R. Bengtsson, H. Uhrenholt and S. Åberg, *Phys. Rev. C* **79** (2009) 064304.
4. J. Randrup, P. Möller and A. J. Sierk, *Phys. Rev. C* **84** (2011) 034613.
5. J. Randrup and P. Möller, Proc. 5th Int. Conf. on Fission and Properties of Neutron-Rich Nuclei, Sanibel Island, Florida, 410 November 2012, (World Scientific, Singapore, 2013) p. 613.
6. J. Randrup and P. Möller, *Phys. Rev. C* **88** (2013) 064606.
7. K.-H. Schmidt, S. Steinhäuser, C. Böckstiegel, A. Grewe, A. Heinz, A. R. Junghans, J. Benlliure, H.-G. Clerc, M. de Jong, J. Müller, M. Pfützner and B. Voss, *Nucl. Phys.* **A665** (2000) 221.
8. P. Möller, J. Randrup and A. J. Sierk, *Phys. Rev. C* **85** (2012) 024306.

9. P. Möller and J. Randrup, *Phys. Rev. C* **91** (2015) 044316.
10. P. Möller and C. Schmitt, *Eur. Phys. J. A* **53** (2017) 7.
11. D. G. Madland and J. R. Nix, *Nucl. Sci. Eng.* **81**, (1982) 213.
12. P. Möller and T. Ichikawa, *Eur. Phys. J. A* **51** (2015) 173.
13. J. Taieb *et al.*, in perparation (2017).
14. D. Ramos *et al.*, in preparation (2017).

Measurements and Simulations of Prompt Neutron Polarization Asymmetries in Photofission

J. M. Mueller

Department of Nuclear Engineering, North Carolina State University, Raleigh, NC 27695, USA
E-mail: jonathan_mueller@ncsu.edu

M. W. Ahmed

Department of Mathematics and Physics, North Carolina Central University, Durham, NC 27707, USA
E-mail: ahmed@tunl.duke.edu

H. R. Weller

Department of Physics, Duke University, Durham, NC 27708, USA
E-mail: weller@tunl.duke.edu

Angular distributions from sub-barrier and near-barrier photofission can yield insights into excited states and level densities at the fission barrier. Photofission experiments were performed on targets of ^{232}Th, ^{233,235,238}U, ^{237}Np, and ^{239,240}Pu using nearly 100% linearly polarized, high intensity, and nearly-monoenergetic γ-ray beams having energies between 5.3 and 7.6 MeV at the High Intensity γ-ray Source (HIγS) located at Duke University and Triangle Universities Nuclear Laboratory[1]. An array of 12–18 liquid scintillators was used to measure prompt fission neutron yields parallel and perpendicular to the plane of beam polarization. Polarization asymmetries, the differences between the in-plane and out-of-plane yields divided by their sums, were measured. Asymmetries close to zero were found for ^{233,235}U, ^{237}Np, and ^{239}Pu while significant asymmetries (~0.2–0.5) were found for ^{232}Th, ^{238}U, and ^{240}Pu. Predictions of the polarization asymmetries based on previously measured photofission fragment angular distributions combined with a model of prompt neutron emission agree well with the experimental results.

In addition to providing insights into the structure of the fission barrier, these polarization asymmetries could be used to measure the fissile versus non-fissile content of special nuclear material[2]. These potential applications provide an incentive to utilize polarization effects in active interrogation of special nuclear material.

Keywords: Fission; Photofission; Angular Distributions; Prompt Neutrons.

1. Introduction

Measurements of photofission fragment angular distributions of even-even actinides for beam energies between 5–8 MeV have shown significant anisotropies[3] while measurements of odd-A actinides have shown nearly isotropic angular distributions[4–6]. The angular distribution of the photofission fragments is expected to be related to the population of excited states at the saddle point of the outer fission barrier[7]. Therefore, the study of photofission fragment angular distributions can yield new insights into nuclear structure and shape evolution at the fission barrier.

Anisotropy in photofission angular distributions can be enhanced with the use of a linearly polarized γ-ray beam to induce photofission[8]. With a linearly-polarized beam typically one measures the polarization asymmetry, which is the difference between the yields in the plane of beam polarization and the yields perpendicular to the plane of beam polarization divided by their sum. In our measurements of photofission polarization asymmetries, we have chosen to measure the prompt neutrons instead of the fission fragments. It has been well established that the majority of the prompt neutrons are emitted after the fragments have nearly fully accelerated and thus are correlated in emission angle with the fission fragments[9]. Therefore, the underlying polarization asymmetries in the fission fragments for even-even photofission should be detectable in the polarization asymmetries of the prompt neutrons.

2. Description of the Experiment

Details of the experiment and the analysis are given elsewhere[1,10], therefore only a short description is provided here. A nearly 100% linearly-polarized, high-intensity ($\sim 5 \times 10^6$ γ/s), pulsed γ-ray beam was generated at the High Intensity γ-Ray Source (HIγS) located at Duke University[11–13]. The γ-ray beam was created by colliding a high-power free electron laser beam with an electron beam in a storage ring. γ-ray beams of energies between 5.3 and 7.6 MeV were generated. The full-width at half maximum at each γ-ray beam energy was approximately 3%. The γ-ray beam was polarized in the horizontal plane.

The γ-ray beams were incident on one of seven gram-sized actinide targets: ^{232}Th, ^{233,235,238}U, ^{237}Np, or ^{239,240}Pu. Prompt fission neutron yields were measured using an array of 12 to 18 liquid scintillator detectors filled with BC-501A. These detectors were placed in the horizontal plane (azimuthal angles of 0° and 180°) and the vertical plane (azimuthal angles of 90° and 270°). When using all 18 detectors, four were placed at each scattering angle of 55°, 90°, and 125° and two were placed at each scattering angle of 72°, 107°, and 142°. Pulse shape discrimination was used to eliminate γ-ray backgrounds. The energies of detected neutrons were determined based on their time-of-flight, and neutron energy thresholds were set at $E_n = 1.5$ MeV to ensure that all detected neutrons came from fission and not the (γ, n) reaction. Possible neutron backgrounds, due to either delayed fission neutrons from earlier events, or due to $^{27}Al(\alpha, n)$ neutrons from α-particles in the actinide sample interacting with the casing of the sample, were subtracted from the prompt neutrons by measuring the out-of-time yield with respect to the beam pulse. The resulting prompt fission neutron yield was integrated over all neutron energies above the energy threshold, and the polarization asymmetry Σ:

$$\Sigma(\theta) = \frac{1}{P}\frac{Y(\theta,\phi=0°)+Y(\theta,\phi=180°)-Y(\theta,\phi=90°)-Y(\theta,\phi=270°)}{Y(\theta,\phi=0°)+Y(\theta,\phi=180°)+Y(\theta,\phi=90°)+Y(\theta,\phi=270°)}$$

was measured as a function of scattering angle θ and beam energy. The fractional beam polarization P was assumed to be 1.0. Experimentally, it was determined that the polarization asymmetry was largest at a scattering angle of 90°, which is in agreement with theory[1] and with previous measurements of unpolarized prompt neutron angular distributions[14]. The polarization asymmetries were corrected for instrumental asymmetries, finite-size effects, and the contamination of each sample with other isotopes.

3. Experimental Results

Figure 1 shows the experimental results for the polarization asymmetry at a scattering angle of 90°. The polarization asymmetries are very different for the even-even and the odd-A actinides. The even-even actinides (^{232}Th, ^{238}U, and ^{240}Pu) generally show large polarization asymmetries, particularly at the lower beam energies studied where the asymmetries are as large as 0.45. However, the odd-A actinides generally show polarization asymmetries much closer to zero. Above beam energies of 6 MeV, the odd-A actinides show polarization asymmetries nearly consistent with zero.

A calculation was performed to predict the values of $\Sigma(90°)$. The primary ingredient in this calculation is the code `FREYA`[15]. `FREYA` is an event-by-event Monte Carlo prompt neutron+γ-ray emission calculation, which simulates the entire fission process using a few parameters specifically tuned to improve the agreement with available experimental data. It was designed primarily for use in neutron-induced and spontaneous fission. For

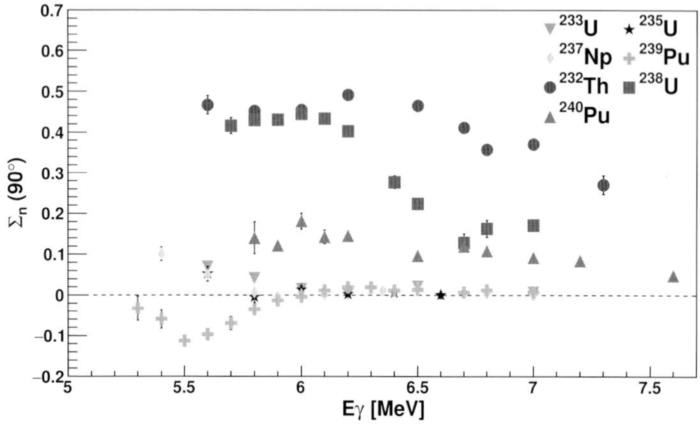

Fig. 1. The measured polarization asymmetry Σ at a scattering angle of 90° is shown as a function of beam energy E_γ. The error bars indicate the combined statistical and systematic uncertainties. Uncertainties not shown are smaller than the size of the data points.

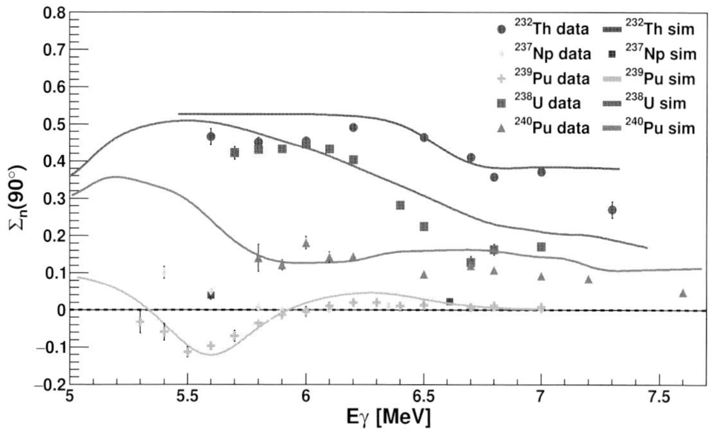

Fig. 2. The predicted polarization asymmetry Σ at a scattering angle of 90° is shown as a function of beam energy E_γ and compared to the experimental measurements.

the present work, it was adapted for use in photofission of ^{232}Th, ^{237}Np, ^{238}U, and ^{239,240}Pu.

The fragment-neutron energy and angular correlations from FREYA were combined with previously measured fragment angular distributions[3–5] to predict the neutron polarization asymmetry. Figure 2 shows the comparison between these predictions and our experimental measurements. Overall, good agreement is observed, especially considering that most of the fragment angular distribution data were originally derived from bremsstrahlung measurements that were unfolded to predict the fragment angular distribution for a monoenergetic γ-ray beam.

4. Applications

4.1. *Enrichment Determination*

As described in Ref. 2, the prompt neutron polarization asymmetries could in principle be used to distinguish even-even from odd-A actinides and to measure the enrichment of special nuclear material (SNM). Figure 3 shows the polarization asymmetry as a function of ^{235}U enrichment. The left-most point is the measurement of the ^{238}U sample, the right-most point is the ^{235}U sample, and the interior points are measurements of combined foils of ^{235}U and ^{238}U. These points were fit as shown by the solid line, and the statistical and systematic uncertainties on the fit are shown by the dashed lines. More information regarding these measurements and the fit can be found in Ref. 16.

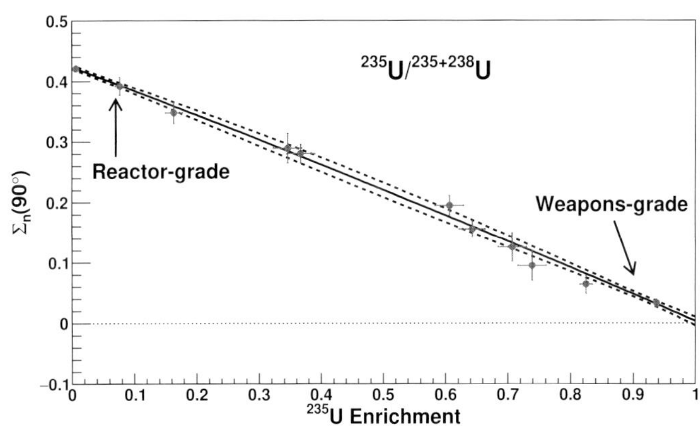

Fig. 3. The polarization asymmetry is shown as a function of ^{235}U enrichment. The points are measurements, the solid line is the result of a fit, and the dashed lines show combined statistical and systematic uncertainties on the fit.

4.2. *Active Interrogation*

It is critical to know the performance of this enrichment assay in the presence of shielding before this assay is used for active interrogation of SNM.

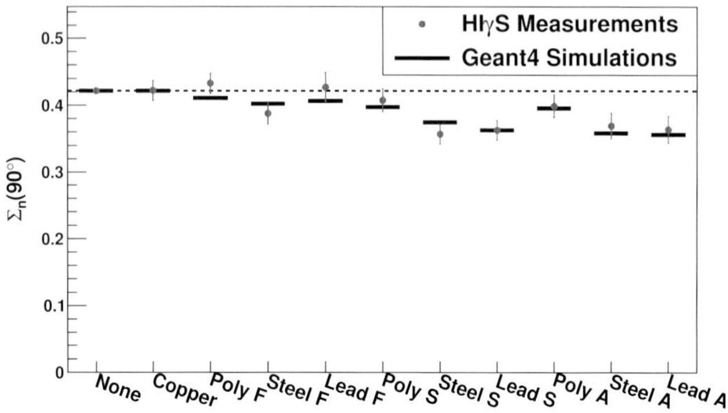

Fig. 4. The polarization asymmetry from ^{238}U photofission is shown as a function of shielding material and location. 'F' refers to shielding between the beam and target, 'S' refers to shielding between the target and detectors, and 'A' refers to 'F' and 'S' combined. The points are experimental measurements and the lines are the results of GEANT4 simulations.

Figure 4 shows the reduction in the polarization asymmetry from ^{238}U photofission for various shielding cases. In all cases, the thickness of the shielding was approximately 2.54 cm. Overall, the impact of the shielding on the polarization asymmetry is relatively small, meaning that measured polarization asymmetries could be used to extract the enrichment of SNM in the presence of shielding up to 2.54 cm. More information regarding these measurements can be found in Ref. 16.

5. Conclusions and Future Work

High-precision measurements of prompt neutron polarization asymmetries in photofission of ^{232}Th, ^{233,235,238}U, ^{237}Np, and ^{239,240}Pu have been performed at the HIγS facility. The prompt neutrons from photofission of the even-even actinides exhibit large polarization asymmetries (~ 0.45), while those from photofission of the odd-A actinides show little to no polarization asymmetries. The differences in asymmetries between the even-even actinides and odd-A actinides are likely due to their different spins and level densities at the saddle point, which for the even-even actinides lead to polarization asymmetries in the fission fragments and corresponding asymmetries in the prompt neutrons. These polarization asymmetries can be used to measure the enrichment of uranium and they are robust in the presence of shielding materials up to 2.54 cm thick.

Future work in photofission angular distributions could benefit from high-precision measurements of the correlation between the fragment mass distribution and the fragment angular distribution. Previous measurements have found a strong correlation between the mass and angular distribution[17,18]. However, no measurements of this effect have yet been performed with highly polarized quasi-monoenergetic γ-ray beams. Such measurements could be increasingly valuable to understanding nuclear structure and shape evolution at the saddle point.

Acknowledgements

The authors would like to acknowledge the following sources of funding: DNDO, Academic Research Initiative (ARI) Grant #2010-DN-077-ARI046-02 and 2008-DN-077-ARI010, LLNL contract with DHS/DNDO HSHQDC-07-X-00213 P00015, DOE Grant #DE-FG02-97ER41033, and the DOE Office of Science Graduate Fellowship.

References

1. J. M. Mueller *et al.*, Prompt neutron polarization asymmetries in photofission of ^{232}Th, ^{233,235,238}U, ^{237}Np, and ^{239,240}Pu, *Physical Review C* **89**, p. 034615 (2014).
2. J. M. Mueller, M. W. Ahmed and H. R. Weller, A novel method to assay special nuclear materials by measuring prompt neutrons from polarized photofission, *Nuclear Instruments and Methods in Physics Research Section A: Accelerators, Spectrometers, Detectors and Associated Equipment* **754**, 57 (2014).
3. N. S. Rabotnov *et al.*, Photofission of ^{232}Th, ^{238}U, ^{238}Pu, ^{240}Pu, ^{242}Pu and the Structure of the Fission Barrier, *Soviet Journal of Nuclear Physics* **11**, p. 285 (1970).
4. A. Soldatov, Y. Tsipenyuk and G. Smirenkin, Angular anisotropy of photofission of ^{239}Pu, *Soviet Journal of Nuclear Physics* **11**, p. 552 (1970).
5. L. P. Geraldo, Angular distribution of the photofission fragments of ^{237}Np at threshold energy, *Journal of Physics G: Nuclear Physics* **12**, 1423 (1986).
6. V. Zhuchko, Y. Ostapenko, G. Smirenkin, A. Soldatov and Y. Tsipenyuk, Angular distribution of fragments from photofission of ^{235}U, *Soviet Journal of Nuclear Physics* **27**, p. 746 (1978).

7. A. Bohr, On the Theory of Nuclear Fission, in *Proc. U.N. Int. Conf. Peaceful Uses At. Energy*, (New York, 1956).
8. J. M. Mueller *et al.*, Measurement of prompt neutron polarization asymmetries in photofission of ^{235,238}U, ^{239}Pu, and ^{232}Th, *Physical Review C* **85**, p. 014605 (2012).
9. J. Fraser, The angular distribution of prompt neutrons emitted in fission, *Physical Review* **88**, 536 (1952).
10. J. M. Mueller, Prompt Neutron Polarization Asymmetries in Photofission of Isotopes of Thorium, Uranium, Neptunium, and Plutonium, PhD thesis, Duke University 2013.
11. V. N. Litvinenko *et al.*, Gamma-ray production in a storage ring free-electron laser, *Physical Review Letters* **78**, 4569 (1997).
12. H. R. Weller *et al.*, Research opportunities at the upgraded HIγS facility, *Progress in Particle and Nuclear Physics* **62**, 257 (2009).
13. A. M. Bernstein, M. W. Ahmed, S. Stave, Y. K. Wu and H. R. Weller, Chiral dynamics in photopion physics: Theory, experiment, and future studies at the HIγS facility, *Annual Review of Nuclear and Particle Science* **59**, 115 (2009).
14. S. Nair, D. B. Gayther, B. H. Patrick and E. M. Bowey, Fission neutron and fragment angular distributions from threshold photofission of ^{232}Th and ^{238}U, *Journal of Physics G: Nuclear Physics* **3**, 965 (1977).
15. J. Randrup and R. Vogt, Calculation of fission observables through event-by-event simulation, *Physical Review C* **80**, p. 024601 (2009).
16. J. M. Mueller *et al.*, Tests of a novel method to assay SNM using polarized photofission and its sensitivity in the presence of shielding, *Nuclear Instruments and Methods in Physics Research Section A: Accelerators, Spectrometers, Detectors and Associated Equipment* **776**, 107 (2015).
17. F. Steiper *et al.*, Mass dependence of fragment angular distributions in the fission of ^{232}Th and ^{236}U induced by polarized photons, *Nuclear Physics A* **563**, 282 (1993).
18. A. Göök, Determination of Photofission Fragment Characteristics of ^{234,238}U and ^{232}Th in the Barrier Region, PhD thesis, Technical University Darmstadt 2012.

Prompt Fission Neutron Spectrum Measurements of ^{252}Cf Using a Multiple Gamma Tagging Method

E. Blain*, B. Casel, R. C. Block and Y. Danon

MANE Department, Rensselaer Polytechnic Institute,
Troy, NY 12180, USA
**E-mail: blaine2@rpi.edu*
www.rpi.edu

Accurate measurement of the prompt fission neutron spectrum is extremely important to several aspects of nuclear engineering. The spectrum for the spontaneous fission of ^{252}Cf is particularly of use since other spectra are often measured as a ratio to Californium and the efficiency of neutron detectors can be measured using it. A measurement was performed at Rensselaer Polytechnic Institute in order to provide a new high accuracy measurement of the spectrum particularly in the regions of little data below 0.5 MeV and above 10 MeV. A measurement was performed using the multiple gamma tagging method and provided a new dataset in the region from 50 keV to 7 MeV which agreed well with previous measurements and theory. Preliminary measurements have also been performed to test the feasibility of using this method to measure the high portion of the spectrum from 5 MeV up to 15 MeV.

Keywords: Fission; Spectrum; Measurement.

1. Introduction

The prompt fission neutron spectrum is an important quantity for several aspects of nuclear engineering including reactor simulations, non-proliferation, and criticality safety [1]. Even the prompt fission spectrum standard of ^{252}Cf has high uncertainties up to 10% in some energy regions. Figure 1 shows several of the current datasets used in the evaluations of ^{252}Cf and from these datasets it is apparent that particularly in the high energy region above 10 MeV and the low energy region below 0.5 MeV there is little data present and the data that does exist is discrepant. A new high accuracy measurement of ^{252}Cf would help to reduce the uncertainties in this dataset in future evaluations. Although ^{252}Cf is typically not of direct importance in reactor simulations it is a key isotope of

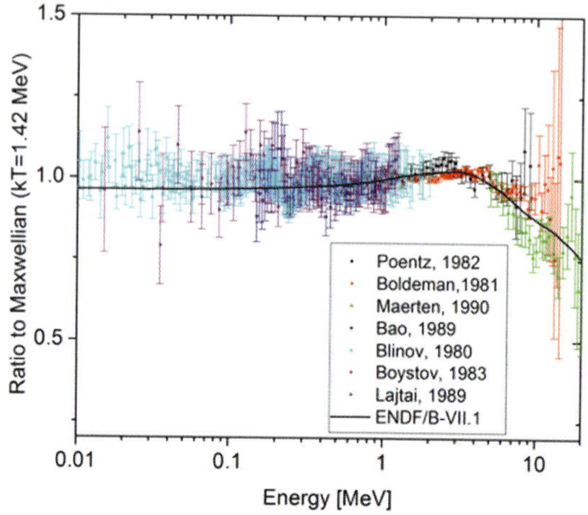

Fig. 1. Previous measurements for the PFNS of the spontaneous fission of ^{252}Cf as well as the ENDF/B-VII.1 evaluation [4] showing large uncertainty and discrepancies between the datasets both above 10 MeV and below 0.5 MeV.

interest for several reasons. Firstly the PFNS of other isotopes such as ^{235}U and ^{239}Pu are often measured as ratios to the ^{252}Cf spectrum in order to minimize systematic uncertainties [2]. Therefore, the uncertainty in these PFNS measurements is directly correlated to the uncertainty in the ^{252}Cf measurement. Additionally, the efficiency of neutron detectors is often calibrated based on the ^{252}Cf PFNS spectrum [3]. Therefore, the accuracy of all of these efficiency measurements is dependent on the accuracy of the ^{252}Cf spectrum. For these reasons a new measurement of ^{252}Cf PFNS would help to reduce the uncertainties in the evaluation and therefore reduce the uncertainties not only in other PFNS measurements of interest but also neutron detector efficiency measurements.

2. Method

2.1. *Multiple Gamma Tagging Method*

In order to measure the PFNS a time-of-flight (ToF) technique was used using the multiple gamma tagging method to determine the fission timing [5]. The multiple gamma tagging method relies on utilizing the prompt gamma multiplicity of fission as measured by an array of gamma detectors. This method allows for much larger sample masses to be used than with conventional fission chambers. Additionally, it does not require the construction of complex multi-plate chambers to achieve larger sample masses. It also has considerable advantages over single gamma tagging. Single gamma tagging methods only require one fission gamma to be measured to determine the fission timing. This method suffers both from high background due to triggering on background gamma rays. And a large time uncertainty due to measurement of late prompt gamma rays [6]. Both of these drawbacks are mitigated by using the multiple gamma tagging method. By requiring a multiplicity of two or more for the gamma rays it greatly reduces the probability of triggering on events from random background or late prompt gamma rays. The multiple gamma tagging method allows for large sample masses to be used which enables higher accuracy measurements in the regions of low count rate, above 10 MeV, and low detection efficiency, below 0.5 MeV without the drawbacks associated with the single gamma tagging method.

2.2. *Determination of Detector Efficiency*

One of the difficulties with directly measuring the PFNS of ^{252}Cf is that the detection efficiency of the neutron detectors needs to be determined without relying on the ^{252}Cf PFNS spectrum. In order to determine the detector efficiency each neutron detector was placed directly in the neutron beam of the Rensselaer Polytechnic Institute (RPI) electron linear accelerator (LINAC). The count rate in the detector is proportional to the incident flux times the detector efficiency, and therefore, if the flux at the detector location is known then the detector efficiency can be solved for. By placing a ^{235}U fission chamber into the neutron beam at the detector location the flux shape can be directly measured with an uncertainty related to the uncertainty of the ^{235}U fission cross section which is a standard [7]. This eliminates the self-referencing by measuring detector efficiency using the ^{252}Cf PFNS. This method was used for the EJ-301 detectors however due to the lack of pulse shape discrimination it could not be used for the EJ-204 detector. Therefore, the SCINFUL code [8] was used to simulate the efficiency of this detector. The uncertainty on the efficiency was found by comparing the measured efficiency for the EJ-301 detector with a corresponding simulation for the EJ-301 detector.

2.3. *Background Correction*

Accurate background correction is extremely important for accurate measurement of the PFNS. There are three separate background components to consider; time independent background, time dependent neutron background and time dependent gamma background. The time independent background is the easiest to quantify and can be found by measuring the ToF spectrum at very long times of flight. Since any counts in this region would not be correlated with a fission event it is taken as the time independent room background and subtracted.

The time dependent gamma background is only present in the EJ-204 plastic detector since pulse shape discrimination can be used to eliminate it from the EJ-301 detector [9]. The time dependent gamma background comes primarily from late prompt gamma rays which originate from the decay of metastable states [10]. These gammas are emitted from fission fragments decaying from isomeric states and typically occur at shorter ToF. In order to quantify this effect an MCNP Polimi [11] simulation was performed to measure the gamma response from a detector as a function of ToF from the fission event. This was done for both the EJ-301 and EJ-204 detectors and the simulation for the EJ-301 detector was compared against experimental data. This comparison showed that the simulation underestimated the time dependent gamma component and a scaling factor was used to scale the simulation to the experiment. Even with these corrections the time dependent gamma background was found to be a less than 1% effect.

The time dependent neutron background is primarily caused by prompt fission neutrons scattering off surrounding materials and into the neutron detectors. This causes the neutrons to arrive at ToFs not corresponding to their incident energy. In order to correct for this effect MCNP Polimi simulations were performed modeling all the surrounding materials and the ToF distribution from scattered neutrons was simulated. This was compared to the experimental values and as can be seen in Figure 2 for the EJ-301 detector. At longer ToF above 100 ns the measured signal is only from scattered neutrons, since the detector does

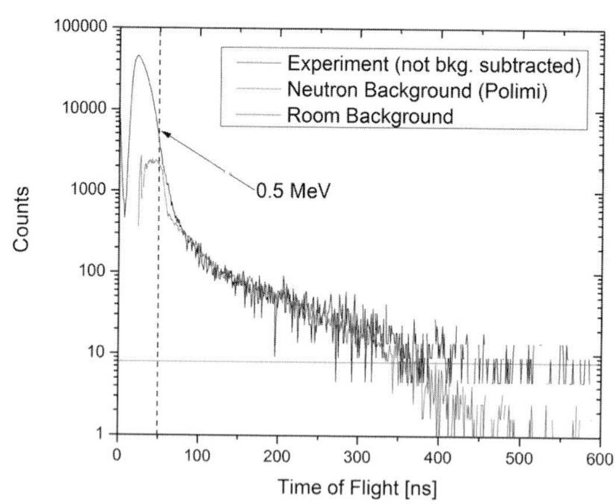

Fig. 2. The simulated background for the spontaneous fission of ^{252}Cf both compared to the experimentally measured ToF spectrum for an EJ-301 detector. The agreement at ToF greater than 100ns where only scattered neutrons exist shows that the simulation can be used to subtract the time dependent neutron background in the PFNS region of interest above 0.5 MeV.

not have the efficiency to detect prompt neutrons in this energy region. The agreement between the simulation and experiment indicates that this method can be used to subtract the neutron background from the measurement.

3. Experimental Setup

For the measurement of the spectrum from 50 keV to 7 MeV a fission chamber previously constructed at RPI was used [5]. This was to verify that there were no inherent biases associated with using the multiple gamma tagging method and the PFNS were comparable using either the multiple gamma tagging method or the fission chamber as the start signal. For these measurements 3 cylindrical EJ-301 liquid scintillators with dimensions 5" diameter by 3" thick and one cylindrical EJ-204 plastic scintillator with dimensions 5" diameter by 0.5" thick were used. The EJ-301 detectors were located at a distance of 50 cm from the fission source and the EJ-204 detector was located at a distance of 48 cm from the fission source. An array of four hexagonal BaF_2 detectors with dimensions 2" by 5" were located 10 cm away from the fission source. Figure 3 shows the experimental setup showing the locations of the neutron and gamma detectors and the fission source. All detector signals were inputted into and set of four Acqiris AP240 8 bit digitizer boards which sample at a Gsample per second sampling rate. This allows for post processing including coincidence timing and pulse shape discrimination to be performed on the data.

For the measurements from 5 MeV to 15 MeV a separate setup was used with the gamma detectors the same distance from the fission source and one EJ-301 detector located 1.28 m from the fission source. In order to increase the counts in

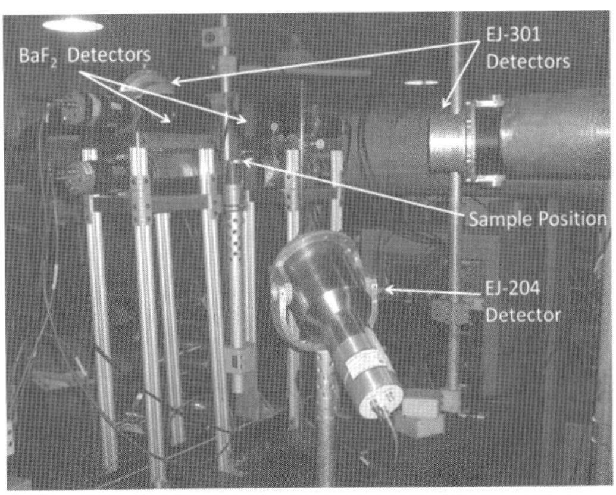

Fig. 3. The expermiental setup for the low Energy measurement of the PFNS of ^{252}Cf spontaneous fission showing the locations of all neutron and gamma detectors.

the high energy region a ^{252}Cf source emitting 10^6 neutrons per second was used instead of the fission chamber. These can be seen in Figure 4. Another modification that was made in this experiment is a new digitizer system was used in order to allow for better coincidence measurements and better pulse height resolution. A 10 bit Struck SIS3305 digitizer was used sampling at a 1.25 Gsample per second sampling rate. This allowed for better timing and pulse height resolution for the high end of the PFNS spectrum.

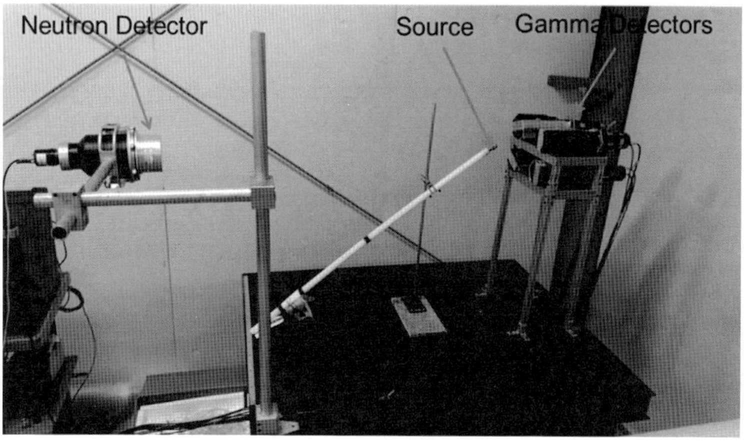

Fig 4. The experimental setup for the high energy spectrum measurement showing the location of the neutron detector, gamma detectors and fission source.

4. Results

Once the data was collected using the digitizer boards the spectra were background subtracted, efficiency corrected and translated from the ToF domain to the incident neutron energy domain. Figure 5 shows the RPI datasets taken with both detectors as well as additional datasets used in the ^{252}Cf evaluation by Manhart [12]. This shows the good agreement between the current measurement and previous datasets with a small uncertainty which will help to reduce the overall uncertainty in future evaluations. Figure 6 shows the most recent data for the high energy PFNS measurement. This shows good separation between the prompt gammas and prompt neutrons as well as a signal to background of 2 at an energy of 16.5 MeV. Additionally, most of this background is gamma background which will be eliminated once pulse shape discrimination has been performed. Initial measurements indicate that less that 5% statistical uncertainty can be obtained for 16.5 MeV for a 40 day measurement of ^{252}Cf.

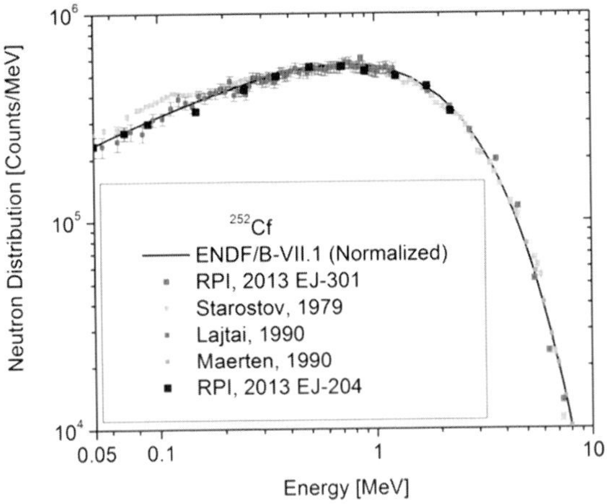

Fig. 5. The PFNS as measured by the multiple gamma tagging method showing the RPI data from 50 keV to 7 MeV showing good agreement with previous datasets and the ENDF/B-VII.1 evaluation. Uncertainties on the RPI data include both statistical and systematic uncertainty.

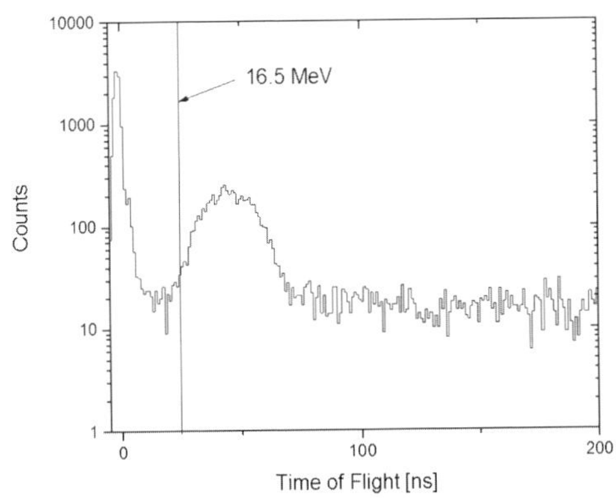

Fig. 6. The measured ToF for prompt fission neutrons from a ^{252}Cf sample highlighting the ability to measure prompt fission neutrons up to 16.5 MeV with good separation from the prompt gamma peak.

5. Conclusions

The PFNS for the spontaneous fission of ^{252}Cf has been successfully measured at RPI using the multiple gamma tagging method in the energy range from 50 keV to 7 MeV. This provides a new high accuracy measurement which will help constrain future evaluations of the PFNS. Additionally, preliminary experiments have been performed to measure the PFNS in the energy range from 5 MeV to 15 MeV. The initial data indicates that under 5% statistical uncertainty can be achieved for these measurements with a measurement time of approximately 40 days. These measurements are currently underway.

Acknowledgements

The authors would like to thank the staff of the LINAC for their assistance and expertise with these measurements. The authors would also like to thank the Stewardship Science Academic Alliance for funding this research through Grant numbers DE-FG52-09NA29453 and DE-NA0001814.

References

1. A. Klein and J. Lance, *Future Direction, Challenges and Opportunities in Nuclear Energy,* Idaho Nat. Lab., Idaho Falls, ID, INL/CON-06-11735, (Jul. 2006).
2. H. Werle and H. Bluhm, Fission neutron spectra measurements of ^{235}U, ^{239}Pu and ^{252}Cf, *J. of Nucl. Energy,* Vol. 26, no. 4, p. 165 (Apr. 1972).
3. G. Knoll, *Radiation Detection and Measurement, Third Edition* (John Wiley and Sons Inc., Hoboken NJ, 2000).
4. M. B. Chadwick *et al.,* ENDF/B-VII.1 Nuclear Data for Science and Technology: Cross Sections, Covariances, Fission Product Yields and Decay Data, *Nuclear Data Sheets,* Vol. 122, issue 12, p. 2887-2996 (Dec. 2011).
5. E. Blain *et al.,* A method to measure prompt fission neutron spectrum using gamma multiplicity tagging, *Nuc. Inst. And Meth. In Physics Research: A,* vol. 805, p. 95-100 (Jan. 2016).
6. T. Granier, Reanalysis of ^{239}Pu prompt fission neutron spectra, *Physics Prodcedia,* Vol. 64, p. 183-189 (2015).
7. J. K. Dickens, *A Monte Carlo Based Computer Program to Determine a Scintillator Full Energy Response to Neutron Detection for En Between 0.1 and 80 MeV: User's Manual and FORTRAN Program Listings,* Oak Ridge Nat. Lab., Oak Ridge TN, Rep. ORNL-6462 (Mar. 1988).
8. A. Carlson *et al.,* International Evaluation of Neutron Cross Section Standars, *Nuclear Data Sheets,* Vol. 110, issue 12, p. 3215-3324 (Dec. 2009).

9. A. M. Daskalakis *et al.,* Quasi-differential neutron scattering from ^{238}U from 0.5 MeV to 20 MeV, *Ann. of Nucl. Energy,* Vol. 73, p. 455-464 (Nov. 2014).
10. C. Wagemans, *The Nuclear Fission Process* (CRC Press, Boca Raton FL, 1991).
11. E. Padovani, S. A Pozzi, *MCNP Polimi User's Manual* (2002).
12. W. Manhart, IAEA-TECDOC-410, p. 158 (1987).

Induced Fission of ^{240}Pu

Aurel Bulgac* and Shi Jin
Department of Physics, University of Washington, Seattle, WA 98195-1560, USA
**E-mail: bulgac@uw.edu*

Piotr Magierski
Faculty of Physics, Warsaw University of Technology, ulica Koszykowa 75, 00-662 Warsaw, POLAND

Kenneth J. Roche
Pacific Northwest National Laboratory, Richland, WA 99352, USA

Ionel Stetcu
Theoretical Division, Los Alamos National Laboratory, Los Alamos, NM 87545, USA

We study the fission dynamics of ^{240}Pu within an implementation of the Density Functional Theory (DFT) extended to superfluid systems and real-time dynamics. We demonstrate the critical role played by the pairing correlations. The evolution is found to be much slower than previously expected in this fully non-adiabatic treatment of nuclear dynamics, where there are no symmetry restrictions and all collective degrees of freedom (CDOF) are allowed to participate in the dynamics.

Keywords: Induced Fission, ^{240}Pu, Time-Dependent Density Functional Theory.

Nuclear fission, discovered in 1939 [1,2], is fast approaching the venerable age of 80 years and proves to be one of the most challenging problems in quantum many-body theory. Nuclear fission is an extremely complex physical phenomenon, starting with the formation of the compound nucleus, the shape evolution until the outer saddle point and eventual slide towards the scission where the fission fragments are formed, accompanied or rather followed by neutron and gamma emissions, followed later on by beta-decay, with times scales of these processes ranging over more than twenty orders of magnitude, see Fig. 1 and Ref. [3]. Likely the most difficult part of this entire process is the descent from the saddle to the scission configuration which over the years has proved quite difficult to define, and the formation of the fission fragments. How a large nucleus with more than 200 nucleons separates into two fragments, how the mass and charge is distributed, how much excitation energy and angular momentum each fragment acquires in this process, how many neutrons and gammas are emitted and at what stage of the fission dynamics, and why and how sometimes even more than two fission fragments are formed? Even tough an enormous body of experimental results exists and a large number of phenomenological models have been developed, the present day microscopic results are far from satisfactory[4]. Nuclear fission is thus unlike another remarkable problem of the quantum many-body problem, namely superconductivity, which since its discovery in 1911[5] was successfully described microscopically in less than 50 years[6].

Qualitatively it was understood a long time ago that fission can be described qualitatively using a hydrodynamic description of the nucleus as a charged liquid drop[2] and that the evolving shape of the fissioning nucleus can be described within a collective model with a deformation potential, see Fig. 1. The independent particle model[7] forced us realize that this smooth deformation potential actually should have quite a lot of roughness, due to the single-particle level crossings as a function of the nuclear deformation, see Fig. 2 and Ref.[8]. Only the lowest A-levels remain mostly occupied while the nuclear shape evolves, as the nucleus does not heat up significantly and the Fermi surface should retain its spherical overall shape. While the nucleus elongates, the Fermi surface becomes oblate, and it can recover its sphericity only if at a level crossing nucleons from the occupied upward going levels jump to

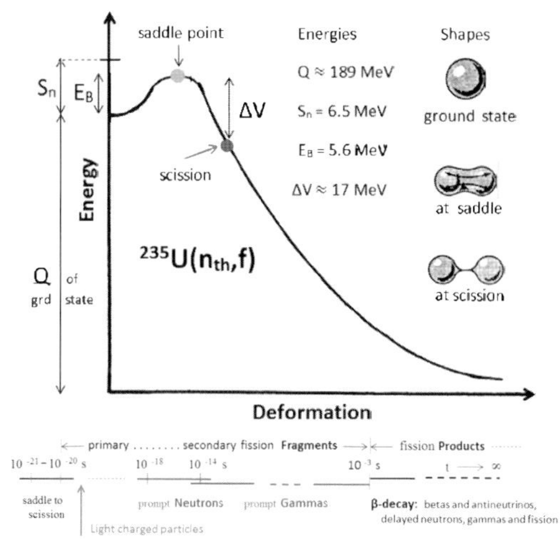

Fig. 1. Qualitative potential energy of a fissioning nucleus versus deformation and characteristic times of various accompanying processes[3].

unoccupied downward going levels, see upper panel of Fig. 2. The total energy of the nucleus, which is to a large extent the sum of the occupied single-particle energies, develops cusps at the level crossings. It was assumed that the residual interactions between the independent particles provide the mechanism for jumping from one level to another at the crossing, and it was also expected that as result also the deformation potential will become smoother. Due to Kramers degeneracy each single-particle level is however double-occupied and the only residual interaction capable of providing an effective mechanism to promote simultaneously two particles from one level to another is the pairing interaction[9].

This requirement can be understood from a different point of view as well[10]. During fission the axial symmetry is typically conserved, and one can expect that the probability distribution of the projections of the angular momenta along the fission axis is also conserved. The initial nucleus has a wider waist than the final fragments and the maximum angular momentum is roughly $p_F R_A$, where p_F is the Fermi momentum and R_A the waist radius. In the final fission products in case of symmetric fission $R_A \to R_{A/2} = R_A/2^{1/3}$ and thus the maximum angular momentum projection is smaller by a factor of $2^{1/3}$ than in the initial nucleus. A dynamics which will conserve the axial symmetry will not be capable to allow for such a dramatic redistribution of angular momenta of the occupied states. This is the main reason while many attempts to describe fission within a time-dependent Hartree-Fock approach failed so far[11]. The pairing interaction is the most effective at coupling nucleon pairs in time reverse quantum states in $(m, -m) \to (m', -m')$ and it preserves the axial symmetry as well. Thus a full microscopic treatment of the pairing interactions in a dynamic approach is crucial in describing fission dynamics, see Fig. 3 and Ref.[13]. The approach adopted in Ref.[13] is based on an extension of the density functional theory to superfluid anytime-dependent phenomena[14]. This approach satisfies all expected symmetries of a nuclear Hamiltonian:

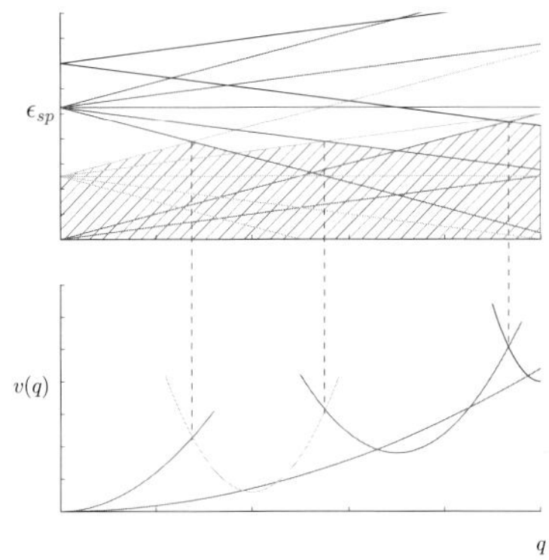

Fig. 2. The qualitative evolution of the single-particle levels (upper panel) and of the total nuclear energy (lower panel) as a function of nuclear deformation[8,9].

translational and rotational invariance, Galilean invariance, isospin invariance up to Coulomb effects and proton/neutron mass difference, gauge symmetry, and renormalizability of the theory. The static and the time dependent formalism has been confronted with a multitude of theoretical tests and with various experimental data in cold atom physics, nuclear physics, and neutron star crust problems.

Even though the nuclear energy density functional is not yet known with high enough accuracy and its origin is mostly phenomenological, its basic properties (volume energy, surface energy, symmetry energy, Coulomb energy, spin-orbit interaction) are such that a very large body of observables (masses, charge radii, collective state) are described rather well. For the normal part of the energy density functional we have chosen a well studied parametrization, the SLy4[15], and for the pairing part we used[16]. This type of parametrizations of the nuclear energy density functional has met with difficulties when describing spontaneous fission lifetimes, since for an under-the-barrier process the life-times depend exponentially on the energy density functional parameters. In the case of induced fission, where the entire dynamics occurs in classically allowed regions, inaccuracies of the order of $\mathcal{O}(1)$ MeV in the total deformation potential energy have a relatively small impact on the observables, such as masses and charges of the fission fragments, total kinetic energy and the excitation

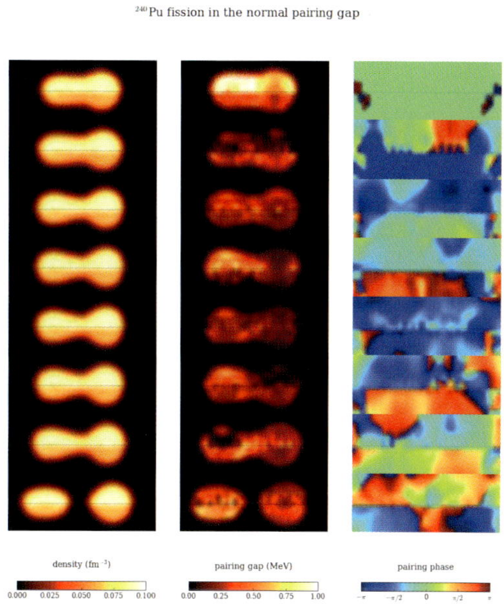

Fig. 3. Induced fission of ^{240}Pu with normal pairing strength last about 14,000 fm/c from saddle-to-scission. The columns show sequential frames of the density (first column), magnitude of the pairing field (second column), and the phase of the pairing field (third column). In each frame the upper/lower part of each frame shows the neutron/proton density, the magnitude of neutron/proton pairing fields, and of the phase of the pairing field respectively[13].

energies of the fragments. The nucleus ^{240}Pu was prepared in a state close in deformation to the outer fission barrier and an equivalent neutron energy in the reaction ^{239}Pu(n,f) of about 1.5 MeV. Our goal was not to describe correctly various fission fragments properties, as for many decades the main difficulty the theory was its inability produce fission starting above but near the top of the fission barrier in a real-time approach.The approximate approaches used widely and based on constructing at first a potential energy surface in a collective space of a typically arbitrary dimension between 2 and 5, which was combined with a recipe to calculate also an appropriate inertia tensor in this collective space, even though they might lead to some reasonable predictions, they do not really prove that a truly microscopic theory is at hand. First of all the choice of collective variables is not rigorous, it is based often on the ability of a specific researcher or group of researchers to solve numerically the problem in the space chosen. It is computationally extremely expensive to construct potential energy surfaces and related inertia tensors in large dimensional spaces. The choice of collective variable is not dictated by a rigorous theory but rather by "intuition." There are also technical difficulties with defining a potential energy surface in a multidimensional space, which is basically a reduction from an infinite dimensional space to a finite dimensional one, a fact well known in mathematics in catastrophe theory even in the case of finite dimensional spaces. Apart from these rather technical difficulties, there are physics problems, as the introduction of collective degrees of freedom implies an almost exact separation of the degrees of freedom into collective and intrinsic with no coupling between them. This implies that during the evolution the intrinsic degrees of freedom are assumed to remain "unexcited," which is never the case, unless one deals with fully integrable models. There is always a coupling between collective and intrinsics degrees of freedom, this is why fragments emerge excited. This aspect is trivial to put in evidence, one can start with a small number of collective degrees of freedom excited, such as quadrupole and octupole deformation, and let the nucleus evolve freely, only to discover that in an unrestricted dynamics other degrees of freedom are immediately excited with significant amplitudes. This is one of the main reasons why the present "microscopic" approaches, based on limited and arbitrarily chosen number of collective degrees of freedom cannot be recognized as a solution of the large amplitude collective nuclear many-body problem. The only viable alternative is to allow all degrees of freedom to be active. Even though this might appear as an insurmountable numerical problem, in fact the problem can be solved with current computers. In a high accuracy simulation of induced fission of ^{240}Pu we integrated in time numerically 256,000 3D time-dependent coupled non-linear complex partial differential equations on a $25^2 \times 50$ fm^3 spatial lattice for about 320,000 time steps using 512 GPUs in about 47 hours or using 1602 GPUs in about 24 hours. The lattice constant in this calculation corresponds to a cutoff momentum of ≈ 500 MeV/c, which is very high and of the same magnitude with the cutoff momenta used in chiral perturbation effective theories of nucleon-nucleon interactions.

The outcome of allowing all collective degrees of freedom to be active and to include time and space dependent pairing fields has been remarkable in several ways. The first surprise was that for the first time an actinide could fission while the dynamics was described with a realistic energy density functional. This could not have happened if pairing correlations would have not been included dynamically, and even if pairing correlations would have been treated approximately at the BCS level either statically or in a time-dependent approach as in Refs.[11,12]. The second surprise was that the properties of the fission fragments came out very close to the observed ones, even though no effort has been put in trying to obtain them. The physics embodied in the nuclear energy density functional is to a large extent accurate and we attribute this agreement to this fact. The third surprise was that the evolution time from the saddle-to-scission was almost an order of magnitude longer than previously predicted[17,18], namely of the order of 10,000 fm/c. It has already been established that in the absence of pairing, or when pairing as a normal

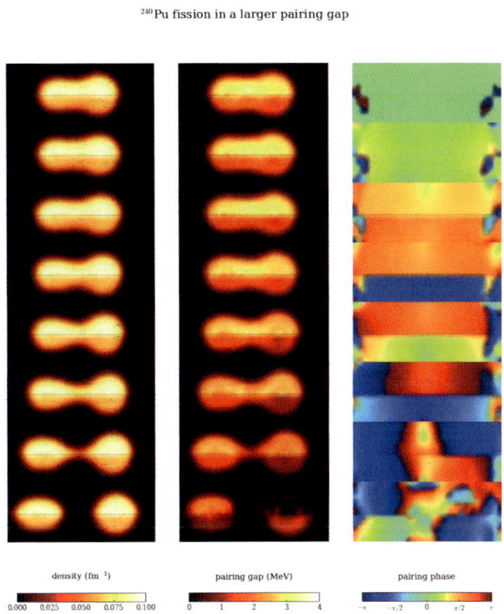

Fig. 4. Induced fission of ^{240}Pu with enhanced pairing strength last about 1,400 fm/c from saddle-to-scission, thus about ten times faster than in the case of normal pairing strength.

strength and is treated in BCS approximation or with frozen initial occupation probabilities, a nucleus will not fission starting at the outer saddle [12], as there is no mechanism to allow for a redistribution of single-particle occupation probabilities. In order to prove convincingly the crucial role played by the pairing correlations in the fission dynamics we have increased artificially the strength of the pairing interaction. The saddle-to-scission time is then reduced dramatically to about 1,400 fm/c, see Fig. 4. The pairing field fluctuates both in magnitude and phase at normal pairing strength, Fig. 3, while these fluctuations are basically absent in case of strong pairing, Fig. 4, when the dynamics is, as expected, similar to the ideal hydrodynamics [18]. The potential energy surface has a lot of "roughness" for normal pairing strength, and the slide down of the nucleus is similar to the motion of an electron in the Drude model of electric conductivity, when the electron is kicked out of the direction of the electric field by elastic collisions with the ions, the length of the trajectory is longer, and the average velocity along the direction of the filed is significantly reduced, even though there is no friction. Similarly, the nucleus from saddle-to-scision remains rather cold and only collective degrees of freedom are significantly excited. Pairing, while not being the engine, it provides the essential "lubricant," without which fission is brought to a "screeching halt."

References

1. O. Hahn and F. Strassmann, Über den Nachweis und das Verhalten der bei der Bestrahlung des Urans mittels Neutronen entstehenden Erdalkalimetalle, Naturwissenschaften **27**, 11 (1939).
2. L. Meitner and O.R. Frisch, Disintegration of Uranium by Neutron: A New Type of Nuclear Reaction, Nature (London), **143**, 239 (1939).
3. F. Gönnenwein, Lectures presented at LANL FIESTA Fission School & Workshop, Sep. 8–12, 2014, Santa Fe, New Mexico, USA, *http://t2.lanl.gov/fiesta2014/*
4. N. Schunck and L.M. Robledo, Microscopic Theory of Nuclear Fission: A review, *Rep. Prog. Phys.* **79**, 116301 (2016).
5. H. Kamerlingh Onnes, Further Experiments with Liquid Helium, KNAW, Proceedings, 13 II, 1910-1911, Amsterdam, 1911, pp. 1093–1113.
6. J. Bardeen, L.N. Cooper and J.R. Schrieffer, Theory of Superconductivity, *Phys. Rev.* **108**, 1175 (1957).

7. O.J. Haxel, H.D. Jensen and H.E. Suess, On the "magic numbers" in nuclear structure, *Phys. Rev.* **75**, 1766 (1949); Maria Goeppert Mayer, On closed shells in nuclei. II, *Phys. Rev.* **75**, 1969 (1949).
8. D.L. Hill and J.A. Wheeler, Nuclear Constitution and the Interpretation of Fission Phenomena, *Phys. Rev.* **89**, 1102 (1953).
9. G.F. Bertsch, The nuclear density of states in the space of nuclear shapes, *Phys. Lett.* **B 95**, 157 (1980); F. Barranco, G.F. Bertsch, R.A. Broglia and E. Vigezzi, Large-Amplitude Motion in Superfluid Fermi Droplets, *Nucl. Phys.* **A512**, 253 (1990); G.F. Bertsch, Large Amplitude Collective Motion, *Nucl. Phys.* **A574**, 169c (1994).
10. G.F. Bertsch and A. Bulgac, Comment on "Spontaneous Fission: A Kinetic Approach," *Phys. Rev. Lett.* **79**, 3539 (1997).
11. C. Simenel and A.S. Umar, Formation and Dynamics of Fission Fragments, *Phys. Rev. C* **89**, 031601(R) (2014); G. Scamps, C. Simenel and D. Lacroix, Superfluid Dynamics of ^{258}Fm Fission, *Phys. Rev. C* **92**, 011602(R) (2015).
12. Y. Tanimura, D. Lacroix and G. Scamps, Collective Aspects Deduced from the Time-Dependent Microscopic Mean-Field with Pairing: Application to the Fission Process, *Phys. Rev. C* **92**, 034601 (2015); P. Goddard, P. Stevenson and A. Rios, Fission Dynamics within Time-Dependent Hartree-Fock: Deformation Induced Fission, *Phys. Rev. C* **92**, 054610 (2015); P. Goddard, P. Stevenson and A. Rios, Fission Dynamics within Time-Dependent Hartree-Fock: Boost Induced Fission, *Phys. Rev. C* **93**, 014620 (2016).
13. A. Bulgac, P. Magierski, K.J. Roche and I. Stetcu, Induced Fission of ^{240}Pu within a Real-Time Microscopic Framework, *Phys. Rev. Lett.* **116**, 122504 (2016).
14. A. Bulgac, Time-Dependent Density Functional Theory and Real-Time Dynamics of Fermi Superfluids, *Ann. Rev. Nucl. Part. Sci.* **63**, 97 (2013).
15. E. Chabanat, P. Bonche, P. Haensel, J. Meyer and R. Schaeffer, A Skyrme parametrization from subnuclear to neutron star densities Part II. Nuclei far from stabilities, *Nucl. Phys.* **A635**, 231 (1998).
16. Y. Yu and A. Bulgac, Energy Density Functional Approach to Superfluid Nuclei, *Phys. Rev. Lett.* **90**, 222501 (2003).
17. K.T.R. Davies, A.J. Sierk and J.R. Nix, Effect of Viscosity on the Dynamics of Fission, *Phys. Rev. C* **13**, 2385, (1976); J. Blocki, Y. Boneh, J.R. Nix, J. Randrup, M. Robel, A.J. Sierk, W.J. Swiatecki, One-body dissipation and the super-viscidity of nuclei, *Ann. Phys.* **113**, 330

(1978); J. Randrup and W.J. Swiatecki, One-body dissipation and nuclear dynamics, *Ann. Phys.* **125**, 193 (1980); J.W. Negele, S.E. Koonin, P. Möller, J.R. Nix and A.J. Sierk, Dynamics of Induced Fission, *Phys. Rev. C* **17**, 1098 (1978).

18. A. Bulgac, M.M. Forbes and S. Jin, Nuclear energy density functionals: What do we really know?, arXiv:1506.09195.

Fission Barriers from a Multidimensionally Constrained Relativistic Mean Field Model*

Bing-Nan Lu[a,b], Jie Zhao[a,c], En-Guang Zhao[a,d] and
Shan-Gui Zhou[a,d,e,f,†]

[a] *CAS Key Laboratory of Frontiers in Theoretical Physics, Institute of Theoretical Physics, Chinese Academy of Sciences, Beijing 100190, China*
[b] *Institut für Kernphysik (IKP-3) and Jülich Center for Hadron Physics, Forschungszentrum Jülich, D-52425 Jülich, Germany*
[c] *Physics Department, Faculty of Science, University of Zagreb, Bijenicka 32,Zagreb 10000, Croatia*
[d] *Center of Theoretical Nuclear Physics, National Laboratory of Heavy Ion Accelerator, Lanzhou 730000, China*
[e] *School of Physics, University of Chinese Academy of Sciences, Beijing 100049, China*
[f] *Synergetic Innovation Center for Quantum Effects and Application, Hunan Normal University, Changsha, 410081, China*
[†] *E-mail: sgzhou@itp.ac.cn*

Many shape degrees of freedom play crucial roles in determining fission barriers. In multidimensionally constrained covariant density functional theories (MDC-CDFTs) which we have developed in recent years, both the axial and the spatial reflection symmetries are broken and all deformations described by $\beta_{\lambda\mu}$ with even μ, including β_{20}, β_{22}, β_{30}, β_{32}, β_{40}, etc., are considered self-consistently. The MDC-CDFTs have been applied to the study of fission barriers and potential energy surfaces of actinide nuclei, third minima in potential energy surfaces of light actinides, shapes and potential energy surfaces of superheavy nuclei, the Y_{32} correlations in $N = 150$ isotones and Zr isotopes, and the shape of hypernuclei. In this contribution we will present an introduction of the MDC-CDFTs and a short review of its applications on fission barriers.

Keywords: Fission Barrier; Third Minimum; Covariant Density Functional.

1. Background

Since the discovery of nuclear fission[1], the quantitative description of the fission process has been very challenging for nuclear theorists. One crucial

*This work was partly supported by the NSF of China (11525524, 11621131001 and 11647601) and the 973 Program of China (2013CB834400). The computation of this work was supported by the HPC Cluster of SKLTP/ITP-CAS and the Supercomputing Center, CNIC of CAS.

ingredient in describing the fission dynamics is the fission barrier prohibiting the dissolving of the fissioning nucleus[2]. Due to the shell effects, many actinide nuclei are characterized by the two-humped fission barrier[3–7]. Consequently any model concerning the fission phenomena should be able to reproduce such a feature and to estimate the two barrier heights under certain assumptions on the nuclear shapes. For example, in the early calculations of the fission barriers, the nuclei are often assumed to be axially symmetric. Later it is found that the occurrence of the triaxial and the octupole deformations has non-negligible consequences in determining the heights of the inner and outer fission barriers. Namely, it has long been known from the macroscopic-microscopic model calculations that the inner fission barrier is usually lowered when the triaxial deformation is allowed, while for the outer barrier, the reflection asymmetric (RA) shape is favored[8,9]. Furthermore, these conclusions were confirmed in the non-relativistic[5] and relativistic[10,11] density functional calculations, respectively. Recently, it is shown in Ref. 12 that the outer barriers are also lowered by the triaxial deformation compared with axially symmetric results. This lowering effect for the reflection-asymmetric outer barrier is $0.5 \sim 1$ MeV, accounting for $10\% \sim 20\%$ of the barrier height. It is thus necessary to include these shape degrees of freedom properly in any systematic calculations.

In recent years, the covariant density functional theory (CDFT)[13] has been employed to study nuclear potential energy surfaces and fission barriers. By breaking both the axial and the spatial reflection symmetries, we have developed multidimensionally constrained (MDC-CDFTs) in which all deformations described by $\beta_{\lambda\mu}$ with even μ are included self-consistently[12,14–16]. In this contribution we will present an introduction of the MDC-CDFTs and a short review of its applications on fission barriers.

2. An introduction of MDC-DCFTs

There are four types of covariant density functionals in the CDFT: the meson exchange or point-coupling nucleon interactions combined with the nonlinear or density dependent couplings (see Ref. 13 for recent reviews). All these four types of functionals have been implemented in our MDC-CDFTs.

In MDC-CDFTs, the nuclear shape is assumed to be invariant under the reversion of x and y axes, i.e., the intrinsic symmetry group is V_4 and all shape degrees of freedom $\beta_{\lambda\mu}$ with even μ (β_{20}, β_{22}, β_{30}, β_{32}, β_{40}, $\ldots$) are included self-consistently. The MDC-CDFTs consist of two types of models: the multidimensionally-constrained relativistic mean field (MDC-RMF) model and the multidimensionally-constrained relativistic Hartree-Bogoliubov (MDC-RHB) model.

In the MDC-RMF model, the BCS approach has been implemented for the particle-particle (pp) channel. This model has been used to study potential energy surfaces and fission barriers of actinides [12,14,17–20], the spontaneous fission of several fermium isotopes [21], the Y_{32} correlations in $N = 150$ isotones [22], and shapes of hypernuclei [23,24], see Refs. 25, 26, 15 for recent reviews.

The Bogoliubov transformation generalizes the BCS quasi-particle concept and provides a unified description of particle-hole (ph) and pp correlations on a mean-field level. In the MDC-RHB model, pairing correlations are treated by making the Bogoliubov transformation and a separable pairing force of finite range [27–31] is adopted. The MDC-RHB model has been used to study the spontaneous fission of fermium isotopes [32] and neutron-rich Zr nuclei [16].

3. Results and discussions

The double-humped fission barriers of actinide nuclei are always used to benchmark the predictive power of theoretical models. As the nucleus evolves from the ground state to the fission configurations, various shape degrees of freedom play important and different roles in determining the heights of the inner and outer barriers. It has been long known that the inner barrier is usually lowered when the triaxial deformation is allowed, while for the outer barrier, the reflection asymmetric shape is favored. In Ref. 12 we have shown that, the reflection asymmetric outer barrier may be further lowered by including the non-axial shape degrees of freedom in the framework of MDC-CDFT.

In Fig. 1 we show the 1-d potential energy curves (PEC) from an oblate shape with β_{20} about -0.2 to the fission configuration with β_{20} beyond 2.0 which are obtained from calculations with different self-consistent symmetries imposed: the axial (AS) or triaxial (TS) symmetries combined with reflection symmetric (RS) or asymmetric cases. The importance of the triaxial deformation on the inner barrier and that of the octupole deformation on the outer barrier are clearly seen: The triaxial deformation reduces the inner barrier height by more than 2 MeV and results in a better agreement with the empirical datum; the RA shape is favored beyond the fission isomer and lowers very much the outer fission barrier. Besides these features,

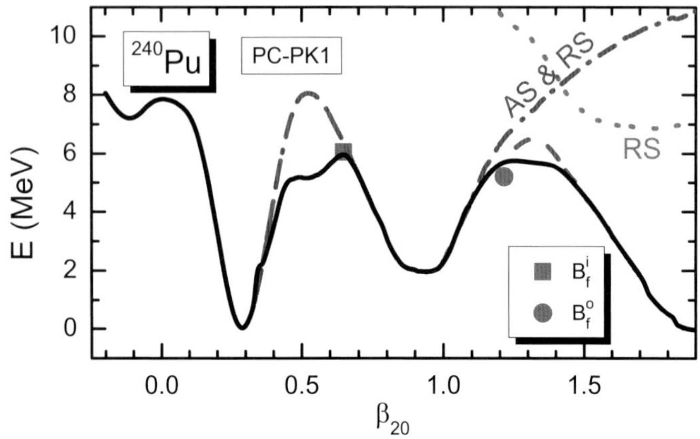

Fig. 1. Potential energy curves of ^{240}Pu with various self-consistent symmetries imposed. The empirical inner (outer) barrier height is denoted by the grey square (circle). The energy is normalized with respect to the binding energy of the ground state. The parameter set used is PC-PK1[33]. Taken from Ref. 12.

it was found for the first time that the outer barrier is also considerably lowered by about 1 MeV when the triaxial deformation is allowed. In addition, a better reproduction of the empirical barrier height can be seen for the outer barrier. It has been stressed that this feature can only be found when the axial and reflection symmetries are simultaneously broken[12].

In Ref. 14 we have applied the MDC-CDFT to investigate systematically the fission barriers of the actinide nuclei with the breaking of axial and reflection symmetries. In Fig. 2 the calculated potential energy curves near the outer barriers of the actinide nuclei are shown. For comparison the results with and without non-axial deformations are both presented. Note that near the second barrier all the results are reflection asymmetric. Because only the β_{20} deformation is constrained, the total binding energy is automatically minimized against other possible deformation parameters. The empirical values of the fission barriers are depicted by dots. It's clearly seen that the axial symmetric calculations overestimate the barrier height. After including the non-axial shapes, the agreement between the calculated values and the experimental ones becomes better.

Besides the two-humped barriers of the PES's in the actinide region, the occurrence of a third one at large deformations beyond the second barrier was predicted by macroscopic-microscopic model (MMM) calculations in the 1970s but contradictory results were later obtained with a number of

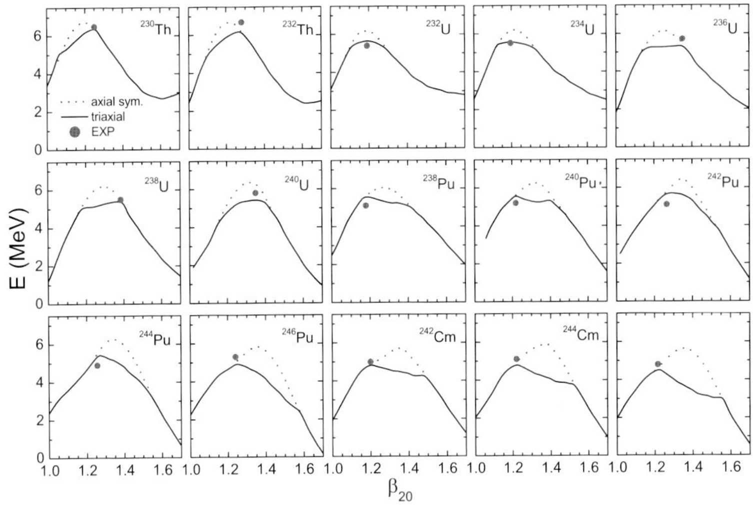

Fig. 2. Potential energy curves of even-even actinides nuclei in the outer barrier regions from the MDC-CDFT calculations. The energy is normalized with respect to the binding energies of the ground states of each nucleus. The experimental data are shown as full dots[34]. Taken from Ref. 14.

different models. In Ref. 20 we have investigated the triple-humped barriers in actinide nuclei in the framework of MDC-CDFT.

Figure 3 displays the self-consistent PES's in the (β_{20}, β_{30}) plane for the light even-even actinides ^{232,234,236,238}U and ^{226,228,230,232}Th, calculated with the relativistic functional DD-ME2. The deformation parameters are restricted to the range $\beta_{20} \sim (1.0, 3.0)$ and $\beta_{30} \sim (0.3, 1.5)$, in which the second barrier together with the third minimum and the third barrier, if they exist, are located. Note that the contour interval for the PES's plotted in Fig. 3 is 0.5 MeV.

In Fig. 3 one finds a third minimum on the PES's of ^{232,234,238}U with the exception of ^{236}U. The minimum is rather shallow for ^{232}U and ^{234}U and very pronounced for ^{238}U. These results are very similar to those obtained in recent calculations based on the MMM[36–39], and the Skyrme Hartree-Fock-Bogoliubov model[40].

In the case of Th isotopes, our calculations predict a pronounced third minimum for ^{226}Th, with a depth of 2–3 MeV. As the neutron number N increases, both the energy of the third minimum and the height of the third barrier decrease, and the depth of the third well is also reduced. This trend has also been predicted in MMM calculations[39]. For ^{230}Th only a shallow

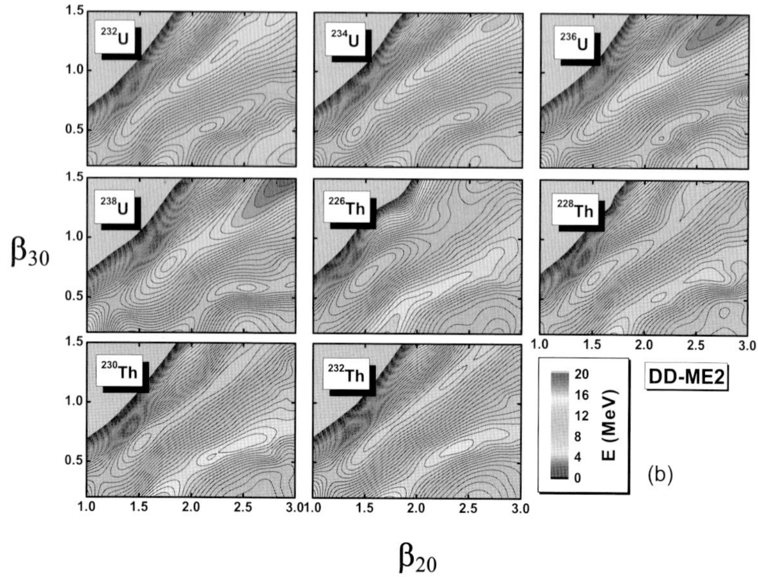

Fig. 3. Self-consistent MDC-CDFT potential energy surfaces of U and Th isotopes in the (β_{20}, β_{30}) plane, calculated with the relativistic functionals DD-ME2[35]. For each nucleus energies are normalized with respect to the binding energy of the absolute minimum. The contours join points on the surface with the same energy, and the energy difference between neighboring contours is 0.5 MeV. Taken from Ref. 20.

minimum of depth less than 1 MeV occurs. The third minimum is very shallow for ^{232}Th when calculated with DD-ME2.

Figure 3 clearly shows that the functional DD-ME2 (lower panel) predicts pronounced third minima. In addition to the lowest fission path on which we have focused in this work, there are other possible paths along which there are more shallow minima and saddle points.

4. Summary

In this contribution we introduce the MDC-CDFTs in which all shape degrees of freedom $\beta_{\lambda\mu}$ deformations with even μ are allowed. We present some applications of the MDC-CDFTs about the potential energy curves and potential energy surfaces of actinide nuclei. It's found that the triaxiality plays important role in determining the fission properties. Particularly, we found that the outer barrier of actinides can be lowered considerably when the triaxiality is included. Consequently the experimental fission barriers are better reproduced. We also studied the large deformation behaviour of the actinide nuclei and examined the existence of the third minimum and third barrier in the corresponding energy surfaces of several nuclei.

References

1. O. Hahn and F. Strassmann, *Naturwissenschaften* **27**, 11 (1939).
2. N. Bohr and J. A. Wheeler, *Phys. Rev.* **56**, 426 (1939).
3. T. Burvenich, M. Bender, J. A. Maruhn and P.-G. Reinhard, *Phys. Rev. C* **69**, 014307 (2004).
4. M. Mirea and L. Tassan-Got, *Cent. Eur. J. Phys.* **9**, 116 (2011).
5. M. Girod and B. Grammaticos, *Phys. Rev. C* **27**, 2317 (1983).
6. V. Blum, J. A. Maruhn, P.-G. Reinhard and W. Greiner, *Phys. Lett. B* **323**, 262 (1994).
7. L. Csige *et al.*, *Phys. Rev. C* **85**, 054306 (2012).
8. V. V. Pashkevich, *Nucl. Phys. A* **133**, 400 (1969).
9. P. Möller and S. G. Nilsson, *Phys. Lett. B* **31**, 283 (1970).
10. H. Abusara, A. V. Afanasjev and P. Ring, *Phys. Rev. C* **82**, 044303 (2010).
11. K. Rutz, J. A. Maruhn, P. G. Reinhard and W. Greiner, *Nucl. Phys. A* **590**, 680 (1995).
12. B.-N. Lu, E.-G. Zhao and S.-G. Zhou, *Phys. Rev. C* **85**, 011301(R) (2012).
13. J. Meng (ed.), *Relativistic Density Functional for Nuclear Structure*, International Review of Nuclear Physics, Vol. 10 (World Scientific Pub Co Pte Ltd, Mar 2016).
14. B.-N. Lu, J. Zhao, E.-G. Zhao and S.-G. Zhou, *Phys. Rev. C* **89**, 014323 (2014).
15. S.-G. Zhou, *Phys. Scr.* **91**, 063008 (2016).
16. J. Zhao, B.-N. Lu, E.-G. Zhao and S.-G. Zhou, *Phys. Rev. C* **95**, 014320 (2017).
17. B.-N. Lu, J. Zhao, E.-G. Zhao and S.-G. Zhou, *J. Phys: Conf. Ser.* **492**, 012014 (2014).
18. B.-N. Lu, J. Zhao, E.-G. Zhao and S.-G. Zhou, *Phys. Scr.* **89**, 054028 (2014).
19. B.-N. Lu, J. Zhao, E.-G. Zhao and S.-G. Zhou, *EPJ Web Conf.* **38**, 05003 (2012).
20. J. Zhao, B.-N. Lu, D. Vretenar, E.-G. Zhao and S.-G. Zhou, *Phys. Rev. C* **91**, 014321 (2015).

21. J. Zhao, B.-N. Lu, T. Nikšić, D. Vretenar and S.-G. Zhou, *Phys. Rev. C* **93**, 044315 (2016).
22. J. Zhao, B.-N. Lu, E.-G. Zhao and S.-G. Zhou, *Phys. Rev. C* **86**, 057304 (2012).
23. B.-N. Lu, E.-G. Zhao and S.-G. Zhou, *Phys. Rev. C* **84**, 014328 (2011).
24. B.-N. Lu, E. Hiyama, H. Sagawa and S.-G. Zhou, *Phys. Rev. C* **89**, 044307 (2014).
25. L.-L. Li, B.-N. Lu, N. Wang, K. Wen, C.-J. Xia, Z.-H. Zhang, J. Zhao, E.-G. Zhao and S.-G. Zhou, *Nucl. Phys. Rev.* **31**, 253 (2014), in Chinese.
26. B.-N. Lu, J. Zhao, E.-G. Zhao and S.-G. Zhou, in *Relativistic Density Functional for Nuclear Structure*, ed. J. Meng, International Review of Nuclear Physics, Vol. 10 (World Scientific Publishing Co. Pte. Ltd., Mar 2016) pp. 171–217.
27. Y. Tian and Z.-Y. Ma, *Chin. Phys. Lett.* **23**, 3226 (2006).
28. Y. Tian, Z.-Y. Ma and P. Ring, *Phys. Lett. B* **676**, 44 (2009).
29. Y. Tian, Z.-Y. Ma and P. Ring, *Phys. Rev. C* **80**, 024313 (2009).
30. T. Nikšić, D. Vretenar and P. Ring, *Phys. Rev. C* **78**, 034318 (2008).
31. T. Nikšić, P. Ring, D. Vretenar, Y. Tian and Z.-Y. Ma, *Phys. Rev. C* **81**, 054318 (2010).
32. J. Zhao, B.-N. Lu, T. Nikšić and D. Vretenar, *Phys. Rev. C* **92**, 064315 (2015).
33. P. W. Zhao, Z. P. Li, J. M. Yao and J. Meng, *Phys. Rev. C* **82**, 054319 (2010).
34. R. Capote, M. Herman, P. Oblozinsky, P. G. Young, S. Goriely, T. Belgya, A. V. Ignatyuk, A. J. Koning, S. Hilaire, V. A. Plujko, M. Avrigeanu, O. Bersillon, M. B. Chadwick, T. Fukahori, Z. Ge, Y. Han, S. Kailas, J. Kopecky, V. M. Maslov, G. Reffo, M. Sin, E. S. Soukhovitskii and P. Talou, *Nucl. Data Sheets* **110**, 3107 (2009).
35. G. A. Lalazissis, T. Nikšić, D. Vretenar and P. Ring, *Phys. Rev. C* **71**, 024312 (2005).
36. M. Kowal and J. Skalski, *Phys. Rev. C* **85**, 061302(R) (2012).
37. P. Jachimowicz, M. Kowal and J. Skalski, *Phys. Rev. C* **87**, 044308 (2013).
38. T. Ichikawa, A. Iwamoto, P. Möller and A. J. Sierk, *Phys. Rev. C* **86**, 024610 (2012).
39. T. Ichikawa, P. Möller and A. J. Sierk, *Phys. Rev. C* **87**, 054326 (2013).
40. J. D. McDonnell, W. Nazarewicz and J. A. Sheikh, *Phys. Rev. C* **87**, 054327 (2013).

Progress on the Description of Low Energyssion Dynamics within the TDGCM+GOA Microscopic Approach

D. Regnier[1], N. Dubray[1], N. Schunck[2] and M. Verriere[1]

[1]*CEA, DAM, DIF, 91297 Arpajon, France*

[2]*Nuclear and Chemical Science Division, Lawrence Livermore National Laboratory, Livermore, CA 94551, USA*

The growing need for nuclear data in applications such as the r-process in astrophysics is an incentive to build theory capable of predicting the ssion yields for a wide range of systems, observables, and input channels. One of the most promising theoretical frameworks is the time dependent generator coordinate method (TDGCM) applied under the Gaussian overlap approximation (GOA). However, the computational cost of this method was up to now a hurdle to perform accurate realistic calculations with two and more collective degrees of freedom. To overcome this limitation, we develop the code FELIX aiming to solve the TDGCM+GOA equation for an arbitrary number of collective variables. In this talk, we will report the recent results related to the calculation of low energy ssion yields within TDGCM+GOA with two collective variables. We will briefly present the numerical methods adopted as well as the status of the latest version of FELIX. Finally, we will emphasize the status of 3-dimensional calculations of the collective ssion dynamics.

Cluster Degrees of Freedom in Fission of Actinides*

A. V. Unzhakova and T. A. Khantuleva
Faculty of Mathematics and Mechanics, Saint Petersburg State University, St. Petersburg, 199034, Russia
E-mail: a.unzhakova@spbu.ru

O. N. Granichin
Faculty of Mathematics and Mechanics, Saint Petersburg State University, St. Petersburg, 199034, Russia
Institute of Problems of Mechanical Engineering, Russian Academy of Sciences, St. Petersburg, 199178, Russia

In this work we study a stabilization of internal cluster structure of the nuclear system under the fluctuations of the asymmetry value used as a constraint in the shell correction calculations. Persistent cluster magic structures play a main role in origination of various fission modes. We get quantitative results for clustering in theoretical description of actinide multimodal fission by means of analyzing the high dimensional potential energy surface's topological complexity. It is shown that traditional adiabatic description is insufficient in the regions of the deformation characterized by the strong internal shell structures rearrangement and therefore it is necessary to take nonequilibrium effects of reclusterization into account.

Keywords: Nuclear Fission; Clustering; State Space Structure.

1. Introduction

1.1. *Cluster degrees of freedom*

Microscopic self-consistent [1–3] and micro-macro shell correction descriptions of a complex phenomenon, such as nuclear fission, produce the potential energy surfaces (PES) with multiple structures. A chain of local minima on PES represents a metastable configuration caused by the magic shell cluster formation. Under the assumption that fission motion is adiabatic until the neck ruptures at some finite radius, the chain of local minima is associated with either symmetrical or asymmetrical fission pathway.

*This work is supported by Russian Science Foundation (project 16-19-00057).

There are still unsolved questions related to nonadiabaticity, decoherence and multiclustering [4, 5] in the theory describing nuclear collective motion in fission.

In this work, we consider a variety of possible fluctuations of the mass-asymmetry collective coordinate. Micro-macro shell correction calculations have been performed for a wide range of the actinide nuclei. We use Cassinian oval shape parameterization and a large deformation space to take all important degrees of freedom into account and to make the cluster structure stabilization possible. The single-particle spectra responsible for the shell structure influence are determined in frame of the realistic mean field with absolutely flexible axial surface [6]. Multidimensional fission PES's have a complex structure, they consist of the several separated sheets and contain the regions of singularities [7]. Local minima in these regions are separated from the other states due to the presence of specific set of discrete quantum numbers because of internal cluster formation. These regions of PES are very sensitive to clustering degrees of freedom. The clustering phenomena introduce a number of different scales into the system and, consequently, causes the large-scale fluctuations around the local minima of PES.

1.2. *Experimental evidence of clustering in fission*

In the last two decades a new kind of nuclear multi-body decay has been observed experimentally by several groups. It was a collinear cluster tripartition (CCT), which is a cluster decay of heavy nuclei with the ruptures of two necks separating the middle cluster. One of these groups worked in Flerov Laboratory of Nuclear Reactions in Dubna and from the very beginning of the multibody decay study was supported by theoretical calculations performed in Bogoliubov Laboratory of Theoretical Physics [8–11]. Series of experiments aimed at searching for CCT have been performed in Dubna. Results of this group for cluster tripartition events in ^{252}Cf(sf) and ^{235}U(n$_{th}$,f) reactions were presented in [12]. An example of the experimental data and its analysis in the framework of the cluster concept is presented in Figure 1.

One can see that generally the agreement between the theoretical results and new experimental data on the fission is improved when clustering degrees of freedom are taken into account. A total yield of the CCT-originated events was found to be not less 4×10^{-3} per binary fission in these experiments. The latest experimental results of the same group show that the

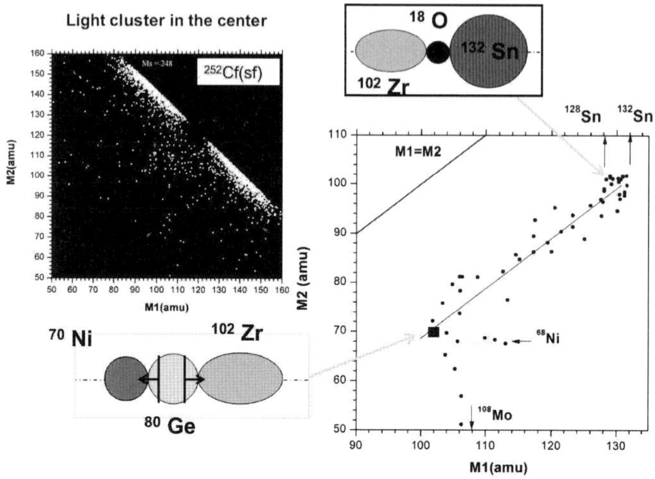

Fig. 1. A mass-mass experimental distribution and an example of the data processing and analysis on the base of clustering concept for the ^{252}Cf(sf) reaction [12].

clustering effects manifest themselves in possible existence of a new kind of the shape isomer states in fission fragments detected [13].

2. Theoretical Background and Models of Clustering

For the first time, in addition to the conventional fissoining shape sequence, two-neck and three-neck shape families have been obtained in Strutinsky's early work [14]. Fission barriers were determined in this paper in the frame of the liquid drop model (LDM) as an extremum of the potential energy, calculated as a sum of surface and Coulomb energies. The nuclear medium was considered in terms of nucleon density, thus the quantum effects of the single-particle motion were eliminated. The main deformation parameter was taken as an adiabatic variable and it was shown that there is no continuous equilibrium pass between different possible nuclear shape families in fission.

Another theoretical approach to the clustering in fission was presented in [15] by introducing the asymmetric two-center shell model. It is one of the rare examples, when phenomenon was predicted before the experimental discovery. The fission valleys in this model are present due to the shell effects, and it was clearly shown why cluster radioactivity was mostly detected in parent nuclei yielding a doubly magic lead daughter.

Similar results were obtained in the paper [11] devoted to the manifestation of clustering effects in the in the structure of actinides. The two-cluster model [9] allows clusters to overlap one another. The clusters may be well separated in momentum space, but have an overlap in coordinate space, what is confirmed by our results for the two-Sn mode. In this mode the compact cluster configuration was identified by varying the deformation parameters. In the dinuclear model [10] clusters are well separated in coordinate space at large elongations of the system, but the strong correlations in momentum space remain. Our calculations results could explain the success of the other cluster models. The clustering is not introduced by the mean field in our model and appears as a result of the system complexity and flexibility of the nuclear surface.

3. Calculations Results

From the very beginning, the shell correction model developed in Dubna was designed to make high-dimensional calculations possible. The practical implementation of this study has been limited before only by the computation capabilities. The nuclear shape variables responsible for the collective motion in fission process are strongly connected with the properties of the nuclear potential. An important feature of the Woods-Saxon type potential [6] is that the thickness of nuclear surface layer remains constant with deformation. Spin-orbit interaction is considered to be proportional to the potential gradient. The Hamiltonian contains the kinetic energy, the potential, the spin-orbit term and the Coulomb potential. Single-particle spectrum of the fissioning system is calculated for every point of the high-dimensional deformation space. The nuclear shape is parameterized in the coordinate system associated with Cassini ovaloids. The deviation from the ovaloid is expressed by expansion into a series of Legendre polynomials. In addition to the single parameter that determines the given Cassini ovaloid, the full set of collective variables includes the expansion coefficients. Thus, it takes from 10 to 20 parameters of deformation in order to determine all possible shapes, which fissioning nucleus may assume on its way to scission one needs. The deformation energy is then minimized as a function of the deformation parameters and the PES structure is determined.

In high-dimensional shell correction calculations the conventional PES splits into several different sheets separated by the singularity regions. The system evolution from the ground state is described as a function of the main elongation (quadrupole moment) taken as an integral constraint. The quadrupole moment and asymmetry value of the strongly deformed nuclear system are determined after energy minimization in respect to the shape deformation parameters is completed. Position of the border between left and right parts of the nuclear shape is determined by taking the derivative of the surface profile function. The neck position is assumed to be located at the point where this derivative is equal to zero. Iterative calculations by asymmetry value have been performed, allowing us to obtain the shapes of the conditional energy minima located higher in energy then conventional fission valleys.

The role of the shell effects in the shape evolution is demonstrated by appearance of the persistent clusters which coincide in the shape and the composition with magic (spherical or deformed) nuclear structures. In terms of energy, the shell effects are most pronounced in ^{252}Cf; in Figure 2 the results of multidimensional calculations for ^{252}Cf are presented as an example of cluster structure formations in heavy nuclei, but their influence extends over a broad region of actinide nuclei. In general, multimodal structure of PES for actinides follows the same pattern. Our cluster states are similar in their nature to the equilibrium ground state deformations and are also obtained as minima of the energy within the continuous adiabatic path on the PES.

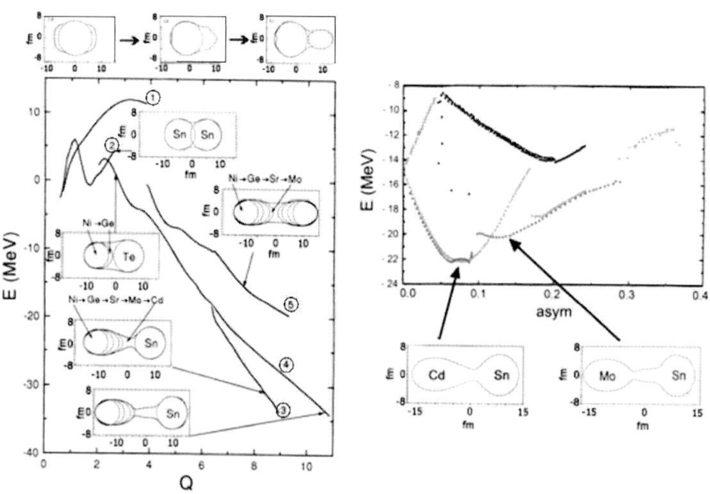

Fig. 2. Potential energy surface sheets obtained as a result of ten-dimensional minimization with integral quadrupole moment Q used as a constraint (left) and conditional energy minima at fixed Q-value obtained by ten-dimensional minimization with asymmetry value used as a constraint (right).

In one of the modes the nuclear shape appears to be close to that of two partially overlapping double magic nuclei ^{132}Sn. This mode, observed experimentally in ^{252}Cf, can be considered as a variation of cluster decay.

4. Cluster Flow Approach to Structure Dynamics

Clustering effects in nuclear fission are related to the nonhomogeneous distribution of nucleons in phase space, whereas the distribution in quasi-classical phenomenological liquid drop component is supposed to be homogeneous. Large amplitude collective motion in fission is usually assumed to be slow and the state of the nuclear system to be equilibrium state. Within the framework of microscopical-macroscopical approach, nucleon shells and other quantum phenomena usually are considered as a small deviations from the uniform distribution and the corresponding correction to the LDM energy are determined. The PES singularities obtained in our multidimensional calculations might be considered as a dramatic change of this uniform distribution, and then they might be treated as change of the cluster state structure. The existence of PES singularities requires the application of nonequilibrium hydrodynamics. The nonlocal model of the unstable state structure was introduced in [16, 17] in terms of correlation function that depends on the characteristics of mesoscale cluster structure. It was shown that the nonequilibrium state can lead to the change of state structure and to the formation of a new cluster.

Therefore, as emerging cluster structures render the conventional hydrodynamics invalid, one could employ the language of the integral kernel $\mathfrak{R}(\vec{r}, \vec{r'}, t, t')$, which is the space-time correlation function averaging the big gradients of the macroscopic fields and forming cluster structures at the mesoscale level. In order to calculate the nonequilibrium influence of the cluster formed inside the system, differential continuous hydrodynamic model is to be substituted by integro-differential model, that is valid under the nonequilibrium conditions. The updated equations have to take the space nonlocality into account.

The model expression of the kernel, responsible for the structure formation:

$$\mathfrak{R}_S(\vec{r}, \vec{r'}, t) = \frac{1}{q(\vec{r}, t)} \exp\left\{ -\frac{\pi(\vec{r} - \vec{r'})}{q(\vec{r}, t)} \right\},$$

involves parameter q that represents a linear size of the structure element. For the structureless continuum medium the value averaging scale tends to zero: $q \to 0$. The generalization of cluster flow approach to the structure dynamics has been done in [18, 19], suggesting to describe the reclusterization process along the fission valley to scission. The main feature of the cluster evolution is the appearance of discrete parameters, responsible for the internal state structure and able to change under rearrangement of the cluster shell structure.

5. Conclusions

In present paper the nuclear fission is described as a dynamical evolution of the nuclear shape in the high-dimensional deformation space. The results of PES calculations for nuclear isotopes in actinide region is examined to trace the development of the magic clusters within the fissioning systems with different nucleon composition. Our calculations show that the organization of complex potential energy surface contains several adiabatic paths to the scission, and the regions of the strong singularities as well. The structure of the state space and corresponding cluster core shell quantum numbers can strongly change at the points of the PES singularities. The sufficient description of the complex nuclear fission phenomenon combines both cluster degrees of freedom and quantum liquid drop aspects.

References

[1] J. D. McDonnell, W. Nazarewicz and J. A. Sheikh, Third minima in thorium and uranium isotopes in a self-consistent theory, *Phys. Rev. C* **87**, p. 054327 (May 2013).

[2] S.-G. Zhou, Multidimensionally constrained covariant density functional theories nuclear shapes and potential energy surfaces, *Phys. Scripta* **91**, p. 063008 (2016).

[3] A. Staszczak, A. Baran, J. Dobaczewski and W. Nazarewicz, Microscopic description of complex nuclear decay: Multimodal fission, *Phys. Rev. C* **80**, p. 014309 (Jul 2009).

[4] W. Nazarewicz and J. Dobaczewski, Dynamical symmetries, multiclustering, and octupole susceptibility in superdeformed and hyperdeformed nuclei, *Phys. Rev. Lett.* **68**, 154 (Jan 1992).

[5] R. V. Jolos, Y. V. Palchikov, V. V. Pashkevich and A. V. Unzhakova, Reflection asymmetric deformation and clustering in heavy nuclei, *Il Nuovo Cimento A* **110**, 941 (1997).

[6] V. V. Pashkevich, On the symmetric deformation of fissioning nuclei, *Nucl. Phys. A* **169** (1971).

[7] A. V. Unzhakova, V. V. Pashkivich and Y. V. Pyatkov, *Potential Energy Calculations for Collinear Cluster Tripartition Fission Events*, in *Fission and Properties of Neutron-Rich Nuclei* (World Scientific, 2013), pp. 652–657.

[8] Y. V. Pyatkov, V. V. Pashkevich, A. V. Unzhakova *et al.*, Nontrivial manifestation of clustering in fission of heavy nuclei at low and middle excitations, *Phys. Atom. Nuclei* **67**, 1754 (2004).

[9] T. M. Shneidman, G. G. Adamian, N. V. Antonenko *et al.*, Cluster approach to the structure of heavy nuclei, *Journal of Physics: Conference Series* **569**, p. 012056 (2014).

[10] S. N. Kuklin, G. G. Adamian and N. V. Antonenko, Description of alpha decay and cluster radioactivity in the dinuclear system model, *Physics of Particles and Nuclei* **47**, 206 (2016).

[11] T. M. Shneidman, G. G. Adamian, N. V. Antonenko *et al.*, Manifestation of cluster effects in the structure of actinides, *EPJ Web of Conferences* **107**, p. 03009 (2016).

[12] Y. V. Pyatkov, D. V. Kamanin, W. von Oertzen *et al.*, The collinear cluster tri-partition (cct) of ^{252}cf (sf): New aspects from neutron gated data, *The European Physical Journal A* **48**, p. 94 (2012).

[13] Y. V. Pyatkov, D. V. Kamanin, A. A. Alexandrov *et al.*, First results on observation of new shape isomers, *Physics Procedia* **74**, 67 (2015).

[14] V. M. Strutinsky, N. Y. Lyashchenko and N. A. Popov, Symmetrical shapes of equilibrium for a liquid drop model, *Nuclear Physics* **46**, 639 (1963).

[15] D. N. Poenaru and W. Greiner, *Cluster Radioactivity*, in *Clusters in Nuclei*, 2010, pp. 1–56.

[16] T. A. Khantuleva, Microstructure formation in the framework of the non-local theory of interfaces, *Mater. Phys. Mech.* **51**, p. 51 (2000).

[17] O. N. Granichin and T. A. Khantuleva, Hybrid systems and randomized measuring in nonequilibrium processes, *Differential Equations and Control Processes* **3**, p. 35 (2004).

[18] T. A. Khantuleva and K. Abdulaev, Internal control in nonequilibrium physical processes, *Cybernetics and Physics* **3**, p. 21 (2014).

[19] O. N. Granichin, What is the actual structure of complex information-control systems?, *Stochastic Optimization in Informatics* **12**, 3 (2016).

Study of Fission Using Multi-Nucleon Transfer Reactions

K. Nishio*[1], K. Hirose[1], R. Léguillon[1], H. Makii[1], R. Orlandi[1], K. Tsukada[1], J. Smallcombe[1,2], S. Chiba[3], Y. Aritomo[4], S. Tanaka[4], T. Ohtsuki[5], I. Tsekhanovich[6], C.M. Petrache[7] and A.N. Andreyev[8,1]

[1] *Advanced Science Research Center, Japan Atomic Energy Agency, Tokai, Ibaraki 319-1195, Japan*

[2] *TRIUMF, Vancouver, British Columbia V6T 2A3, Canada*

[3] *Laboratory for Advanced Nuclear Energy, Tokyo Institute of Technology, 2-12-1, Ookayama, Meguro-ku, Tokyo, 152-8550 Japan*

[4] *Kindai University, Higashi-Osaka, 577-8502, Japan*

[5] *Research Reactor Institute, Kyoto University, Kumatori-cho, Osaka 590-0494, Japan*

[6] *University of Bordeaux, 351 Cours de la Libration, 33405 Talence Cedex, France*

[7] *Centre des Sciences Nucléaire et des Sciences de la Matière, Université Paris-Saclay, CNRS/IN2P3, 91406 Orsay, France*

[8] *Department of Physics, University of York, York, YO10 5DD, United Kingdom*

It is shown that the multi-nucleon transfer reactions is a powerful tool to study fission of exotic neutron-rich actinide nuclei, which cannot be accessed by particle-capture or heavy-ion fusion reactions. In this work, multi-nucleon transfer channels of the reactions of ^{18}O+^{232}Th, ^{18}O+^{238}U and ^{18}O+^{248}Cm are used to study fission for various nuclei from many excited states. Identification of fissioning nuclei and of their excitation energy is performed on an event-by-event basis, through the measurement of outgoing ejectile particle in coincidence with fission fragments. Fission fragment mass distributions are measured for each transfer channel. The experimental data are compared to a calculation based on the fluctuation-dissipation model. Role of multi-chance fission in fission fragment mass distributions is discussed.

Keywords: Multi-Nucleon Transfer Reaction; Fission; Fragment Mass Distribution.

1. Introduction

Nuclear fission is largely influenced by the shell structures appearing in a deformed nuclear shape. Fission-fragment mass distributions (FFMDs) are one of the most fundamental data. Neutron- and charged particle capture reactions have been used to populate low-excited compound nuclei for fission study. Around 2000 GSI developed a Coulex-induced fission of relativistic RIBs in inverse kinematics, where comprehensive fission studies were performed for several tens of nuclei in the neutron-deficient Ac-U region[1]. Recently, β/EC delayed fission was investigated for the very proton-rich nucleus using radioactive beams, and ^{180}Hg was found to show an asymmetric fission[2].

Multi-nucleon transfer (MNT) reaction is another unique approach to populate neutron-rich nuclei, which cannot be accessed by other reactions. Furthermore, excited states of compound nuclei (CN) range widely from around the fission barrier to higher energies, allowing us to measure the excitation energy dependence of FFMDs. Recently, an inverse kinematics technique was used in the MNT channels of the ^{238}U+^{12}C reaction to study fission[3,4]. In these experiments, sufficiently-high A and Z resolution for FFs was achieved due to their kinematic boost, allowing the complete mass- and atomic-number identification for fission fragments.

At the tandem accelerator facility of the Japan Atomic Energy Agency (JAEA), we are studying fission using MNT reactions in the reactions using several target nuclides, e.g. ^{18}O + ^{232}Th[5]. An obvious advantage of this method is a relatively easy possibility to change the projectile and/or the target nuclei. In this report some experimental data are shown.

2. Experimental Methods

An ^{18}O beam was supplied by the JAEA-tandem accelerator at a typical beam intensity of about 0.5 pnA. Beam energies were 157-162 MeV. Targets were prepared by electrodeposition of oxide-target material on a thin Ni backing. Thickness of the target-material layer was 30–148 μg/cm^2.

For the event-by-event identification of the transfer channel (thus, of the fissioning nucleus) and of respective coincident FFs, a detection system consisting of a ΔE - E silicon detector telescope and four multiwire proportional chambers (MWPC) were used, see Fig. 1. Specific transfer channels were identified by detecting projectile-like (ejectile) nuclei in twelve 75 μm-thick trapezoidal ΔE silicon detectors which were mounted in a cone around the beam axis. After passing through the ΔE detector, the ejectiles impinged on the 300μm-thick annular silicon strip detector (E-detector). From the peak of elastic scattering in the sum spectrum E_{tot}=ΔE + E_{res}, the energy resolution was obtained to be 0.9–1.0 MeV (FWHM), which also determines the precision for the excitation energy of the compound nucleus. Figure 2 shows the ΔE - E_{tot} spectrum for ejectiles obtained in the ^{18}O+^{232}Th reaction, where the parabolic lines correspond to different transfer channels, including a clear separation of specific isotopes.

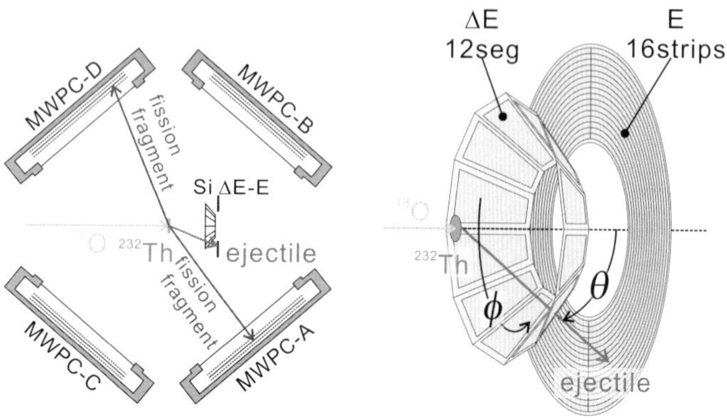

Fig. 1. Schematic detection set-up (left) and expanded view of the silicon ΔE-E detector telescope (right). See text for details.

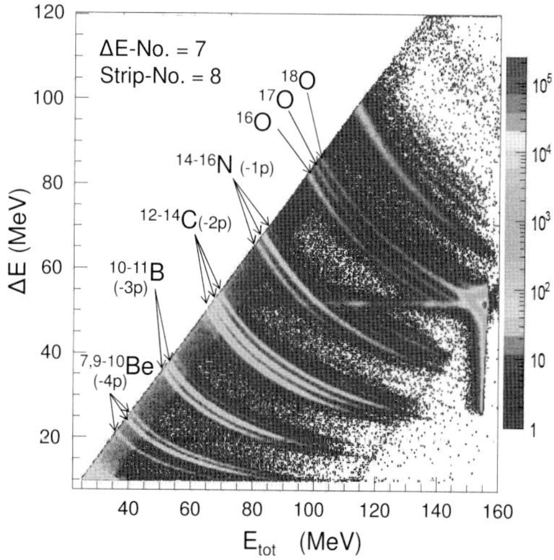

Fig. 2. ΔE-$E_{\rm tot}$ spectrum for ejectiles measured by one pair of the ΔE-E detectors obtained in the ^{18}O+^{232}Th reaction[5]. The curves corresponding to different ejectiles are marked with the respective isotopes. The scattered ^{18}O is also seen in the plot.

The coincident FFs resulting from the fission of excited nuclei after the MNT reaction are detected by four $200\times200\,\mathrm{mm}^2$ position-sensitive MWPCs, see in Fig. 1. Time difference, ΔT, between two coincident MWPCs were measured to determine the masses of both fragments.

3. Results

Figure 3 shows the FFMDs for nuclei of $^{231-234}$Th*, $^{232-236}$Pa* and $^{234-238}$U*, populated in the MNT reactions of ^{18}O + ^{232}Th. The FFMDs of the ^{231,234}Th*, ^{234,235,236}Pa* nuclei were obtained for the first time in this experiment. For the other nuclei, the known FFMD data were systematically extended to excitation energies as high as 60 MeV. It follows from Fig. 3 that mass-asymmetric fission dominates at low excitation energies for all measured nuclei. The yield in the mass-symmetric fission region increases with excitation energy and the double-humped shapes tend to become structureless.

In the results of ^{18}O+^{238}U, we obtained the FFMDs for totally 19 nuclei by expanding the analysis of ejectiles of oxygen (uranium compound nuclei), nitrogen (protoactinium), carbon (plutonium), boron (americium), and beryllium (curium) isotopes[6]. In the ^{18}O+^{238}U reactions, new FFMDs data were obtained for nine nuclei of ^{240}U*, ^{240,241,242}Np*, ^{241,243}Pu*,

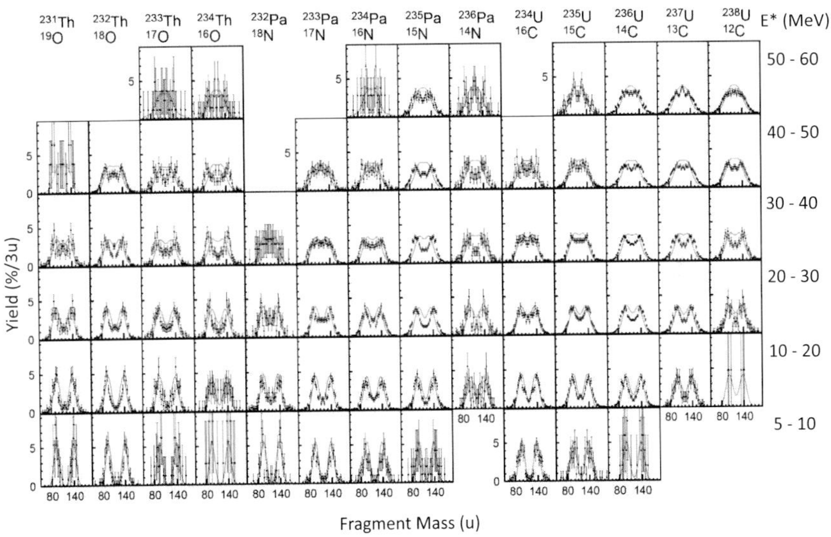

Fig. 3. FFMDs obtained in the multinucleon transfer channels of the reaction ^{18}O+^{232}Th (data points with error bars). The fissioning nucleus and the corresponding ejectile are indicated on the top of the plot. The data are shown for sequential 10 MeV excitation-energy E^* intervals, indicated on the right side. The curves are the results of Langevin calculations (see text)[5] after a convolution with the experimental mass resolution, where the initial CN excitation energy is used to start the calculation, but effects of multi-chance fission (neutron evaporation prior to fission) were not included as discussed in the context of Fig. 5.

^{245,246}Am*, ^{247}Cm*. In the recent measurement of ^{18}O+^{248}Cm, new FFMDs of eleven nuclei are further generated,^{247,249}Cm, ^{249,250,251,252}Bk, ^{251,253}Cf, ^{254,256}Es, and ^{255}Fm[6]. It was found in these experiments that data of low-excited states were not obtained with enough statistics when more than two protons are transferred from the projectile to the target nucleus, due to the small probability to populate low excited states of $E^* < \sim 20$MeV.

The evolution of the center of the light- and heavy-fragment groups ($\bar{A}_{\rm L}$ and $\bar{A}_{\rm H}$) with the mass of the CN in low energy fissions of $10 < E^* < 20$ MeV is shown in Fig. 4, where data were obtained from the three MNT reactions of ^{18}O+^{232}Th, ^{18}O+^{238}U and ^{18}O+^{248}Cm. It is found that the $\bar{A}_{\rm H}$ value are nearly kept constant around 141, whereas center of the light fragment increases linearly with mass of fissioning nucleus.

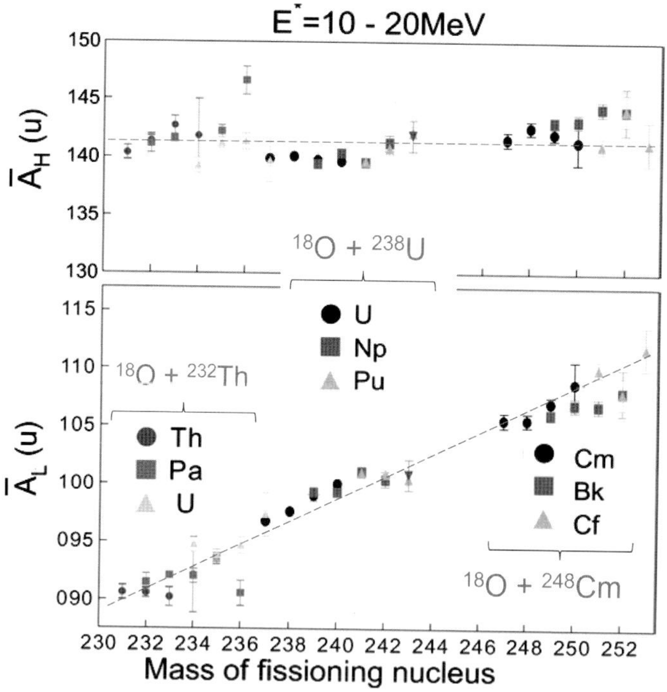

Fig. 4. Center of the light and heavy fragment groups ($\bar{A}_{\rm L}$ and $\bar{A}_{\rm H}$) as a function of mass of the fissioning nuclei in low-excitation fission of $10 < E^* < 20$ MeV. Data are obtained from the reactions of ^{18}O+^{232}Th, ^{238}U and ^{248}Cm using the same setup shown in Fig. 1.

4. Discussions

The measured FFMDs from the MNT reactions are compared with calculations based on the fluctuation-dissipation model[7]. The nuclear shape is defined by three parameters (distance between two potential centers, deformation of fragments, and mass-asymmetry), and the corresponding energy is given by a sum of the liquid-drop energy $V_{\rm LD}$ and the shell correction energy $V_{\rm shell}$. The latter term is represented as $V_{\rm shell}(0)\exp(-E^*/E_{\rm d})$ using the shell correction energy at the zero temperature $V_{\rm shell}(0)$ and shell damping parameter $E_{\rm d}$, where $E_{\rm d}$ =20 MeV was chosen as in[7]. For simplicity, we assumed that the total excitation energy of the system after the MNT reactions is given to the excitation energy (E^*) of the fissioning nucleus.

The calculated results are shown in in Fig. 3, where the original theoretical curves were broadened with the experimental resolution. One can see that the calculated FFMDs reproduce reasonably well both the global shape of the experimental distributions and also the positions of the light and heavy-fragment peaks for most of studied nuclides, at the excitation energies below $\sim$30 MeV. Furthermore, observation that lighter element (thorium) has a pronounced peak-to-valley ratio than the heavier element (uranium) was also reproduced by the calculation. These agreement demonstrates the reasonable treatment of the shell correction energy at these excitation energies, and confirms the validity of the shell-damping energy of $E_{\rm d}$=20 MeV originally introduced in[8].

However, significant deviations of the calculation in Fig. 3 are recognized, especially at higher excitation energies. For example, calculated spectra at E^* =30−40 MeV and higher energies show a tendency to favor the symmetric fission mode for most of the studied nuclei, whereas experimentally several nuclei still exhibit a clear mass asymmetry especially for the most neutron-rich isotopes (for example ^{234}Th*, ^{236}Pa*, ^{238}U*). This can be interpreted as the effects of multi-chance fission that excited compound nucleus emits neutrons prior to fission. With increasing the excitation energy, the contribution from multi-chance fission becomes increasingly important. Thus, the fission observable, in particular FFMDs, become a superposition of several contributions, originating from fission at different excitation energies. These features are demonstrated by Fig. 5, which compares the experimental data for fission of ^{238}U* at the initial excitation energy $E^* = 35$ MeV with the Langevin calculation taking into account multi-chance fission. According to calculations, at this initial energy, the 1st- and 2nd-chance fission occurs with somewhat lower

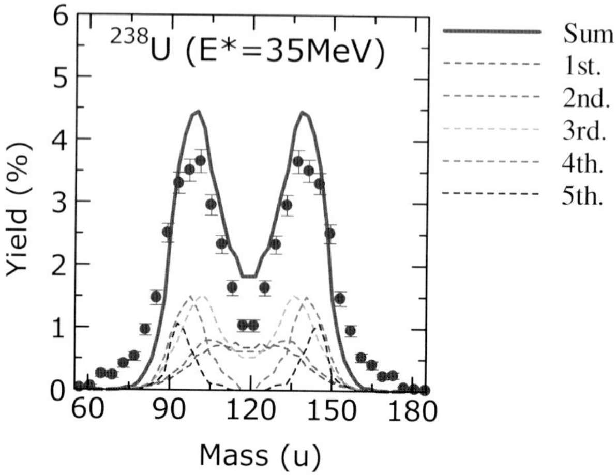

Fig. 5. Experimental FFMD of $^{238}U^*$ (data points with error bars) measured at the initial excitation energy of 35 MeV, obtained from the inelastic scattering channel of the reaction $^{238}U(^{18}O,^{18}O)^{238}U^*$, is compared with the Langevin calculation[7] taking into account multi-chance fission. The solid curve shows the finally calculated FFMD, where the contributions from every multi-chance fission are shown by the dashed curves with different colors.

probabilities (calculated by the GEF code[9]) from respective higher excitation energies, which results in more symmetric-like fission. On the contrary, the higher-chance fissions, after emission of several neutrons (3–5, in this case), occur at lower excitation energies, thus they lead predominantly to an asymmetric mass split. The finally calculated FFMD shown by the solid line is the sum of the FFMDs over the possible multi-chance fissions; it is seen that the measured asymmetric FFMD predominantly comes from the higher-chance fissions. Systematic calculation by extending nuclei and excitation-energies is under way[6].

5. Summary and Outlook

It is shown that the multi-nucleon transfer reactions are the powerful tool to investigate fission for nuclei which cannot be accessed by particle-capture and heavy-ion fusion reactions. In addition to FFMDs, fission barrier height can be determined. Important advantage in the normal kinematics is that the nuclei to be studied can be significantly expanded by using available high-purity radioactive-target. The MNT reactions study using ^{237}Np, ^{234}Am, ^{231}Pa, ^{249}Cf, and ^{254}Es are planned at the JAEA tandem facility.

In addition to investigate the fission-fragment properties, a measurement of prompt neutrons in coincidence with fission fragments has stated to obtain neutron multiplicity from individual fragment $\bar{\nu}(A)$ and its excitation energy dependence.

Acknowledgement

Special thanks are due to the crew of the JAEA tandem facility for their beam operation. Present study is supported by Comprehensive study of delayed-neutron yields for accurate evaluation of kinetics of high-burn up reactors and "Development of prompt-neutron measurement in fission by surrogate reaction method and evaluation of neutron-energy spectra" by the Ministry of Education, Culture, Sports, Science and Technology of Japan (MEXT).

References

1. K.-H. Schmidt *et al.*, *Nucl. Phys. A* **665**, 221 (2000).
2. A.N. Andreyev *et al.*, *Phys. Rev. Lett.* **105**, 252502 (2010).
3. M. Caamaño *et al.*, *Phys. Rev. C* **88**, 024605 (2013).
4. C. Rodríguez-Tajes *et al.*, *Phys. Rev. C* **89**, 024614(2014).
5. R. Léguillon *et al.*, *Phys. Lett. B* **761**, 125 (2016).
6. K. Hirose *et al.*, Preparing for publication.
7. Y. Aritomo and S. Chiba, *Phys. Rev. C* **88**, 044614 (2013).
8. A.N. Ignatyuk *et al.*, *Sov. J. Nucl. Phys.* **21**, 255 (1975).
9. K.-H. Schmidt, B. Jurado, C. Amouroux and C. Schmitt, *Nuclear Data Sheets* **131**, 107 (2016).

Low-Energy Fission and Beta-Delayed Fission with Radioactive Beams*

A.N. Andreyev

Department of Physics, University of York, York, UK
Advanced Science Research Center, Japan Atomic Energy Agency, Tokai, Japan

In the last decade, through technological, experimental and theoretical advances, the situation in experimental low-energy fission studies has changed dramatically. With the use of advanced production and detection techniques, much more detailed fission information can be obtained for traditional regions of fission research and, very importantly, new regions of nuclei have become accessible for fission studies.

The talk will give a review of recent low-energy fission experiments in very proton-rich nuclei in the lead region, with an emphasis on the exotic phenomenon of beta-delayed fission (βDF). This decay mode, which ideally suits to study low-energy fission, is experimentally accessible only in the well-studied uranium region and in the scarcely-studied (by fission) neutron-deficient lead region. Our pioneering βDF study of ^{180}Tl at ISOLDE(CERN) showed an unexpected asymmetric fission-fragment mass split for ^{180}Hg [1], despite a symmetric split in two semi-magic ^{90}Zr nuclei was expected. This result also identified a new region of asymmetric fission in addition to the well-known asymmetry in trans-uranium nuclei.

More recently, new βDF data on the neutron-deficient nuclides ^{194,196}At and ^{200,202}Fr were obtained at ISOLDE. A mixture of asymmetric and

---

*On behalf of the IS466/IS534 collaboration: RILIS-ISOLDE(CERN) — University of York (UK) — IKS, KU Leuven (Belgium) — Comenius University, Bratislava (Slovakia), JAEA, Tokai (Japan) — SCK•CEN, Mol (Belgium).

symmetric fission is clearly seen in respective daughter isotopes (after precursor's beta decay) $^{194,196}Po$ and ^{202}Rn [2]. This establishes a phenomenon of multimodal fission in this region of nuclei. The results will be compared with predictions from several current fission models.

As a conclusion, further perspective of low-energy fission studies with radioactive beams will be presented.

References

1. A. N. Andreyev *et al.*, "New type of asymmetric fission in proton-rich nuclei", *Phys. Rev. Lett.* **105**, 252502 (2010).
2. L. Ghys *et al.*, "Evolution of fission-fragment mass distributions in the neutron-deficient lead region", *Phys. Rev. C* **90**, 041301(R) (2014).

Experimental Measures of Fission Time Scales

B. B. Back

Argonne National Laboratory,
Argonne, IL 60439, USA
E-mail: back@anl.gov

The nuclear fission process involves a drastic rearrangement of the nuclear matter of the fissioning system as it undergoes shape changes from a single, compact nucleus to two separated fission fragments. Understanding the dynamics of this process has been the focus of much research for decades, both theoretical and experimental, using a number of different approaches. Early on, Kramers [1] proposed that the fission rate is not well described by the statistical counting of transition states as proposed by Bohr and Wheeler [2] but that the friction or viscosity of the nuclear matter can play a substantial role in slowing down the rate by large factors. Aside from the theoretical estimates of fission time scales, a number of different approaches have been used to obtain experimental information on this quantity. Although the divergence of results obtained by the different experimental approaches have been known for years, these discrepancies have not yet been resolved and they have recently received renewed attention [3]. In this paper I will recount some of the experimental approaches to estimate the fission time scales and discuss their respective strengths and weaknesses without coming close to provide a resolution of the discrepancies.

Keywords: Fission, Quasi-Fission, Time Scales, Nuclear Dissipation.

1. Introduction

Soon after the discovery of the production of chemical element Ba in neutron bombardment of U by Hahn and Strassman [4] in late 1938 it was realized by Meitner and Frisch [5] and proven by physical measurements performed by Frisch [6] that these products are produced by fission of the U nucleus. A detailed and inside account of this development is given by Frisch [7]. In June of 1939 an in-depth theoretical explanation of this phenomenon was given by Bohr and Wheeler [2], who estimated the fission rate in terms of the number of transition states at the fission barrier in a purely statistical manner. A somewhat different approach to the theory of fission was put forth by Kramers the following year [1], in which he pointed out that the dynamics of the fission process may come into play for a certain range of the strength of nuclear viscosity. However, this possibility was largely ignored until detailed studies of the competition between fission and neutron emission showed that the fission branch appeared to the suppressed in hot nuclei populated in heavy-ion reactions [8, 9]. With the appearance of dynamical effects in the fission process, the fission time scale became of interest and the development of experimental methods for determination of the fission time scale became a central focus of fission research. The dynamical time scale becomes particularly evident in cases where the transition state approach would predict very short fission times because a large number of transition states are available. This occurs for example at high excitation energy, or in purely dynamic reactions, such as quasi-fission, which may not encounter a 'barrier' in the evolution of the nuclear shapes along the path to scission. The time-scales for this type of reactions are therefore more sensitive to the dynamics of the matter re-arrangements taking place during the process. Recently, it has, however, become evident that there are substantial discrepancies between some of the methods for obtaining time scales for this type of reactions [3]. Experimental measurements of the time scales for fission, *i.e.* the time between the initial capture and formation of a di-nuclear system until the final fragments separate at scission all rely on different, but well understood processes that can take place during this evolution. These include, di-nuclear rotation of the system, emission of neutrons, Giant Dipole Resonance γ-rays, the distance traveled by the di-nuclear system and the emission of K X-rays. A discussion of these methods and the results will be addressed in this paper along with some of the discrepancies that have been observed between them.

2. Di-nuclear rotation

In some quasifission reactions one observes a pronounced correlation between the mass and angle of emission of the fragments, which is inconsistent with compound nucleus fission. This behavior indicates that the di-nuclear system rotates only by a finite angle before the system breaks apart at scission as demonstrated and discussed in the following publications [10-17] and illustrated in Fig. 1. The central point in determining the dynamical time scale in these reactions is to relate the angular velocity, ω, of the system

$$\omega = L / J \,, \tag{1}$$

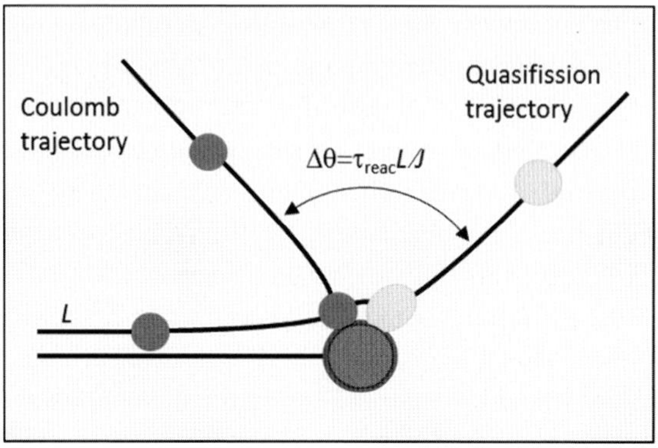

Fig. 1. Illustration of the difference in Coulomb and quasifission trajectories and the rotation angle of the di-nuclear system during contact.

where, L is the orbital angular momentum and J is the moment of inertia of the combined system, to the observed rotation angle of the system during the reaction time, τ_{reac}, which is then given by

$$\Delta\theta = \omega\tau_{reac} = \tau_{reac}L/J, \text{ or } \tau_{reac} = \Delta\theta J/L. \quad (2)$$

Here, the average angular momentum, L, associated with a given final mass range of the fragments may be obtained from the measured total capture cross section.

Using this method, Tōke *et al.* [11] find reaction times in the $(2.5\text{-}7.5)\times 10^{-21}$s range for quasifission in the reaction of 6 MeV/u ^{64}Ni+^{238}U. A further analysis [12] shows that the time scale for the observed mass transfer in this case is in good agreement with those expected on the basis of the one-body dissipation mechanism given in terms of the wall-and-window formula [18], which gives a characteristic mass transfer time of $\tau = 4.8\times 10^{-21}$s compared to $\tau = (5.3\pm 1)\times 10^{-21}$s resulting from the analysis of U-induced quasifission reactions [11]. A recent analysis and simulation of the quasifission process of 6.01 MeV/u ^{40}Ca+^{238}U give similar time scales [16]. These observations also agree well with time-dependent Hartree-Fock calculations that are presented in the same paper.

3. Pre-scission neutron emission

Reaction time scales can also be derived from a comparison of the branching ratio of different decay modes *e.g.* neutron emission and (quasi) fission. In this approach it is assumed that the decay rate of one of the branches, *i.e.* neutron emission, is well described by theory. In this case, the measurement is somewhat complicated by the fact that neutrons are also emitted by the fragments after scission, but since this source of neutrons is moving a high velocities, it is possible to separate the pre- and post-scission neutrons on the basis of kinematics. The number of pre-scission neutrons observed in the reaction thus relates to the time-delay or reduced rate of the fission process. Studies of this nature have been carried out by Hinde *et al.* [19-22] for a range of different reactions including the $^{64}Ni+^{238}U$ system [21], which was studied at 6.5 MeV/u. Here a fission delay time of $\tau_d = 9 - 19 \times 10^{-21}$s was found for fragments in the 120<A<220 u mass range as shown in Fig. 2. One notices that a larger value was found for the A=100u mass bin, which because of symmetry consideration should have the same value as the A=200u mass bin.

The time scale derived in this manner is slightly longer, but consistent with the rotation angle results mentioned above. As mentioned in [21], these measurements may be slightly skewed toward longer times because of the neutron emission from the fragments during the post-scission acceleration, which may be counted as pre-scission neutrons as they appear to originate from a source moving at close to the compound nucleus velocity.

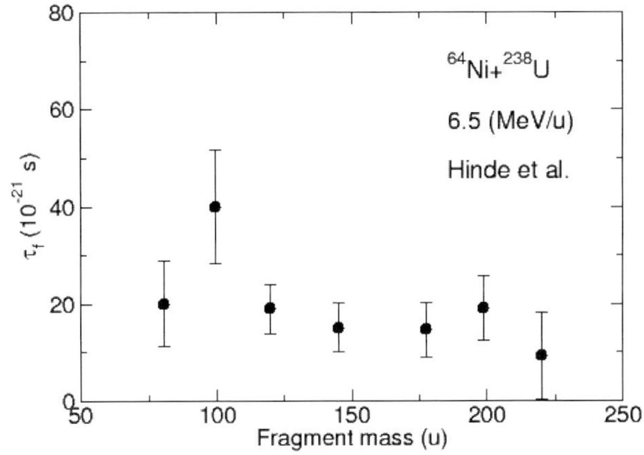

Fig. 2. Fission times are shown as a function of fragment mass. The data were obtained from measurements of the multiplicity of pre-scission neutrons by Hinde *et al.* [21].

4. GDR gamma-ray emission

A third method, based on the emission of γ-rays from the Giant Dipole Resonance (GDR) in competition with fission, has been pursued in a range of different systems [23-25]. This method relies on statistical model calculations of the γ-ray spectra from the accelerated fission fragments and the possible observation of compound nucleus γ-emission prior to scission from the GDR expected be centered in the $E_\gamma \sim 80/A^{1/3}$ MeV region. In a study of the $^{58}Ni+^{165}Ho$ system at 6.35 MeV/u Nestler *et al.* [25] found no indication of a pre-scission GDR component in the γ-ray spectrum measured in coincidence with quasi-fission products, which allowed the authors to set an upper limit of $\tau_{reac} \leq 11 \times 10^{-21}$s. This time scale is good agreement with those derived the $^{64}Ni+^{238}U$ system by the methods discussed above.

5. Crystal blocking

In the crystal blocking technique [26] one takes advantage of the fact that the prompt emission of fission fragments from lattice sites are substantially suppressed, whereas a delay in the fission process allows the system to travel to locations in between these where the fission fragments can freely emerge along 'channels' in the crystal. The sensitivity of this method is typically in the range of 10^{-18} to 10^{-14} s depending chiefly on the recoil velocity of the fissioning system and the temperature of the crystal. In a recent experiment, Morjean *et al.* studied the reaction of a ^{238}U beam impinging onto a Ni crystal at 6.62 MeV/u [26] and found that the observed yield in the direction of the $\langle 110 \rangle$ crystal axis, as depicted in Fig. 3, was consistent with a fission time scale of $\sim\tau_{reac} = 1 \times 10^{-18}$ s, a result that appears to be at variance with those obtained from the techniques discussed above.

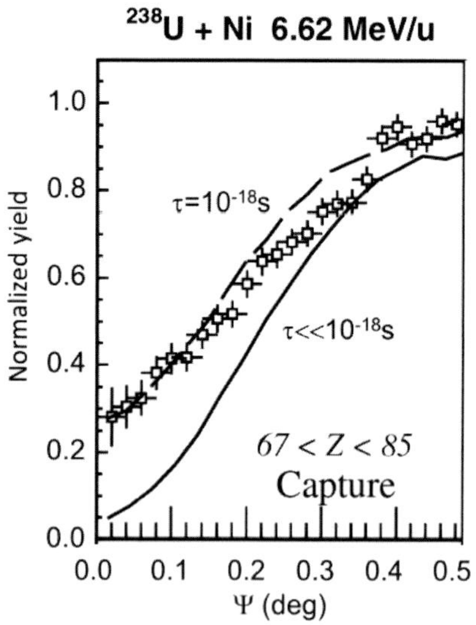

Fig. 3. Blocking dip for fission fragments in the range 67<Z<85 relative to the crystal $\langle 110 \rangle$ axis (open squares) is compared with calculations for fission life times of 10^{-18} s (dashed curve) and $\ll 10^{-18}$ s (solid curve). Adapted from Ref. [26].

6. K X-ray emission

Inner shell vacancies created during the fusion process and their subsequent filling and the associated emission of characteristic K X-rays, a process that occurs on a short time scale, offers another method of assessing the life-time of the combined system. An advantage of this method is that the K X-ray spectrum and the life-time of the K-shell vacancies are reliably calculable using standard methods. This approach has been taken in a recent study of the $^{238}U+^{64}Ni$ reaction at 6.6 MeV/u by Frégeau *et al.* [27]. In coincidence with quasi-fission fragments in the range $50 < Z < 91$, these authors observe a peak in the X-ray spectrum that is interpreted as K X-ray emission from the combined system during its lifetime. From the analysis of the experimental evidence, these authors arrive at the conclusion that composite Z = 120 system has a lifetime of $\tau_{Z=120} \geq 2.5 \times 10^{-18}$ s. Again, this result is in strong disagreement with the lifetimes derived from the mass-angle distribution of fission fragments and from neutron and GDR γ-ray emission discussed in Sect. 2-4.

7. Comparison and discussion

Four different measurements of the lifetime of the composite system formed in the $^{238}U+Ni$ collision system have been reported in the literature [11, 21, 26, 27]. Although the bombarding energy differ by up to 10% between the experiments and in one case non-separated Ni target material was used, it is argued that the experimental conditions are sufficiently similar that a direct comparison of the results has validity. The four experiments and the resulting lifetimes are listed in Table.1.

One observes that there is reasonable agreement between the methods using the di-nuclear rotation and the emission of pre-scission neutrons to estimate the lifetimes of the fissioning system. The remaining discrepancy of a factor of 2-3 may very well be associated with the different methods of analysis of the data or the fact that the di-nuclear rotation experiment was performed at a 10% lower beam energy. Although this discrepancy may be reduced in a consistent re-analysis of the two data sets, it is not considered essential in this context.

Likewise, one finds that the results from the experiments using crystal blocking and K X-ray emission are in reasonable agreement on a time scale of the order $\tau = 1 - 2.5 \times 10^{-18}$ s, almost three order of magnitudes longer than what was found by the first two methods. In face of such a large discrepancy it is of course relevant to inquire about its origin, whether it's faulty or misinterpreted data. In this connection it is probably worth to point out that there is a relatively large body of work based on the methods of di-nuclear rotation and pre-scission emission of either neutrons or GDR γ-rays [10-17, 19-25] whereas the method of crystal blocking and K X-ray emission are much less studied [26, 27] for this type of reaction. Only brief reports on the latter two methods for this heavy system are available in the literature with no follow-up full papers that could go into further detail about the experiment and data analysis have been published. This situation makes it difficult to assess the full validity of the results. Ideally, new experiments using both methods would be very beneficial for reaching a satisfactory understanding of these puzzling results.

Table 1. Comparison of the characteristics and the results obtained in four different experimental measurements of the fission lifetime.

System	E_{beam} (MeV/u)	Method	Fragment range	Lifetime (10^{-21} s)	Reference
$^{238}U+^{64}Ni$	6.0	Di-nuclear rotation	180<A<222	2.5-7.5	Tōke *et al.* [11]
$^{238}U+^{64}Ni$	6.5	Neutron emission	120<A<220	9-19	Hinde *et al.* [21]
$^{238}U+^{nat}Ni$	6.62	Crystal blocking	67<Z<85	~1000	Morjean *et al.* [26]
$^{238}U+^{64}Ni$	6.6	K X-ray emission	35<Z<90	>2500	Frégeau *et al.* [27]

8. Conclusions

On the basis of the analysis and discussion presented above, it is of course tempting to dismiss the two outlier results, namely those obtained from crystal channeling and K X-ray emission. Although this author may have personal opinions about the validity of these results, this is based only on the fact that they are at variance with those of previous measurements and various accepted theoretical models of the fission dynamics. In this paper, only the fact that there is a substantial discrepancy between the various results has been pointed out, but no resolution has been offered. This is clearly an unsatisfactory situation scientifically, and additional effort should be devoted to resolve the discrepancy. Additional experiments using the two methods that have produced outlier results would probably help to clarify the situation, especially if detailed accounts of the experimental conditions and data analysis were made available in the literature.

Acknowledgments

This work was supported by the US Department of Energy, Office of Nuclear Physics, under contract No. DE-AC02-06CH11357.

References

1. H. A. Kramers, *Physica* **7**, 284 (1940).
2. N. Bohr and J. A. Wheeler, *Phys. Rev.* **56**, 426 (1939).
3. A. K. Sikdar, A. Ray and A. Chatterjee, *Phys. Rev. C* **93**, 041604 (2016).
4. O. Hahn and F. Strassmann, *Nature* **27**, 11 (1939).
5. L. Meitner and O. R. Frisch, *Nature* **143**, 239 (1939).
6. O. R. Frisch, *Nature* **143**, 276 (1939).
7. O. R. Frisch, *What little I remember*, Cambridge University Press, Cambridge (1979).
8. A. Gavron *et al.*, *Phys. Rev. Lett.* **47**, 1255 (1981).
9. P. Grangé *et al.*, *Phys. Rev. C* **34**, 209 (1986).
10. R. Bock *et al.*, *Nucl. Phys. A* **388**, 334 (1982).
11. J. Tõke *et al.*, *Nucl. Phys. A* **440**, 327 (1985).
12. W. Q. Shen *et al.*, *Phys. Rev. C* **36**, 115 (1987).
13. B. B. Back *et al.*, *Phys. Rev. C* **53**, 1734 (1995).
14. C. Simenel *et al.*, *Phys. Lett. B* **710**, 607 (2012).
15. R. du Rietz *et al.*, *Phys. Rev. C* **88**, 054618 (2013).
16. A. Wakhle *et al.*, *Phys. Rev. Lett.* **113**, 182502 (2014).
17. E. Prasad *et al.*, *Phys. Rev. C* **93**, 024607 (2016).
18. J. Randrup and W. J. Swiatecki, *Nucl. Phys. A* **429**, 105 (1984).
19. D. J. Hinde *et al.*, *Nucl. Phys A* **452**, 550 (1986).
20. D. J. Hinde *et al.*, *Nucl. Phys. A* **472**, 318 (1987).
21. D. J. Hinde *et al.*, *Phys. Rev. C* **45**, 1229 (1992).
22. J. Cabrera *et al.*, *Phys. Rev. C* **68**, 034613 (2003).
23. R. Butsch *et al.*, *Phys. Rev. C* **44**, 1515 (1991).
24. D. J. Hofman *et al.*, *Phys. Rev. Lett.* **72**, 470 (1994).
25. J. Nestler *et al.*, *Phys. Rev. C* **51**, 2218 (1995).
26. M. Morjean *et al.*, *Phys. Rev. Lett.* **101**, 072701 (2008).
27. M. O. Fregeau *et al.*, *Phys. Rev. Lett.* **108**, 122701 (2012).

Posters

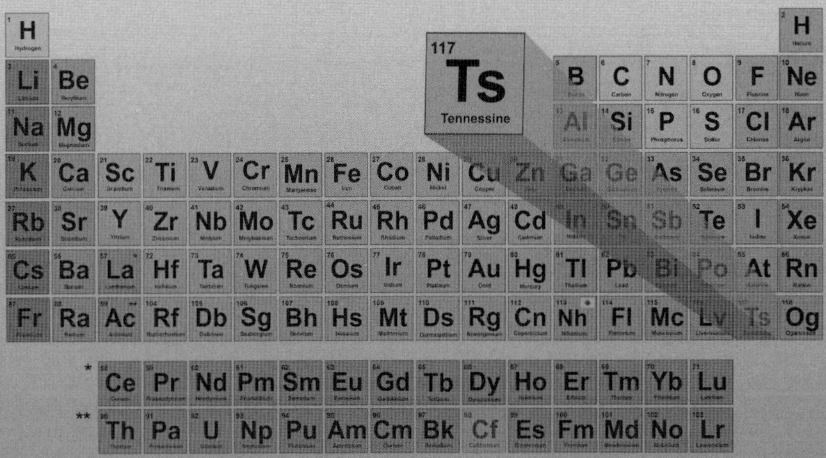
117
Ts
Tennessine
H He
Li Be B C N O F Ne
Na Mg Al Si P S Cl Ar
K Ca Sc Ti V Cr Mn Fe Co Ni Cu Zn Ga Ge As Se Br Kr
Rb Sr Y Zr Nb Mo Tc Ru Rh Pd Ag Cd In Sn Sb Te I Xe
Cs Ba La Hf Ta W Re Os Ir Pt Au Hg Tl Pb Bi Po At Rn
Fr Ra Ac Rf Db Sg Bh Hs Mt Ds Rg Cn Nh Fl Mc Lv Ts Og
* Ce Pr Nd Pm Sm Eu Gd Tb Dy Ho Er Tm Yb Lu
** Th Pa U Np Pu Am Cm Bk Cf Es Fm Md No Lr

Multipole Mixing Ratios of Gamma Band to Ground Band Transitions in ^{102,104,106,108}Mo, ^{108,110,112}Ru, and ^{112,114,116}Pd

J. M. Eldridge*, B. Fenker, C. Goodin, C. J. Zachary, J. H. Hamilton, E. Wang and A. V. Ramayya

Department of Physics and Astronomy, Vanderbilt University
Nashville, TN 37240, United States of America
**E-mail: jonathan.m.eldridge@vanderbilt.edu*

B. Fenker

Physics Department, Texas A&M
College Station, TX 77843, United States of America

A. V. Daniel, G. M. Ter-Akopian and Yu. Ts. Oganesson

Flerov Laboratory of Nuclear Reactions, JINR, Dubna, Rusia

Multipole mixing ratios have been measured, using ^{252}Cf spontaneous-fission, γ-ray coincidence data, for transitions from states in the γ-vibrational-bands to states in the ground state bands of deformed, neutron-rich isotopes, ^{102,104,106,108}Mo, ^{108,110,112}Ru, ^{112,114,116}Pd. These mixing ratios have been found to be pure, or nearly pure, E2, in agreement with theory.

1. Introduction

The multipole mixing ratio of γ-rays emitted from excited nuclei can give much information about the spin and parity of nuclear energy levels. When a nucleus decays from an excited state, the angular momentum selection rule for the emitted γ-ray is $\Delta J = |J_i - J_f| \leq L$. Additionally, parity must be conserved in any electromagnetic nuclear decay producing additional selection rules. Electric radiation of multipolarity L has parity $\pi_e = (-1)^L$ while magnetic radiation has parity $\pi_m = (-1)^{L+1}$. Using these selection rules, a γ-decay from a 2^+ state into a 1^+ state permits $M1$, $E2$, or $M3$ radiation but only the $M1$ and $E2$ components are ever observed experimentally. Furthermore, electric multipole radiation is favored over its magnetic counterpart in deformed nuclei leading to ML, $E(L+1)$ mixtures but leaving EL, $M(L+1)$ mixtures extremely rare.

Because the γ vibrational band represents, in deformed nuclei, vibrations of a quadrupole state ($I^{\pi} = 2^{+}$), and subsequent rotations of that vibrational state, electromagnetic radiation emitted when a nucleus decays from the γ-band to the ground-state rotational band is expected to be pure quadrupole. [1] Thus, only an $E2$ component should be observed in $\Delta I = 0, 1$, γ-band to ground-band decays which would normally permit a $M1 + E2$ mixing. This work seeks to confirm this prediction in ^{102,104,106,108}Mo, ^{108,110,112}Ru, and ^{112,114,116}Pd.

2. Experiment and Methods

The measurement of these multipole mixing ratios was done by the angular correlation method detailed in [2, 3]. Our data are coincidence measurements of a ^{252}Cf spontaneous fission source. A fit of the relative intensity of two gammas with respect to the cosine of the angle between them, yields two parameters, A_2 and A_4 from which one can determine the spins and multipole mixing ratio, δ, for one of the two gammas, provided δ is known for the other. Thus, we paired the γ-band to ground-band transitions with a $\Delta I = 2$ transition within the ground-band, which are assumed to be pure E2 in this work. The program "Delta" provided by NNDC [4] was used to aid in these calculations.

3. Results

Table 1 shows all the mixing ratios found in the present work. Results marked with a "^" indicate that the work is adopted from [5] or [3]. A mixing ratio in parentheses indicates that it falls outside 1.5σ for A_2^{exp}, A_4^{exp}. In some cases there were two solutions for the mixing ratio. In these cases the value listed first is slightly closer to A_2^{exp}, A_4^{exp}, and thus is mathematically preferred. Occasionally, both (or the only) values are in parentheses, this means that no solution fell within he uncertainty of A_2^{exp}, A_4^{exp}, the best solution(s) are still listed. "^{108}Mo*" was measured using the $4_{\gamma}^{+} \to 2_{\gamma}^{+} \to 2^{+}$ cascade, rather than the $2_{\gamma}^{+} \to 2^{+} \to 0^{+}$ cascade.

4. Conclusion

The multipole mixing ratios of γ-band to ground-band transitions in ^{102,104,106,108}Mo, ^{108,110,112}Ru, and ^{112,114,116}Pd have been measured by an angular correlation method from ^{252}Cf spontaneous fission data. In most cases these gammas have been found to be pure E2, in agreement with theory. Those transitions which showed possible E2/M1 mixing or significant deviation of δ from A_2^{exp}, A_4^{exp} are the subjects of ongoing research.

Table 1: A Table of δ Values Found in This Work

Isotope	A_2^{exp}, A_4^{exp}	$\delta(M1/E2)$
	$2_\gamma^+ \to 2^+$	
^{102}Mo	-0.17(1),0.30(2)	7^{+2}_{-1}
^{104}Mo	-0.16(2),0.4(1)	8^{+3}_{-2}
^{106}Mo	-0.16(2),0.21(7)	$(8^{+3}_{-2}), (0.59(4))$
^{108}Mo	-0.11(4),0.10(8)	$0.51(7), (19^{+\infty}_{-12})$
^{108}Mo*	0.00(4),-0.01(7)	$-0.34^{+0.17}_{-0.20}, 11^{+\infty}_{-7}$
^{108}Ru^	-0.108(23),0.32(3)	23^{+68}_{-10}
^{110}Ru^	-0.09(2),0.33(3)	$90^{+\infty}_{-50}$
^{112}Ru^	-0.05(2),0.31(3)	-30^{+10}_{-50}
^{112}Pd	0.020(6),0.311(9)	$-7.9^{+0.3}_{-0.4}$
^{114}Pd	-0.046(7),0.30(1)	-25^{+3}_{-2}
^{116}Pd	-0.06(1),0.33(2)	-70^{+40}_{-280}
	$3_\gamma^+ \to 2^+$	
^{102}Mo	-0.29(2),-0.05(2)	$-9^{+1}_{-2}, (-0.28(2))$
^{104}Mo	-0.19(2),-0.12(4)	50^{+460}_{-20}
^{106}Mo	-0.08(1),-0.08(3)	$6.1^{+0.7}_{-0.6}$
^{108}Mo^	-0.12(7),0.1(1)	$-0.06(8), (10^{+273}_{-5})$
^{108}Ru	-0.22(3),-0.08(6)	$-40^{+30}_{-\infty}$
^{110}Ru^	-0.23(1),-0.10(3)	-30^{+10}_{-30}
^{112}Ru^	-0.16(2),-0.11(4)	20^{+20}_{-10}
^{112}Pd	-0.37(1),-0.05(2)	$(-4.3^{+0.1}_{-0.3}), (-0.46^{+0.05}_{-0.02})$
^{114}Pd	-0.372(7),-0.08(1)	$-4.2(2)$
^{116}Pd	-0.25(1),-0.09(3)	-15^{+3}_{-5}
	$4_\gamma^+ \to 4^+$	
^{102}Mo	-0.18(2),0.13(3)	2^{+3}_{-1}
^{104}Mo	-0.15(1),0.14(2)	9^{+7}_{-3}
^{106}Mo	-0.17(1),0.13(1)	$1.8^{+0.5}_{-0.3}, 4(1)$
^{108}Mo	-0.18(2),0.00(4)	$(2.6^{+0.9}_{-0.6})$
^{108}Ru	-0.20(2),0.08(4)	$2.5^{+0.6}_{-0.4}$
^{110}Ru	-0.17(2),0.11(4)	4^{+5}_{-3}
^{112}Pd	-0.08(3),0.13(5)	-9^{+3}_{-10}
^{114}Pd	-0.15(6),0.1(1)	$8^{+\infty}_{-7}, 1.3^{+2.6}_{-0.4}$

Table 1: δ Values Found in This Work (Continued)

Isotope	A_2^{exp}, A_4^{exp}	$\delta(M1/E2)$
	$5_\gamma^+ \to 4^+$	
^{104}Mo^	-0.10(1),-0.07(2)	30^{+20}_{-10}
^{106}Mo^	0.03(1),-0.05(2)	4.1(3)
^{110}Ru	-0.09(3),0.07(5)	$-0.03(4)$,(20^{+130}_{-10})
^{112}Pd	-0.17(5),0.06(8)	$-18^{+8}_{-81}, -0.15(6)$
^{114}Pd	-0.16(7),-0.1(1)	$-20^{+10}_{-\infty}$
	$6_\gamma^+ \to 6^+$	
^{106}Mo	-0.107(28),0.02(4)	$(1.4^{+0.4}_{-0.3})$,(-20^{+8}_{-26})
	$7_\gamma^+ \to 6^+$	
^{104}Mo^	0.08(3),-0.05(4)	$3.4^{+0.7}_{-0.5}$
^{106}Mo	0.08(3),-0.08(5)	$3.2^{+0.8}_{-0.6}$
	$9_\gamma^+ \to 8^+$	
^{104}Mo	0.10(6),-0.15(9)	$3(1), (0.31^{+0.10}_{-0.09})$

Acknowledgements

The work at Vanderbilt University and Lawrence Berkley National Laboratory are supported by the US Department of Energy under Grant No. DE-FG0588ER40407 and Contract No. DE-AC03-76SF00098. The Work at Tsinghua University was supported by the National Natural Science Foundation of China under Grant No. 11175095.

References

[1] A. Bohr and B. R. Mottelson, 1975, *Nuclear Structure* (Benjamin, New York), Vol. II.

[2] A. V. Daniel *et al.*, Nuclear Instruments and Methods in Physics Research Section B **262**, 399 (2007).

[3] C. Goodin, *Angular Correlations of Prompt Gamma-Rays from the Spontaneous Fission of ^{252}Cf*, Ph.D. Dissertation, 2010, accessible online at `http://etd.library.vanderbilt.edu/available/etd-07232008-152352/`.

[4] P. Ekstrom, DELTA, `http://www.nndc.bnl.gov/nndcscr/ensdf_pgm/analysis/delta`.

[5] B. Fenker, *Multipole Mixing Ratios of Transitions out of High Spin Gamma Vibrational States in Neutron Rich Mo, Ru Isotopes*, Senior Honors Thesis, 2010, accessible online at `http://hdl.handle.net/1803/4106`.

Deriving Prompt Neutron Emission from Fission Product Yields

P. Jaffke

Los Alamos National Laboratory
Los Alamos, NM 87545, USA
pjaffke@lanl.gov

A natural connection exists between the fission product yields and the prompt neutron emission. The average neutron emission for a fission of a particular nucleus $\bar{\nu}$ is typically known to a very high level of certainty. In addition, the average neutron emission as a function of the fragment mass $\bar{\nu}(A)$ also provides a connection between not only average quantities, such as $\bar{\nu}$, but also the prompt neutron emission distribution. In this way, one can utilize established prompt neutron data as a constraint on fission product yields.

1. Introduction

Both the fission product yields and the prompt neutron emission from fission have been widely studied in numerous fissioning systems. Typically, these measurements are conducted separately as they demand different detector qualities. Fission product yields (FPY) and the prompt neutron emission (PNE) are critical for reactor safety and nuclear physics. As the applications of the FPY and PNE are far-reaching, it is imperative that the experimental data or theoretical models are self-consistent. We present new work to evaluate the consistency between data sets of FPY and PNE.

2. Consistency Checks

We utilize two constraints, charge and mass conservation, on the FPY in order to relate it to the PNE. Charge conservation can be realized using the charge distributions of the FPY for a specific fragment mass $Y_A(Z)$. This distribution is known to be Gaussian in shape[1] and normalizing each for a given fragment mass yields the fractional independent yield $\mathcal{I}_A(Z)$:

$$\mathcal{I}_A(Z) = \frac{N \times F(A,Z)}{\sqrt{2\pi\sigma^2}} \exp[-(Z - Z_p)^2/(2\sigma^2)], \tag{1}$$

where Z_p and σ^2 are the mean and variance in the Gaussian distribution and $F(A, Z)$ represents a correction for even-odd shell effects, thus requiring a renormalization constant N. We take FPY data from three major data sets: ENDFB-VII.1[2], JEFF-3.1.1[3], and JENDL-4.0[4]. We fit each charge distribution for a given A with Eq. 1 and plot the best-fit value of Z_p in Fig. 1. One can see that the function $Z_p(A)$ is semi-linear in nature and that the three databases display very similar functions within error bars. We also find that our derived Z_p values follow the Wahl systematics[1]. In addition, the inset of Fig. 1 shows the result of summing Z_p values for two fragments exactly complementary in mass, that is, $Z_p(A_L) + Z_p(A_H)$ with $A_L + A_H = A_C$. We note that this summation clearly neglects prompt neutron emission and the fact that $Z_p(A_L) + Z_p(A_H) \neq Z_C$ illustrates the need for prompt neutron emission. Instead, one can demand charge conservation in fission and require $Z_p(A) + Z_p(A_c) = Z_C$ which determines a particular mass pairing between A and some complement A_c. Then, the difference $A_C - A_c$ is the average total neutrons emitted by A and its complement A_c, $\bar{\nu}_T(A) = \bar{\nu}(A) + \bar{\nu}(A_c)$.

Using the PNE data, specifically $\bar{\nu}(A')$ with A' the pre-neutron mass, one can formulate the same $\bar{\nu}_T(A)$ parameter by iteratively solving

$$\bar{\nu}(A') + A' = A_C - \bar{\nu}(A'_c) - A'_c, \tag{2}$$

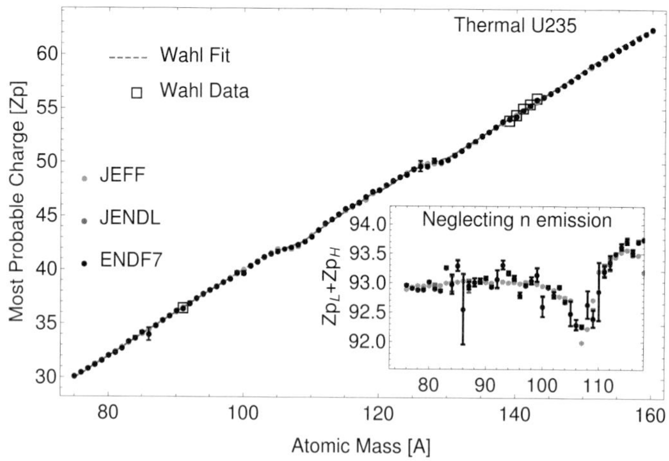

Fig. 1. Best-fit values of mean charge Z_p for the Gaussian distributions (see text) as a function of fragment mass. Results for three major fission databases are provided along with Wahl's systematics and a few data points from Wahl. The inset plot shows the summed Z_p for fragments complementary in mass should neutron emission be neglected.

which amounts to a description of mass conservation. Thus, with Eq. 2 an A-A_c pair is established and the resulting $\bar{\nu}_T(A)$ is plotted against the derived values from the FPY data in Fig. 2. The results of Fig. 2 illustrate the connection between the FPY and PNE. In addition, several trends in $\bar{\nu}(A')$ can be derived from the $\bar{\nu}_T(A)$ curve above: the relative slopes of the sawtooth rises in $\bar{\nu}(A')$ determine the slope of the $\bar{\nu}_T(A)$ wings and the peaks of each curve are related. One can also calculate the average $\bar{\nu}$ from the $\bar{\nu}_T(A)$ curve by weighting over one of the fragment peaks. We present the resulting values of $\bar{\nu}$ in Table 1.

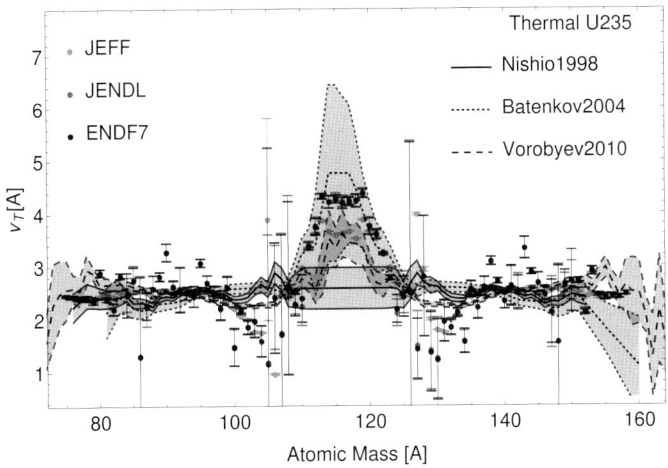

Fig. 2. Calculated $\bar{\nu}_T(A)$ using the charge conservation of the FPY and the method described above, with three major databases, compared with the derived values from experimental data for $\bar{\nu}(A)$[5–7]. Shaded regions denote error bands.

Table 1. Prompt neutron emission $\bar{\nu}$ as derived from FPY compared with experimental data.

	^{235}U	^{239}Pu	^{241}Pu	^{238}U	^{252}Cf
JEFF-3.1.1	2.42 ± 0.06	2.82 ± 0.07	3.13 ± 0.08	4.63 ± 0.11	4.61 ± 0.12
JENDL-4.0	2.43 ± 0.09	2.77 ± 0.12	2.96 ± 0.16	4.27 ± 0.21	3.67 ± 0.19
ENDF7	2.43 ± 0.10	2.78 ± 0.12	2.96 ± 0.15	4.30 ± 0.20	3.68 ± 0.18
Exp.[8]	2.4251 ± 0.0034	2.879 ± 0.006	2.924 ± 0.007	4.24 ± 0.69[9]	3.7675 ± 0.004

One can see from Table 1 that the derived $\bar{\nu}$ values are close to the experimental values. A significant difference is noted in the ^{252}Cf spontaneous fission data from JEFF-3.1.1, which is caused by a systematic shift of ~ 0.5

units in the Z_p values for the light fragments. In this way, our consistency check illustrates a known discrepancy in the major FPY libraries.

3. Conclusion

The work presented here outlines a consistency check between FPY and PNE. As both data sets are integral for physics and nuclear safety calculations it is imperative that the variables are self-consistent. This procedure is applied to several fissioning systems and highlights disagreements, such as the JEFF-3.1.1 ^{252}Cf spontaneous fission FPY, which shows a Z_p shift in the light fragments that results in an erroneously high value for $\bar{\nu}$.

References

1. A. C. Wahl, *Systematics of fission-product yields*, tech. rep., Los Alamos Nat. Lab. (2002).
2. M. B. Chadwick *et al.*, ENDF/B-VII.1 Nuclear Data for Science and Technology: Cross Sections, Covariances, Fission Product Yields and Decay Data, *Nucl. Data Sheets* **112**, 2887 (2011).
3. M. A. Kellett *et al.*, *The JEFF-3.1/-3.1.1 radioactive decay data and fission yields sub-libraries* (Nuclear Energy Agency, 2009).
4. S. Keiichi *et al.*, Jendl-4.0: A new library for nuclear science and engineering, *Journal of Nuclear Science and Technology* **48**, 1 (2011).
5. K. Nishio *et al.*, Multiplicity and energy of neutrons from ^{235}U (nth, f) fission fragments, *Nuclear Physics A* **632**, 540 (1998).
6. O. I. Batenkov *et al.*, Comparison of Experimental and Calculated Mass Distributions of Fission Fragments in Proton-Induced Fission of ^{232}Th, ^{235}U, ^{238}U and ^{237}Np in the Intermediate Energy Region, p. 625 (2004).
7. A. Vorobyev *et al.*, Investigation of the prompt neutron emission mechanism in low energy fission of ^{235,233}U (nth, f) and ^{252}Cf (sf), **8**, p. 03004 (2010).
8. S. F. Mughabghab, Atlas of Neutron Resonances: Resonance Parameters and Thermal Cross Sections. Z= 1-100 (2006).
9. T. Ethvignot *et al.*, Neutron Multiplicity in the Fission of ^{238}U and ^{235}U with Neutrons up to 200 MeV, *Phys. Rev. Lett.* **94**, p. 052701 (Feb 2005).

New Determination of the Mo-Ba Yield Matrix for ^{252}Cf

A.H. Thibeault[1], E. H. Wang[1], C.J. Zachary[1], J. H. Hamilton[1], A.V. Ramayya[1], B. Musangu[1,2], J. O. Rasmussen[3], Y. X. Luo[1,3], G. M. Ter-Akopian[4], Yu. Ts. Oganessian[4] and S. J. Zhu[5]

[1]*Department of Physics, Vanderbilt University, Nashville, TN 37235, USA*
[2]*Department of Physics, Furman University, Greenville, SC 29613, USA*
[3]*Lawrence Berkeley National Laboratory, Berkeley, CA 94720, USA*
[4]*Joint Institute for Nuclear Research, RU-141980 Dubna, Russian Federation*
[5]*Department of Physics, Tsinghua University, Beijing 100084, China*

Using quadruple coincidence events of prompt fission gamma rays from Gammasphere data on spontaneous fission of ^{252}Cf, we made a careful analysis of the yield matrix of coincident pairs of barium (Z = 56) and molybdenum (Z = 42) fission fragments. The accuracy of previously determined yield matrices is improved upon with the use of high accuracy quadruple coincidences, the increased statistics of the most recent Gammasphere data, and improved level schemes for barium and molybdenum isotopes. The previously proposed extra hot fission mode (up to ten neutron evaporated) has been confirmed in our reanalysis. Our results are well in agreement with the results from the 1995 Gammasphere data analysis of the Ba-Mo yields.

In our first analysis of the Mo-Ba fission yield data from the spontaneous fission of ^{252}Cf, an extra hot fission mode that significant amounts of 8, 9 and 10 neutrons was reported [1], see Fig. 1. This mode was attributed to hyper deformed ^{144}Ba at scission. This analysis was repeated with our second set of Gammasphere data with the mode still present but with reduced intensity (Fig. 2). In 2000 a new experiment was carried out with Gammasphere in which 5.7×10^{11} γ-γ-γ and 1.9×10^{11} γ-γ-γ-γ and higher fold gamma coincidences were collected in the spontaneous fission of ^{252}Cf. Because of the importance of understanding this extra hot fission mode, the Mo-Ba yield data were analyzed using both the triple and quadruple data. Also more recent level schemes of the Mo and Ba nuclei likewise improved the accuracy of the analysis. Radware software was used to analyze the quadruple coincidence data. The relative intensities of all transitions feeding the ground states of the barium isotopes for

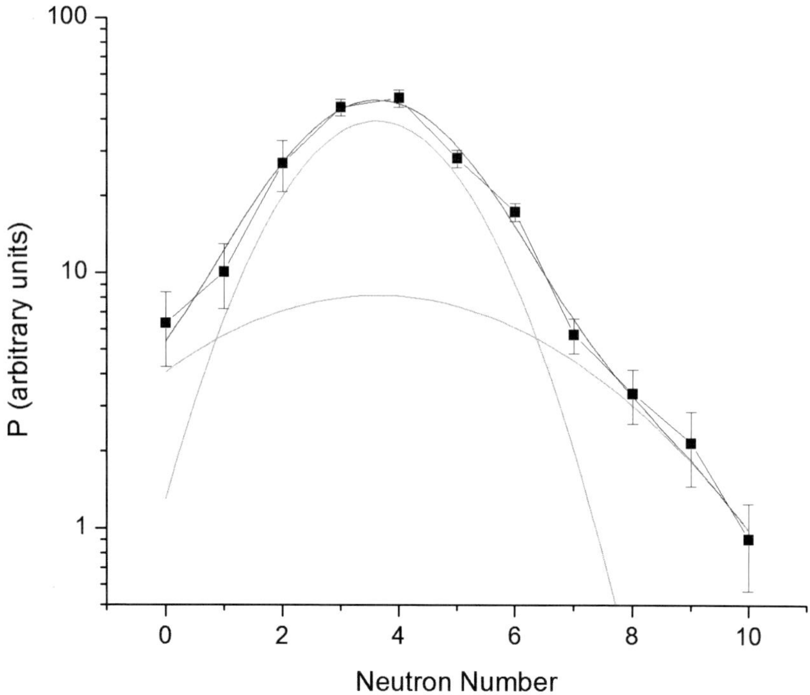

Fig. 1. Relative intensities of the 0 to 10 neutron emissions in the Mo-Ba matrices from analysis of 1992 ORNL Compton Suppression Spectrometer data [1]. The smooth lines are Gaussian fits to the 0-6 and 8-10 neutron emission data.

each molybdenum isotope were determined based on new level schemes [3-5] to produce a new set of relative yields.

The new yields of Mo-Ba are given in Table I and Fig. 3. The errors are significantly reduced because of the improved statistics, the use of quadruple data to further remove accidental coincidences and improved knowledge of level schemes. As seen in Fig. 3, a similar deviation from a Gaussian fit to the data for the 0 to 6 neutron emission channels is seen at neutron numbers 7, 8, 9, 10. There is a larger error in the 8 neutron yield from accidental coincidences. Also shown in Fig. 3 is a Gaussian fit the 8, 9, 10 neutron data with the Gaussian fit to the 0-6 neutron emission subtracted for the data. This new analysis, coupled with the new analysis of the Zr-Ce yields that does not exhibit 8, 9, 10 neutron emission [6] confirms the existence of this extra hot mode only in Mo-Ba yield. The Mo-Ba yield data also are given in Table 1.

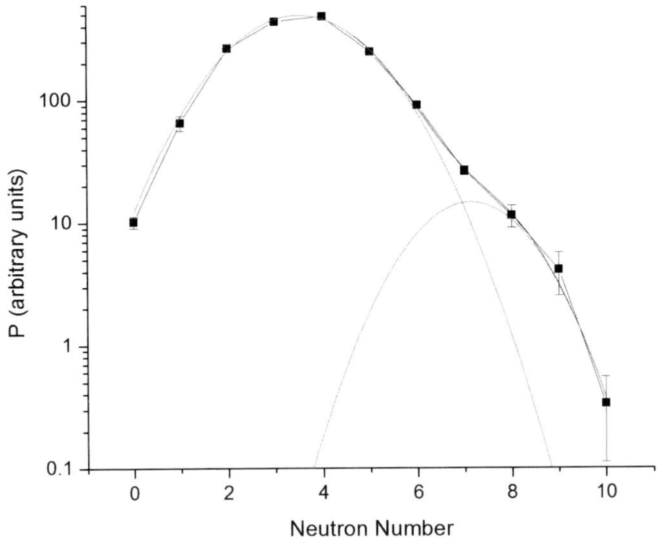

Fig. 2. Relative intensities of the 0 to 10 neutron emissions in the Mo-Ba matrices from analysis of 1995 Gammasphere data [2]. The smooth lines are Gaussian fits to the 0-6 and 8-10 neutron emission data.

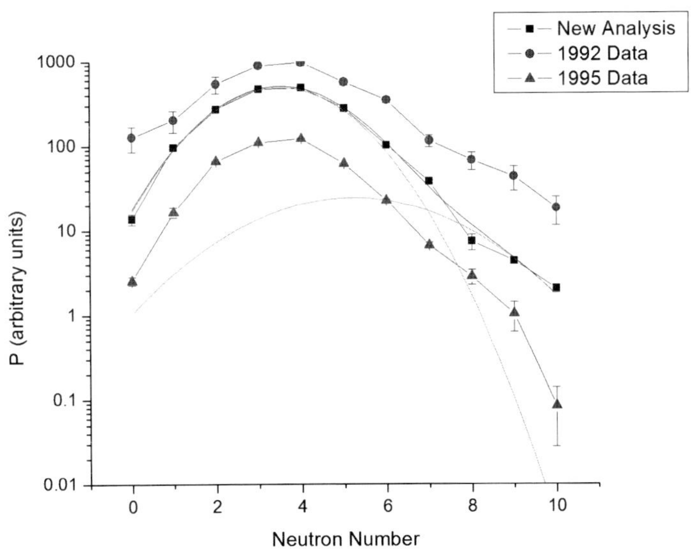

Fig. 3. The experimental Mo-Ba yield curves for the 1992 data [1] and 1995 data [2] are shown along with the present analysis. A smooth Gaussian fit to the 0-6 neutron emission data and a smooth Gaussian fit to the 8, 9, 10 data with the 0-6 neutron Gaussian fit subtracted are shown.

Table 1. New yield matrix for Ba-Mo from the spontaneous fission of ^{252}Cf.

Yield	^{138}Ba	^{139}Ba	^{140}Ba	^{141}Ba	^{142}Ba	^{143}Ba	^{144}Ba	^{145}Ba	^{146}Ba	^{147}Ba	^{148}Ba
^{102}Mo						0.044(6)	0.128(13)	0.243(24)	0.280(26)	0.194(28)	0.045(5)
^{103}Mo			0.008(1)		0.079(7)	0.132(10)	0.384(24)	0.503(33)	0.323(20)	0.045(6)	0.005(1)
^{104}Mo	0.016(2)	0.009(1)	0.022(9)	0.055(4)	0.211(13)	0.462(28)	0.925(53)	0.666(40)	0.327(20)	0.087(12)	
^{105}Mo	0.017(2)	0.016(2)	0.053(6)	0.141(10)	0.467(32)	0.782(53)	0.965(64)	0.577(40)	0.150(11)		
^{106}Mo	0.020(8)	0.032(5)	0.127(14)	0.324(24)	0.731(51)	0.800(54)	0.607(41)	0.245(18)	0.084(15)		
^{107}Mo	0.035(5)	0.060(7)	0.302(31)	0.396(39)	0.616(58)	0.413(40)	0.207(20)				
^{108}Mo	0.009(1)	0.021(2)	0.232(21)	0.141(13)	0.131(12)	0.068(6)	0.024(2)				

Conclusion

A new yield matrix was determined for Ba-Mo fission partners from the spontaneous fission of ^{252}Cf. A similar deviation from the Gaussian fit to the normal fission mode was found for the 8, 9, and 10 neutron channels as found in previous analyses, to confirm the existence of the proposed "extra-hot-fission" mode. Errors are reduced in this newest analysis compared to previous studies because of the greater number of statistics of the latest Gammasphere experiment, the use of quadruple coincidences in the analysis and improved level schemes. The larger error in the 8 neutron channel yield relative to other errors introduces some uncertainty as to the strength of the "extra-hot-fission" mode.

Acknowledgments

From the Department of Physics and Astronomy at Vanderbilt University I would like to thank Joseph H. Hamilton, Akunuri V. Ramayya, Enhong H Wang, and Christopher J. Zachary for their guidance on this project. I would also like to thank Brooks Musangu from Furman University for his help and support. He was also supported by the Furman Advantage program. The work at Vanderbilt University and Lawrence Berkeley National Laboratory are supported by the US Department of Energy under Grant No. DE-FG05-88ER40407 and Contract No. DE-AC03-76SF00098. The work at Tsinghua University was supported by the National Natural Science Foundation of China under Grant No. 11175095. The work at JINR was supported by the Russian Foundation for Basic Research Grant No. 08-02-00089 and by the INTAS Grant No. 03-51-4496.

References

1. G. M. Ter-Akopian *et al.*, *Phys. Rev. C* **55**, 1146 (1997).
2. S.-C. Wu *et al.*, *Phys. Rev. C* **62**, 041601(R) (2000).
3. M. R. Bhat, Evaluated Nuclear Structure Data File (ENSDF).
4. Experimental Unevaluated Nuclear Data List, www.nndc.bnl.gov/xundl.
5. Private communication from Vanderbilt University.
6. B.M. Musangu *et al.*, talk at 6th International Conference on Fission and Properties of Neutron-rich Nuclei.

Reinvestigation of Octupole Correlations in ^{146,147}La

E. H. Wang,[1] W. Lewis,[1,2] C. J. Zachary,[1] J. H. Hamilton,[1] A. V. Ramayya,[1] J. K. Hwang,[1] S. H. Liu,[1] N. T. Brewer,[1] Y. X. Luo,[1] J. O. Rasmussen,[3] S. J. Zhu,[4] G. M. Ter-Akopian[5] and Yu. Ts. Oganessian[5]

[1]*Department of Physics, Vanderbilt University, Nashville, TN 37235, USA*
[2]*Department of Physics, Furman University, Greenville, SC 29613, USA*
[3]*Lawrence Berkeley National Laboratory, Berkeley, CA 94720, USA*
[4]*Department of Physics, Tsinghua University, Beijing 100084, China*
[5]*Joint Institute for Nuclear Research, RU-141980 Dubna, Russian Federation*

High spin states of neutron rich ^{146,147}La have been reinvestigated by γ-γ-γ and γ-γ-γ-γ coincidence data from a ^{252}Cf spontaneous fission experiment by using Gammasphere. Two new bands in ^{146}La have been established. One of them is proposed to be the octupole parity partner of the previously known band. The ground state band of ^{147}La has been established with a proposed $5/2^+$ band-head. Angular correlations of cascades have been used to study the spins and parities of the states. The B(E1)/B(E2) ratios and dipole moments between the proposed octupole bands in ^{146,147}La have been measured showing a decreasing trend from ^{144}La to ^{146}La, and from ^{145}La to ^{147}La.

Keywords: ^{146}La; ^{147}La; Octupole Correlation.

1. Introduction

Previous theoretical predictions suggested an island of static octupole deformation located at Z = 56 and N = 88 [1, 2]. Octupole deformation and/or correlations have been reported in Ba (Z = 56) and Ce (Z = 58) in experiments [3–13]. For La isotopes with odd Z = 57 lying between the Ba (Z = 56) and Ce (Z = 58), octupole correlations are expected and evidence was reported in $^{143-145,147}$La from ^{248}Cm and ^{252}Cf spontaneous fission (SF) experiments [14–17].

For odd-odd ^{146}La with Z = 57 and N = 89, some levels were investigated by β-decay of ^{146}Ba [18], ^{252}Cf SF [19] (and the earlier publications therein), ^{235}U(n,F) experiments [20], by our group previously [21]. In the present work, the high spin levels and previous bands of ^{146,147}La have been reinvestigated and expanded from the high statistics data of our ^{252}Cf SF experiment. Octupole correlations in ^{146}La are proposed and the ground state band in ^{147}La has been established for the first time by combining with previous β-decay data [22, 23].

2. Experimental Method

The experiment with ^{252}Cf was carried out at the Lawrence Berkeley National Laboratory(LBNL). A 62 μCi ^{252}Cf source was sandwiched between two Fe foils of thickness 10mg/cm^2. By using 101 Ge detectors of Gammasphere, the data were sorted into 5.7×10^{11} γ-γ-γ and higher fold γ events and 1.9×10^{11} γ-γ-γ-γ and higher fold γ coincident events. These γ coincident data were analyzed by RADWARE software packages [24].

3. Experimental Results

The level schemes of $^{146,147}La$ are shown in Fig. 1. Transitions and levels above the 619.9+x keV level in ^{146}La form three band structures. Our measured halflife for this state is 14(1) ns. Note the ground state of the ^{147}Ba parent was assigned as $5/2^-$ in Ref. [27] and this may change the previous low lying spin assignments.

Note that low lying transitions in ^{146}La are also observed in the Coulomb excitation experiment [25], which provides evidence for the mass determination.

Angular correlations for γ transitions in $^{146,147}La$ were measured. Details of the technique can be found in Ref. [26]. The result confirms E2 multipolarity of the 270.3, 397.1 and 509.8 keV transitions in ^{146}La, the E2 transitions in band (1), the E2 297, 414, 421 keV transitions in band (3) and E1 linking transitions from band (2) to band (1) in ^{147}La.

4. Discussion

The octupole bands in ^{146}La are analogous to the bands reported in ^{144}Cs [28, 29], ^{148}La [30], ^{148}Pr [31] and ^{143}Cs[32]. Such systematic comparison supports our spin and parity assignments in ^{146}La.

The average B(E1)/B(E2) ratio for the bands (1) and (2) in ^{146}La is $0.05(1)\times(10^{-6}fm^{-2})$, compared to $0.41(6)\times(10^{-6}fm^{-2})$ in ^{144}La. The average Q_0 over the four neighbouring even-even Ba Ce nuclei [33,34], is used to determine the Q_0 value of $^{144,146}La$, and yields a 0.045 e fm D_0 of ^{146}La and 0.11 e fm for ^{144}La.

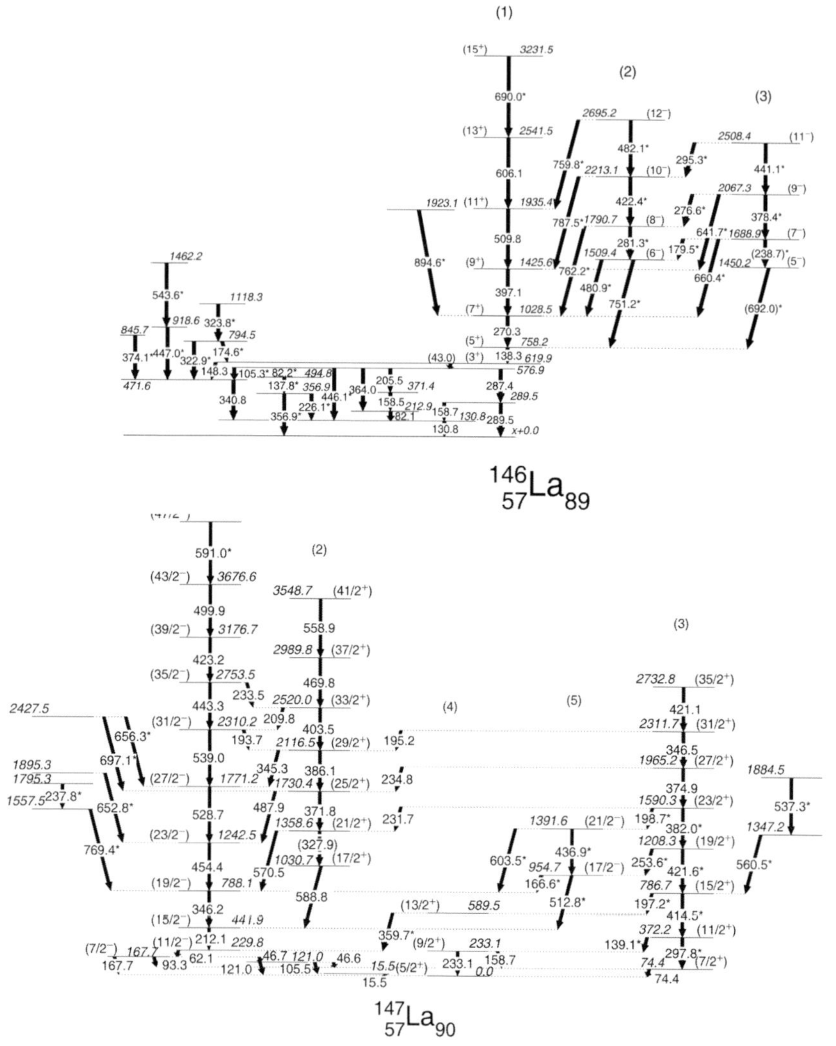

Fig. 1. The level scheme of ^{146,147}La in the present work. New transitions are labeled with an asterisk.

The average B(E1)/B(E2) ratio for the octupole bands in ^{147}La is $0.40\times(10^{-6}\text{fm}^{-2})$, while such value in ^{145}La is $1.0\times(10^{-6}\text{fm}^{-2})$. The D_0 values of ^{145,147}La are 0.18 and 0.13 e fm.

We note that the recent measurement of B(E3) in ^{144}Ba [35] shows direct evidence for octupole correlations in this region.

5. Conclusion

In conclusion, high spin states in ^{146,147}La have been reinvestigated. The s = −1 octupole bands in ^{146}La have been observed. The previous assignment of s = +i octupole doublet has been confirmed by angular correlation measurements. The experimental B(E1)/B(E2) ratios and D_0 dipole moments exhibit a decreasing trend from ^{144}La to ^{146}La, and from ^{145}La to ^{147}La.

Acknowledgments

The work at Vanderbilt University and Lawrence Berkeley National Laboratory are supported by the US Department of Energy under Grant No. DE-FG0588ER40407 and Contract No. DE-AC03-76SF00098. The work at Tsinghua University was supported by the National Natural Science Foundation of China under Grant No. 11175095. W. Lewis was an undergraduate doing a summer research experience supported by Furman and Vanderbilt. The author would also give special thank to the private communication with Brian Bucher at LLNL.

References

1. W. Nazarewicz *et al.*, *Nucl. Phys. A* **429**, 269 (1984).
2. G.A. Leander *et al.*, *Phys. Lett. B* **152**, 284 (1985).
3. W.R. Phillips *et al.*, *Phys. Rev. Lett.* **57**, 3257 (1986).
4. W.R. Phillips *et al.*, *Phys. Lett. B* **212**, 402 (1988).
5. S.J. Zhu *et al.*, *Phys. Lett. B* **357**, 273 (1995).
6. J.H. Hamilton *et al.*, *Prog. Part. Nucl. Phys.* **35**, 635 (1995).
7. S.J. Zhu *et al.*, *J. Phys. G* **23**, L77 (1997).
8. S.J. Zhu *et al.*, *Phys. Rev. C* **60**, 051304 (1999).
9. W. Urban *et al.*, *Nucl. Phys. A* **613**, 107 (1997).
10. M.A. Jones *et al.*, *Nucl. Phys. A* **605**, 133 (1996).
11. H.J. Li *et al.*, *Phys. Rev. C* **90**, 047303 (2014).
12. Y.J. Chen *et al.*, *Phys. Rev. C* **73**, 054316 (2006).
13. S.J. Zhu *et al.*, *Phys. Rev. C* **85**, 014330 (2012).
14. W. Urban *et al.*, *Phys. Rev. C* **54**, 945 (1996).
15. S.J. Zhu *et al.*, *Phys. Rev. C* **59**, 1316 (1999).
16. J.G. Wang *et al.*, *Phys. Rev. C* **75**, 064301 (2007).
17. Y.X. Luo *et al.*, *Nucl. Phys. A* **818**, 121 (2009).
18. C. Chung *et al.*, *Phys. Rev. C* **31**, 2199 (1985).

19. L.K. Peker and J.K. Tuli, *Nucl. Data Sheets* **82**, 187 (1997).
20. T.A. Khan, D. Hofmann and F. Horsch, *Nucl. Phys. A* **205**, 488 (1973).
21. J.K. Hwang *et al.*, *Phys. Rev. C* **58**, 3252 (1998).
22. E. Der Mateosian and L.K. Peker, *Nucl. Data Sheets* **66**, 705 (1992).
23. N. Nica, *Nucl. Data Sheets* **110**, 749 (2009).
24. D.C. Radford, *Nucl. Instrum. Methods Phys. Res. A* **361**, 297 (1995).
25. B. Bucher *et al.*, *Aps. Bull.* **60** (13), DC00009; B. Bucher *et al.* (private communication).
26. A.V. Daniel *et al.*, *Nucl. Instrum. Methods B* **262**, 399 (2007).
27. T. Rzaca-Urban *et al.*, *Phys. Rev. C* **87**, 031305(R) (2013).
28. T. Rzaca-Urban *et al.*, *Phys. Rev. C* **80**, 064317 (2009).
29. Y.X. Luo *et al.*, *Nucl. Phys. A* **838**, 1 (2010).
30. Y.X. Luo *et al.*, *Nucl. Phys. A* **825**, 1 (2009).
31. E.H. Wang *et al.*, *Phys. Rev. C* **92**, 034317 (2015).
32. W. Urban *et al.*, *Phys. Rev. C* **69**, 017305 (2004).
33. S. Raman *et al.*, *At. Data Nucl. Data Tables* **36**, 1 (1987).
34. B. Pritychenko, M. Birch, B. Singh, M. Horoi, *At. Data Nucl. Data Tables* **107**, 1 (2016).
35. B. Bucher *et al.*, *Phys. Rev. Lett.* **116**, 112503 (2016).

New Measurement of Zr-Ce Yield Matrix for ^{252}Cf

B. Musangu,[1,2] E. H. Wang,[1] J. H. Hamilton,[1] C. J. Zachary,[1] A. Thibeault,[1] J. M. Eldridge,[1] A. V. Ramayya,[1] W. H. Brantley,[2] J. O. Rasmussen,[3] Y. X. Luo,[1,3] G. M. Ter-Akopian,[4] Yu. Ts. Oganessian[4] and S. J. Zhu[5]

[1]*Department of Physics, Vanderbilt University, Nashville, TN 37235, USA*
[2]*Department of Physics, Furman University, Greenville, SC 29613, USA*
[3]*Lawrence Berkeley National Laboratory, Berkeley, CA 94720, USA*
[4]*Joint Institute for Nuclear Research, RU-141980 Dubna, Russian Federation*
[5]*Department of Physics, Tsinghua University, Beijing 100084, China*

The Zr-Ce fission fragment yield matrix has been determined from the spontaneous fission of ^{252}Cf using Gamasphere data. About $1.9*10^{11}$ quadruple-coincidence events of fission gamma rays were used to reconstruct the yield matrix of coincident pair of zirconium [Z=40] and cerium [Z=58] fission fragments from 0 to 8 neutron multiplicities. The data used in the current analysis have better statistics than that used in the previous study on the spontaneous fission of ^{252}Cf. Also, the previous work only had triple gamma coincidence events. Thus, the present study was expected to yield more accurate results. However, the results from this study were similar to those from the previous study, with smaller error limits.

1. Introduction

In the spontaneous fission of ^{252}Cf, Zirconium ($_{40}$Zr) and Cerium ($_{58}$Ce) are fission partners. This means that whenever ^{252}Cf fissions resulting in an isotope of Zr, an isotope of Ce is also a product. On average, each fission of ^{252}Cf produces 3.7 neutrons, however, as few as 0, and as many as 6 or more neutrons are possible [1]. This means that for any particular product of ^{252}Cf spontaneous fission, either the three neutron or four neutron partners (also called 3n or 4n channel) should be the most frequent partner. The study of relative yields gives us information on the excitation, deformation, and tip distance for fragments of given mass and nuclear charge [1]. In this paper we will describes the recent work done involving the re-evaluation of neutron multiplicity and yields of the correlated Zr-Ce fragment pair in earlier spontaneous fission of ^{252}Cf [1]. We analyzed the coincidence gamma-gated spectra by measuring the area under the lowest $2^+ \rightarrow 0^+$ ground state band transitions observed in the de-excitation of even-even fission products and all the transitions feeding the ground states in odd A products.

It was shown [2] and confirmed in earlier work [1] that the total intensities of the lowest $2+ \rightarrow 0+$ ground-state band transitions observed in the de-excitation of even-even fission products reflect, to a high degree of accuracy (< 5%), the total independent yields of these isotopes. Additionally, it was also shown that the values of relative yields of pairs of fission fragments, as they appear after neutron evaporation, can be extracted from the analysis of prompt γ–γ–γ and γ–γ–γ–γ coincidences detected with large detector arrays. Newer level schemes published in the National Nuclear Data Center and private communication with Vanderbilt University were used to reduce the errors in the analysis.

2. Proceedings and Discussion

Triple-coincidence and quadruple-coincidence data used in the present work was taken by 101 Hyper-Pure Gammasphere, at Lawrence Berkley National Laboratory in 2000. The Gammasphere consists of 110 Hyper-Pure-Germanium (HPGe) detectors arranged in a spherical configuration. Each HPGe crystal also has seven Bismuth Germanate (BGO) Compton shields. These Compton shields detect gamma rays that escape the HPGe crystal after undergoing Compton scattering, thus reducing many of the background detections. About $5.7*10^{11}$ γ–γ–γ and higher fold, and $1.9*10^{11}$ γ–γ–γ–γ and higher fold coincidence events were collected by Gammasphere. For our purposes, both γ–γ–γ and γ–γ–γ–γ coincidence data were analyzed using the RADWARE software package, details can be found in [1].

In the case of determining the Zr-Ce yield matrix, measurements of multiple γ–rays emitted by the Zr-Ce fission fragment pair formed in spontaneous fission of ^{252}Cf were used to extract the yields. A two-dimensional matrix of γ–γ–γ–γ coincidences was created from the initial data by selecting the γ ray coincidences occurring within 50ns. The peaks observed in this two-dimensional spectrum arise from the coincidences between γ ray emitted promptly by both complementary fission fragments of different fragment pairs. The area under the peak was estimated using the method described in the previous work [1]. This fission pair involves even-even fragments therefore; the yields were determined from the area of the peaks corresponding to the coincidences between 2^+ 0^+ transitions in both fragments. The relative yields of the fragment pair were obtained after corrections for the known detection efficacies and internal conversion probabilities. The internal conversion coefficients are from [4].

The table below displays the relative yield data that were collected. These yields of the correlated fragment pair of Zr-Ce in the spontaneous fission of ^{252}Cf were next scaled according to Ter-Apkopian's independent yield [1] and summed for each isotope of Ce. This summation and Ter-Akopian's calculated data for Zr-Ce, were both normalized such that ^{146}Ce had a value of 100. Then these two data sets were compared to see if Ter-Akopian's calculations could be verified. As can be seen below in Fig. 2, the present relative yield data are in agreement with the previous ones, and thus experimentally confirm the results from the Zr-Ce study [1] but with smaller error limits.

Table 1. Yields of correlated fragment pairs of Zr-Ce in the spontaneous fission of ^{252}Cf.

	^{144}Ce	^{146}Ce	^{147}Ce	^{148}Ce	^{149}Ce	^{150}Ce
^{98}Zr		0.023(3)		0.072(5)	0.067(13)	0.224(24)
^{99}Zr		0.052(3)		0.13(14)	0.134(26)	0.187(9)
^{100}Zr		0.139(4)	0.249(6)	0.552(21)	0.221(5)	0.418(11)
^{101}Zr		0.317(12)	0.542(21)	0.621(24)	0.133(6)	0.073(16)
^{102}Zr	0.022(2)	0.400(6)	0.404(7)	0.349(13)		0.058(3)
^{103}Zr	0.059(11)	0.214(19)	0.175(47)	0.1399(8)		
^{104}Zr		0.07(7)				

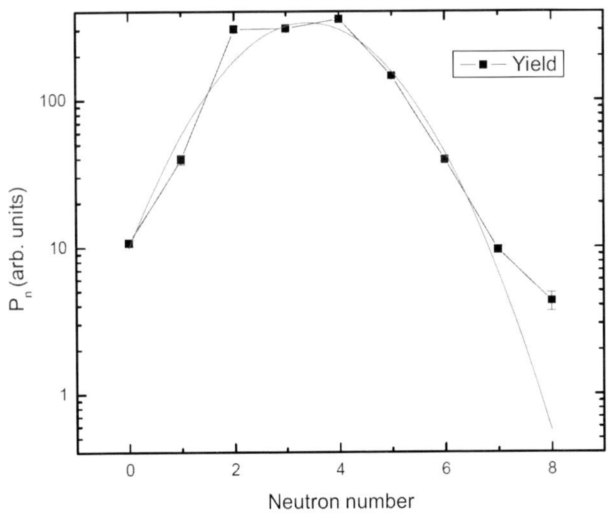

Figure 1. New relative yield curve with a simple Gaussian fit to the data.

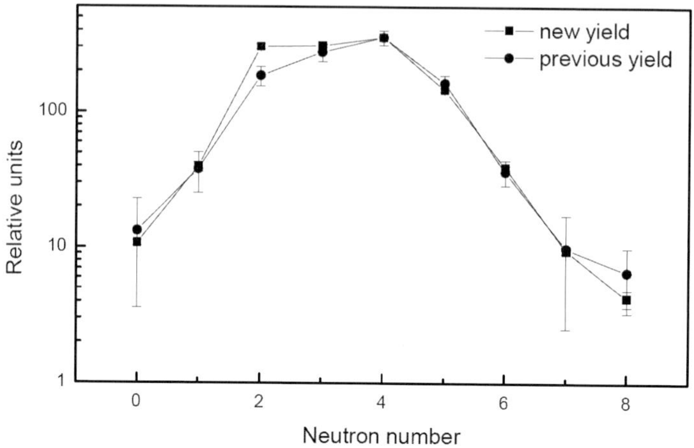

Figure 2. Combined relative yields. The black line represents the curve from the current experiment while the circle points represent the Ter-Akopian's, *et al.*, curve [1]

Acknowledgments

Firstly, I would like to thank Dr. J. H. Hamilton, Dr. A. V. Ramayya, Dr. E. Wang, C. Zachery, A. Thibeault from Vanderbilt University and Dr. W. H. Brantley from Furman University for giving me the opportunity to work on this project as well as helping me with this project. Secondly, I would like to thank my family and friends for their support. The work at Vanderbilt University and Lawrence Berkeley National Laboratory are supported by the US Department of Energy under Grant No. DE-FG05-88ER40407 and Contract No. DE-AC03-76SF00098. The work at Tsinghua University was supported by the National Natural Science Foundation of China under Grant No. 11175095. The work at JINR was supported by the Russian Foundation for Basic Research Grant No. 08-02-00089 and by the INTAS Grant No. 03-51-4496. In the summer of 2016 Brooks Musangu at Vanderbilt University was also supported by the Furman University through Furman Advantage.

References

1. G. M. Ter-Akopian *et al.*, *Phys. Rev. C* **55**, 1146 (1997).
2. C. Goodin *et al.*, *Phys. Rev C* **74**, 017309 (2006).
3. M. R. Bhat, Evaluated Nuclear Structure Data File (ENSDF), and Experimental Unevaluated Nuclear Data List (XUNDL).
4. bricc.anu.edu.au.

Studying the Photon Strength Function of ^{97}Zr Using the ^{96}Zr(n,g) and ^{96}Zr(d,p) Reactions

J. Winkelbauer[1], S. Mosby[1], A. Couture[1], H.Y. Lee[1], S. Kuvin[3], C.R. Hoffman[2], J. Ullmann[1], A. Spyrou[4], S.N. Liddick[4], A.C. Larsen[5], B.B. Back[2], M.L. Avila[2], B.V. Kheswa[5], T. Renstroem[5], G. Tveten[5], R. Talwar[2] and D. Santiago-Gonzales[6]

[1]*Los Alamos National Laboratory, Los Alamos NM 87545 USA*
[2]*Argonne National Laboratory, Lemont IL 60439 USA*
[3]*University of Connecticut, Storrs CT 06269 USA*
[4]*Michigan State University/National Superconducting Cyclotron Laboratory, East Lansing MI 48824 USA*
[5]*Physics Department, University of Oslo, Oslo Norway*
[6]*Louisiana State University, Baton Rouge, LA 70803*

A major barrier in the study of neutron-induced nuclear reactions, which play a critical role in stockpile stewardship and astrophysics, is the impossibility of direct measurements with short-lived radioactive isotopes. Theoretical models can be used to describe these reactions, although the nuclear structure inputs such as the Photon Strength Function (PSF) for these exotic nuclei are often poorly constrained. Recently, a program to investigate the PSF for medium-mass nuclei has begun as a collaboration between Los Alamos National Laboratory (LANL) and Argonne National Laboratory (ANL), combining unique experimental capabilities from both laboratories. At the Manuel J. Lujan Neutron Scattering Center at LANL, The Detector for Advanced Neutron Capture Experiments (DANCE) provides direct measurements of gamma ray cascades from neutron capture reactions on stable or long-lived radioactive nuclei. At the Argonne Tandem Linear Accelerator System (ATLAS) facility at ANL, single neutron transfer reactions in inverse-kinematics provide complementary data on short-lived radioactive nuclei. The Helical Orbit Spectrometer (HELIOS) is a device that was designed to study transfer reactions in inverse kinematics by detecting the charged ejectiles inside of a large-bore solenoidal magnet. The APOLLO array was designed and built at LANL to be placed inside the magnetic field of HELIOS to measure gamma ray cascades from the nuclear states populated in neutron transfer reactions. As a test case for this research program, the ^{96}Zr(n,γ) reaction was measured using DANCE and the ^{96}Zr(d,p) reaction was measured using HELIOS+APOLLO. ^{96}Zr lies near the light mass peak in the ^{239}Pu fission spectrum, so neutron capture rates on the neutron-rich Zr isotopes are important for fission applications. While DANCE provides Multi-Step Cascade spectra from ^{97}Zr as well as cross section information, HELIOS+APOLLO will provide more detailed nuclear structure information about the intermediate states in ^{97}Zr. Initial results from the ^{96}Zr(d,p) and ^{96}Zr(n,γ) measurements will presented. This work benefited from the use of the LANSCE accelerator facility. Work was performed under the auspices of the US Department of Energy by Los Alamos National Security, LLC under contract DEAC52-06NA25396.

In memory of our long-time friend Prof. Walter Greiner, who wrote this in 2002 for our Sanibel Conference.

Sanibel Poem by Walter Greiner

"Sanibel, Sanibel,
what a beautiful island
rich in shells,
and beaches with fine sand.

This is one of Gods best creations
ideally suited for recreation
and also, as we can see
for conferences with high fee.

Joe and Jannelle
once felt particularly well,
discovered many years ago:
Sanibel is the place to go!

And every year again,
to escape storm and rain
they come down from Tennessee
and enjoy the beauty of land, sun and sea.

Having friends world-wide,
they enjoyed with pride,
to move forward to share
tranquility, beauty, wildlife and air.

This happened 10 years ago
celebrating 60 years of Joe
and since every fifth years
we come together there.

Now in 2002
we have assembled too,
and Joe is seventy of age
we congratulate and state:

Continue Joe, for many years!
You look still young
and may become
near hundred: we shall cheer!

It is within your genes
Inherited from your father.
And, if I read: that means,
we may meet here often after;

At least for 6 more times
we shall enjoy progress in our science,
as we did in the past.
Long might it last!

Your father still wrote papers
At 98 of years.
Sure, you will match him later.
After all, you are his peer.

Jannelle was always on your side,
She gives you strength, she is your guide.
Treasure her, as you always did,
So that both of you are here indeed.

When we celebrate the next occasions,
streaming in from many nations
during 30 years ahead,
enjoying in '32 the greatest of all fest.

It will be interesting to see
where nuclear physics will then be.
Will the driplines be explored?
And the superheavies still adored?

Is the vacuum then understood,
will antinuclei, strange clusters have set foot
and arouse general attention
with their structure, I just mention.

That's all important to understand
galaxies, stars and distant land
at the edge of our universe
can we ever end this verse?

The future is full of promise
and indeed, we have great wish.
It's like looking into dark sky
and your thoughts do fly by
far beyond where we cannot see
and we ask, can we there be?!

We are in the hands of the Lord
And ask him for us to afford
happiness, health and insight:
May our all future be quite bright!

Let me now thank you all,
organizers, secretaries, here in this hall.
Enjoy still further Sanibel
and have safe journey home; keep well!"

Participant List

Dieter Ackermann
GANIL
dieter.ackermann@ganil.fr

Navin Alahari
GANIL
navin@ganil.fr

Sergio Almaraz-Calderon
FSU
salmarazcalderon@fsu.edu

Andrei Andreyev
University of York
Andrei.Andreyev@york.ac.uk

Ani Aprahamian
University of Notre Dame
aapraham@nd.edu

Thomas Aumann
TU Darmstadt
t.aumann@gsi.de

Juha Aysto
Helsinki Inst. of Physics
juha.aysto@helsinki.fi

Birger Back
Argonne National Laboratory
back@anl.gov

George Bertsch
University of Washington
bertsch@uw.edu

Carrol Bingham
University of Tennessee
cbingham@utk.edu

Ezekiel Blain
RPI
ezekial.blain@gmail.com

Klaus Blaum
Max-Planck Inst. for Nuclear Phys
sekretariat.blaum@mpi-hd.mpg.de

Michael Block
GSI
m.block@gsi.de

Maria Borge
CERN
mgb@cern.ch

William Brantley
Furman University
bill.brantley@furman.edu

Alex Brown
Michigan State University
brown@nscl.msu.edu

Brian Bucher
Idaho National Lab
brian.bucher@inl.gov

Aurel Bulgac
University of Washington
aurel.bulgac.phys@gmail.com

Ken Carter
Retired
kencarter@bellsouth.net

Bradley Cheal
The University of Liverpool
bradley.cheal@liverpool.ac.uk

Satoshi Chiba
Tokyo Institute of Technology
chiba.satoshi@nr.titech.ac.jp

Jerry Cole
Retired
jeraldcole@verizon.net

David Dean
Oak Ridge National Laboratory
deandj@ornl.gov

Giacomo DeAngelis
INFN LNL
giacomo.deangelis@lnl.infn.it

Matthew Devlin
LANL
devlin@lanl.gov

Maria Diakaki
NTUA
mdiakaki@cern.ch

Jens Dilling
TRIUMF
jdilling@triumf.ca

Sergey Dmitriev
JINR
dmitriev@jinr.ru

Jacek.Dobaczewski
University of York
jacek.dobaczewski@york.ac.uk

Jerry Draayer
Louisiana State University
draayer@lsu.edu

Christoph Duellmann
GSI Darmstadt
c.e.duellmann@gsi.de

Dana Duke
LANL
dduke@lanl.gov

Jonathan Eldridge
Vanderbilt University
jonathan.m.eldridge@vanderbilt.edu

Paul Fallon
LBNL
pfallon@lbl.gov

Aleksandra Fijakowska
Rutgers University
Aleksandra.kuzniak@gmail.com

Stefan Frauendorf
University Notre Dame
sfrauend@nd.edu

Alexandra Gade
NSCL (MSU)
gade@nscl.msu.edu

Jacklyn Gates
LBL
jmgates@lbl.gov

Konrad Gelbke
FRIB/NSCL (Retired)
gelbke@nscl.msu.edu

Jaime Gomez
LANL
jaime.gomez@lanl.gov

Katherine Grzywacz Jones
University of Tennessee
kgrzywac@utk.edu

Robert Haight
LANL (Retired)
robert.haight87501@gmail.com

Franz-Josef Hambsch
EC-JRC-Dir G
franz-josef.hambsch@ec.europa.eu

Joseph Hamilton
Vanderbilt University
j.h.hamilton@vanderbilt.edu

Kalee Hammerton
NSCL/MSU
hammerto@nscl.msu.edu

Heiko Hergert
NSCL/FRIB
hergert@nscl.msu.edu

Sigurd Hofmann
GSI
s.hofmann@gsi.de

Yan Huang
Tsinghua University
h-y12@mails.tsinghua.edu.cn

Paul Inman
Vanderbilt University
paul.inman@vanderbilt.edu

Mikhail Itkis
JINR
itkis@jinr.ru

Patrick Jaffke
LNL
pjaffke@lanl.gov

Norris Johnson
Retired
norrisndj@gmail.com

Ari Jokinen
University of Jyvaskyla
ari.jokinen@jyu.fi

Matthias Junker
LNGS
junker@lngs.infn.it

Roman Kezerashvili
City Tech, City Univ. of NY
rkezerashvili@citytech.cuny.edu

Jonathon Koglin
LLNL
jkoglin89@gmail.com

Hisaaki Kudo
Niigata University
hkudo@sc.niigata-u.ac.jp

Walter Kutschera
University of Vienna
walter.kutschera@univie.ac.at

Sean Liddick
NSCL/MSU
liddick@nscl.msu.edu

Walter Loveland
Oregon State University
lovelanw@onid.orst.edu

Yixiao Luo
Vanderbilt University/LBNL
yxluo@lbl.gov

Wenchao Ma
Retired
wm14@msstate.edu

Brett Manning
LANL
bmanning@lanl.gov

Scott Marley
Louisiana State University
smarley@lsu.edu

Joachim Maruhn
Frankfurt University
jamaruhn@me.com

Peter Moller
LANL
moller@lanl.gov

Jonathan Mueller
North Carolina State Univ.
jonathan_mueller@nscu.edu

Brooks Musangu
Furman University
brooks.musangu@furman.edu

Witek Nazarewicz
FRIB/NSCL
witek@frib.msu.edu

Katsuhisa Nishio
Japan Atomic Energy Agency
nishio.katsuhisa@jaea.go.jp

Yuri Oganessian
JINR
oganessian@jinr.ru

Steven Pain
ORNL
stevenpain@nuclearemail.org

Stanley Paulauskas
Univ. of Tennessee
stanpaulauskas@gmail.com

Heikki Penttilä
University of Jyvaskyla
penttila@jyu.fi

William Peters
Univ. of TN/ORNL
wapeters.delien@gmail.com

Dorin Poenariu
Horia Hulubei National Inst.
poenaru@nipne.ro

Akunuri Ramayya
Vanderbilt University
a.v.ramayya@vanderbilt.edu

Jorgen Randrup
LBNL
jrandrup@lbl.gov

John Rasmussen
LBNL
oxras1@gmail.com

Rene Reifarth
Goethe University Frankfurt
reifarth@physik.uni-frankfurt.de

David Regnier
IPNO
regnier@ipno.in2p3.fr

Lee Riedinger
University of Tennessee
lrieding@utk.edu

Peter Ring
Technical University Munich
ring@ph.tum.de

James Roberto
Oak Ridge National Lab
robertojb@ornl.gov

Reginald Ronningen
NSCL/FRIB Retired
ronninge@msu.edu

Gencho Rusev
LANL
rusev@lanl.gov

Guy Savard
ANL/University of Chicago
savard@anl.gov

Kevin Siegl
University of Notre Dame
ksiegl@nd.edu

Mallory Smith
Univ. of Dame
msmith40@nd.edu

Nichols Stone
Univ. of Tennessee
nick.stone@physics.ox.ac.uk

Mark Stoyer
LLNL
mastoyer@llnl.gov

Julien Taieb
CEA DAM
julian.taieb@cea.fr

Kemper Talley
Bredesen Center
kemper.talley@gmail.com

Patrick Talou
LANL
talou@lanl.gov

Steven Taylor
Univ. of Tennessee
staylo65@vols.utk.edu

Andrew Thibeault
Vanderbilt University
andrew.h.thibeault@vanderbilt.edu

Diego Torres
Universidad Nacional de Colombia
datorresg@unal.edu.co

Fredrik Tovesson
LANL
tovesson@lanl.gov

Tomohiro Uesaka
RIKEN
uesaka@riken.jp

Sait Umar
Vanderbilt University
sait.a.umar@vanderbilt.edu

Anna Unzhakova
St. Petersburg University
annunzh@gmail.com

Vladimir Utyonkov
JINR
utyonkov@jinr.ru

Marc Verriere
CEA/DAM/DIF
verriere.m@gmail.com

Ramona Vogt
LLNL and UC Davis
vogt2@llnl.gov

Takahiro Wada
Kansai University
wadataka@kanusai-u.ac.jp

David Walter
Rutgers University
dwalter@physics.rutgers.edu

Enhong Wang
Vanderbilt University
enhong.wang@vanderbilt.edu

Yutaka Watanabe
WNSC, IPNS, KEK
yutaka.watanabe@kek.jp

Carl Wheldon
University of Birmingham
c.wheldon@bham.ac.uk

Ingo Wiedenhoever
Florida State University
iwiedenhoever@fsu.edu

Jeff Winger
Mississippi State Univ.
j.a.winger@msstate.edu

Jack Winkelbauer
LANL
jackwinkelbauer@gmail.com

Furong Xu
Peking University
frxu@pku.edu.cn

Walid Younes
LLNL
younes1@llnl.gov

Christopher Zachary
Vanderbilt University
christopher.j.zachary@vanderbilt.edu

Shan-Gui Zhou
Chinese Academy of Sciences
sgzhou@itp.ac.cn

Feng-Shou Zhang
Beijing Normal University
fszhang@bnu.edu.cn

Shaofei Zhu
Argonne National Lab
zhu@anl.gov

Sixth International
Fission and
Neutron-rich